Access Denied

The Information Revolution and Global Politics
William J. Drake and Ernest J. Wilson III, editors
mitpress.mit.edu/IRGP-series

The Information Revolution and Developing Countries
Ernest J. Wilson III, 2004

Human Rights in the Global Information Society
edited by Rikke Frank Jørgensen, 2006

Mobile Communication and Society: A Global Perspective
Manuel Castells, Mirela Fernández-Ardèvol, Jack Linchuan Qiu, and Araba Sey, 2007

Access Denied: The Practice and Policy of Global Internet Filtering
edited by Ronald Deibert, John Palfrey, Rafal Rohozinski, and Jonathan Zittrain, 2008

Access Denied

The Practice and Policy of Global Internet Filtering

edited by Ronald Deibert, John Palfrey, Rafal Rohozinski, and
Jonathan Zittrain

The MIT Press Cambridge, Massachusetts London, England

For information about special quantity discounts, please e-mail special_sales@mitpress.mit .edu.

This book was set in Swis721 on 3B2 by Asco Typesetters, Hong Kong.
Printed and bound in the United States of America.

Library of Congress Cataloging-in-Publication Data

Access denied : the practice and policy of global Internet filtering / edited by Ronald Deibert . . . [et al.].
p. cm. — (The information revolution & global politics series)
Includes bibliographical references and index.
ISBN 978-0-262-54196-1 (pbk. : alk. paper) — ISBN 978-0-262-04245-1 (hardcover : alk. paper)
1. Computers—Access control. 2. Internet—Censorship. 3. Internet—Government policy.
I. Deibert, Ronald.
QA76.9.A25.A275 2008
005.8—dc22 2007010334

10 9 8 7 6 5 4 3 2

Contents

Foreword

The Internet is the operating system of global politics. Ideas, messages, news, information, and money ricochet around the world in minutes, crossing time zones and borders in real time. Charities, banks, corporations, governments, nongovernmental organizations, and terrorist organizations all use the Internet to do business, to organize, and to speed communications. Internet technology is implicated in almost everything done in world politics today.

But the Internet is not the free operating zone that its early proponents expected. Contrary to conventional wisdom, states have shown an increased willingness to intervene to control communication through the Internet. And they have done so with precision and effectiveness.

At the beginning of the decade, few were aware of the scale of the problem. Advocacy and rights organizations charged that a handful of countries were blocking access to Web sites, but they had little evidence to support their claims. Good empirical knowledge of the scope of the problem did not exist.

Four years ago, a group of scholars at the University of Toronto, Harvard, and Cambridge (Oxford joined later) came together to begin systematic research on patterns of Internet censorship and surveillance worldwide. At the time, the project seemed very ambitious. The researchers proposed to put together a combination of contextual political and legal research and technical interrogations of the Internet in the countries under investigation. It relied heavily on the work of partners working in the countries where governments were engaged in active censorship. The project was extraordinarily challenging; in almost every case, the research implied a direct threat to national security and put researchers' personal safety at risk.

The project was ambitious in other ways as well. A transatlantic collaboration among four universities is difficult to manage at the best of times, but the ONI includes dozens of researchers and collaboration with nongovernmental, rights, and advocacy organizations all over the world. The project is also truly interdisciplinary. It involves sociologists, lawyers, international relations scholars, political scientists, and some of the world's most skilled computer programmers.

From 2003 to 2006, the ONI collaboration paid handsome dividends. It has produced eleven major country reports, reports that revealed a startling trend. States were aggressively finding ways to filter and control access to information for citizens within their borders. The reports were detailed, supported by strong evidence that had an immediate impact on policy worldwide. The ONI's China report was delivered before two U.S. congressional committees and was featured in newspapers and on television around the world. The reports highlighted the embarrassing evidence that major U.S. corporations were implicated in Internet censorship practices. Once, the best and brightest of Silicon Valley were wiring the world; now, they were profiting from their collaboration with governments who were censoring and blocking websites. The ONI's dogged investigations called into question the conventional wisdom about the Internet's open architecture.

The significance of the research that ONI has conducted goes beyond its analysis of Internet surveillance and censorship. It speaks to fundamental questions of world politics, its structure, its power relationships, and its new forms of global control and resistance. The essays in this volume engage with all these issues. The editors of *Access Denied* present not only detailed overviews of their country investigations, but several incisive chapters that probe the legal, theoretical, and political implications of the growth of Internet-content-filtering practices worldwide.

Access Denied tells us unmistakably that the Internet is one of the most important—and most contested—terrains of global politics. It is being fought over by states, civil society organizations, and corporations. The essays in this volume do a superb job of educating us about the new battlefield of global politics.

Janice Gross Stein
Director, Munk Centre for International Politics

Preface

This book is a testament to collaboration. About five years ago, it became clear to several of us—at the University of Cambridge, Harvard Law School, and the University of Toronto—that we might accomplish more by working together, across institutions and continents, than we could by going it alone. Since that time, the Oxford Internet Institute has joined our team, along with more than fifty researchers around the globe. Collaboration is not easy; we have had our share of struggles along the way to keep our partnership functioning effectively. Neither the analytical chapters of this volume nor the new global data set that we have compiled, on which our analytical work relies, would be possible without the partnership that joins us.

The insight that brought us together as collaborators was the sense that the architecture of the Internet was changing rapidly—and that these changes would have far-reaching implications. One of the forces at work is that states are using technical means, in addition to other kinds of controls, to block access to sites on the Web that their citizens seemed to wish to access. We set out, together, to enumerate these technical restrictions as they emerged, to track them over time and across states and regions, and to set them into a broader context. Though we have published many of our findings to our Web site (http://www.opennet.net) and will continue to do so, this book is our first effort to tie the many strands of our shared work together into a single fabric.

Just as we shared a sense of the importance of this area of inquiry, we realized also that this phenomenon could not properly be understood without bringing to bear a series of academic disciplines to analyze it and to set it into a fulsome context. The way we have approached our work, which begins with technical enumeration, required technologists among us to develop a new methodology for testing for choke points in the Internet. Political scientists and international relations theorists hold another piece of the puzzle, as do those with expertise in regional studies. Those of us who study and practice international law and how it relates to information technologies understand another part of the whole. Our shared view is that interdisciplinary research is the only way truly to understand our field in all its complexity.

Most important of all, there are those people on the ground, in the places where the state is seeking to impose control over the Internet, who have shed particular light upon what is happening in the places we are studying. Many of these people take risks every day in the interest of promoting human rights, the rule of law, and other universally good causes. Many of these people have put themselves in harm's way, in one fashion or another, to help make this book possible. It is to these heroes, scattered about the globe and about the Internet, that we dedicate this book.

Many good people deserve explicit acknowledgement for their contribution to this book. We each have been blessed by extraordinary teams at our respective institutions and our networks in the field. Some of these contributors are not listed here, at their request; they know who they are.

The Advanced Network Research Group at the Cambridge Security Programme could not have done its work without the support of some key individuals within the University. Rafal Rohozinski, the director of the research effort and ONI Principal Investigator, would like to thank Professors James Mayall and Christopher Hill at the Centre of International Studies, who made available the fellowship under which much of the ONI's work over the past three years took place. Professor Yezid Sayegh was key to paving the way for the project and has been a constant supporter of the work, providing intellectual insight and encouragement. Peter Cavanaugh, the executive director of the Cambridge Security Programme, and Leslie Fettes were patient and willing to provide support, even when we were forced, by necessity, to make payments to our partners in the Commonwealth of Independent States and Middle East via transfers to questionable financial institutions or, at times, in small currency stuffed into plain paper envelopes. Professor Ross Anderson, and the Security Group at the Cambridge Computer Laboratory, was extraordinarily supportive and brought to our project Dr. Steven Murdoch, who has gone on to become the ONI's chief technology officer. Steven's quiet and diligent manner has led to some of the ONI's more interesting findings, and he continues to spearhead the development of tools and methods that will keep our work ahead of the emerging trends.

The work done by Cambridge in mapping and contextualizing emerging information controls in the Commonwealth of Independent States could not have happened without special partnership with the Eurasia I-Policy Network (EIPN), in particular its dynamic regional coordinator Tattu Mambetalieva (Kyrgyzstan). Under Tattu's leadership, EIPN members, who represent NGOs from nine CIS countries, went well beyond the requirements of the yearbook and engaged policymakers, security actors, academia, and businessmen in examining the emerging governance and policy of the Internet in their countries. Their commitment not only led to great research but also helped reverse policies in some countries. Some unfortunately paid the price for speaking too loudly; during the course of our work over the past three years, members of our team have been harassed, arrested, and in one case died under questionable circumstances. Special mention goes out to our country coordinators, only some of whom we can name: Emin Akhndov (Azerbaijan), Vadim Dryganov (Belarus), Alexsei Marcuic and Vladislav Spirlenko (Institute for Information Policy, Moldova), Dr. Alexandra Belyaeva (Russian Federation), and Andriy Paziuk (Privacy Ukraine). For those whom we cannot, thanks goes out to the Civil Initiative for Internet Policy in Kazakhstan and the public foundation "GIPI" in Tajikistan. Extraspecial mention goes out to our team in Uzbekistan, who toil under great personal risk and in total anonymity. Cambridge and EIPN also are supported by a fantastic in-field administration and technology team from the Civil Initiative for Internet Policy in Kyrgyzstan, who make working in the CIS seem easy: Alexsei Bebinov, Lira Samykbaeva, and Zlata Shramko.

Cambridge also would like to recognize the engagement of the Institute of Information Security Problems, Moscow State University, for its willingness to engage with the Advanced Network Research Group around two NATO-sponsored roundtables examining Internet controls, and to bring to the table representatives from the Russian National Security Council as well as major security organizations and businesses. This engagement has started an important public-policy process around these critical issues between representatives of Russian state institutions, business, and civil society.

In the Middle East, Cambridge partnered with Palestinians and Israelis to conduct testing in what can be termed "a highly complex political and security environment." Special thanks go out to Dr. Michael Dahan (Hebrew University) for his insights on Israeli information society. Especial thanks to our Palestinian partners, Engineer Wassim Abdullah, Dr. Mashour Abudaka, His Excellency Dr. Sabri Saidam, and Sam Bahour and the technical staff at the Centre for Continuing Education, Bir Zeit University, without whom the work in the West Bank and Gaza would not have been possible.

Finally, the Cambridge team benefited from some excellent past and present researchers: Dr. David Mikosz, Deirdre Collings, and Joanna Michalska, all of whom undertook much of the grounded foundational research upon which our present work in the CIS and Middle East depends.

Dr. Robert Faris at the Berkman Center for Internet & Society at Harvard Law School has led the research staff, at Harvard Law School and also across all institutions, with grace and poise. Rob deserves as much credit as anyone for the quality and integrity of the research that underlies this work, as well as for a great deal of the text in this book.

Rob Faris has been joined and supported by an unusually strong group of research fellows on the Berkman Center's team. Among these Berkman fellows, Derek Bambauer, now a law professor, stands out. Derek spent more than two years, as a student and as a research fellow, developing the methodology, gathering earlier versions of these data, and drafting reports that form the core of much of what we conclude in this book. Jeffrey Engerman, now a lawyer in private practice, contributed a great deal of wisdom as to our methods and the way we handle and analyze our data. Derek and Jeff also coordinated a generation of research assistants who helped us to produce the first versions of many of the state-specific reports on which our work is grounded. Stephanie Wang, a terrific lawyer and researcher, brought exceptional regional understanding to our work in East Asia. Vesselina Haralampieva lent similar expertise to our work in the region encompassing the Commonwealth of Independent States. Helmi Noman and Elijah Zarwan ably led our work in the Gulf and North Africa regions, respectively. Our partners in the Cyberlaw and International Human Rights Clinics at Harvard Law School—fellows Phil Malone, Matt Lovell, and Bonnie Docherty, and Professor Jim Cavallaro—have co-led missions with exceptional students from our respective clinics to Southeast Asia and Russia as we gathered data for this project.

An extraordinary cadre of student researchers from Harvard Law School and the surrounding academic community has been responsible for pulling together much of the detail that has gone into this project. Kevin O'Keefe, a graduate student in East Asian studies, is first among equals. The first student to work on Internet filtering at the Berkman Center, Benjamin Edelman, now a professor at Harvard Business School, deserves thanks for his important role in the early days of this research.

The country profiles were produced under the guidance and authorship of principal investigator Rafal Rohozinski and Vesselina Haralampieva for the Commonwealth of Independent States, Helmi Noman and Elijah Zarwan for the North Africa and Middle East region, and Stephanie Wang and Kevin O'Keefe for Asia. Many people contributed to the research, writing, and editing of these profiles, including: James Ahlers, Aisha Ahmad, Anna Brook, Chris Conley, Evan Croen, Matthieu Desruisseaux, Charles Frentz, Anthony Haddad, Christina Hayes, Joanna Huey, Samuel Hwang, Sajjad Khoshroo, Jehae Kim, Saloni Malhotra, Katie Mapes, Miriam Simun, Tobias Snyder, Elisabeth Theodore, and Christina Xu. The following individuals made important contributions to the research in the field: Shahzad Ahmad, Shanti Alexander, Tatyana Bezuglova, Srijana Bhattarai, Alexander Blank, Matt Boulos, Xiao Wei Chen, Yee Yeong Chong, Lino Clemente, Kathleen Connors, Peter Daignault, Shubhankar Dam, Elliott Davis, Siddharth Dawara, Charles Duan, Bipin Gautam, Nah Soo Hoe, Tina Hu, Ang Peng Hwa, Mary Joyce, Randy Kluver, David Levenson, Eitan Levisohn, Saloni Malhotra, Efrat Minivitski, Ron Morris, Caroline Nellemann, Jeff Ooi, Sai Rao, David Rizk, Sajan Sangraula, Katie Smith, Amine Taha, Lokman Tsui, Allison Turbiville, Neha Viswanathan, Dinesh Wagle, Sally Walkerman, Naaman Weiss, Aaron Williamson, K. H. Yap, and Jeffrey Yip. We are grateful to those who took the time to read and comment on our work, including: Markus Breen, Silke Ernst, Peyman Faratin, Daniel Haeusermann, Nancy Hafkin, Luis Muñoz, Eric Osiakwan, Russel Southwood, and James Thurman. We also would like to offer our thanks to the following individuals for their valuable guidance and help with our research: Ananta Agrawal, Roby Alampay, Cherian George, Tyler Giannini, Chandrachoodan Gopalakrishnan, Rishikesh Karra, Sudhir Krishnaswamy, Arun Mehta, Parishi Sanjanwala, Xiao Qiang, and Zaw Zaw.

Hope Steele expertly edited each of the country profiles and regional overviews for this book with great care, grace, and patience. Ha Nguyen designed the country profiles and regional overviews, performing multiple miracles on short notice with true poise and artistic skill.

A number of people participated in the writing, editing, research, and testing anonymously. We undoubtedly have not included others who deserve our thanks.

The Berkman Center's work on this project drew upon many within the Berkman Center's community for whom the OpenNet Initiative is not their sole obsession. Colin Maclay, the Center's managing director, contributed both substantive insights and a steady hand. Catherine Bracy and Seth Young, with the backing of the Center's wonderful administrative staff, kept the relevant trains running on time, despite plenty of events that could have thrown them off the rails. Andrew Heyward, Peter Emerson, Evan Croen, Amanda Michel, Andrew Solomon, and Patrick McKiernan—along with a group of volunteer advisors—have assisted us in shaping the way that we communicate the findings of our study. Wendy Seltzer and Urs Gasser, fellows of the Center and also professors of law, each challenged our thinking at many stages of this research and offered helpful feedback on various drafts that became parts of this book. Research fellows Ethan Zuckerman, Michael Best, David Weinberger, and Rebecca MacKinnon (now a professor of journalism) went out of their way, as did many other Berkman fellows, to lend hands and contacts, along with welcome critiques of our methods and our conclusions. A group of our colleagues from around Harvard (Joseph Nye) and at neighboring MIT (Eric von Hippel) also reviewed drafts and participated in an informal peer review session. We also have learned much from the participants in the global process to develop a set of ethical guidelines for corporations operating in regimes that practice censorship and surveillance. Dunstan Hope and Aron Cramer of Business for Social Responsibility; Leslie Harris of CDT; Andrew McLaughlin and Bob Boorstin of Google; Michael Samway of Yahoo!; Ira Rubinstein of Microsoft; Orville Schell, Xiao Qiang, Deirdre Mulligan, and Roxanna Altholz at the University of California-Berkeley; and others have offered valuable commentary and guidance.

Jonathan Zittrain and I thank especially our faculty colleagues associated with the Berkman Center and Harvard Law School, who in many respects are the reasons we do what we do for a living. We are grateful to Charlie Nesson, the founder of the Berkman Center; Terry Fisher, the Center's faculty director; Jack Goldsmith, one of the most insightful contributors to our field; Larry Lessig, whose ideas about the regulation of cyberspace through code infuse all our work; and Yochai Benkler, who keeps reminding us why this all matters.

At the Oxford Internet Institute at the University of Oxford, Sangamitra Ramachander contributed helpful research assistance, and Bill Dutton and the Institute's research staff participated in a number of workshop sessions that helped us test and refine our hypotheses. The Institute generously has hosted two ONI-related conferences, and its investment of intellectual capital in the project is much appreciated.

At the Citizen Lab at the Munk Centre at the University of Toronto, a dynamic team of extraordinary "hacktivists" has contributed immensely to the technical and other research work of the ONI. Nart Villeneuve's pioneering methods of remote network interrogation laid the basis for the ONI's technical methodology. His dogged pursuit of network anomalies, questionable practices, and seemingly intractable problems helps drive the engine of the ONI on a daily basis. Michelle Levesque worked alongside Nart in the early years of the ONI to develop and refine the ONI's suite of testing tools. Both of them have approached their responsibilities with great enthusiasm and spirit to hunt down and document patterns of Internet content filtering and surveillance worldwide. They are truly Net Ninjas.

As a Citizen Lab senior research fellow responsible for the ONI's "deep dives" into Asia, Dr. Francois Fortier has helped convene and lead a dynamic group of researchers in the region. Although relatively new to the project, his tremendous organizational and intellectual skills already have contributed invaluably, and we look forward to his ongoing and expanding role in the project in the years to come.

Over the years, numerous programmers and researchers have worked at the Citizen Lab, bringing ingenuity and dogged determination to the ONI's forensic investigations. These include, in no particular order, Graeme Bunton, Sarah Boland, James Nicholas Tay, Eugene Fryntov, Anton Fillipenko, Michael Hull, Pat Smith, Tim Smith, Oliver Day, Julian Wolfson, Stian Haklev, Konstantin Kilibarda, David Wade-Farley, Peter Wong, and Liisa Hyyrylainen.

Jane Gowan, of Agent5 design, has brought her remarkable creativity and artistic sensibilities to help enrich and enliven the ONI's presentation of its work, including our 2006 poster of Internet censorship, many of the graphics and other visualizations included herein, and the striking cover art that frames this volume. Her professionalism, enthusiasm and creativity are much appreciated.

As Director of the Citizen Lab, Ron Deibert would like to thank the staff at the Munk Centre for International Studies for providing such a supportive environment for the Lab and the ONI's research activities, in particular its director, Janice Stein, and Marketa Evans, Wilhelmina Peters, and Penny Alford, as well as the Munk Centre's technical support staff. Thanks also to the University of Toronto's Computing and Network Services, in particular Eugene Sicunius, for tolerating and supporting our (at times) unconventional methods.

As the list of the contributors makes plain, the OpenNet Initiative is an expensive project to operate. There would be no global data set and no book were it not for the vision of our program officers and the willingness to take risks of the boards of their foundations. We owe deep thanks to all at the John D. and Catherine T. MacArthur Foundation for a multiyear, $3 million grant that has provided the bulk of the funding for this book project. In particular, the foundation's president, Jonathan Fanton, its vice president Elspeth Revere, and program officer John Bracken have provided invaluable counsel and, of course, financial support. The Open Society Institute of the Soros Foundation provided the ONI its first grant; it was Jonathan Peizer, then the OSI's chief technology officer, who connected us—fittingly enough, by e-mail—in the first place. Darius Cuplinskas and Vera Franz of the OSI's Information Program have earned our unending thanks for their loyal support of the ONI and its work. Ron Deibert and Rafal Rohozinski owe an enormous debt of gratitude to Anthony Romero and the Ford Foundation for seed funding that helped contribute to the realization of the Citizen Lab and the Advanced Network Research Group. We are very grateful to the International Development Research Centre (IDRC) of Canada for providing funding for ONI's continuing engagement in the Asia, Africa, and Middle East regions, and support for the ONI's mapping and other visualization projects.

Most of all, we each thank our families and friends who have supported us as we have traveled the world to compile these data and spent long hours away in writing them up.

John G. Palfrey
on behalf of the OpenNet Initiative Principal Investigators

OpenNet Initiative
opennet.net

Citizen Lab, Munk Centre for International Studies, University of Toronto

Berkman Centre for Internet & Society, Harvard Law School

Advanced Network Research Group, Cambridge Security Programme, University of Cambridge

Oxford Internet Institute, Oxford University

Introduction

Jonathan Zittrain and John Palfrey

A Tale of Two Internets

Tens of thousands of international travelers descended upon the Tunis airport for the World Summit on the Information Society in 2005. The summit brought together policy-makers, journalists, nongovernmental organization (NGO) leaders, academics, and others to consider the present and future of information and communications technologies. Polite Tunisian handlers in crisp, colorful uniforms guided arriving summit attendees to buses that took those with credentials to one of several sites nearby.

The capital, Tunis, hosted the main conference facilities. The seaside town of Yasmine-Hammamet, with boardwalks, theme parks, casinos, and breathtaking sunsets, housed delegates who could not find lodging in the city. Within the main conference facilities in Tunis, they would experience the Internet as though they were in a Silicon Valley start-up: unfettered access to whatever they sought to view or write online.

But those by the sea in Yasmine-Hammamet, outside the United Nations–sponsored conference facilities, encountered a radically different Internet—the one that is commonplace for Tunisians. If attendees sought to view a site critical of the summit's proceedings or mentioning human rights—for instance, a site called Citizen's Summit, at www.citizens-summit.org/—they would see a page indicating that a network error had occurred. Among other curious things, the page was written in French, not the native Arabic. The blockpage is partially accurate: something in the network had caused that information never to reach the surfer's laptop.[1] But it was not an error.

The blockage is intentional, one of thousands put in place daily by the government of Tunisia. The ad hoc filtering of information underway in Tunisia is flatly at odds with the ideals touted by World Summit participants. Tunisia's filtering system was implemented long before the World Summit kicked off, and it was unaffected by the attention the summit brought to Tunisia.

A filtering system is meant to stop ordinary citizens from accessing some parts of the Internet deemed by the state to be too sensitive, for one reason or another. The information blocked ranges from politics to sexuality to culture to religion. As user-generated content has gained in popularity and new tools have made it easier to create and distribute it, filtering regimes have pivoted to stop citizens from publishing undesirable thoughts, images, and sounds, whether for a local or an international audience. The system that facilitates a state's Internet filtering can also be configured to enable the state to track citizens' Web surfing or to listen in on their conversations, whether lawful or unlawful.

A Tale of Many Internets

Tunisia is not a special case. More than three dozen states around the world now filter the Internet. This book contains the results of the first systematic, academically rigorous global study of all known state-mandated Internet filtering practices. Previously, the OpenNet Initiative and others have reported only anecdotally or sporadically on the scope of Internet filtering. Our first goal in writing this book is to present the data from this global study, allowing others to make use of it in their own empirical work, or to place it within a normative framework. Second, in addition to state-by-state test results, we have commissioned a series of essays analyzing these test results and related findings from a variety of perspectives—what this emerging story means from the standpoint of technology, as a matter of international law, in the context of corporate ethics, and for the vibrant activist and political communities that increasingly rely upon Internet technologies as a productivity enhancer and essential communications tool.

For this first global study, we have sought to find those places in the world that practice state-mandated technical filtering. The definition of what we are and are not covering here is important to set forth at the outset: we seek to describe technical blockages of the free flow of information across the Internet that states put in place or require others to institute. To determine where to test for such blockages, we have drawn upon our own technical probes and forensic analyses of networks, published reports of others who track these matters, and credible unpublished reports that we received either through interviews or over the transom. Our emphasis on state-mandated technical filtering underscores our own sense that "West Coast Code," in Lawrence Lessig's terms (computer code), is more malleable, more subtle, more effective in many contexts, and less easily noted, changed, or challenged than "East Coast Code" (ordinary law and regulation), which is typically less opaque in its operation.[2] Straightforward state regulation of speech without technological components can, of course, result in censorship; our work here is designed to focus on regulation that, when implemented through code, seems more a force of nature than an exercise of political or physical power.

Thus it is entirely possible that a state that does not require or inspire technical filtering can possess a set of regulations or social norms or market factors that render its information environment less free than a state with fairly extensive technical filtering. A rich and comprehensive picture of what a truly "free" or "open" information environment looks like can rely only in part on conclusions about Internet filtering. The essays that accompany our presentation of the data are intended to provide some, though by no means all, of the relevant context. A shrewd observer might well make a case that the extensive regulatory regimes for speech in Canada, the United States, and the United Kingdom—from which states the majority of our researchers hail—result in a more constrained information environment than a state with technical filtering but little else by way of law, norms, or markets to constrain an Internet user. We map out filtering practices, and the law and regulation behind them, so that they may take their place within a larger mosaic of assessing and judging the flow of information within and across the world's jurisdictions.

The states that practice state-mandated filtering are predominantly clustered in three regions of the world: east Asia, the Middle East and North Africa, and central Asia. A handful of states outside these regions also encourage or mandate certain forms of filtering. Someone in the United States, for instance, may encounter state-mandated Internet filtering on some computers in libraries and schools. A citizen in northern Europe might find child pornography blocked online. In France and Germany, content that includes imagery related to Nazism or Holocaust denial is blocked in various ways and at various levels. The emerging trend points to more filtering in more places, using more sophisticated techniques over time. This trend runs parallel to the trajectory of more people in more places using the Internet for more important functions in their lives.

We find that filtering implementations, and their respective scopes and levels of effectiveness, vary widely among the states that filter. China institutes by far the most extensive filtering regime in the world, with blocking occurring at multiple levels of the network and spanning a wide range of topics. Singapore, by contrast, despite a widely publicized filtering program, in fact blocks access to only a handful of sites. Each of the sites blocked in Singapore is pornographic in nature. Several states, including some in central Asia, filter only temporarily when elections or other key moments make the control of the information environment most important to the state. Most states implement filtering regimes that fall between the poles of China and Singapore, with significant variation from one to the next. Each of these state-mandated filtering regimes can be understood only in the political, legal, religious, and social context in which they arise. It is just this context that we seek to provide in the chapters that follow our presentation of the data.

Our aim in this volume is to document, with the greatest degree of precision possible, technical Internet filtering wherever we have been able to find it, and to set it in a context that acknowledges the nuances and complexity of this matter. We have relied upon an extensive network of researchers in each of the regions of the world that we have studied, as well as area-studies experts based outside those regions. We chose to study and report on the states covered in this volume, as well as other states that appear not to be filtering but are on our "watch list," because our researchers, members of the press, or others in this field—Reporters Sans Frontières or Human Rights Watch, for instance—have identified these states as potentially carrying out state-level filtering. The lists used in the testing that forms the core of our set of findings are the product of study of the political, social, cultural, and religious issues in each of the states we have reviewed. While there is no doubt filtering underway in places around the world that we have yet to uncover, our goal in this volume is to be as comprehensive as possible.

The core of the data we present is found in short reports covering each state that we have studied in depth, with an overview for each of the three regions—east Asia, the Middle East and North Africa, and central Asia—identifying themes and trends across states. The section on testing results for each state sets forth the types of content blocked by category and includes documentation of the most noteworthy content-specific findings.

We intend to update this study annually. Our intention is to develop a publicly accessible online database of filtering test results worldwide over time. Taken together, these reports represent a starting point in understanding the nature and future of global Internet filtering.

In addition to the state-specific data, we present a series of chapters that builds arguments grounded in our empirical findings about Internet filtering. The first short chapter, by Robert Faris and Nart Villeneuve, includes a set of issues that emerge from the data: trends and themes from a global perspective. Our chapter 2 gives an overview of the politics and practice of Internet filtering. The third chapter, by Ross Anderson and Steven Murdoch, considers the technology that powers the Internet filtering and highlights its strengths and limitations. The fourth chapter, by Mary Rundle and Malcolm Birdling, takes up the extent to which international law might bear on Internet filtering. Our chapter 5 examines the ethical issues for corporations seeking to avail themselves of markets in states that filter. The final chapter, by Ronald Deibert and Rafal Rohozinski, looks in depth at the impact of Internet filtering upon the activist community that increasingly relies upon the Internet for mission-critical activities.

While we bring our own normative commitments to this work—those of us who have contributed to this work tend to favor the free flow of bits as opposed to proprietary control of information, whether by states or companies or both—our goal is not to point fingers or assign blame, but rather to document a trend that we believe to be accelerating and to set that trend in context. We seek to prompt further conversation across cultures and disciplines about what changes in Internet filtering practices mean for the future of the Internet as well as the future of markets, social norms, and modes of governance around the world. We look forward to the conversations as others put these data into the proper, broader context—into the larger mosaic of political and cultural freedom—into which they belong.

Notes

1. For one of many contemporaneous accounts, see John Palfrey, On Being Filtered in Tunisia, or, What WSIS Should Really Focus On, http://blogs.law.harvard.edu/palfrey/2005/11/14/on-being-filtered-in-tunisia-or-what-wsis-should-really-focus-on/.

2. Lawrence Lessig, *Code and Other Laws of Cyberspace* (New York: Basic Books, 1999), 53–54.

1

Measuring Global Internet Filtering

Robert Faris and Nart Villeneuve

The Scope and Depth of Global Internet Filtering

In this chapter, we set out to provide an overview of the data regarding Internet filtering that the OpenNet Initiative[1] has gathered over the past year. Empirical testing for Internet blocking was carried out in forty countries in 2006. Of these forty countries, we found evidence of technical filtering in twenty-six (see table 1.1). This does not imply that only these countries filter the Internet. The testing we carried out in 2006 constitutes the first step toward a comprehensive global assessment. Not only do we expect to find more countries that filter the Internet as we expand our testing, but we also expect that some of the countries that did not show signs of filtering in 2006 will institute filtering in subsequent years.[2]

Conceptually, the methodology we employ is simple. We start by compiling lists of Web sites that cover a wide range of topics targeted by Internet filtering. The topics are organized into a taxonomy of categories that have been subject to blocking, ranging from gambling, pornography, and crude humor to political satire and Web sites that document human rights abuses and corruption. (See table 1.2.) Researchers then test these lists to see which Web sites are available from different locations within each country.[3]

The states that filter the Internet must choose which topics to block (the scope of filtering) and how much of each topic to filter (the depth of filtering). The results of these decisions are summarized in figure 1.1, comparing the breadth and depth of filtering for the countries where evidence of filtering was found.

The number of different categories in which Internet filtering was found to occur is shown on the horizontal axis. We put this forward as a measure of the scope of Internet filtering in each country. (The categories are shown in table 1.2.)

The vertical axis depicts the comprehensiveness of filtering efforts as measured by the highest degree of content blocked in any of the topical categories. This captures a markedly different angle on filtering. If the breadth of filtering represents the ambition of censors to limit information related to a range of topics, the depth of filtering measures the success in actually blocking content. This might correspond to the level of sophistication of the filtering regime and amount of resources devoted to the endeavor, or it may be a reflection of the resolve and political will to shut down large sections of the Internet.

The countries occupying the upper right of figure 1.1, including Iran, China, and Saudi Arabia, are those that not only intercede on a wide range of topics but also block a large amount of content relating to those topics. Myanmar and Yemen cover a similarly broad scope, though with less comprehensiveness in each category. South Korea is in a league of its own. It has opted to filter very little, targeting North Korean sites, many of which are hosted in Japan. Yet South Korea's thoroughness in blocking these sites manifests a strong desire to eliminate access to them. There is a cluster of states occupying the center of the plot that

Table 1.1
Filtering by state

Evidence of filtering	Suspected filtering	No evidence of filtering
Azerbaijan	Belarus	Afghanistan
Bahrain	Kazakhstan	Algeria
China		Egypt
Ethiopia		Iraq
India		Israel
Iran		Kyrgyzstan
Jordan		Malaysia
Libya		Moldova
Morocco		Nepal
Myanmar		Russia*
Oman		Ukraine
Pakistan		Venezuela
Saudi Arabia		West Bank/Gaza
Singapore		Zimbabwe
South Korea		
Sudan		
Syria		
Tajikistan		
Thailand		
Tunisia		
United Arab Emirates		
Uzbekistan		
Vietnam		
Yemen		

*Testing in Russia was limited to a selection of ISPs in Moscow; these preliminary results may not extend beyond this sample.

Table 1.2
Categories subject to Internet filtering

Categories subject to Internet filtering
Free expression and media freedom
Political transformation and opposition parties
Political reform, legal reform, and governance
Militants, extremists, and separatists
Human rights
Foreign relations and military
Minority rights and ethnic content
Women's rights
Environmental issues
Economic development
Sensitive or controversial history, arts, and literature
Hate speech
Sex education and family planning
Public health
Gay/lesbian content
Pornography
Provocative attire
Dating
Gambling
Gaming
Alcohol and drugs
Minority faiths
Religious conversion, commentary, and criticism
Anonymizers and circumvention
Hacking
Blogging domains and blogging services
Web hosting sites and portals
Voice over Internet Protocol (VOIP)
Free e-mail
Search engines
Translation
Multimedia sharing
P2P
Groups and social networking
Commercial sites

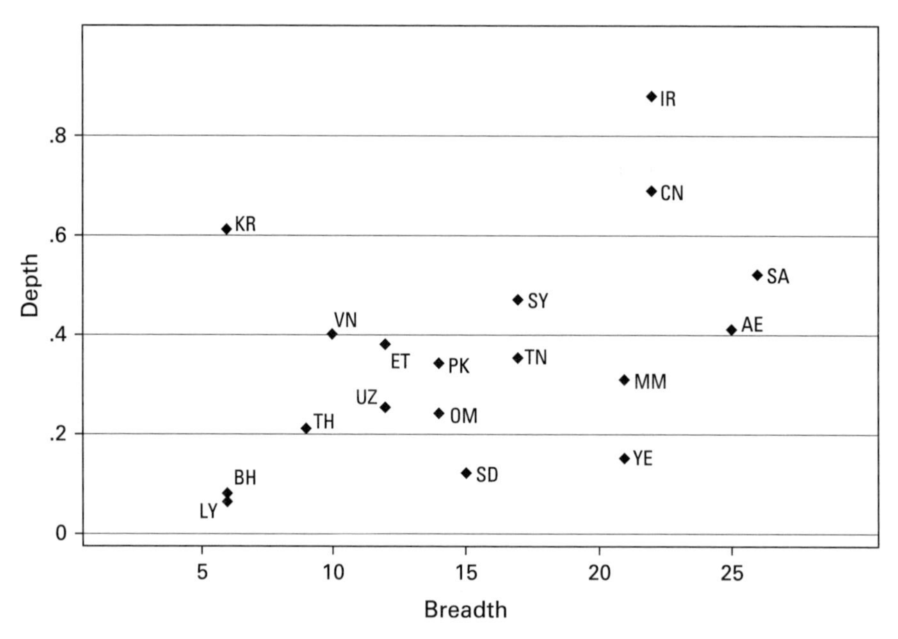

Figure 1.1
Comparing the breadth and depth of filtering. AE—United Arab Emirates; BH—Bahrain; CN—China; ET—Ethiopia; IR—Iran; JO—Jordan; KR—South Korea; LY—Libya; MM—Burma/Myanmar; OM—Oman; PK—Pakistan; SA—Saudi Arabia; SD—Sudan; SY—Syria; TH—Thailand; TH—Tunisia; UZ—Uzbekistan; VN—Vietnam; YE—Yemen. A number of countries that filter a small number of sites are omitted from this diagram, including Azerbaijan, Belarus, India, Jordan, Kazakhstan, Morocco, Singapore, and Tajikistan.

are widely known to practice filtering. These countries, which include Syria, Tunisia, Vietnam, Uzbekistan, Oman, and Pakistan, block an expansive range of topics with considerable depth. Ethiopia is a more recent entrant into this category, having extended its censorship of political opposition into cyberspace.

Azerbaijan, Jordan, Morocco, Singapore, and Tajikistan filter sparingly, in some cases as little as one Web site or a handful of sites. The evidence for Belarus and Kazakhstan remains inconclusive at the time of this writing, though blocking is suspected in these countries.

Of equal interest are the states included in testing in 2006 in which no evidence of filtering was uncovered (see table 1.1). We make no claims to have proven the absence of filtering in these countries. However, our background research supports the conclusion drawn from the technical testing that none of these states are currently filtering Internet content.[4]

Later in the book we turn our attention to the question of why some countries filter and others do not, even under similar political and cultural circumstances.

The Principal Motives and Targets of Filtering

On September 19, 2006, a military-led coup in Thailand overthrew the democratically elected government headed by Prime Minister Thaksin Shinawatra. Thailand is not unfamiliar with such upheavals. There have been seventeen coups in the past sixty years. This time, however, Internet users noticed a marked increase in the number of Web sites that were not accessible, including several sites critical of the military coup.[5] A year earlier in Nepal, the king shut down the Internet along with international telephone lines and cellular communication networks when he seized power from the parliament and prime minister. In Bahrain, during the run-up to the fall 2006 election, the government chose to block access to a number of key opposition sites. These events are part of a growing global trend. Claiming control of the Internet has become an essential element in any government strategy to rein in dissent—the twenty-first century parallel to taking over television and radio stations.

In contrast to these exceptional events, the constant blocking of a swath of the Internet has become part of the everyday political and cultural reality of many states. A growing number of countries are blocking access to pornography, led by a handful of states in the Persian Gulf region. Other countries, including South Korea and Pakistan, block Web sites that are perceived as a threat to national security.

Notwithstanding the wide range of topics filtered around the world, there are essentially three motives or rationales for Internet filtering: politics and power, social norms and morals, and security concerns. Accordingly, most of the topics subject to filtering (see table 1.2) fall under one of three thematic headings: political, social, and security. A fourth theme—Internet tools—encompasses the networking tools and applications that allow the sharing of information relating to the first three themes. Included here are translation tools, anonymizers, blogging services, and other Web-based applications categorized in table 1.2.

Protecting intellectual property rights is another important driver of Internet content regulation, particularly in western Europe and North America. However, in the forty countries that were tested in 2006, this is not a major objective of filtering.[6]

Figure 1.2 compares the political and social filtering practices of these same twenty-seven countries. On one extreme is Saudi Arabia, which heavily censors social content. While there is also substantial political filtering carried out in Saudi Arabia, it is done with less scope and depth. On the other fringe are Syria and China, focusing much more of their extensive filtering on political topics. Myanmar and Vietnam are also notable for their primary focus on political issues, which in the case of Vietnam contradicts the stated reason for filtering the Internet.[7] Iran stands out for its pervasive filtering of both political and social material.

Filtering directed at political opposition to the ruling government is a common type of blocking that spans many countries. Politically motivated filtering is characteristic of authoritarian and repressive regimes. The list of countries that engage in substantial political blocking includes Bahrain, China, Libya, Iran, Myanmar, Pakistan, Saudi Arabia, Syria, Tunisia,

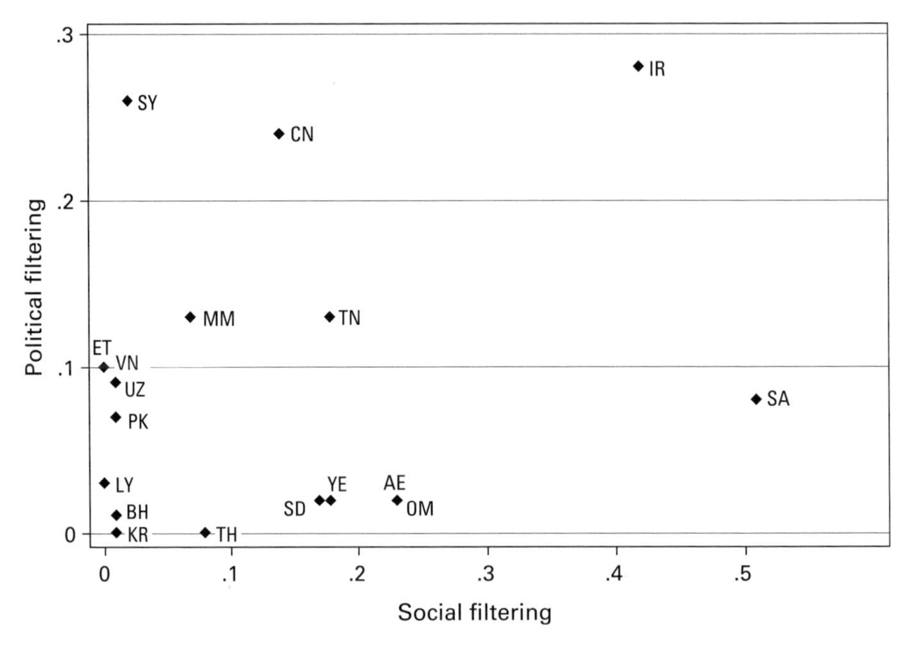

Figure 1.2
Political and social filtering. AE—United Arab Emirates; BH—Bahrain; CN—China; ET—Ethiopia; IR—Iran; JO—Jordan; KR—South Korea; LY—Libya; MM—Burma/Myanmar; OM—Oman; PK—Pakistan; SA—Saudi Arabia; SD—Sudan; SY—Syria; TH—Thailand; TH—Tunisia; UZ—Uzbekistan; VN—Vietnam; YE—Yemen. A number of countries that filter a small number of sites are omitted from this diagram, including Azerbaijan, Belarus, India, Jordan, Kazakhstan, Morocco, Singapore, and Tajikistan.

Uzbekistan, and Vietnam.[8] Thailand and Ethiopia are the most recent additions to this group of countries that filter Web sites associated with political opposition groups. Yet in other countries with an authoritarian bent, such as Russia and Algeria, we have not uncovered filtering of the Internet.

The perceived threat to national security is a common rationale used for blocking content. Internet filtering that targets the Web sites of insurgents, extremists, terrorists, and other threats generally garners wide public support. This is best typified by South Korea where pro–North Korean sites are blocked, or by India where militant and extremist sites associated with groups that foment domestic conflict are censored. In Pakistan, Web sites devoted to the Balochi independence movement are blocked. Similarly, the Web sites of separatist or radical groups such as the Muslim Brotherhood are blocked in some countries in the Middle East.

Social filtering is focused on those topics that are held to be antithetical to accepted societal norms. Pornographic, gay and lesbian, and gambling-related content are prime examples

Box 1.1
Identifying and documenting Internet filtering

Measuring and describing Internet filtering defies simple metrics. Ideally, we would like to know how Internet censorship reduces the availability of information, how it hampers the development of online communities, and how it inhibits the ability of civic groups to monitor and report on the activities of the government, as these answers impact governance and ultimately economic growth. However, this is much easier to conceptualize at an abstract level than to measure empirically. Even if we were able to identify all the Web sites that have been put out of reach due to government action, the impact of blocking access to each Web site is far from obvious, particularly in this networked world where information has a habit of propagating itself and reappearing in multiple locations. Nevertheless, every obstacle thrown into the path of citizens seeking out information bears a cost or, depending on how one views the contribution of a particular Web site to society, a benefit. With this recognition of the inherent complexity of evaluating Internet censorship, we set out with modest goals—to identify and document filtering.

Two lists of Web sites are checked in each of the countries tested: a global list and a local list. The global list is a standardized list of Web sites that cover the categories listed in table 1.1. The global list of Web sites is comprised principally of internationally relevant Web sites with English content. The same global list is checked in each of the countries in which we have tested. A separate local list is created for each of the countries tested; it includes Web sites related to the specific issues and context of the study country.

These testing lists encompass a wide variety of content including political topics such as human rights, political commentary and news, religion, health and sex education, and Web sites sponsored by separatists and militant organizations. Pornography, gambling, drugs, and alcohol are also represented in the testing lists. The lists embody portions of the Web space that would be subject to Internet filtering in each of the countries being tested. They are designed to unearth filtering and blocking behavior where it exists. Background research is focused on finding sites that are likely to be blocked. In countries where Internet censorship has been reported, the lists include those sites that were alleged to have been blocked. These are not intended to be exhaustive lists of the relevant subject matter, nor do we presume to have identified all the Web sites that are subject to blocking.

The actual tests are run from within each country using software specifically designed for this purpose. Where appropriate, the tests are run from different locations to capture the differences in blocking behavior across Internet service providers (ISPs). The tests are run across multiple days and weeks to control for normal connectivity problems.

The completion of the initial accessibility testing is just the first step in the evaluation process. Additional diagnostic work is required to separate normal connectivity errors from intentional tampering. As described in further detail later, there are a number of technical alternatives for filtering the Internet, some of which are relatively easy to discover. Others are difficult to detect and require extensive diagnostic work to confirm.

of what is filtered for social and cultural reasons. Hate speech and political satire are also the target of Internet filtering in some countries. Web sites that deny the Holocaust or promote Nazism are blocked in France and Germany. Web sites that provide unflattering details related to the life of the king of Thailand are censored in his country.

An emergent impetus for filtering is the protection of existing economic interests. Perhaps the best example is the blocking of low-cost international telephone services that use Voice-over Internet Protocol (VoIP) and thereby reduce the customer base of large telecommunications companies, many of which enjoy entrenched monopoly positions. Skype, a popular and low-cost Internet-based telephone service, has been blocked in Myanmar and United Arab Emirates, which heavily block VoIP sites. The Web sites of many VoIP companies are also blocked in Syria and Vietnam.

Many countries seek to block the intermediaries: the tools and applications of the Internet that assist users in accessing sensitive material on the Internet. These tools include translation sites, e-mail providers, Weblog hosting sites, and Web sites that allow users to circumvent standard blocking strategies. Blogging services such as Blogspot are often targeted; eight countries blocked blogs hosted there, while Syria, Ethiopia, and Pakistan blocked the entire domain, denying access to all the blogs hosted on Blogspot. Fourteen countries blocked access to anonymity and censorship circumvention sites. Both SmartFilter, used in Sudan, Tunisia, Saudi Arabia, and UAE, and Websense, used in Yemen, have filtering categories—called "Anonymizers" and "Proxy Avoidance," respectively—used to block such sites.

A handful of countries, including China, Vietnam, and states in the MENA region (the Middle East and North Africa), block Web sites related to religion and minority groups. In China, Web sites that represent the Falun Gong and the Tibetan exile groups are widely blocked. In Vietnam, religious and ethnic sites associated with Buddhism, the Cao Dai faith, and indigenous hill tribes are subject to blocking. Web sites that are aimed at religious conversion from Islam to Christianity are often blocked in the MENA region. Decisively identifying the motives of filtering activity is often impossible, particularly as the impact of filtering can simultaneously touch a host of social and political processes. That being said, it probably would be a mistake to attribute the filtering of religious and ethnic content solely to biases against minority groups, as these movements also represent a political threat to the ruling regimes.

A Survey of Global Filtering Strategies, Transparency, and Consistency

There are many techniques used to block access to Internet content. Each of these techniques can be used at different levels of Internet access within a country. Internet filtering is most commonly implemented at two levels: at the ISPs within the country and on the Internet backbone at the international gateway. These methods may overlap; an ISP may filter content using one particular technique while another technique is used at the international gateway.

Pakistan is an example of a country that blocks at both the international gateway and at the ISP level.

There are a few principal techniques used for Internet filtering including IP blocking, DNS tampering, and proxy-based blocking methods. (For blocking behavior by country, see table 1.3.) These techniques are presented in further detail by Anderson and Murdoch in chapter 3.

IP blocking is effective in blocking the intended target and no new equipment needs to be purchased. It can be implemented in an instant; all the required technology and expertise is

Table 1.3
Blocking techniques

	IP blocking	DNS tampering	Blockpage	Keyword
Azerbaijan	X		X	
Bahrain		X	X	
China	X			X
Ethiopia	X			
India	X	X		
Iran			X	X
Jordan	X			
Libya	X			
Myanmar			X	
Oman			X	
Pakistan	X	X		
Saudi Arabia			X	
Singapore			X	
South Korea	X	X	X	
Sudan			X	
Syria			X	
Thailand			X	
Tunisia			X	
United Arab Emirates			X	
Uzbekistan*			X	
Vietnam		X	X	
Yemen			X	X

Blocking behavior included in this table may include international gateway level filtering, and filtering techniques used by different ISPs.
* In Uzbekistan, the blockpage does not clearly indicate that filtering is occurring but rather redirects users to a third-party Web site.

readily available. Depending on the network infrastructure within the country it may also be possible to block at or near the international gateways so that the blocking is uniform across ISPs.

Countries new to filtering will generally start with IP blocking before moving on to more expensive filtering solutions. ISPs most often respond quickly and effectively to blocking orders from the government or national security and intelligence services. Therefore they block what is requested in the cheapest way using technology already integrated into their normal network environment. Blocking by IP can result in significant overblocking as all other (unrelated) Web sites hosted on that server will also be blocked.

China uses IP blocking to obstruct access to at least three hundred IP addresses. This blocking is done at the international gateway level affecting all users of the network regardless of ISP. The IPs blocked among the two backbone providers, China Netcom and ChinaTelecom, are remarkably similar.[9]

The ISP ETC-MC in Ethiopia uses IP blocking to block, among other sites, Google's Blogspot blogging service. This results in all Blogspot blogs being blocked in Ethiopia. Pakistan implements IP blocking at the international gateway level. In addition to blocking the IP for Blogspot, they also block Yahoo's hosting service, which results in major overblocking. For example, in targeting www.balochvoice.com they are actually blocking more than 52,000 other Web sites hosted on that same server.

DNS tampering is achieved by purposefully disrupting DNS servers, which resolve domain names into IP addresses. Generally, each ISP maintains its own DNS server for use by its customers. To block access to particular Web sites, the DNS servers are configured to return the wrong IP address. While this allows the blocking of specific domain names, it also can be easily circumvented by simple means such as accessing an IP address directly or by configuring your computer to use a different DNS server.

In Vietnam, the ISP FPT configures DNS to not resolve certain domain names, as if the site does not exist. The ISP Cybernet in Pakistan also uses this technique. The ISP Batelco in Bahrain uses this technique for some specific opposition sites. Batelco did not, however, completely remove the entry (the MX record for e-mail still remains). In India, the ISP BHARTI resolves blocked sites to the invalid IP address 0.0.0.0 while the ISP VSNL resolves blocked sites to the invalid IP address 1.2.3.4. The South Korean ISP, Hananet, uses this technique but makes the blocked Web site resolve to 127.0.0.1. This is the IP address for the "localhost." Another South Korean ISP, KORNET, makes blocked sites resolve to an ominous police Web site. This represents an unusual case in which DNS tampering resolves to a blockpage.[10]

Our tests revealed that there is often a combination of IP blocking and DNS tampering. It may be a signal that countries are responding to the outcry concerning the overblocking associated with IP blocking and moving to the targeting of specific domain names with DNS tampering. In India, for example, the Internet Service Providers Association of India reportedly has sent instructions to ISPs showing how to block by DNS instead of by IP.[11]

In proxy-based filtering strategies, Internet traffic passing through the filtering system is reassembled and the specific HTTP address being accessed is checked against a list of blocked Web sites. These can be individual domains, subdomains, specific long URL paths, or keywords in the domain or URL path. When users attempt to access blocked content they are subsequently blocked. An option in this method of filtering is to return a *blockpage* that informs the user that the content requested has been blocked.

Saudi Arabia uses SmartFilter as a filtering proxy and displays a blockpage to users when they try to access a site on the country's block list. The blockpage also contains information on how to request that a block be lifted. Saudi Arabia blocks access to specific long URLs. For example, www.humum.net/ is accessible, while www.humum.net/country/saudi.shtml is blocked. United Arab Emirates, Oman, Sudan, and Tunisia also use SmartFilter. Tunisia uses SmartFilter as a proxy to filter the Internet. But instead of showing users a blockpage indicating that the site has been blocked, they have created a blockpage that looks like the Internet Explorer browser's default error page (in French), presumably to disguise the fact that they are blocking Web sites.

A proxy-based filtering system can also be programmed such that Internet traffic passing through the filtering system is reassembled and the specific HTTP address requested is checked against a list of blocked keywords. No country that ONI tested blocked access to a Web site as a result of a keyword appearing in the body content of the page, however, there are a number of countries that block by keyword in the domain or URL path, including China, Iran, and Yemen.

China filters by keywords that appear in the host header (domain name) or URL path. For example, while the site http://archives.cnn.com/ is accessible, the URL http://archives.cnn.com/2001/ASIANOW/east/01/11/falun.gong.factbox/ is not. When this URL is requested, reset (RST) packets are sent that disrupt the connection, presumably because of the keyword *falun.gong*. Iran uses a filtering proxy that displays a blockpage when a blocked Web site is requested. On some ISPs in Iran, such as Shatel and Datak, keywords in URL paths are blocked. This most often affects search queries in search engines. For example, here is a query run on Google for *naked* in Arabic (www.google.com/search?hl=fa&q=%D9%84%D8%AE%D8%AA&btnG=%D8%A8%D9%8A%D8%A7%D8%A8) that was blocked. Ynet in Yemen blocks any URL containing the word *sex*. The domain www.arabtimes.com is blocked in Oman and the UAE but the URL for the Google cached version (http://72.14.235.104/search?q=cache:8utpDVLa1yYJ:www.arabtimes.com/+arabtimes&hl=en&ct=clnk&cd=1) is also blocked because *www.arabtimes.com* appears in the URL path.

Filtering systems can also be configured to redirect users to another Web site. In most cases, redirection is identical to blockpage filtering, the only difference being the route used to produce the blockpage. ISPs in Iran, Singapore, Thailand, and Yemen all use redirection to a blockpage. Uzbekistan uses redirection but instead of redirecting to a blockpage the filters send users to Microsoft's search engine at www.live.com, suggesting that the government wishes to conceal that fact that blocking has taken place.

There are thus various degrees of transparency in Internet filtering. Where blockpages are used, it is clearly apparent to users when a requested Web site has been intentionally blocked. Other countries give no indication that a Web site is blocked. In some cases, this is a function of the blocking technique being used. Some countries, such as Tunisia and Uzbekistan, appear to deliberately disguise the fact that they are filtering Internet content, going a step farther to conceal filtering activity beyond the failure to inform users that they are being filtered.

Another subset of countries, including Bahrain and United Arab Emirates, employ a hybrid strategy, indicating clearly to users that certain sites are blocked while obscuring the blocking of other sites behind the uncertainty of connection errors that could have numerous other explanations. In Bahrain, users normally receive a blockpage. However, for the specific site www.vob.org, Bahrain uses DNS tampering that results in an error. In United Arab Emirates all blocked sites with the exception of www.skype.com returned a blockpage. There is an apparent two-tiered system in place. They are willing to go on the record as blocking some sites, and not for others.

Providing a blockpage informing a user that their choice of Web site is not available by action of the government is still short of providing a rationale for the blocking of that particular site, or providing a means for appealing this decision. Very few countries go this far. A small group of countries, including Saudi Arabia, Oman, and United Arab Emirates, and some ISPs in Iran, allow Internet users to write to authorities to register a complaint that a given Web site has been blocked erroneously.

Centralized filtering regimes require all Internet traffic to pass through the same filters. This results in a consistent view of the Internet for users within the country; all users experience the same degree of filtering. This is most commonly implemented at the international gateway. When filtering is delegated to the ISP level, and hence decentralized, there may be significant differences among ISPs regarding the filtering techniques used and the content that is filtered. In this case, access to Web sites may vary substantially depending on the blocking choices of individual ISPs. (Table 1.4 presents the use of centralized and/or decentralized filtering strategies across the countries in the study, and the resulting consistency in filtering within each country.) In Iran there is considerable variation in the blocking among ISPs. For example, one ISP blocks considerably less political content than the other six ISPs tested. Only one ISP out of the five tested in Azerbaijan, AzNet, blocks access to a considerable amount of social content, most of which is pornographic, while the others block access to only a single IP address. In Myanmar, there is substantial variation in the filtering between the two ISPs tested. One filters much more pornography, while the other blocks a significantly greater portion of politically oriented Web sites. In the United Arab Emirates, an ISP that serves primarily the free-trade

Table 1.4
Comparing filtering regimes

	Locus	Consistency	Concealed filtering	Transparency and accountability
Azerbaijan	D	Low		Medium
Bahrain	C	High	Yes	Low
China	C and D	Medium	Yes	Low
Ethiopia	C	High	Yes	Low
India	D	Medium		High
Iran	D	Medium		Medium
Jordan	D	High		Low
Libya	C	High	Yes	Low
Morocco	C	High	Yes	Low
Myanmar	D	Low		Medium
Oman	C	High		High
Pakistan	C and D	Medium	Yes	High
Saudi Arabia	C	High		High
Singapore	D	High		High
South Korea	D	High		High
Sudan	C	High		High
Syria	D	High		Medium
Tajikistan	D	Low		Medium
Thailand	D	Medium		Medium
Tunisia	C	High	Yes	Low
United Arab Emirates	D	Low		Medium
Uzbekistan	C and D	High	Yes	Low
Vietnam	D	Low	Yes	Low
Yemen	D	High		Medium

The **Locus** of filtering indicates where Internet traffic is blocked. **C** indicates that traffic is blocked from a central location, normally the Internet backbone, and affects the entire state equally. **D** indicates that blocking is decentralized, typically implemented by ISPs. (Note that this study does not include filtering at the institutional level, for example, cybercafés, universities, or businesses.)
Consistency measures the variation in filtering within a country across different ISPs where applicable.
Concealed filtering reflects either efforts to conceal the fact that filtering is occurring or the failure to clearly indicate filtering when it occurs.
Transparency and accountability corresponds to the overall level of openness in regard to the practice of filtering. It also considers the presence of concealed filtering, the type of notice given to users regarding blocking, provisions to appeal or report instances of inappropriate blocking, and public acknowledgement of filtering policies.

zone has not historically filtered the Internet, while the predominant ISP for the rest of the country has consistently filtered the Internet.

Modifications can be made to the blocking efforts of a country by the authorities at any time. Sites can be added or removed at their discretion. For example, during our tests in Iran the Web site of the *New York Times* was blocked, but for only one day. Some countries have also been suspected of introducing temporary filtering around key time periods such as elections.

Hosting modifications can also be made to a blocked site resulting in it becoming accessible or inaccessible. For example, while Blogspot blogs were blocked in Pakistan due to IP blocking, the interface to update one's blog was still accessible. However, Blogspot has since upgraded its service and the new interface is hosted on the blocked IP, making the interface to update one's blog inaccessible in Pakistan. The reverse is also possible. For example, if the IP address of a Web site is blocked, the Web site may change its hosting arrangement in order to receive a new IP address, leaving it unblocked until the new IP address is discovered and blocked.

Summary Measures of Internet Filtering

To summarize the results of our work, we have assigned a score to each of the countries we studied. This score is designed to reflect the degree of filtering in each of the four major thematic areas: 1) the filtering of political content, 2) social content, 3) conflict- and security-related content, and 4) Internet tools and applications. Each country is given a score on a four-point scale that captures both the breadth and depth of filtering for content of each thematic type (see table 1.5).

- Pervasive filtering is defined as blocking that spans a number of categories while blocking access to a large portion of related content.
- Substantial filtering is assigned where either a number of categories are subject to a medium level of filtering in at least a few categories or a low level of filtering is carried out across many categories.
- Selective filtering is either narrowly defined filtering that blocks a small number of specific sites across a few categories, or filtering that targets a single category or issue.
- Suspected filtering is assigned where there is information that suggests that filtering is occurring, but we are unable to conclusively confirm that inaccessible Web sites are the result of deliberate tampering.

The scores in table 1.5 are subjective evaluations based upon the quantitative information gathered during a year of testing and research. In 2006, we tested thousands of Web sites across more than 120 ISPs in 40 countries, creating a database with close to 200,000 observations. Each observation is in turn based on the conclusion of an average of ten accessibility tests. Despite the breadth of this data, a purely quantitative reporting might be

Table 1.5
Summary of filtering

	Political	Social	Conflict and security	Internet tools
Azerbaijan	●	—	—	—
Bahrain	●●	●	—	●
Belarus	○	○	—	—
China	●●●	●●	●●●	●●
Ethiopia	●●	●	●	●
India	—	—	●	●
Iran	●●●	●●●	●●	●●●
Jordan	●	—	—	—
Kazakhstan	○	—	—	—
Libya	●●	—	—	—
Morocco	—	—	●	●
Myanmar	●●●	●●	●●	●●
Oman	—	●●●	—	●●
Pakistan	●	●●	●●●	●
Saudi Arabia	●●	●●●	●	●●
Singapore	—	●	—	—
South Korea	—	●	●●●	—
Sudan	—	●●●	—	●●
Syria	●●●	●	●	●●
Tajikistan	●	—	—	—
Thailand	●	●●	—	●
Tunisia	●●●	●●●	●	●●
United Arab Emirates	●	●●●	●	●●
Uzbekistan	●●	●	—	●
Vietnam	●●●	●	—	●●
Yemen	●	●●●	●	●●

●●● Pervasive filtering; ●● Substantial filtering; ● Selective filtering; ○ Suspected filtering; — No evidence of filtering.

misleading unless we were able to effectively measure the relative importance of each Web site. For example, the blocking of BBC or Wikipedia represents far more than the blocking of a less prominent Web site. Similarly, blocking a social networking site or a blogging server could have a profound impact on the formation of online communities and on the publication of user-generated content. While Internet users will eventually provide alternatives to recreate these communities on other sites hosted on servers that are not blocked, the transition of a wide community is unlikely given the time, effort, and coordination required to reconstitute a community in another location. At the other extreme, the blocking of one pornographic site will have a minor impact on Internet life if access to thousands of similar sites remains unimpeded. For these reasons, we have decided to summarize the results of testing categorically, considering both the scope and depth of the quantitative testing results, in conjunction with expert opinion regarding the significance of the blocking of individual Web sites.

It is tempting to aggregate the results by summing up the scores in each category. Yet doing so would imply that the blocking of political opposition is equivalent to filtering that supports conservative social values or the fear of national security risks. These competing sets of values suggest that a number of different weighting schemes might be appropriate. In any case, the results are generally quite clear, as the most pervasive filtering regimes tend to filter across all categories.

Country-specific and Global Filtering

A comparison between the blocking of country-specific sites and the blocking of internationally relevant Web sites provides another view of global filtering. Not surprisingly, we found that

Box 1.2
Where we tested

The decision where to test was a simple pragmatic one—where were we able to safely test and where did we have the most to learn? Two countries did not make the list this year because of security concerns: North Korea and Cuba. Learning more about the filtering practices in these countries is certainly of great interest to us. However, we were not confident that we could adequately mitigate the risks to those who would collaborate with us in these countries.

A number of other countries in Europe and North America that are known to engage in filtering to varying degrees were not tested this year. This decision again was a fairly easy practical choice. The filtering practices in these countries are better understood than in other parts of the world and we therefore had less to contribute here. Many of the countries in Europe focus their Internet filtering activity on child pornography. This is not a topic that we will test for ethical and legal reasons.

the incidence of blocking Web sites in our testing lists was approximately twice as high for Web sites available in a local language compared to sites available only in English or other international languages. Figure 1.3 shows that many countries focus their efforts on filtering locally relevant Web content. Ethiopia, Pakistan, Syria, Uzbekistan, and Vietnam are examples of countries that extensively block local content while blocking relatively few international Web sites. China and Myanmar also concentrate more of their filtering efforts on country-specific Internet content, though they block somewhat more global content. Middle Eastern filtering regimes tend to augment local filtering with considerably more global content. This balance mirrors the use of commercial software, generally developed in the West, to identify and block Internet content.

Table 1.6 shows an alternative view of filtering behavior, looking at the blocking of different types of content providers rather than content. The apparent prime targets of filtering are blogs, political parties, local NGOs, and individuals. In the case of blogs, a number

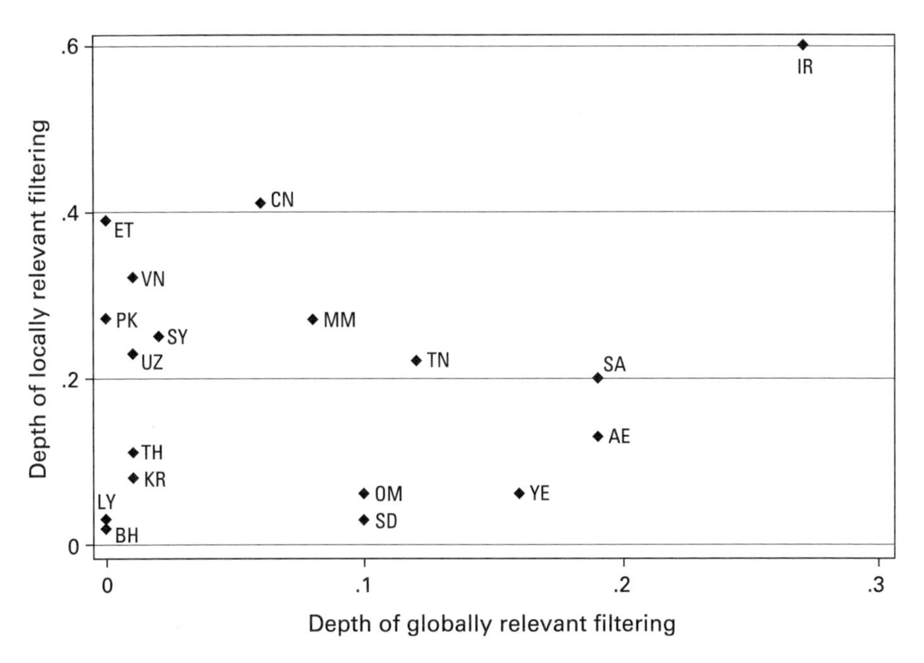

Figure 1.3
Filtering targeted at local sites and global sites. AE—United Arab Emirates; BH—Bahrain; CN—China; ET—Ethiopia; IR—Iran; JO—Jordan; KR—South Korea; LY—Libya; MM—Burma/Myanmar; OM—Oman; PK—Pakistan; SA—Saudi Arabia; SD—Sudan; SY—Syria; TH—Thailand; TH—Tunisia; UZ—Uzbekistan; VN—Vietnam; YE—Yemen. A number of countries that filter a small number of sites are omitted from this diagram, including Azerbaijan, Belarus, India, Jordan, Kazakhstan, Morocco, Singapore, and Tajikistan.

of countries, including Pakistan and Ethiopia, have blocked entire blogging domains, which inflates these figures. Logically, these assessments represent more accurately the result of filtering rather than the intention. Establishing the intention of blocking is never as clear. The blocking of this wide array of blogs could be the result of a lack of technical sophistication or a desire to simultaneously silence the entire collection of blogs hosted on the site.

The other prominent target of filtering is political parties, followed by NGOs focused on a particular region or country, and Web sites run by individuals. The implications of targeting civic groups and individual blogs are addressed by Deibert and Rohozinski in chapter 6 of this volume.

First Steps Toward Understanding Internet Filtering

In this chapter, we summarize what we have learned over the past year regarding the incidence of global Internet filtering. Taking an inventory of filtering practices and strategies is a necessary and logical first step, though still far from a thorough understanding of the issue. The study of Internet filtering can be approached by asking why some states filter the Internet or by asking why others do not. The latter question is particularly apt in countries that maintain a repressive general media environment while leaving the Internet relatively open. This is not

Table 1.6
Blocking by content provider

Content provider type	Portion of content filtered
Academic	0.02
Blogs	0.20
Chat and discussion boards	0.05
Government	0.03
Government media	0.02
International governmental organizations	0.00
Independent media	0.06
Individual	0.09
International NGOs	0.02
Labor groups	0.05
Locally focused NGOs	0.09
Militant groups	0.01
Political parties	0.19
Private businesses	0.06
Religious groups	0.02
Regional NGOs	0.04

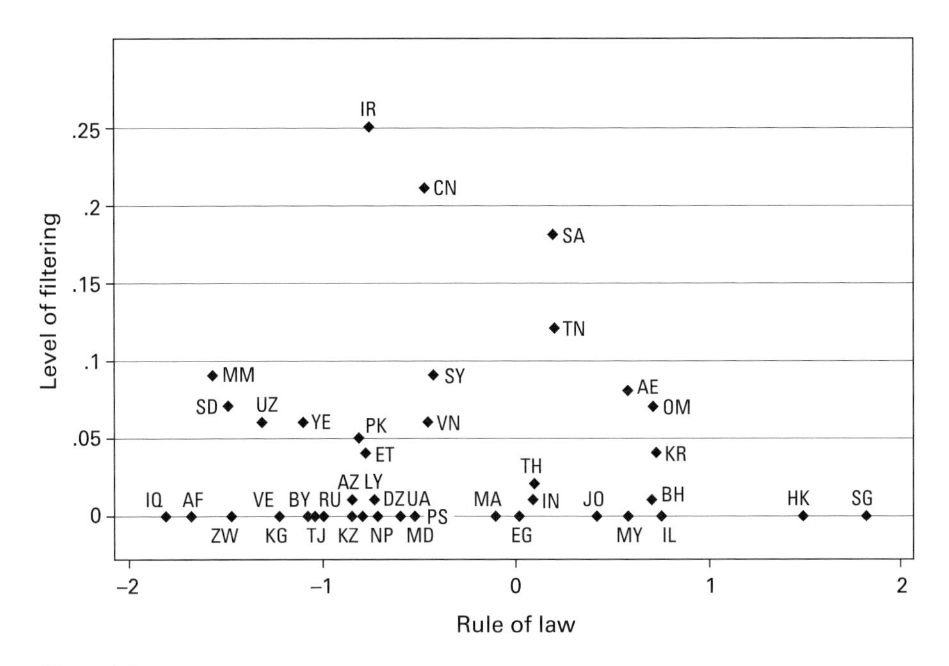

Figure 1.4

Filtering and the rule of law. AE—United Arab Emirates; AF—Afghanistan; AZ—Azerbaijan; BH—Bahrain; BY—Belarus; CN—China; DZ—Algeria; EG—Egypt; ET—Ethiopia; HK—Hong Kong; IL—Israel; IN—India; IR—Iran; IQ—Iraq; JO—Jordan; KG—Kyrgyzstan; KR—South Korea; KZ—Kazakhstan; LY—Libya; MA—Morocco; MD—Moldova; MM—Burma/Myanmar; MY—Malaysia; NP—Nepal; OM—Oman; PK—Pakistan; PS—Gaza/West Bank; RU—Russia; SA—Saudi Arabia; SD—Sudan; SG—Singapore; SY—Syria; TH—Thailand; TH—Tunisia; TN—Tunisia; TJ—Tajikistan; UA—Ukraine; UZ—Uzbekistan; VE—Venezuela; VN—Vietnam; YE—Yemen; ZW—Zimbabwe.

an uncommon circumstance. Pointing simply toward the absence of a solid rule of law does not seem promising. As seen in figure 1.4, there is no simple relationship between the rule of law and filtering, at least not as rule of law is defined and measured by the World Bank.[12] A country can maintain a better-than-average rule of law record and still filter the Internet. Similarly, many countries suffer from a substandard legal situation while maintaining an open Internet.

Comparing filtering practices with measures of voice and accountability is more telling. The countries that actively engage in the substantial filtering of political content also score poorly on measures of voice and accountability. This is true for both political and social Internet blocking, as shown in figures 1.5 and 1.6. Yet many of the anomalies persist. We are still far from explaining why some countries resort to filtering while others refrain from taking this step. This does stress the diversity of strategies and approaches that are being taken to regulate

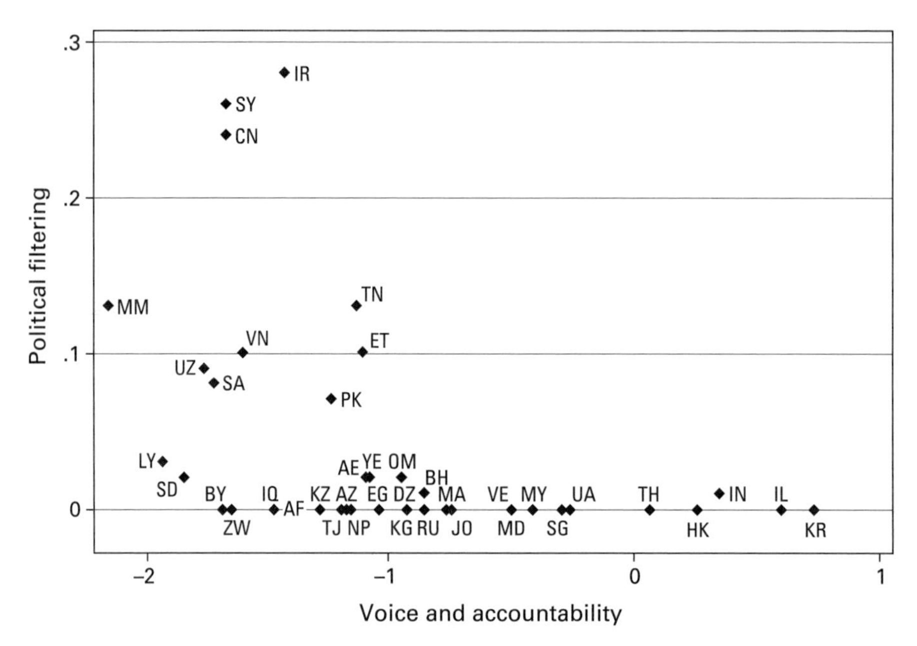

Figure 1.5

Political filtering and voice and accountability. AE—United Arab Emirates; AF—Afghanistan; AZ—Azerbaijan; BH—Bahrain; BY—Belarus; CN—China; DZ—Algeria; EG—Egypt; ET—Ethiopia; HK—Hong Kong; IL—Israel; IN—India; IR—Iran; IQ—Iraq; JO—Jordan; KG—Kyrgyzstan; KR—South Korea; KZ—Kazakhstan; LY—Libya; MA—Morocco; MD—Moldova; MM—Burma/Myanmar; MY—Malaysia; NP—Nepal; OM—Oman; PK—Pakistan; PS—Gaza/West Bank; RU—Russia; SA—Saudi Arabia; SD—Sudan; SG—Singapore; SY—Syria; TH—Thailand; TH—Tunisia; TN—Tunisia; TJ—Tajikistan; UA—Ukraine; UZ—Uzbekistan; VE—Venezuela; VN—Vietnam; YE—Yemen; ZW—Zimbabwe.

the Internet. We are also observing a recent and tremendously dynamic process. The view we have now may change dramatically in the coming years.

The link between repressive regimes and political filtering follows a clear logic. However, the link between regimes that suppress free expression and social filtering activity is less obvious. Part of the answer may reside in that regimes that tend to filter political content also filter social content.

Figure 1.7 demonstrates that few states restrict their activities to one or two types of content. Once filtering is implemented, it is applied to a broad range of content. These different types of filtering activities are often correlated with each other, and can be used as a pretense for expanding government control of cyberspace.

Vietnam, for example, uses pornography as its publicly stated reason for filtering, yet blocks little pornography. It does, however, filter political Internet content that opposes one-party rule

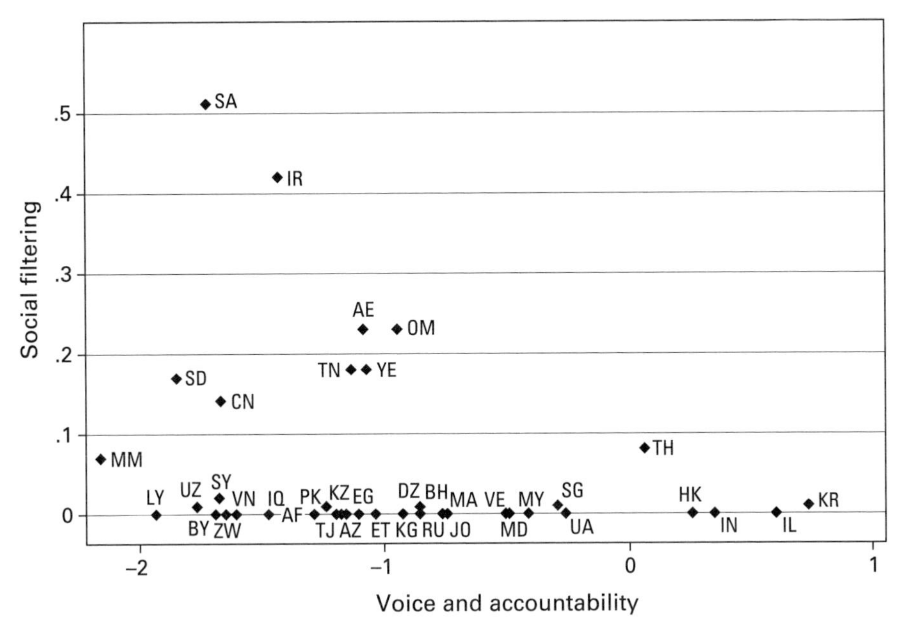

Figure 1.6

Social filtering and voice and accountability. AE—United Arab Emirates; AF—Afghanistan; AZ—Azerbaijan; BH—Bahrain; BY—Belarus; CN—China; DZ—Algeria; EG—Egypt; ET—Ethiopia; HK—Hong Kong; IL—Israel; IN—India; IR—Iran; IQ—Iraq; JO—Jordan; KG—Kyrgyzstan; KR—South Korea; KZ—Kazakhstan; LY—Libya; MA—Morocco; MD—Moldova; MM—Burma/Myanmar; MY—Malaysia; NP—Nepal; OM—Oman; PK—Pakistan; PS—Gaza/West Bank; RU—Russia; SA—Saudi Arabia; SD—Sudan; SG—Singapore; SY—Syria; TH—Thailand; TH—Tunisia; TN—Tunisia; TJ—Tajikistan; UA—Ukraine; UZ—Uzbekistan; VE—Venezuela; VN—Vietnam; YE—Yemen; ZW—Zimbabwe.

in Vietnam. In Saudi Arabia and Bahrain, filtering does not end with socially sensitive material such as pornography and gambling but expands into the political realm.

Once the technical and administrative mechanisms for blocking Internet content have been put into place, it is a trivial matter to expand the scope of Internet censorship. As discussed in subsequent chapters, the implementation of filtering is often carried by private sector actors—normally the ISPs—using software developed in the United States. Filtering decisions are thus often made by selecting categories for blocking within software applications, which may also contain categorization errors resulting in unintended blocking. The temptation and potential for mission creep is obvious. This slope is made ever more slippery by the fact that transparency and accountability are the exception in Internet filtering decisions, not the norm.

In the following chapter, Zittrain and Palfrey probe in further detail the political motives and implications of this growing global phenomenon, with subsequent chapters elaborating on technical, legal, and ethical considerations.

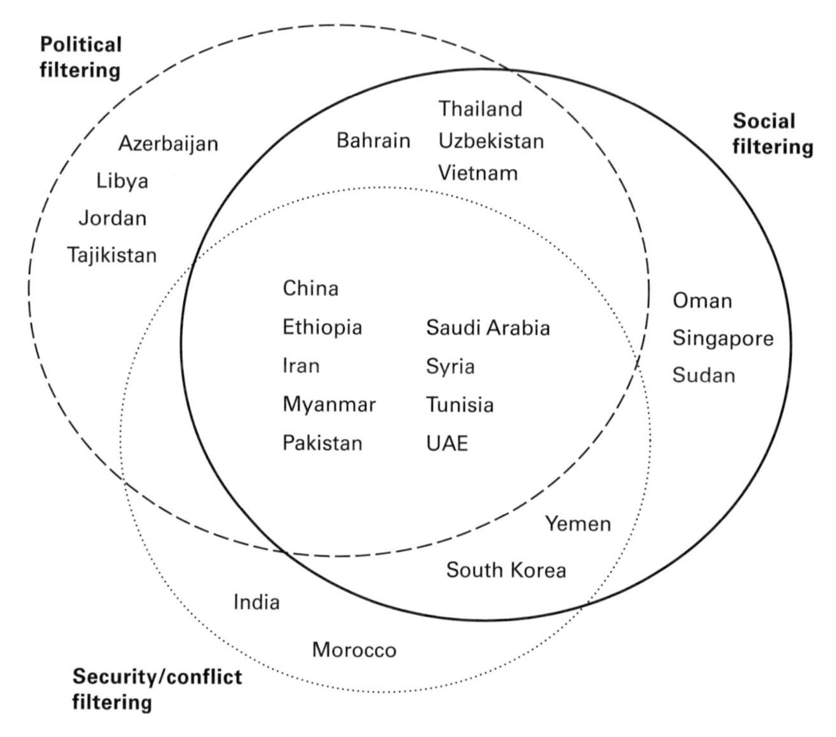

Figure 1.7
Content filtering choices.

Notes

1. The OpenNet Initiative is a collaboration of four institutions: the Citizen Lab at the University of Toronto, the Oxford Internet Institute at Oxford University, the Berkman Center for Internet & Society at Harvard Law School, and the University of Cambridge. More information is available at http://www.opennetinitiative.net.
2. A number of countries are currently debating strategies and legislation to filter the Internet, including Norway, Russia, and many countries in Latin America.
3. Each set of tests is performed on different Internet service providers within the country.
4. The Internet filtering tests carried out in Russia in 2006 were limited to ISPs accessible in Moscow. These results therefore do not necessarily reflect the situation in other areas of the country.
5. The blocking of two sites garnered most of the attention: one devoted to opposition to the September 19 coup (http://www.19sep.com/) and another hosted by Thai academics (http://www.midnightuniv.org/).
6. The strategies for addressing alleged intellectual property rights violations can vary significantly from standard Internet filtering. Rather than blocking Web sites that continue to be available from other locations, efforts generally focus on taking down the content from the Web sites that have posted the material and on removing the sites from the results of search engines. Moreover, takedown efforts are often instigated by private parties with the threat of subsequent legal action rather than being initiated by government action. See www.chillingeffects.org for more information.

7. The ONI Vietnam report is available at http://www.opennetinitiative.net/studies/vietnam/ONI_Vietnam_Country_Study.pdf.
8. We were not able to test in Cuba or North Korea. Both countries are reported to engage in pervasive filtering in addition to curtailing access to the Internet. See ''Going Online in Cuba: Internet under Surveillance,'' http://www.rsf.org/IMG/pdf/rapport_gb_md_1.pdf, and Tom Zeller, ''The Internet Black Hole That Is North Korea,'' *New York Times*, 23 October 2006.
9. There are two principal ISPs in China—one that covers the north and one the south. The smaller ISPs in China that serve Internet users connect to the Internet backbone through one of these large ISPs.
10. It also demonstrates that the use of DNS tampering does not necessitate a lack of transparency in filtering. If it were deemed important, users could be informed that the Web site they were seeking was being intentionally blocked.
11. See Shivam Vij, ''Blog Blockade Will Be Lifted in 48 Hours,'' Rediff India Abroad, http://www.rediff.com/news/2006/jul/19blogs.htm.
12. Information on the compilation and estimation of the ''rule of law'' and ''voice and accountability'' measures are available at the World Bank Governance and Anti-Corruption Web site, www.worldbank.org/wbi/governance. Their definitions of these indicators are: ''Voice and Accountability includes in it a number of indicators measuring various aspects of the political process, civil liberties, political and human rights, measuring the extent to which citizens of a country are able to participate in the selection of governments.'' ''Rule of Law includes several indicators which measure the extent to which agents have confidence in and abide by the rules of society. These include perceptions of the incidence of crime, the effectiveness and predictability of the judiciary, and the enforceability of contracts.''

2

Internet Filtering: The Politics and Mechanisms of Control

Jonathan Zittrain and John Palfrey

It seems hard to believe that a free, online encyclopedia that anyone can edit at any time could matter much to anyone. But just as a bee can fly despite its awkward physiognomy, Wikipedia has become wildly popular and enormously influential despite its unusual format. The topics that Wikipedians write about range more broadly than any other encyclopedia known to humankind. It has more than 4.6 million articles comprising more than a billion words in two hundred languages.[1] Many Google search queries will lead to a Wikipedia page among the top search results. Articles in Wikipedia cover the Tiananmen Square protests of 1989, the Dalai Lama, the International Tibet Independence Movement, and the Taiwan independence movement. Appearing both in the English and the Chinese language versions of Wikipedia—each independently written—these articles have been written to speak from what Wikipedia calls a "neutral point of view."[2] The Wikipedians' point of view on some topics probably does not seem so neutral to the Chinese authorities.

Wikipedia has grown so influential, in fact, that it has attracted the attention of China's censors at least three times between 2004 and 2006.[3]

The blocking and unblocking of Wikipedia in China—as with all other filtering in China, without announcement or acknowledgment—might also be grounded in a fear of the communal, critical process that Wikipedia represents. The purpose of Wikipedia is "to create and distribute a multilingual free encyclopedia of the highest quality to every single person on the planet in their own language,"[4] and the means of creating it is through engagement of the public at large to contribute what it knows and to debate in earnest where beliefs differ, offering sources and arguments in quasiacademic style.

While its decentralization creates well-known stability as a network, this decentralization reflected at the "content layer" for the purpose of ascertaining truth might give rise to radical instability at the social level in societies that depend on singular, official stories for their legitimacy. Wikipedia makes it possible for anyone to tell one's own story about what happened—and, more threateningly to a regime intent on controlling the information environment, to compare notes with others in a process designed to elicit truth from competing perspectives. The once stable lock of a regime accustomed to telling its citizenry how things happened—where states have controlled their media environments for a long time—is threatened.

Wikipedia is the poster-story of a new iteration of the Internet, known as the *read-write Web*, or *Web 2.0* in Silicon Valley terms, or the *semantic Web* in MIT terms. This phenomenon—in which consumers of information can also easily be creators—threatens to open and to destabilize political environments that were previously controlled tightly by those in power. In a world where Wikipedia is accessible, citizens not only can read different versions of the story than the version that the state would have them read, they can help to create them. And not just in their own language: as automated translation tools come into their own, they can interact in many languages.

This version of the Internet also continues the process of breaking down geographic barriers between states by allowing information to flow easily from one jurisdiction to another. The editors of a Wikipedia article are telling the story not just for the benefit of their neighbors in their own country, but for anyone in any place to see. The destabilization that Wikipedia makes possible is also a threat to the ability of a given state to control how its own "brand" is perceived internationally.

So why and how would the Chinese state block—and then unblock—Wikipedia repeatedly? That question lies at the heart of this book. Internet filtering is a complex topic, easy to see from many vantage points but on which it is hard to get a lasting fix. As a practical matter, it is easy for a state to carry out technical Internet filtering at a simple level, but very tricky—if not plain impossible—for a state to accomplish in a thorough manner. As a policy matter, are states putting in place filtering regimes because they are concerned that their own citizens will learn something they should not learn? Or that their citizens will say something they should not say? Or that someone in another state will read something bad about the state that is filtering the Net? Where is filtering merely a ministerial task, taken on because the state bureaucracy feels it must at least look like it is making an effort, and where is it a central instrument of policy, initiated if not orchestrated at the highest levels of power?

As a normative matter, broad, informal Internet filtering seems like an infringement of the civil liberties—or, put more forcefully, the human rights—of all of us who use the free, public, unitary, global network of networks that the Internet constitutes. But states have a strong argument that they have the right to control domestic matters, whether or not they occur in cyberspace, and there is often little that other states can do to influence them. The future of the Internet, if not all geopolitics, hangs in the balance.

We are still in the early stages of the struggle for control on the Internet. Early theorists, reflecting the libertarian streak that runs deep through the hacker community, suggested that the Internet would be hard to regulate. Cyberspace might prove to be an alternate jurisdiction that the long arm of the state could not reach. Online actors, the theory went, need pay little heed to the claims to sovereignty over their actions by traditional states based in real space.

As it turns out, states have not found it so very hard to assert sovereignty when and where they have felt the need to do so. The result is the emergence of an increasingly balkanized

Internet. Instead of a World Wide Web, as the data from our study of Internet filtering makes plain, it is more accurate to say we have a Saudi Wide Web, an Uzbek Wide Web, a Pakistani Wide Web, a Thai Wide Web, and so forth. The theory of "unregulability" no longer has currency, if ever it did. Many scholars have described the present reality of the reassertion of state control online, despite continued hopes that the Internet community itself might self-regulate in new and compelling ways.[5]

A key aspect of control online—and one that we prove empirically through our global study of Internet filtering—is that states have, on an individual basis, defied the cyberlibertarians by asserting control over the online acts of their own citizens in their home states. The manner in which this control is exercised varies. Sometimes the law pressures citizens to refrain from performing a certain activity online, such as accessing or publishing certain material. Sometimes the state takes control into its own hands by erecting technological or other barriers within its confines to stop the flow of bits from one recipient to another. Increasingly, though, the state is turning to private parties to carry out the control online. Many times, those private parties are corporations chartered locally or individual citizens who live in that jurisdiction. In chapter 5, we describe a related, emerging problem, in which the state requires private intermediaries whose services connect one online actor to another to participate in online censorship and surveillance as a cost of doing business in that state.

The dynamic of control online has changed greatly over the past ten years, and it is almost certain to change just as dramatically in the ten years to come. The technologies and politics of control of the Internet remain in flux. As one example of this continued uncertainty, participants in the Internet Governance Forum (IGF), an open global body chartered via the process that produced two meetings of the World Summit on the Information Society (WSIS), continue to wrestle with a broad set of unanswered questions related to control of the online environment. At a simple level, the jurisdictional question of who can sue whom (and where that lawsuit should be heard, and under the law of which jurisdiction decided, for that matter), remains largely unresolved, despite a growing body of case law. A series of highly distributed problems—spam, spyware, online fraud—continues to vex law enforcement officials and public policy–makers around the world. Intellectual property law continues to grow in complexity, despite some degree of harmonization underway among competing regimes. Each of these problems leaves many unresolved issues of global public policy in its wake. Internet filtering, the core focus of this book, and the related matter of online surveillance, present an equally, if not more, fraught set of issues for global diplomats to address.

Suppressing and Controlling Information on the Internet

The idea that states would seek to control the information environment within their borders is nothing new. Freedom of expression has never been absolute, even in those liberal democracies that hold these freedoms most dear. The same is true of the related freedoms of association, religion, and privacy. Most states that have been serious about controlling the information environment have done so by holding on to the only megaphones—whether it takes the form of a printing press, a newspaper, a radio station, or a television station—and banning anyone from saying anything potentially subversive.

The rise of the Internet, initially seen as little more than an information delivery mechanism, put pressure on this strategy of control. Early in the twenty-first century, the Saudi state was one of the first to grapple publicly with what the introduction of the Internet might mean. Rather than introducing the Internet in its unfettered—and fundamentally Western—form, the Saudi authorities decided to establish a system whereby they could stop their citizens from accessing certain materials produced and published from elsewhere in the world. As an extension of its longstanding traditional media controls, the Saudis set up a technical means of filtering the Internet, buttressed by a series of legal and normative controls. All Internet traffic to and from Saudi citizens had to pass through a single gateway to the outside world. At that gateway, the Saudi state established a technical filtering mechanism. If Saudi citizens sought to access a Web page that earlier had been found to include a pornographic image, for instance, the computers at the gateway would send back a message telling them, in Arabic, that they have sought to access a forbidden page—and, of course, not rendering the requested page. At a fundamental level, this basic form of control was initially about blocking access to information that would be culturally and politically sensitive to the state.

The issue that the Saudi state faces, of desiring to keep its citizens from accessing subversive content online, is an issue that more and more states are coming to grips with as the Internet expands. The network now joins more than one billion people around the world. At the same time, new issues are arising that are prompting states to establish Internet filtering mechanisms. The read-write Web, exemplified by Wikipedia and the phenomena of blogging, YouTube, podcasting, and so forth, adds a crucial dimension—and additional complexity—as states now grapple with the ease with which their own citizens are becoming publishers with local, national, and international audiences.

How Internet Filtering Works: Law, Technologies, and Social Norms

When states decide to filter the Internet, the approach generally involves establishing a phalanx of laws and technical measures to block their citizens from accessing or publishing information online. The laws are ordinarily extensions of pre-existing media or telecommunications regulatory regimes. Occasionally, these laws take the form of Internet-specific statutes and regulations. These laws rarely explicitly establish the technical filtering regime, but more commonly establish a framework for restricting certain kinds of content online and banning certain online activities.

There are at least five levels of Internet legal control with respect to content control online.[6] States have employed content restrictions, which disallow citizens from publishing or accessing certain online content. Licensing requirements call for intermediaries to carry out certain

Internet filtering, as well as surveillance activities. Liability placed on Internet service providers and Internet content providers can ensure that intermediaries affirmatively carry out filtering and surveillance without a license requirement. Registration requirements establish the need to gather data about citizens accessing the Internet from a certain IP address, user account, cybercafé location, and so forth. And self-monitoring requirements—coupled with the perception, real or imagined, of online surveillance—prompt individual, corporate, and other users to limit their own access and publication online. At the same time, some states are experiencing international pressure to pass Internet-related laws, including omnibus cybercrime statutes that include reference to eliminating access to certain types of banned sites.[7]

The interplay among these types of regulations is a key aspect of this narrative. China, for instance, bundles Internet content restrictions with its copyright laws. This set of regulations sets a daunting web of requirements in front of anyone who might access the Internet or provide a service to another Internet user. These rules create a pretext that can be used to punish those who exchange undesirable content, even though the law may not be invoked in many instances it might cover—including copyright infringement. Vietnam has taken a similar approach, assigning a number of different relevant ministries and agencies a piece of the responsibility to limit what can be done and accessed online. Much of the legal regulation that empowers state agencies to carry out filtering and surveillance tends to be very broadly and vaguely stated, where it is stated at all.

A theme that runs through this book is that Internet regulation takes many forms—not just technical, not just legal—and that regulation takes place not just in developing economies but in some of the world's most prosperous regimes as well. Vagueness as to what content is banned exists not just in China, Vietnam, and Iran, but also in France and Germany, where the requirement to limit Internet access to certain materials includes a ban on "propaganda against the democratic constitutional order."[8] Often, these local legal requirements strike a dissonant chord when set alongside international human rights standards, a topic covered in greater detail, and from two different perspectives, in chapters 5 and 6 of this book.

As our global survey shows, and as Faris and Villeneuve set forth in chapter 1 of this volume, several dozen states have gone beyond a legal ban on citizen publication or access of undesirable material online and have set up technical means of filtering its citizens' access to the Internet. In establishing a technical filtering regime, a state has several initial options: domain name system (DNS) filtering, Internet protocol (IP) address filtering, or URL filtering.[9] Most states with advanced filtering regimes implement URL filtering, as this method can be the most accurate (see "Filtering and Overbreadth" section later in this chapter).[10]

To implement URL filtering, a state must first identify where to place the filters. If the state directly controls the Internet service provider(s), the answer is clear. Otherwise, it may require private or semiprivate ISPs to implement the blocking as part of their service. The technical complexities presented by URL filtering become nontrivial as the number of users grows to millions rather than tens of thousands. Some states appear to have limited the number of people who can access the Internet, as Myanmar has, in order to keep URL filtering manageable—or to be able to shut off access to the network entirely, as the military junta appears to have done in September 2007.

Technical Internet filtering is not perfect in any jurisdiction. Even the most sophisticated technical filtering regimes can have difficulty covering those cases where users are intent on getting or publishing certain information, and willing to invest effort and risk to do so. Every system suffers from at least two shortcomings: a technical filtering system either underblocks or overblocks content, and technically savvy users can circumvent the controls with a modicum of effort. Citizens with technical knowledge can generally circumvent any filters that a state has put in place. Some states acknowledge as much: the overseer of Saudi Arabia's filtering program admits that technically savvy users simply cannot be stopped from accessing blocked content.[11] While no state will ultimately win the game of cat-and-mouse with those citizens who are resourceful and dedicated enough to employ circumvention measures, many users will never do so.

For some states, like Singapore, the state's bark is worse than the bite of the filtering system. The widely publicized Singaporean filtering system blocks only a small handful of pornographic sites. The Singapore system is more about sending a message, one that underscores the substantial local self-censorship that takes place there, than it is about blocking citizens from accessing or publishing anything specific. For other states, like those with the most thorough and sophisticated filtering regimes—for instance, China, Iran, and Uzbekistan—the undertaking is far more substantial and has far-reaching consequences.

Locus of Filtering

Most state-mandated filtering is effected by private ISPs that offer Internet access to citizens under licenses to operate in that jurisdiction. These licenses tend to include requirements, explicit or implicit in nature, that the ISPs implement filtering at the behest of the state. Some states partially centralize the filtering operation at private Internet exchange points (IXPs)—topological crossroads for network traffic—or through explicit state-run clearing points established to serve as gatekeepers for Internet traffic. Some states implement filtering at public Internet access points, such as the computers found within cybercafés or in public libraries and schools, as in the United States. Such filtering can take the form of software used in many American libraries and schools for filtering purposes, or *normative* filtering—government-encouraged social pressure by shop proprietors, librarians, and others as citizens surf the Internet in common public places.

The exercise of traditional state powers can have a powerful impact on Internet usage without rendering all content in a particular category inaccessible. China, Vietnam, and Iran, among others, have each jailed "cyber-dissidents."[12] Against this backdrop, the blocking of Web pages may be intended to deliver a message to users that the government monitors Internet usage. This message is reinforced by methods allowing information to be gathered about what sites a particular user has visited after the fact, such as the requirement that a user provide passport information to set up an account with an ISP and tighter controls of users at cybercafés, as in Vietnam. The on-again, off-again blocking of Wikipedia in China might well be explained, in part, by this mode of sending a message that the state is watching in order to prompt self-censorship online.

While our research can tell us what Web sites a regime has targeted for filtering, the real extent to which the information environment is "free" and "open" is sharply limited. It is not as easy to determine, for instance, the extent of citizens' attempts to reach blocked sites, the degree to which citizens are deterred by the threat of arrest or detection, and how much the invisibility of specific content actually affects the regime's internal dialogue. Our research provides the data to draw conclusions about the choices made by states as to the content to be filtered, how these decisions are affected by the mechanisms for filtering they have employed, and how these governments attempt to balance the overblocking or underblocking that is today inherent in any filtering regime.

Types of Content Filtered

As Faris and Villeneuve describe in chapter 1, states around the world are blocking access to Internet content for its political, religious, and social connotations. Sensitivities related to specific content within these broad categories vary greatly from state to state, tracking, to large extent, local conflicts. The Internet content blocked for social reasons—commonly pornography, information about gay and lesbian issues, and information about sex education—is more likely to be the same across countries than the political and religious information to which access is blocked.

Web content is constantly changing, of course, and no state we have studied in the past five years seems able to carry out its Web filtering in a comprehensive manner—at least not through technical means. In other words, no state has been able to consistently block access to a range of sites meeting specified criteria. The most thorough job of blocking might be the high rate of blocking online pornography in Saudi Arabia and other Gulf states—a fairly stunning achievement given the amount of pornographic material available online—and one managed with the ongoing help of a U.S. firm, Secure Computing, that also assists schools in keeping children away from such Web sites. China has the most consistent record of responding to the shifting content of the Web, likely reflecting a devotion of the most resources to the filtering enterprise. Our research shows changes among sites blocked over time in some states, such as Iran, Saudi Arabia, and China. As we repeat this global survey in future years, we expect to be able to describe changes over time with greater certainty.

The Reality and Perception of Internet Surveillance and Other Soft Controls

Just as these several dozen states use technical means to block citizens from accessing content on the Internet, each state also employs legal and other "soft" means of control. Most states use a combination of media, telecommunications, national security, and Internet-specific laws and regulatory schemes to restrict the publication of and access to information on the Internet. Most states that filter require Internet service providers to obtain licenses before providing Internet access to citizens. Some states place pressure on cybercafés and ISPs to monitor Internet usage by their customers. With the exception of a few places, no state seems to communicate much at all with the public about its process for blocking and unblocking content on the Internet. Most states, instead, have only a series of broad laws that cover content issues online. The effect of these regimes is to put citizens on notice that they should not publish or access content online that violates certain norms and to create a sense that someone might be paying attention to their online activity.

Our global survey of Internet filtering in 2006 turned up instances where the Internet is not subject to online filtering, but where the state manages to dampen online dissidence through other means. In Egypt, for instance, the Internet is not filtered (reports suggest that Egypt at one time blocked the Muslim Brotherhood site, but did not appear to do so during our testing),[13] but security forces have detained people for their activities online.[14] The perception that the online space is subject to extensive state surveillance leads to a broad fear of accessing or publishing information online that may be perceived to be subversive—though bloggers and other online activists cross the perceived lines with regularity.[15]

The Spectrum from Manual to Automatic Filtering

Most of the filtering regimes we studied, with the exception of parts of the Chinese filtering regime, appear to rely on the preidentification and categorization of undesirable Web sites. As the Web grows in scope as well as form, it is likely that states with an interest in filtering will attempt to develop or obtain technology to automatically review or generalize about the content of a Web page as it is accessed, or other Internet communication as it happens. The Web 2.0 phenomenon only makes this challenge harder, as citizens have the ability to publish online content on the fly and to syndicate that content for free.

The job of the censor in a Web 2.0 world might or might not be accomplished by looking for certain keywords in the title of the page or on its *link*—its URL. While URLs are clearly not as determinative of content as the name of a television channel or a newspaper, they may be in some situations (consider what generalizations one might draw about a URL of www.google.com/search?q=tiananmen). In others, a URL may serve as an adequate proxy—URLs containing particular obscene words are more likely to have obscene content. If the goal is to block all content coming from a particular state, the top-level domain structure makes this remarkably simple. On the other end of the spectrum, however, are blogging sites or generic free Web-hosting sites like www.geocities.com, where the presence of a page within the general site provides little information about the content or authorship of other pages on the site.

Despite the obvious imperfections of filtering via URLs, we have found little evidence that the states in which we tested, with the exception of China, attempted a dynamic assessment of the content of a Web page instead of the URL at the time of request by a user. China may be the sole exception to this rule. Our research has documented an elaborate network of controls including keyword-based URL filtering; it may be the case that Chinese filtering systems can be triggered based upon the presence of keywords within a Web page's content. OpenNet Initiative researcher Steven Murdoch along with his colleagues Richard Clayton and Robert N. M. Watson have published a paper that describes in detail the workings of the "Great Firewall of China," including this dynamic filtering based on Web page content. As Clayton, Murdoch, and Watson note, "We have demonstrated that the 'Great Firewall of China' relies on inspecting packets for specific content."[16] Yemen and Iran also have the capacity to block sites based on keywords in URLs. Commercial software packages such as SmartFilter make such URL filtering trivial. As a general rule, with the possible exception of China, access to a site is based on its URL; if a URL has triggered a block, one could take down the offensive content within the page and replace it with the most innocuous material possible, and the original link will continue to trigger a block.

URL-based blocking does not, however, require the identification of every page that is to be blocked. Our research indicates that the most prevalent form of blocking is at the domain level. Once a state has identified www.playboy.com as undesirable, the logical step is to deny all requests to that domain, whether http://playboy.com/playmates/2003/may.html or playboy.com/articles/interviews/index.html.

The parallel between the URL-based approach with the approach of the traditional censor is that the domain is deemed on the whole undesirable, and the censor makes no effort to disaggregate the content within. The decision is most complicated when a single domain hosts truly disparate content, such as free hosting sites like Geocities or Angelfire, blogging domains like Blogspot or Blogger, community sites like Google or Yahoo! groups, or university sites like mit.edu that can include student home pages about subjects like Tibet. Within these realms, our research found ample evidence of both blocking of the entire domain and selected blocking of *subsites*, or pages within the domain. Such blocking is discussed in the respective state reports in the appendix to this book. The Berkman Center's Web site at http://cyber.law.harvard.edu was blocked in China in 2002 after our first report of Internet filtering was placed there. (The powers that be in Harvard's central university administration declined to repost the study at www.harvard.edu.)

We have also observed several other means of URL-based filtering. As the results presented in chapter 1 show, several states—including China, Iran, Myanmar, and Yemen—block access to all URLs containing particular strings of letters (such as "ass"), whether such banned terms appear in the domain or in superfluous characters at its end. Those sites' IP addresses are independently blocked, as blocked domains could otherwise be accessed via this method.

Some blocking approaches are cruder still. We observed that the United Arab Emirates and Syria blocked every site found within the Israeli top-level country code domain: no pages from any domain ending in ".il" were accessible there.[17]

This last example demonstrates the dramatic difference between URL-based filtering and content-based filtering. The structure of the Internet makes it very easy to block all sites ending in .il, but extremely hard, if not impossible, to block all sites containing content about Israel, a project that our data indicates was never seriously undertaken within the country. It may be that the purpose of blocking ".il" was more a statement about the Syrian and UAE view on Israel, rather than an attempt to prevent its citizens from discovering particular information. Also, the block likely operates in both directions: someone from a ".il" address may have a hard time accessing content in the UAE as a result of the filtering there.

The Role of Commercial Software in State-Mandated Internet Filtering

Commercial services, including the U.S.-based Secure Computing's SmartFilter, Websense, and Fortinet, appear to assist, or to have assisted, states that filter with the implementation and management of block lists. These services provide extensive lists of URLs categorized using proprietary methods. The commercial services typically fall in the middle of a spectrum between manual and automated filtering. The URL for a site found to contain content related to gambling will be offered as a digital update to the "gambling" block list of those states subscribing to the filtering services' lists.

For topics such as pornography or drugs, few states appear to invest the resources required to maintain active block lists where they can procure a list from commercial Internet filtering companies. The challenge in doing so is compounded by multiple means of typically accessing such Web sites—http://www.norml.com, http://norml.com, http://www.norml.org, and http://209.70.46.40 all bring the user to the home page of the National Organization for the Reform of Marijuana Laws. Each of these means of accessing the site (and others) must be added to the block list in order to block citizens' access in a thorough manner. The task is further complicated when site operators realize that they are being filtered and attempt to evade simple filtering techniques by changing their URLs; for instance, Iran has blocked the site at www.pglo.org, but in a subsequent test, it had not blocked the same content on www.pglo1.org.[18] Additionally, since filtering on a national scale requires complex infrastructure, making sure that the same list of blocked sites is present on each machine performing the filtering, at either a centralized or ISP-specific level, is no simple task.

A state subscribing to such a service is limited to the categories made available by the commercial software providers. While generally useful for content targeted according to the common desires of parents, schools, and companies in the West (such as "pornography," "drugs," or "dating"), these products also include broad categories such as "religion" and "politics" that are not fine-tuned enough to match state-specific goals. These categories will not, in the off-the-shelf version of the software, include filtering of content critical of Islam or opposing the government of Vietnam. To account for the generality of these categories, each of the installations of filtering software we observed appears to allow a state to augment a commercial block list with its own URLs.

Aside from such fine tuning, however, states using commercial filtering services must choose between allowing or blocking all URLs within a category. For example, a previous version of the SmartFilter service provides the choice of blocking or allowing all URLs in the ''anonymizer/translator'' category. Even though a state may wish to block anonymizers in order to prevent circumvention, that same state may wish to preserve access to translators as a useful tool.[19] Language presents an additional problem, as all the commercial filtering software we observed is produced by American companies. The blocking in states using these commercial filters therefore tilts heavily toward evaluating—and in turn prompting blocks of—English-language sites. This tilt leans precisely the opposite way from the tilt of those states that develop their own blocking systems, which generally seek to block content in local languages more than content in foreign languages. In some of the states using commercial filtering software, we have observed heavy filtering of English language content in some categories, while the same content appeared to be freely available in the local language—likely the inverse of what the state was seeking to accomplish through its filtering regime.[20]

When commercial filtering software is in use, a given second-level domain—for instance, cnn.com—may include some sites that are blocked and other sites that are unblocked. Our testing of SmartFilter has determined that the software attempts a more exact match first, and in its absence falls back to categories assigned more generally to areas of a domain or the domain itself. For instance, SmartFilter categorizes the *Sports Illustrated* home page at sportsillustrated.cnn.com as ''sports.'' The default categorization for any Web page located within this site, as shown by the category SmartFilter assigns to a request for http://sportsillustrated.cnn.com/does_not_exist, will also be ''sports.'' However, the page for the most recent swimsuit edition, at http://sportsillustrated.cnn.com/features/2006_swimsuit, is categorized as ''provocative attire/mature.'' Thus, it appears that any Web page within the *Sports Illustrated* site will, logically enough, be assigned to the ''sports'' category whether or not SmartFilter has analyzed the content of the page, unless this default has been overridden with a page-specific categorization.[21]

Commercial filtering software may alleviate some of the difficulties of filtering presented by the technical structure of the Internet. Our data show that states using such software are much less likely to miss alternative means of accessing blocked sites, for instance, visiting http://ifex.org to get around a block of http://www.ifex.org, as was possible in Vietnam during our testing. Commercial software companies have refined their filtering techniques to anticipate, detect, and prevent these relatively simple methods of evading blocking. There are others, as Deibert and Rohozinski note in chapter 6 of this volume, who seek to achieve just the opposite. A game of cat and mouse is well underway.

Filtering and Transparency

As Faris and Villeneuve document in chapter 1 of this book, states adopt a range of practices in terms of how explicitly they discuss their filtering regime with the public and the amount that citizens can learn about it. No state that we have studied in the past five years makes its block list generally available, though partial information has found its way to the surface in a few instances. In India, through freedom of information filings, some citizens have obtained information as to the list of those sites filtered. In Bahrain, citizens have compiled a partial block list and posted it to the Internet. In Thailand, prior to the 2006 coup, a list of many thousand Web sites had been posted online, plausibly leaked by the state, but not mapping closely to the facts we have observed. A combination of citizen efforts and the circulation of Pakistan Telecommunication Authority blocking orders on the Internet have resulted in a partial list of blocked sites in Pakistan coming to our attention.[22]

Saudi Arabia is the most transparent state in terms of Internet filtering. The Saudi state sets forth the rationale and practices related to filtering on an easily accessible Web site in both Arabic and English. (In our first round of testing, in 2002, Saudi Arabia enabled us to run tests directly against its system, but would not show us the list that it was using to determine which sites it was blocking at any given moment; since publication of our first report on this topic, the Saudis have disallowed us such easy and direct access to their system.) In Saudi Arabia, citizens may suggest sites for blocking or for unblocking, in either Arabic or English, via a public Web site. Access to most of these sites prompts a blockpage to appear, indicating to those seeking access to a Web site that they have reached a disallowed site. Most states have enacted laws that support the filtering regime and provide citizens with some context for why and how it is occurring, though rarely with any degree of precision. However, among the states we studied, some of the central Asian states that practice just-in-time filtering on sensitive topics—as well as China, whose officials sometimes deny the presence of Internet filtering—obscure the nature and extent of their filtering regimes to the greatest extent.[23]

Some states, such as Saudi Arabia and UAE, make an effort to suggest that their citizens are largely in support of the filtering regime, particularly when it comes to blocking access to pornographic material. For instance, the agency responsible for both Internet access and filtering in Saudi Arabia conducted a user study in 1999 and reported that 45 percent of respondents thought "too much" was blocked, 41 percent thought the amount blocked was "reasonable," and 14 percent found it "not enough."[24] We have not delved into the veracity of these findings.

Citizens may, in some instances, participate in the decision making as to whether a site may be filtered or not. Three of the states in which we tested (Saudi Arabia, UAE, and Yemen) respond to a request for a blocked site with a page that includes a mechanism for suggesting that the particular URL may be blocked in error. However, to make such a suggestion requires the user to have knowledge of the content of the Web page not able to be visited—and the confidence, perhaps not well-placed, that such self-identification would not put the user in jeopardy of state sanction.

Trends in Internet Filtering

Researchers associated with the OpenNet Initiative have been collecting empirical data on Internet filtering since 2001. Our methodology circa 2006 is far more sophisticated than it has been in the past. The coverage of our research is far broader, now covering every state known or credibly suspected to carry out Internet filtering. During this five-year period, we have observed the following trends:

- The overall trend in Internet filtering is toward more states adopting filtering regimes. The states with the most extensive filtering practices fall primarily in three regions: east Asia, the Middle East and North Africa, and central Asia. State-mandated, technical filtering does occur in other parts of the world, but in a more limited fashion, such as the Internet filtering common in libraries and schools in the United States, child pornography filtering systems in northern Europe, and the filtering of Nazi paraphernalia and Holocaust denial sites in France and Germany.
- Some of the newest filtering regimes, such as those coming online in the Commonwealth of Independent States (CIS), appear to be more sophisticated than the first-generation systems still in place in some states. The early means of filtering—such as Saudi Arabia's early system, with a heavy emphasis on pornography and offering citizens a clear blockpage—are no longer the only ways to accomplish Internet filtering. The net result is greater variation in what it means to filter content online.
- In the Commonwealth of Independent States and in parts of the Middle East and North Africa, the filtering we are seeing is highly targeted in nature and carried out "just-in-time" to block access to information during sensitive time periods. ONI principal investigator Rafal Rohozinski and his coauthor Deirdre Collings predicted such an eventuality: "In democratically-challenged countries, we are likely to see increasing constraints on the 'openness' of the Internet during election periods, and these constraints may be more subtle than outright filtering and blocking."[25]

  The ONI has monitored three elections to date, one in Kyrgyzstan (2005), one in Belarus (2006), and one in Nigeria (2007). As Rohozinski and Collings wrote,

  > The February 2005 elections in Kyrgyzstan marked the ONI's first foray into election monitoring. During the Kyrgyz elections ONI researchers were able to document two major Denial of Service (DoS) attacks directed against ISPs hosting major opposition newspapers. The attacks were commissioned from a commercial "bot herder" and traced back to a group of Ukrainian hackers-for-hire. ONI was not able to identify who was ultimately responsible for these attacks. Direct links to the Kyrgyz authorities could not be established. Thus, while no direct filtering took place, the DoS attack resulted in the indirect censorship of websites while exonerating the Kyrgyz authorities of any direct responsibility. The Kyrgyz case also raised the issue of who benefits most from this kind of indirect filtering. In Kyrgyzstan, the target of the DoS attacks—opposition

newspaper websites—continued to publish print editions while claiming that they were being "censored" by the government.

Of the Belarus election, Rohozinski and Collings wrote,

> [T]he quality and consistency of access to some sites varied considerably, and on critical days, up to 37 opposition and independent sites across 25 different ISPs were inaccessible from within the state-owned Beltelecom network. On election day and after the website of the main opposition candidate (Aleksandr Milinkevich) was "dead," as was another opposition site—Charter 97. On the day that the police cleared the last remaining protesters from October Square (25 March) Internet connectivity by way of Minsk telephone dial-up services failed.
>
> And, there were three instances of confirmed "odd DNS errors" affecting opposition websites. While no case yielded conclusive evidence of government inspired tampering, the pattern of failures as well as the fact that mostly opposition and independent media sites were affected, suggests that something other than chance was afoot.[26]

The just-in-time filtering phenomenon has reared its head in the Middle East region as well. Bahrain blocked several Web sites in the run-up to the country's parliamentary elections in 2006 and Yemen banned access to several media and local politics Web sites ahead of the country's 2006 presidential elections. Bahrain also briefly blocked access to Google Earth in 2006, citing security reasons. For about a month in 2006, Jordan blocked access to the VoIP Web site skype.com, also citing security concerns.

- Our most recent data, collected in 2006 and 2007, suggest that we may also be seeing, for the first time, the emergence of in-stream filtering. This process involves entities based in large states—possibly including Chinese, Russian, and Indian ISPs—that provide Internet service to other states, passing along the filtering practices to their customer states. While the data are inconclusive that such in-stream filtering is taking place extensively, the hallmarks of such activity are present in our recent findings. We will continue to monitor closely for the emergence of this phenomenon, as it might point to a new series of security concerns.
- There is a continued growth in the creation of online information by citizens, including citizen journalism, in many parts of the world, but filtering is having an impact on how people carry it out. In some cases, the existence of a filtering regime leads these citizen journalists to limit the topics that they cover. For instance, environmental activists writing online in China have tended to stick closely to the issues related to the environment, which tend not to be blocked, while steering clear of related political topics that are censored. In other cases, such as the Middle East region, citizens banter with the censors. In the Commonwealth of Independent States we may be witnessing a backlash, in the form of Internet filtering, because of the perceived influence of citizen media on the outcome of elections there.[27]
- Citizens and citizen journalists practice self-censorship. For example, moderators of online discussion forums remove contributions that could lead to the blocking of the forums. On the other hand, cyberactivists exploit alternative technologies to circumvent filtering systems. Many Web sites that discuss sensitive issues use online groups such as Yahoo! Groups as part of their contingency plans, so once a Web site is blocked, users continue the discussion and the exchange of content via the group e-mails.

- We have evidence of more filtering at the edges than in a centralized manner, especially in the Commonwealth of Independent States. One might also consider the cybercafé-based controls in China, say, as compared to the approach of setting up the "Great Firewall" at the state's geopolitical boundaries. Those states that have not developed centralized filtering systems may find it more effective to build them at the edges. This phenomenon suggests that those who lobby against network blockages may have to expand their view of the network to include the devices that attach to it.
- We have observed an increase in alternative modes of filtering, both in engineering technique and through increased licensing, registration, and reporting requirements in some states.
- We have uncovered evidence of filtering undertaken by some Internet sites depending on where they believe their users to be located. In these instances, the entity that is publishing the sites—rather than the state where the person accessing or publishing the information is located—is limiting who can access its site. This process, combined with geolocation of the source of a request for a Web page, has occasionally been prompted in the past by a legal proceeding, such as the French insistence that Yahoo! not provide its citizens with access to certain Nazi-related items in the Yahoo! auction sites. More recently, our data show that gambling sites, U.S. military Web sites in the ".mil" domain, and some dating sites are filtered from the server side.
- States continue to be most concerned with blocking of sites in the local language, as opposed to sites in nonlocal languages—even though commercial filtering software sometimes accomplishes the inverse. In the Commonwealth of Independent States, blocking is almost exclusively of local-language sites. In the Middle East and North Africa, much of the blocking focuses on local-language sites, with some blockage of English sites—especially where commercial filtering systems developed in the United States are in use.
- Internet filtering is increasingly being used to block access to certain online applications beyond Web sites accessed by Web browsers. This trend is particularly important as software transitions toward more and more of an online service model. Google Earth and Skype, among other Voice-over Internet Protocol services, are blocked in some states. Other online applications, such as non-Web-based anonymizers that allow anonymous Internet usage, are consistently blocked in many places.

Normative Analysis of Internet Filtering

Few would condemn all those who would seek to filter Internet content; in fact, nearly every society filters Internet content in one way or another. Certainly all states regulate the information environment in some fashion, as Jack Goldsmith and Tim Wu's work makes plain.[28] The purpose of this research is to provide the empirical data needed to understand this form of state control online, what it means for the future of the Internet, and what choices are involved in a state's decision to filter the Internet.

The perspective in support of state-mandated Internet filtering is straightforward. States have the sovereign right to carry out Internet filtering as they see fit. The same goes for

Internet-based surveillance. Internet filtering and surveillance, this argument goes, is no more a matter for international decision making than any other domestic policy concern. The nature of the network and its potential uses are irrelevant to the analysis. The Internet is not exceptional.

There are several possible critiques of Internet filtering. First, one might argue that technical filtering is fatally flawed from the outset; because it cannot be carried out in a manner that is not over- or under-broad, it cannot be done in a way that is sufficiently protective of civil liberties. Second, as a related critique, Internet filtering implicates human rights concerns, particularly the freedom of expression, and extends to the freedom of association, of religion, and of privacy in some instantiations. Finally, one might conclude that Internet filtering is unwise on public policy grounds because it is anathema to the good things to which ICTs can give rise, such as innovation, creativity, and stronger democracies.

The hardest cases are those that some would argue are acts of law enforcement while others contend that they are clear violations of international norms. Consider a sovereign, jealous of the opposition's power, that disables access to opposition Web sites in the lead-up to an election—and then relents once the threat of losing control is abated—as some of our findings from central Asia would suggest happens. Or a state that routinely uses censorship and surveillance as a key element of a campaign to persecute a religious minority group. Or a state that relies upon online surveillance for the purpose of jailing political dissidents whose acts the state has committed to respect pursuant to international human rights norms. What about when a state is trying to protect public morals by keeping citizens from looking at garden-variety online pornography, but in so doing also block information on culturally sensitive matters, such as HIV/AIDS prevention or gay and lesbian outreach efforts? We set forth three primary critiques here. These cases, each real, put the normative problem of Internet filtering into sharp relief.

The Argument in Favor of Internet Filtering: Legitimate State Control Online

The need for states to be able to exercise some measure of control online is broadly accepted. Likewise, states ought to be able to provide rights of action—ordinarily, the right to sue someone—to their citizens to enable them to seek redress for harms done in the online environment. Though one might disagree, these core presumptions are not challenged in this book. The easiest, perhaps most universal case is the common abhorrence of child pornography. Most societies share the view that imagery of children under a certain age in a sexually compromising position is unlawful to produce, possess, or distribute. The issue in the context of child pornography is less whether the state has the right to assert control over such material, but rather the most effective means of combating the problem it represents, and the problems to which it leads, without undercutting rights guaranteed to citizens. The prevention of online fraud or other crimes, often targeting the elderly or disadvantaged, likewise represents a common purpose for some measure of state control of bits online. Some would argue that intellectual property protection represents yet another such example, though the merits of that proposition are hotly contested.

One of the key findings of our research is the extent to which states cannot do the job of content control alone, which in turn adds another layer of complexity to the analysis of Internet filtering as a public policy matter. Where the state cannot effectively carry out its mandate in these legitimate circumstances, the state reasonably turns to those best positioned to assert control of bits. Often, though not always, the state turns to Internet service providers of one flavor or another. The law enforcement officer, for instance, calls upon the lawyers representing ISPs to turn over information about users of the online service who are suspected of committing a common crime, such as online fraud. As criminals use the Internet in the course of wrongdoing, states need to be able to access the increasingly useful store of evidence collected online.

The strongest form of this argument is that online censorship and surveillance is a legitimate expression of the sovereign authority of states. As we have described, Saudi Arabia, which implements one of the most extensive and longest-running filtering regimes, did not introduce Internet access to its citizens until the state authorities were comfortable that they could do so in a manner that would not be averse to local morals or norms. In particular, the Saudi regime has concerned itself with blocking access to online pornography, which it has done with a startlingly high degree of effectiveness over the past five years—though the scope of its filtering has grown over time, now including more political information than when we first began testing there in 2002.

A state has a right to protect the morality of its citizens, the argument goes, and unfettered access to and use of the Internet undercuts public morality in myriad ways. Many regimes, including those in Western states (including the United States), have justified online surveillance of various sorts on the grounds of ordinary law enforcement activities, such as the prevention and enforcement of domestic criminal activity. Most recently, states have begun to justify online censorship and surveillance as a measure to counteract international terrorism. Put more simply, Internet filtering and surveillance, in an environment where the Internet is considered a form of territory alongside land or sea or air, are an expression of the unalterable right of a state to ensure its national security.

Counterarguments: The Infirmities of Technical Filtering

One of the enduring facts of technical filtering of the Internet is that no state has managed to implement a perfect system. The primary deficiency of any technical filtering system is that the censor must choose between two shortcomings: either the system suffers from overbreadth, that is, sites that are not meant to be filtered are filtered, or underbreadth, that is, not all sites meant to be filtered are filtered. In most instances, the filtering regime suffers from a combination of these two deficiencies. Coupled with the extent to which savvy Internet users can evade the filtering regime, state authorities undertaking technical filtering know that they cannot succeed completely.

The public policy questions to which these problems give rise are many and complex. If a filtering regime cannot be implemented in an accurate manner, should it be undertaken at all? Under U.S. law, these shortcomings make any such system constitutionally suspect, if not outright infirm, but other legal systems would likely draw a different conclusion. Is overbreadth or underbreadth preferable in a filtering regime? States often respond by turning more and more to intermediaries—search engine providers, ISPs, cybercafé owners, and so forth—to make these decisions on the fly.

Filtering and Overbreadth Internet filtering is almost impossible to accomplish with any degree of precision. A country that is deciding to filter the Internet must make an "overbroad" or "underbroad" decision at the outset. The filtering regime will either block access to too much or too little Internet content. Very often, this decision is tied to whether the state opts to use a home-grown system or whether to adopt a commercial software product, such as SmartFilter or Websense, two products made in the United States and licensed to some states that filter the Internet. Bahrain, for instance, has opted for an underbroad solution for pornography; its ISPs appear to block access to a small and essentially fixed number of blacklisted sites. Bahrain may seek to indicate disapproval of access to pornographic material online, while actually blocking only token access to such material, much as Singapore does. United Arab Emirates, by contrast, seems to have made the opposite decision by attempting to block much more extensively in similar categories, thereby sweeping into its filtering basket a number of sites that appear to have innocuous content by any metric.

Most of the time, states make blocking determinations to cover a range of Web content, commonly grouped around a second-level domain name or the IP address of a Web service (such as www.un.org or 66.102.15.100), rather than based on the precise URL of a given Web page (such as www.un.org/womenwatch/), or a subset of content found on that page (such as a particular image or string of text). This approach means that the filtering process will often not distinguish between permissible and impermissible content so long as any impermissible content is deemed "nearby" from a network standpoint. In the case of the above example, the WomenWatch site was unavailable in Vietnam not because of the state attempts to block all sites relating to gender equality issues (judged by the availability of all other similar sites we tested), but because of a block placed on the entire www.un.org domain.

Because of this wholesale acceptance or rejection of a particular site—which may or may not correspond to a given speaker or related group of speakers—it becomes difficult to know exactly what speech was deemed unacceptable for citizens to access. Bahrain, a state in which we have found a handful of blocked sites, has blocked access to a discussion board at www.bahrainonline.org. The message board likely contains a combination of messages that would be tolerated independently and those that are explicitly meant to be subject to filtering. Likewise, we found minimal blocking for internal political purposes in UAE, but the state did block a site that essentially acted as a catalog of criticism of the state. Our tests cannot determine whether it was the material covering human rights abuses or discussion of historical border disputes with Iran, but in as much as the discussion of these topics is taking place within a broad dissention-based site, the calculation we project onto the censor looks significantly different than that for a site with a different ratio of "offensive" to approved content.

For those states using commercial filtering software and update services to try to maintain a current list of blocked sites matching particular criteria, we have noted multiple instances where such software has mistaken sites containing gay and lesbian content for pornography. For instance, the site for the Log Cabin Republicans of Texas has been blocked by the United States–based SmartFilter as pornography, apparently the basis for its blocking by United Arab Emirates. Our research suggests that gay and lesbian content is itself often targeted for filtering; one might surmise that, even when it is not explicitly targeted, states that implement related filters are not overly concerned with its unavailability due to overbreadth.

As content changes increasingly quickly on the Web and generalizations become more difficult to make by URL or domain name—thanks in part to the rise of simpler, faster, and aggregated publishing tools such as Weblogging (blogging) services—accurate filtering is likely to get trickier for filtering regimes to address over time unless they want to take the step of banning nearly everything. For example, free Web-hosting domains tend to group an enormous array of changing content and thus provoke very different responses from state governments. In 2004, Saudi Arabia blocked every page we tested on http://freespace.virgin.net and www.erols.com.[29] However, our research indicated the www.erols.com sites had been only minimally blocked in 2002, and the http://freespace.virgin.net sites had been blocked in 2002, but accessible in 2003 before being reblocked in 2004. In all three tests, Saudi Arabia practiced URL blocking on www.geocities.com (possibly through SmartFilter categorization), blocking only 3 percent of more than one thousand sites tested in 2004. Vietnam blocked all sites we tested on the www.geocities.com and http://members.tripod.org domains.

Contrast this last example with Yahoo! Groups, which Vietnam appears to filter on a group-by-group basis. We found that the state blocks access to the pages of two groups discussing the Cao Dai religion in general, but our testers were able to access the page of a California Cao Dai youth group. Two factors may play a role in this decision. Groups may provide more "benefit" to the censor, due to their interactive nature, and thus implicate the social and possibly economic impacts of the Internet. Groups, too, may have a limited, albeit large, number of possibilities—a single group could, in theory be monitored at the group level where there is much more metadata about the content contained therein, whereas Geocities could be grouped by user, but a particular user may offer large numbers of pages on very varied topics.

In our 2005 testing, we located 115 Weblogs within 3 large blogging domains (blogsky, blogspot, and persianblog) that were blocked in Iran. This blocking corresponded to only 24 percent of all blogs tested within those domains, and our testing was designed to locate blocked sites. Clearly, Iran desires to block access to some blogs, but has not seen fit to block all blogs. Our empirical data do not help to explain why filtering authorities in Iran made this decision, but it clearly was the result of a deliberate action. Also note that the site for www.movabletype.org, an application designed to allow blogging to take place on any domain, was blocked. Perhaps this indicates a policy of containing the blogs by restricting them to the large blogging domains, where they can then be reviewed and potentially filtered on a one-by-one basis.

China's response to the same problem provides an instructive contrast. When China became worried about bloggers the state shut down the main blogging domains for a period of weeks—much as they have, periodically, for Wikipedia. When the domains came back online, they contained filters that would reject posts containing particular keywords.[30] In effect, China moved to a content-based filtering system, but determined that the best place for such content evaluation was not the point of Web page access but the point of publication, and possessed the authority to force these filters on the downstream application provider. Most of these providers coded these restrictions into the software provided to bloggers. This approach is similar to that taken with Google to respond to the accessibility of disfavored content via Google's caching function. Google was blocked in China until a mechanism was put in place to prevent cache access.[31] In the fall of 2005, Saudi Arabia was reported to have blocked access to all blogs on the Blogger network, which plainly represented an overbroad set of blocks. These examples make clear the length to which regimes can go to preserve "good" access instead of simply blocking an entire service.

Alternate approaches that demand a finer-grained means of filtering, such as the use of automated keywords to identify and expunge sensitive information on the fly, or greater manual involvement in choosing individual Web pages to be filtered, are possible so long as a state is willing to invest the time and resources necessary to render them effective. China in particular appears to be prepared to make such an investment, one mirrored by choices made by the Chinese state in the context of traditional media. For example, China allows CNN to be broadcast within the country with a form of time delay so the feed can be temporarily turned off when, in one case, stories about the death of political reformer Zhao Ziyang were broadcast.

Filtering and Underbreadth One of the primary surprises in our data over the past several years is the infrequency with which plainly sensitive pages were blocked within otherwise acceptable sites. For instance, we found no cases where specific articles were blocked on major news sites, except in China. In fact, the regimes in which we tested very rarely made an attempt to block www.cnn.com, www.nytimes.com, http://bbc.co.uk, or others. (Exceptions to this rule include the Voice of America news site at www.voanews.com. It was blocked in both Iran and China, and China also blocks the entire BBC news site.) In fact, not only was CNN's international news site at edition.cnn.com generally accessible in our China testing, a page within that domain dedicated to the massacre in Tiananmen Square was also not filtered. Several factors might be at work here—the sheer volume of news stories produced by major outlets may make thorough review impossible, or the speed at which new stories are posted may simply be too quick for an update across all the necessary filtering technology.

One instance where such URL-specific blocking had been applied in the past was Saudi Arabia's treatment of Amnesty International's Web site. In 2002, we tested twenty-five hundred pages within the amnesty.org domain and found nineteen blocked; all were within the directory www.amnesty.org/ailib/intcam/saudi, corresponding to a report entitled "Saudi Arabia: A Secret State of Suffering." However, these same pages were tested in 2003, 2004, and 2006 and were accessible in each instance.

Human Rights Concerns Related to Internet Filtering

Internet censorship and surveillance prompt legitimate legal and normative concerns. Some state-mandated acts of online control are not straightforward acts of local law enforcement. As the practice of Internet filtering—and its close cousin, Internet surveillance—become more commonplace and more sophisticated, human rights activists and academics tracking this activity have begun to question whether some regimes of this sort violate international laws or norms. Quite often, the states that carry out online censorship and surveillance are signatories to international human rights convenants or have their own rules that preserve certain civil liberties for their citizens. The United States is home to a controversy of this sort as well, as the Electronic Frontier Foundation and others have filed a class-action lawsuit against telecommunications giant AT&T for collaborating with the National Security Agency in a wiretapping program.

The most straightforward of the critiques of Internet filtering and surveillance are grounded in concerns for individual civil liberties against the encroachment of overbearing states. The online environment is increasingly a venue in which personal data is stored. Personal communications increasingly flow across the wires and airwaves that compose the Internet. The basic rights of freedom of expression and individual privacy are threatened by the extension of state power, aided by private actors, into cyberspace. When public and private actors combine to restrict the publication of and access to online content, or to listen in on online conversations, the hackles of human rights activists are understandably raised. As Mary Rundle and Malcolm Birdling argue in chapter 4 of this book, one might contend that the right of free association is likewise violated by certain Internet-censorship and surveillance regimes that are emerging around the world. Most complaints cite the Universal Declaration of Human Rights or the International Covenant on Civil and Political Rights as grounding ideals—if not binding commitments—to which many states have agreed to hold themselves.

Concerns about Imposing Restrictions on the Internet

Even if one agrees with the strong form of the state sovereignty argument and sets aside objections based on international laws and norms, one might still contend that these filtering regimes are unwise from a public policy vantage point. Internet censorship and surveillance, the technologist might argue, violate the so-called end-to-end principle of network design and therefore risk stunting the future growth of the network and the innovation that might derive from it. This argument is typically grounded in adherence to the end-to-end principle. The end-to-end principle stands for the proposition that the ''intelligence'' in the network should not be placed in the middle of the network, but rather at the end-points. Technologists often chalk up the extraordinarily rapid growth of Internet throughout the world to this simple idea. By imposing control in the middle of the network—say, at the ''Great Firewall'' that surrounds China, proxy servers in Iran, or ISPs in dozens of states around the world—rather than at the user level, the censors are stymieing the further growth of the network.

The importance of ''generative'' information platforms also counsels against unwarranted state intrusion into the online environment.[32] Rather than hewing to the original design of the network, the decision-maker should favor those technical decisions that enable acts of innovation on top of the existing layers in the ecosystem—including not just the middle of the network, but also at the edges. The kinds of individual creativity made possible by the personal computer (PC), including self-expression in the form of the creation of user-generated content, might be thwarted by the presence of a censorship and surveillance regime. The on-again, off-again blockage of the user-generated encyclopedia, Wikipedia, makes this case clearly. The sporadic use of filtering regimes to block the use of Voice-over Internet Protocol (VoIP), often to protect the monopoly in voice communications of a local incumbent, also stands for this proposition.

These filtering regimes, along with surveillance practices that often go hand in hand with them, pose a danger in terms of having an adverse impact on the emergence of democracies around the world. The Internet has an increasing amount to do with the shape that democracies are taking in many developing states. As Ronald Deibert and Rafal Rohozinski argue in chapter 6 of this book, activists make use of the Internet in ways that are having a substantial impact on their societies.

The Internet is a potential force for democracy by increasing means of citizen participation in the regimes in which they live. The Internet is increasingly a way to let sunlight fall upon the actions of those in power—and providing an effective disinfectant in the process. The Internet can give a megaphone to activists and to dissidents who can make their case to the public, either on the record or anonymously or pseudonymously. The Internet can help make new networks, within and across cultures, can be an important productivity tool for otherwise underfunded activists, and can foster the development of new communities built around ideas. The Internet can open the information environment to voices other than the organs of the state that have traditionally had a monopoly on the broadcast of important stories and facts, which in turn gives rise to what William Fisher refers to as "semiotic democracy."[33] Put another way, the Internet can place the control of cultural goods and the making of meaning in the hands of many rather than few. The Internet is increasingly an effective counterweight to the consolidation in big media, whether the Internet is controlled by a few capitalists or the state itself.

The Internet also can be a force for economic development, which is most likely the factor holding back some states from filtering the Internet more extensively or from imposing outright bans on related technologies. The Internet is widely recognized as a tool that is helping to lead to the development of technologically sophisticated, empowered middle classes. Entrepreneurship in the information technology sector can lead to innovation, the growth of new firms, and more jobs.

This critique of Internet filtering boils down to a belief in the value of a relatively open information environment because of the likelihood that it can lead to a beneficial combination of greater access to information, more transparency, better governance, and faster economic growth. The Internet, in this sense, is a generative network in human terms. In the hands of the populace at large, the Internet can give rise to a more empowered, productive citizenry.

An Alternate Viewpoint: The "Slope of the Freedom Curve"

As our colleague Charles Nesson has pointed out, another vantage point altogether might lead to the best conclusion about Internet filtering. The point is not whether a single snapshot of an Internet filtering regime reveals a "bad" or a "good" system. Two jurisdictions, after all, could filter in exactly the same way, yet one could be moving toward freedom and the other toward further control of the online environment. In Professor Nesson's articulation, the issue is not the absolute extent of filtering at a given moment but rather the "slope of the freedom curve" that is most relevant. If the value at issue is whether an ICT environment is relatively open or relatively closed, then the key fact is whether a state is headed toward a more open system or a more closed system. The extent to which the Internet filtering picture is in constant flux lends further appeal to this vantage point.

Looking Ahead: The Future of Filtering, Weblogs, and Wikis

Regardless of whether states are right or wrong to mandate filtering and surveillance, the slope of the freedom curve favors not the censor but the citizens who wish to evade the state's control mechanisms. Most filtering regimes have been built on a presumption that the Internet is like the broadcast medium that predates it: each Web site is a "channel," each Web user a "viewer." Channels with sensitive content are "turned off," or otherwise blocked, by authorities who wish to control the information environment. But the Internet is not a broadcast medium. As the Internet continues to grow in ways that are not like broadcast, filtering is becoming increasingly difficult to carry out effectively. The extent to which each person using the Internet can at once be a consumer and a creator is particularly vexing to the broadcast-oriented censor. Combined with the absence of scarcity in terms of the number of channels or spectrum and the fast-dropping cost of accessing Internet from a wide range of devices including shared terminals and mobile devices, the changes in the online environment give an edge to the online publisher against the state's censor in the medium- to long-run.

Along with Wikipedia, Weblogs offer a poignant example of these growing challenges for the censor. No current filtering regime appears designed to address content developed on blogs, podcasts, and Wikis and accessed via Really Simple Syndication (RSS) feeds in aggregators, next-generation peer-to-peer networks, BitTorrent, and so forth. The most effective model demonstrated to date may be China's moves in the past few years to require blog service publishers to block keywords in blog posts, though even this approach can be only a partial means of blocking subversive content over time. Chinese bloggers routinely turn to broadly understood code words to evade the censorship built into the tools.

As online content changes very quickly and can be accessed through new means, the process of prescreening content and establishing a blockpage—akin to updating one's static virus definitions as new viruses are isolated and defined—breaks down. The process must become an heuristic one to function properly, if at all. Multimedia content, which is harder to screen and is accessed in different ways than through the World Wide Web, poses similar challenges for filtering regimes. Those states that are intent on filtering the Internet will have to adapt quickly if they intend to keep up. These adaptations might take the form of more aggressive filtering, or a shift to surveillance of user behavior with legal sanction for those who receive or transmit forbidden material.

In light of the prevalence of structural-based blocking in the states we studied, the trajectory of the Internet to a more dynamic environment will continue to create new problems for filtering regimes. The use of Weblogs by citizens—human rights activists, for instance—as a means of self-publishing is sharply on the rise in many cultures around the world. The general trend on the Internet is the divorcing of content from structure through the syndication of blogs via RSS and similar technologies. Syndication allows the text of a blog to be easily reproduced on other Web sites anywhere[34] in a way that circumvents filtering—since the retrieval of content from a blocked URL is done by the site the user is visiting, potentially located in a country with little or no filtering, instead of by the user's machine. While such mirroring of content has always been possible on the Internet, syndication represents a dramatic decrease in the amount of time and level of technical skill required to easily replicate content. In many ways, this freeing of content from structure mirrors how large sites are internally managed. The reason that CNN can easily display the same article at multiple URLs is that the text of the article can be retrieved from a single location, eliminating the need to separately create each HTML page on which it displays. Through this means, the acceptance or rejection of a large site in its entirety may in itself be a partial reaction to the problem created for URL-based structural filtering when content is not strictly tied to location.

Consider the implications for the censor of this technological change. The rise of publishing through blogs has caused concern in China, Iran, and Saudi Arabia at a minimum, judging from the reaction of their filtering regimes to block some blog-hosting services wholesale for a period of time. Assume that all blogs within the persianblog domain are available via RSS feed. The publisher could create a Web site specifically for the purpose of evading blocking, listing and displaying all such blocked blogs. This site itself could become a target for blocking by the Iranian government, since any mechanism for making this site known to users would also make it known to the filtering authorities.[35]

But using widely available aggregation tools, a user who wants to read this information does not need to go to a single URL to access the information published there. Instead, the user only needs to know the place where the XML feed is located at any given moment—which need not, ultimately, be at a stable location, so long as the user has a means of being updated as to its location at any given moment. In this version of the Web—trivial, using today's technologies—anyone can make any such blogs they choose available on any Web page or in an e-mail in-box or on a mobile device.

Another approach that citizen journalists might take would be to seek to bury the blocked blogs within a much larger number of blogs. The publisher could then establish a site or a feed that aggregates this larger number of blogs. Then, still using simple technologies, the readers could either read the full set of aggregated information or could run a filter of their own against the aggregated group of blogs to distill the information that the publisher wanted them to be able to access. Though these methods add a layer of complexity that would no doubt dissuage some Internet users, the net effect would be a publication mode that would be extremely difficult for the state to filter using current methods.

The state's censor would still have several options for responding to syndication methods of dissemination. First, the state could attempt to ban syndication, aggregation, and peer-to-peer technologies that might make these circumvention efforts easy to carry out. States have not, however, tended to pursue such a heavy-handed mode of regulation. Second, the state could seek to block the sites where the information is published and where the aggregation takes place. However, the potentially unlimited proliferation of such blogs and aggregator sites makes this unfeasible. A last option could involve a fallback to more traditional forms of state coercion—threatening both bloggers, readers, and those who provide them services with sanction. The difficulty of anonymous access leaves open the alternative of identifying users after they have accessed banned content. It is this last option that seems most in keeping with previous filtering and surveillance practices, especially since intermediaries closer to the user can be pressed into service to help.

The enduring point of this glimpse not so far into the future is that as Internet technologies continue to evolve, so too will state censors have to evolve their methods of Internet filtering if they wish to keep up. ONI's early election monitoring efforts in Kyrgyzstan and Belarus, combined with some of the most recent test results from the Commonwealth of Independent

States, suggest that some states are already seeking to turn on and off the Internet-filtering spigot at key moments. The simple proxy-based model, with a corresponding blockpage, will soon look as dated as a 1980s mainframe computer in a peer-to-peer world. If states persist in mandating filtering of the Internet, the narrative of China's on-again, off-again blocking of Wikipedia will be played out over and over as more citizens of the world build upon the generative Internet.

Notes

1. "Wikipedia:Size comparisons," Wikipedia, http://en.wikipedia.org/wiki/Wikipedia:Size_comparisons (last accessed December 28, 2006).
2. "Wikipedia:Overview FAQ," Wikipedia, http://en.wikipedia.org/wiki/Wikipedia:Overview_FAQ (last accessed December 28, 2006).
3. "Blocking of Wikipedia in mainland China," Wikipedia, http://en.wikipedia.org/wiki/Blocking_of_Wikipedia_in_mainland_China (last accessed December 28, 2006).
4. *Ibid.*
5. See Jack L. Goldsmith and Tim Wu, *Who Controls the Internet: Illusions of a Borderless World*, (New York: Oxford University Press, 2006) 65–86; Jack L. Goldsmith, *Against Cyberanarchy*, 65 U. Chi. L. Rev. 1199 (1998); Jack L. Goldsmith, *The Internet and the Abiding Significance of Territorial Sovereignty*, 5 Ind. J. Global Legal Stud. 475 (1998); Jack Goldsmith and Timothy Wu, *Digital Borders*, Legal Affairs, Jan.–Feb. 2006, available at http://www.legalaffairs.org/issues/January-February-2006/feature_goldsmith_janfeb06.msp; Jonathan Zittrain, *Internet Points of Control*, 44 B.C. L. Rev. 653 (2003); Lawrence Lessig, *The Zones of Cyberspace*, 48 Stan. L. Rev. 1403 (1996).
6. ONI researcher Stephanie Wang deserves particular credit for her work on this taxonomy.
7. Consider the draft Cyber Crime Act, under consideration in the Kingdom of Thailand, on file with authors. Proposed section 14 would provide that "service providers" face criminal sanctions if they are aware of various offenses and "do not manage to immediately erase such computer data."
8. § 86 of the German Criminal Code. According to § 4(1) JMStV, online content involving the following is prohibited, irrespective of whether it is otherwise prohibited under criminal law or other law: 1) Propaganda against the democratic constitutional order (§ 86 of the Criminal Code). 2) Use of symbols of unconstitutional organizations (such as the swastika, § 86a of the Criminal Code) (an "unconstitutional organization" is essentially an organization that is against or threatens the principles of a free democratic state). 3) Incitement to hatred and violence against segments of the population or defamation of distinct groups. 4) Denial of the Holocaust or other specific acts perpetrated under the Nazi regime. 5) Depictions of cruel or inhuman violence against humans in a way that glamorizes such acts or makes light of their gravity—this includes virtual representations. 6) Depictions that instigate or incite the commission of certain crimes (namely, those defined in § 126(1) of the Criminal Code [Disturbance of the Peace through the Threat of Perpetrating Criminal Acts]). 7) Glorification of war. 8) Violation of human dignity through the depiction of human death or mortal suffering. 9) Erotic depictions of minors, including virtual depictions. 10) Pornography involving children, animals, or violence, including virtual depictions. 11) Content that has been blacklisted by the Federal Commission for Media Harmful to Young Persons or is similar in nature thereto. Translation and research assistance by Daniel Hausermann and James Thurman of the University of St. Gallen, Switzerland.
9. "Why Block by IP Address?" Internet Censorship Explorer, http://ice.citizenlab.org/index.php?p=78 (last accessed January 15, 2007).
10. For instance, IP filtering forces the choice of blocking all sites sharing an IP address. A recent ONI bulletin found over 3,000 Web sites blocked in an attempt to prevent access to only 31 (see http://www.opennetinitiative.net/bulletins/009/ [last accessed January 15, 2007]). DNS blocking requires an entire domain and all subdomains to be either wholly blocked or wholly unblocked (http://ice.citizenlab.org/index.php?p=78 [last accessed January 15, 2007]).

11. See "Internet Filtering in Saudi Arabia in 2004," OpenNet Initiative, http://opennet.net/studies/saudi at n23 (stating the director of Saudi Arabia's Internet Services Unit director "knows that anyone with much knowledge of the Internet and computers can blow right by the Saudi content filters" and "sees the filtering as a way to protect children and other innocents from Internet evils, and not much more than that").
12. "Appeal Court Confirms Prison for Cyber-Dissident While Blogger is Re-imprisoned," Iran: Reporters sans Frontières, http://www.rsf.org/article.php3?id_article=12564 (accessed February 15, 2005) ("Javad Tavaf, a student leader and the editor of the popular news website Rangin Kaman, which for a year had been criticising the Guide of the Islamic Revolution, was arrested at his home on 16 January 2003 by people who said they were from the military judiciary, which later denied it had arrested him."). Internet—China, China: Reporters Sans Frontières, http://www.rsf.org/article.php3?id_article=10749. Internet—Vietnam, Vietnam: Reporters Sans Frontières, http://www.rsf.org/article.php3?id_article=10778.
13. See "Blocking Web Sites," Human Rights Watch, http://hrw.org/reports/2005/mena1105/4.htm#_Toc119125716 (last accessed February 15, 2007).
14. See Elijah Zarwan, "False Freedom: Online Censorship in the Middle East and North Africa," Human Rights Watch, http://hrw.org/reports/2005/mena1105/ (last accessed January 15, 2007). In addition to his work for HRW, Zarwan contributed research to the ONI work in 2006.
15. See Amnesty International, "Egypt: New Concerns about Freedom of Expression," http://www.amnestyusa.org/news/document.do?id=ENGMDE120182006 (last accessed January 15, 2007).
16. Richard Clayton, Steven J. Murdoch, and Robert N. M. Watson, "Ignoring the Great Firewall of China," http://www.cl.cam.ac.uk/~rnc1/ignoring.pdf (last accessed January 15, 2007).
17. See, for instance, "Syria," Human Rights Watch, http://hrw.org/reports/2005/mena1105/6.htm#_Toc119125750 (last accessed February 15, 2007), a report on Syrian blocking for which ONI provided technical testing support.
18. The effectiveness of such a maneuver will be limited by the need to publicize to users that the content can be found at the alternate site without bringing it to the attention of those managing the filtering.
19. This appears to have been the choice facing Saudi Arabia, where our research indicates that the category was generally allowed but the URLs for anonymizers were added to its local block list. Subsequent versions of SmartFilter have instituted separate categories for these two types of sites.
20. UAE has elected to block sites falling within the "dating" category, resulting in very high levels of blocking for English-language dating sites, but we found no blocking of Arabic-language dating sites.
21. The same default structure will also be applied to the "2006_swimsuit" directory: SmartFilter assigns the category of "provocative attire/mature" to a request for sportsillustrated.cnn.com/features/2006_swimsuit/does_not_exist. Again, the assignment of a default is always subject to a more granular categorization. We checked URL categorizations via the SmartFilterWhere tool at http://www.securecomputing.com/sfwhere (last accessed December 29, 2006).
22. See "Internet Filtering in Pakistan," Internet Censorship Explorer, http://ice.citizenlab.org/?p=204 (last accessed February 15, 2007).
23. Chinese official Yang Xiaokun stated at the 2006 Internet Governance Forum (IGF): "In China, we don't have software blocking Internet sites. Sometimes we have trouble accessing them. But that's a different problem. . . . We do not have restrictions at all." See transcript for October 31, 2006 "Openness" session, at http://www.intgovforum.org/IGF-Panel2-311006am.txt. The Uzbek government has also denied filtering, calling it "impossible." See Ferghana.Ru news agency, "Foreign Minister Eljer Ganiyev: We lack the capacities to restrict access to Internet," June 3, 2005, at http://enews.ferghana.ru/article.php?id=969; RadioFreeEurope/RadioLiberty, "Rights Group Lists 'Enemies of Internet' at UN Summit," November 15, 2005, at http://www.rferl.org/featuresarticle/2005/11/2fdba63a-153a-4268-af4b-e6ebcf54e9ef.html (last accessed September January 3, 2007).
24. "The Old User Survey Results," Internet Services Unit, http://www.isu.net.sa/surveys-&-statistics/user-survey.htm (last accessed September 7, 2004), reporting an online survey of 260 users from July through September 1999.
25. Rafal Rohozinski and Deirdre Collings, "The Internet and Elections: The 2006 Presidential Election in Belarus (and its implications)," OpenNet Initiative Internet Watch Report 001, April 2006, http://www.opennetinitiative.net/studies/belarus/ONI_Belarus_Country_Study.pdf (last accessed January 15, 2007).
26. Ibid.

27. See Regional Overview for the Middle East and North Africa and the Regional Overview for the Commonwealth of Independent States.
28. Goldsmith and Wu, *supra* note 5.
29. Except for the root page at http://www.erols.com itself, potentially indicating a desire to manage perceptions as to the extent of the blocking.
30. "Filtering by Domestic Blog Providers in China," OpenNet Initiative, http://www.opennetinitiative.net/bulletins/008/ (last accessed January 15, 2007).
31. The mechanism for so doing turned out to be extremely rudimentary, as outlined in a previous ONI bulletin. See "Google Search & Cache Filtering Behind China's Great Firewall," http://www.opennetinitiative.net/bulletins/006/ (last accessed January 15, 2007).
32. Jonathan L. Zittrain, *The Future of the Internet—and How to Stop It* (New Haven: Yale University Press, 2007).
33. William W. Fisher III, *Promises to Keep* (Stanford: Stanford University Press, 2004).
34. RSS also allows the user to access a blog via an application other than a Web browser such as a desktop aggregator. While we have not yet performed any testing in this area, it seems that a request issued directly to a blocked blog for an RSS feed would also be blocked, as it would be subject to the same filtering mechanisms despite the use of a different application. The RSS feed may be accessed at a different URL, but it stands to reason that, even if this were the case, the process of blocking a blog would simply be expanded to include this URL.
35. VOA has taken an interesting approach to a similar problem in its creation of a general anonymizer that fully circumvents government filtering. The site has a new address every day, which is broadcast via radio (and other means). ONI has not tested how quickly the Iranian filters are able to update and block these changing sites (see http://opennetinitiative.net/advisories/001/).

3

Tools and Technology of Internet Filtering

Steven J. Murdoch and Ross Anderson

Internet Background

TCP/IP is the unifying set of conventions that allows different computers to communicate over the Internet. The basic unit of information transferred over the Internet is the Internet protocol (IP) *packet*. All Internet communication—whether downloading Web pages, sending e-mail, or transferring files—is achieved by connecting to another computer, splitting the data into packets, and sending them on their way to the intended destination.

Specialized computers known as *routers* are responsible for directing packets appropriately. Each router is connected to several communication *links*, which may be cables (fiber-optic or electrical), short-range wireless, or even satellite. On receiving a packet, the router makes a decision of which outgoing link is most appropriate for getting that packet to its ultimate destination. The approach of encapsulating all communication in a common format (IP) is one of the major factors for the Internet's success. It allows different networks, with disparate underlying structures, to communicate by hiding this nonuniformity from application developers.

Routers identify computers (hosts) on the Internet by their IP address, which might look like 192.0.2.166. Since such numbers are hard to remember, the domain name system (DNS) allows mnemonic names (domain names) to be associated with IP addresses. A host wishing to make a connection first looks up the IP address for a given name, then sends packets to this IP address. For example, the Uniform Resource Locator (URL) www.example.com/page.html contains the domain name "www.example.com." The computer that performs the domain-name-to-IP-address lookup is known as a DNS resolver, and is commonly operated by the Internet service provider (ISP)—the company providing the user with Internet access.

During connection establishment, there are several different ways in which the process can be interrupted in order to perform censorship or some other filtering function. The next section describes how a number of the most relevant filtering mechanisms operate. Each mechanism has its own strengths and weaknesses and these are discussed later. Many of the blocking mechanisms are effective for a range of different Internet applications, but in this chapter we concentrate on access to the Web, as this is the current focus of Internet filtering efforts.

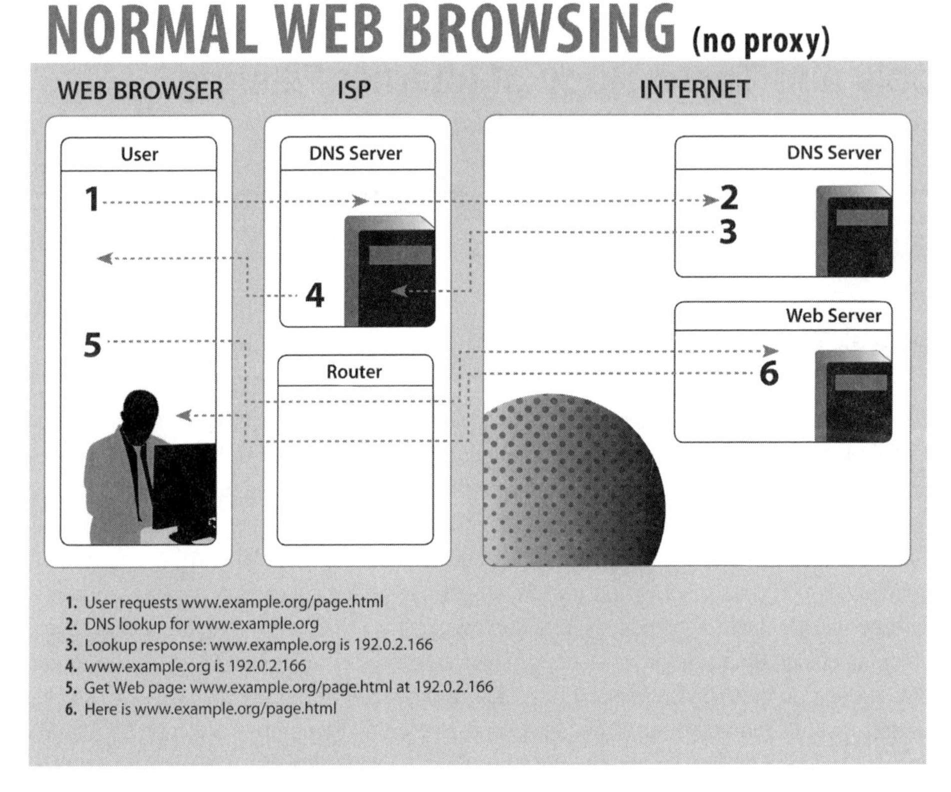

Figure 3.1
Steps in accessing a Web page via normal Web browsing without a proxy.

Figure 3.1 shows an overview of how a Web page (http://www.example.com/page.html) is downloaded. The first stage is the DNS lookup (steps 1–4), as mentioned above, where the user first connects to their ISP's DNS resolver, which then connects to the Web site's DNS server to find the IP address of the requested domain name—"www.example.com." Once the IP address is determined, a connection is made to the Web server and the desired page—"page.html"—is requested (steps 5–6).

Filtering Mechanisms

The goals of deploying a filtering mechanism vary depending on the motivations of the organization deploying them. They may be to make a particular Web site (or individual Web page) inaccessible to those who wish to view it, to make it unreliable, or to deter users from even attempting to access it in the first place. The choice of mechanism will also depend upon the capability of the organization that requests the filtering—where they have access to, the people against whom they can enforce their wishes, and how much they are willing to spend. Other considerations include the number of acceptable errors, whether the filtering should be overt or covert, and how reliable it is (both against ordinary users and those who wish to bypass it). The next section discusses these trade-offs, but first we describe a range of mechanisms available to implement a filtering regime.

Here, we discuss only how access is blocked once the list of resources to be blocked is established. Building this list is a considerable challenge and a common weakness in deployed systems. Not only does the huge number of Web sites make building a comprehensive list of prohibited content difficult, but as content moves and Web sites change their IP addresses, keeping this list up-to-date requires a lot of effort. Moreover, if the operator of the site wishes to interfere with the blocking, the site could be moved more rapidly than it would be otherwise.

TCP/IP Header Filtering

An IP packet consists of a *header* followed by the data the packet carries (the *payload*). Routers must inspect the packet header, as this is where the destination IP address is located. To prevent targeted hosts being accessed, routers can be configured to drop packets destined for IP addresses on a blacklist. However, each host may provide multiple services, such as hosting both Web sites and e-mail servers. Blocking based solely on IP addresses will make all services on each blacklisted host inaccessible.

Slightly more precise blocking can be achieved by additionally blacklisting the *port number*, which is also in the TCP/IP header. Common applications on the Internet have characteristic port numbers, allowing routers to make a crude guess as to the service being accessed. Thus, to block just the Web traffic to a site, a censor might block only packets destined for port 80 (the normal port for Web servers).

Figure 3.2 shows where this type of blocking may be applied. Note that when the blocking is performed, only the IP address is inspected, which is why multiple domain names that share the same IP address will be blocked, even if only one is prohibited.

TCP/IP Content Filtering

TCP/IP header filtering can only block communication on the basis of where packets are going to or coming from, not what they contain. This can be a problem if it is impossible to establish the full list of IP addresses containing prohibited content, or if some IP address contains enough noninfringing content to make it unjustifiable to totally block all communication with it. There is a finer-grained control possible: the content of packets can be inspected for banned keywords.

As routers do not normally examine packet content but just packet headers, extra equipment may be needed. Typical hardware may be unable to react fast enough to block the infringing packets, so other means to block the information must be used instead. As packets

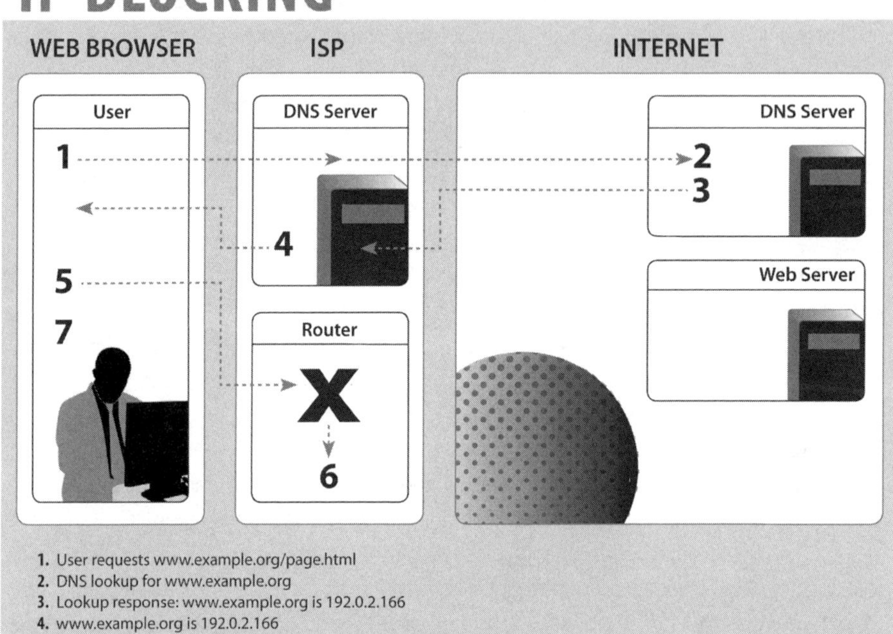

Figure 3.2
IP blocking.

have a maximum size, the full content of the communication will likely be split over multiple packets. Thus while the offending packet will get through, the communication can be disrupted by blocking subsequent packets. This may be achieved by blocking the packets directly or by sending a message to both of the communicating parties requesting they terminate the conversation.[1]

Another effect of the maximum packet size is that keywords may be split over packet boundaries. Devices that inspect each packet individually may then fail to identify infringing keywords. For packet inspection to be fully effective, the stream must be reassembled, which adds additional complexity. Alternatively, an HTTP proxy filter can be used, as described later.

DNS Tampering

Most Internet communication uses domain names rather than IP addresses, particularly for Web browsing. Thus, if the domain name resolution stage can be filtered, access to infringing

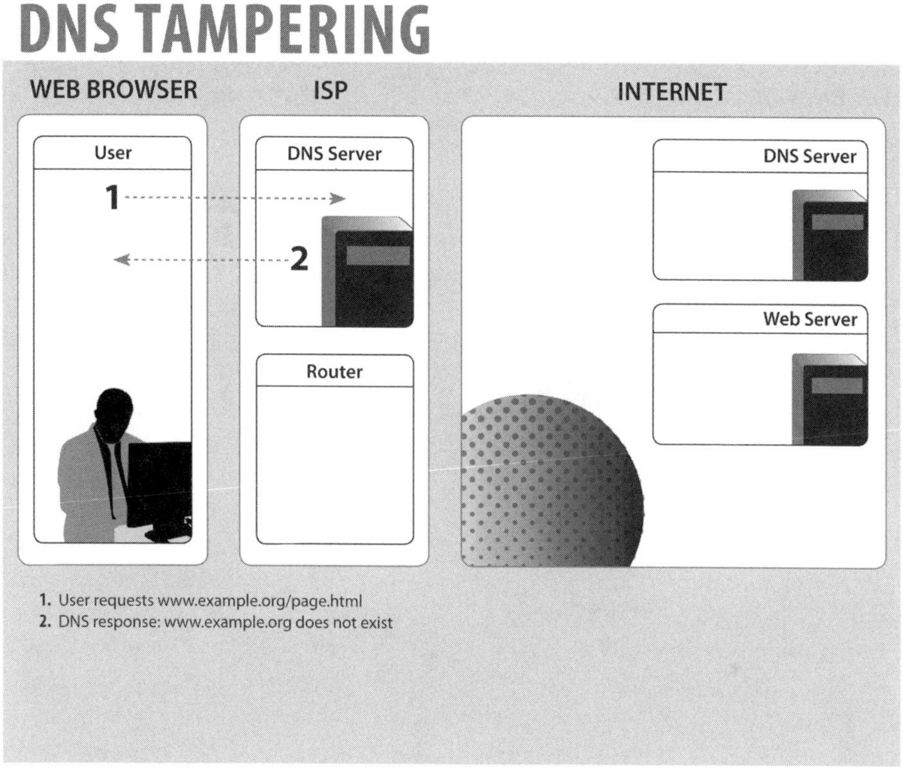

Figure 3.3
DNS tampering via filtering mechanism.

sites can be effectively blocked. With this strategy, the DNS server accessed by users is given a list of banned domain names. When a computer requests the corresponding IP address for one of these domain names, an erroneous (or no) answer is given. Without the IP address, the requesting computer cannot continue and will display an error message.[2]

Figure 3.3 shows this mechanism in practice. Note that at the stage the blocking is performed, the user has not yet requested a page, which is why all pages under a domain name will be blocked.

HTTP Proxy Filtering

An alternative way of configuring a network is to not allow users to connect directly to Web sites but force (or just encourage) all users to access Web sites via a *proxy server*. In addition to relaying requests, the proxy server may temporarily store the Web page in a *cache*. The advantage of this approach is that if a second user of the same ISP requests the same

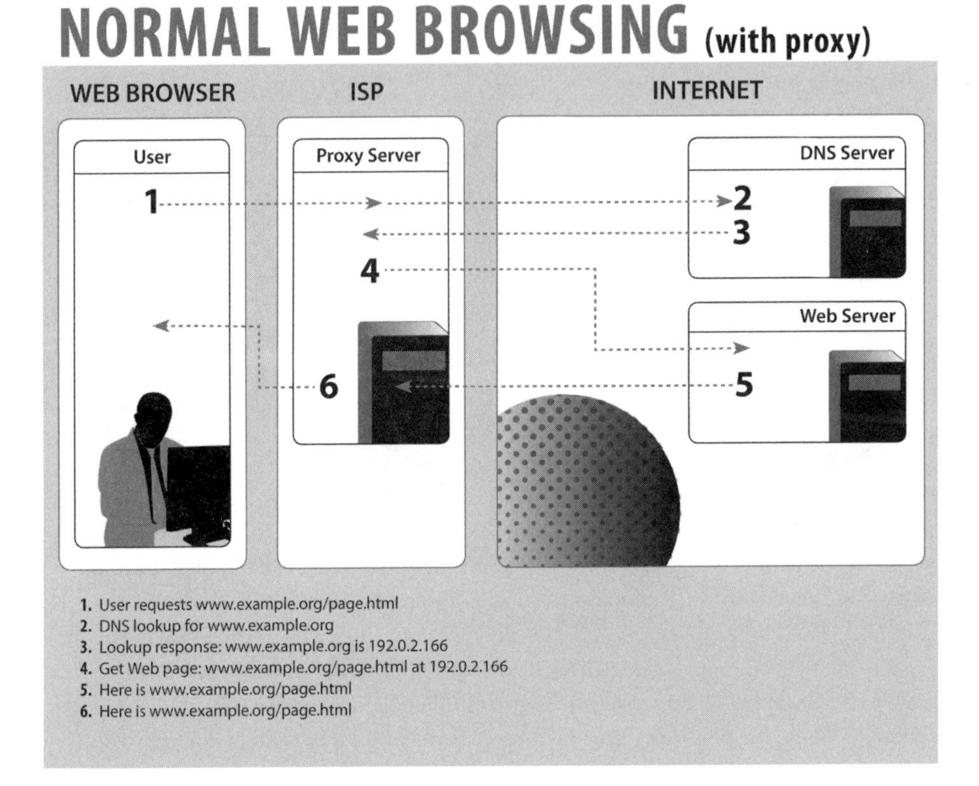

Figure 3.4
Normal Web browsing with a proxy.

page, it will be returned directly from the cache, rather than connecting to the actual Web server a second time. From the user's perspective this is better since the Web page will appear faster, as they never have to connect outside their own ISP. It is also better for the ISP, as connecting to the Web server will consume (expensive) bandwidth, and rather than having to transfer pages from a popular site hundreds of times, they need only do this once. Figure 3.4 shows how the use of a proxy differs from the normal case.

However, as well as improving performance, an HTTP proxy can also block Web sites. The proxy decides whether requests for Web pages should be permitted, and if so, sends the request to the Web server hosting the requested content. Since the full content of the request is available, individual Web pages can be filtered, not just entire Web servers or domains.

An HTTP proxy may be nontransparent, requiring that users configure their Web browsers to send requests via it, but its use can be forced by deploying TCP/IP header filtering to block normal Web traffic. Alternatively, a transparent HTTP proxy may intercept outgoing Web

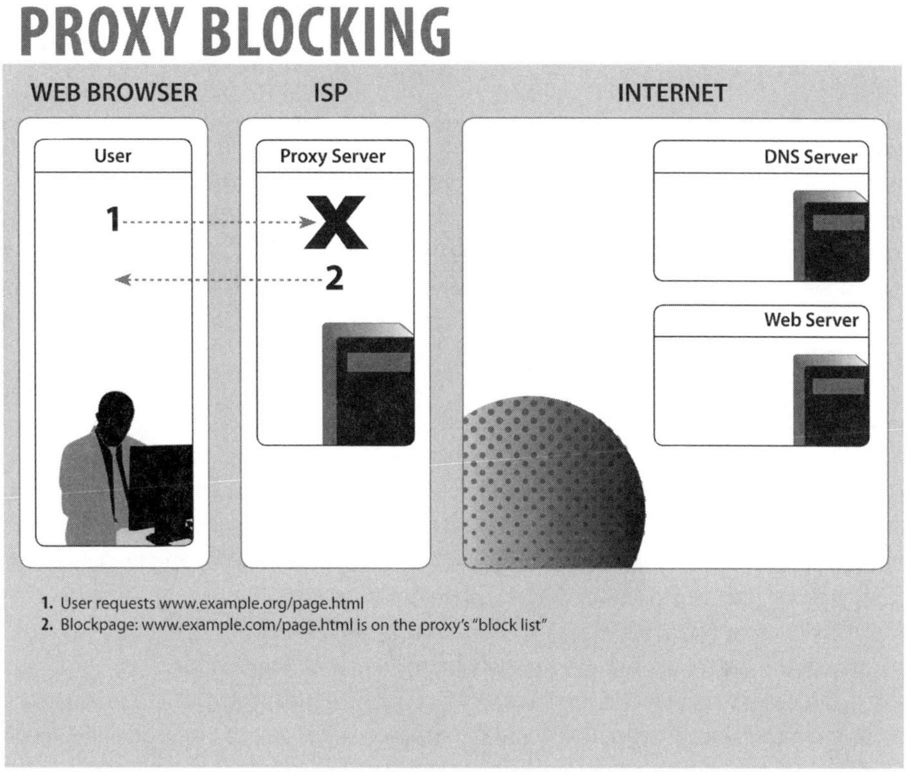

PROXY BLOCKING

WEB BROWSER ISP INTERNET

User

Proxy Server

DNS Server

Web Server

1

2

1. User requests www.example.org/page.html
2. Blockpage: www.example.com/page.html is on the proxy's "block list"

Figure 3.5
HTTP proxy blocking.

requests and send them to a proxy server. While being more complex to set up, this option avoids any configuration changes on the user's computer.

Figure 3.5 shows how HTTP proxy filtering is applied. The ISP structure is different from figure 3.1 because the proxy server must intercept all requests. This gives it the opportunity of seeing both the Web site domain name and which page is requested, allowing more precise blocking when compared to TCP/IP header or DNS filtering.

Hybrid TCP/IP and HTTP Proxy

As the requests intercepted by an HTTP proxy must be reassembled from the original packets, decoded, and then retransmitted, the hardware required to keep up with a fast Internet connection is very expensive. So systems like the BT Cleanfeed project[3] were created, which give the versatility of HTTP proxy filtering at a lower cost. It operates by building a list of the IP addresses of sites hosting prohibited content, but rather than blocking data flowing to these servers, the traffic is redirected to a transparent HTTP proxy. There, the full Web address is inspected and if it refers to banned content, it is blocked; otherwise the request is passed on as normal.

Denial of Service

Where the organization deploying the filtering does not have the authority (or access to the network infrastructure) to add conventional blocking mechanisms, Web sites can be made inaccessible by overloading the server or network connection. This technique, known as a Denial-of-Service (DoS) attack, could be mounted by one computer with a very fast network connection; more commonly, a large number of computers are taken over and used to mount a distributed DoS (DDoS).

Domain Deregistration

As mentioned earlier, the first stage of a Web request is to contact the local DNS server to find the IP address of the desired location. Storing all domain names in existence would be infeasible, so instead so-called recursive resolvers store pointers to other DNS servers that are more likely to know the answer. These servers will direct the recursive resolver to further DNS servers until one, the "authoritative" server, can return the answer.

The domain name system is organized hierarchically, with country domains such as ".uk" and ".de" at the top, along with the nongeographic top-level domains such as ".org" and ".com." The servers responsible for these domains delegate responsibility for subdomains, such as example.com, to other DNS servers, directing requests for these domains there. Thus, if the DNS server for a top-level domain deregisters a domain name, recursive resolvers will be unable to discover the IP address and so make the site inaccessible.

Country-specific top-level domains are usually operated by the government of the country in question, or by an organization appointed by it. So if a site is registered under the domain of a country that prohibits the hosted content, it runs the risk of being deregistered.

Server Takedown

Servers hosting content must be physically located somewhere, as must the administrators who operate them. If these locations are under the legal or extra-legal control of someone who objects to the content hosted, the server can be disconnected or the operators can be required to disable it.

Surveillance

The above mechanisms inhibit the access to banned material, but are both crude and possible to circumvent. Another approach, which may be applied in parallel to filtering, is to monitor which Web sites are being visited. If prohibited content is accessed (or attempted to be accessed) then legal (or extra-legal) measures could be deployed as punishment.

If this fact is widely publicized, it will discourage others from attempting to access banned content, even if the technical measures for preventing it are inadequate. This type of publicity has been seen in China with Jingjing and Chacha,[4] two cartoon police officers who inform Internet users that they are being monitored and encourage them to report suspected rule-breakers.

Social Techniques

Social mechanisms are often used to discourage users from accessing inappropriate content. For example, families may place the PC in the living room where the screen is visible to all present, rather than somewhere more private, as a low-key way of discouraging children from accessing unsuitable sites. A library may well situate PCs so that their screens are all visible from the librarian's desk. An Internet café may have a CCTV surveillance camera. There might be a local law requiring such cameras, and also requiring that users register with government-issue photo ID. There is a spectrum of available control, ranging from what many would find sensible to what many would find objectionable.

Comparison of Mechanisms

Each mechanism has different properties of who can deploy systems based around them, what the cost will be, and how effective the filtering is. In this section we compare these properties.

Positioning of System and Scope of Blocking

No single entity has absolute control of the entire Internet, so those who wish to deploy filtering systems are limited in where they can deploy the required hardware or software. Likewise a particular mechanism will block access only to the desired Web site by a particular group of Internet users.

In-line filtering mechanisms (HTTP proxies, TCP/IP header/content filtering, and hybrid approaches) may be placed at any point between the user and the Web server, but to be reliable they must be at a *choke point*—a location that all communication must go through. This could be near the server to block access to it from all over the world, but this requires access to the ISP hosting the server (and they could simply disconnect it completely).

More realistically, these mechanisms are deployed near or in the user's ISP, thereby blocking content from users of its network. For countries with tightly controlled Internet connectivity, these measures can also be placed at the international gateway(s), which makes circumvention more difficult and avoids ISPs being required to take any special action. The positioning of surveillance mechanisms share the same requirements.

DNS tampering is more limited, in that it must be placed at the recursive resolver used by users and is normally within their ISP. The actual list of blocked sites could, however, be managed on a per-country basis by mandating that all ISPs look up domain names through the government-run DNS server.

Server takedown must be done by the ISP hosting the server and domain deregistration by the registry maintaining the domain use by the Web site. This will usually be a country top-level domain and so be controlled by a government. The physical location of the server need not correspond to the country code used.

Denial-of-Service attacks are the most versatile in terms of location, in that the attacker may be anywhere and an effective attack will prevent access from anywhere.

Finally, social influence is most effectively applied by the country that can impose legal sanctions on the people who are infringing the restrictions, be that people accessing banned Web sites or people publishing banned content.

Error Rate

All the mechanisms suffer from the possibility of errors that may be of two kinds: "false positives"—where sites that were not intended to be blocked are inaccessible, and "false negatives"—where sites are accessible despite the intention that they be blocked. There is commonly a trade-off between these two properties, which are also known as overblocking and underblocking. The trade-off between false positives and false negatives is a pervasive issue in security engineering, appearing in applications from biometric authentication to electronic warfare. The *Receiver Operating Characteristic* (ROC) is the term given to the curve that maps the trade-off between false negative and false positive. Tweaking a parameter typically moves the operating point of the system along the curve; for example, one may obtain fewer false negatives but at the cost of more false positives. In general, the way to improve this trade-off is to devise more precise ways of discriminating between desired and undesired results. This will, in general, shift the ROC curve, so that false negatives and false positives may be reduced at the same time.

TCP/IP header filtering is comparatively crude and must block an entire IP address or address range, which may host multiple Web sites and other services. Taking into account the port number makes the discrimination more precise in that it might limit the blocking to only Web traffic, but this still will often include several hundred Web sites.[5] Server takedown makes the discrimination less precise, in that it will also make all content on the server inaccessible (including content not served over the Web at all).

DNS tampering and domain deregistration will allow individual Web sites to be blocked but, with the exception of e-mail, which may be handled differently at the DNS level, all services on that domain will be made inaccessible. Both may be more precise than packet header filtering, as multiple servers may be hosted on one machine, and blacklisting that machine may take down many Web sites other than the target site.

TCP/IP content filtering allows particular keywords to be filtered, allowing individual Web pages to be blocked. It does run the risk of missing keywords that are split over multiple packets, but this would be unusual for standard Web browsers.

HTTP proxy and hybrid approaches give the greatest flexibility, allowing blocking both by full Web page URL and by Web page content.

Denial-of-Service attacks are the most crude of the options discussed. Since they normally make sites inaccessible by saturating the network infrastructure, rather than the server itself, many servers could be blocked unintentionally, and perhaps the entire ISP hosting the prohibited content.

Surveillance and the threat of legal measures can be effective, as the human element allows much greater subtlety. Even if the authorities have not discovered a site that should be blocked, self-censorship will still discourage users from attempting to access it. However, such measures are also likely to result in overblocking by creating a climate of fear.

Detectability

Given adequate access to computers that are being blocked from accessing certain Web sites, it is possible to reliably detect most of the mechanisms already discussed. Mechanisms at the server side are more difficult. For example, although the server being blocked can detect Denial of Service, it may be difficult to differentiate from a legitimate "flash crowd." Similarly, a server that has been taken down, or whose domain name has been deregistered for reasons of blocking, appears the same as one that has suffered a hardware failure or DNS misconfiguration.

Surveillance is extremely difficult to detect technically if it has been competently implemented. However, the results of surveillance (arrests or warnings) are often made visible in order to deter future infringement of the rules. So it may be possible to infer the existence of surveillance, but law enforcement agencies may choose to hide precisely how they obtained the information used for targeting.

Circumventability

Although the mechanisms discussed will block access to prohibited resources to users who have configured their computers in a normal way, the protections may be circumvented. However, the effort and skills required vary.

DNS filtering is comparatively easy to bypass by the user selecting an alternative recursive resolver. This type of circumvention may be made more difficult by blocking access to external DNS servers, but doing so would be disruptive to normal activities and could also be bypassed.

TCP/IP header filtering, HTTP proxies, and hybrid proxies may all be fooled by redirecting traffic through an open proxy server. Such servers may be set up accidentally by computer users who misconfigure their own computers. Alternatively, a proxy could be specifically designed for circumventing Internet filtering. Here, the main challenge is to discover an open proxy as many are shut down rapidly due to spammers abusing them, or blocked by organizations that realize they are being used for circumvention.

TCP/IP content filtering will not be resisted by a normal HTTP proxy as the keywords will still be present when communicating with the proxy server. However, encrypted proxy servers may be used to hide what is being accessed through them.

Server takedown, Denial of Service, and domain deregistration are more difficult to resist and require effort on the part of the service operator rather than those who access the Web site. Moving the service to a different location is comparatively easy, as is changing the domain name—particularly if the service has planned for this possibility. More difficult is to notify their users of the new address before the attack is repeated.

Reliability

Even where users are not attempting to circumvent the system, they may still be able to access the prohibited resource. Provided they are implemented correctly and the hardware is capable of handling the required processing, all except Denial of Service and social techniques will block all accesses. The problem with Denial-of-Service attacks is that when systems are overloaded, they will drop some requests at random. This results in some connections, which the censor intended to block, getting through. With social techniques, if someone is simply unaware of the risks they may visit the banned site regardless.

Organizations implementing technical filtering systems must also build a list of sites and pages to block. This is a considerable undertaking if the content to be blocked is a type of content, such as pornography, rather than a specific site, such as an opposing political party. There are commercial filtering products that contain a regularly updated list of material commonly objected to, but even this is likely to miss significant content. Keyword filtering (whether at TCP/IP packet level or by HTTP proxy) mitigates this partially, as only the prohibited keywords need to be listed, rather than enumerating all sites that contain them, but sites aware of this technique can simply not use the offending keyword and select an equivalent term.

Cost and Speed

The cost of deploying a filtering mechanism depends on the complexity of the hardware required to implement it. Also, due to the limited market, specialized Internet filtering equipment is comparatively expensive, so if general purpose facilities can be used to implement filtering, the cost will be lower.

Both of these factors result in TCP/IP header filtering being the cheapest option available. Routers already implement logic for redirecting packets based on destination IP address and adding so-called null routing entries, which discard packets to banned sites, is fairly easy. However, routers can only handle up to a maximum number of rules at a time, so this could become a problem in routers working near their limit. Adding port numbers to these rules requires some additional facilities within the router, but as only the header needs to be inspected, the speed penalty of enabling this is small.

TCP/IP content filtering requires inspecting the payload of the IP packet, which is not ordinarily done by routers. Additional hardware may be required, which, for the data rates found on high-speed Internet links, would be expensive. A cheaper option, which reduces reliability but would considerably decrease cost, is for the filter to examine IP packets as they pass, rather than stopping them for the duration of the examination. Now the filtering equipment is not a bottleneck and may be slower, at the cost of missing some packets. When an infringement of policy is detected, the filtering hardware could send a message to both ends of the connection, requesting that they terminate.

DNS tampering is also very inexpensive as recursive resolvers need not respond particularly rapidly and existing configuration options in DNS servers can be used to implement filtering.

HTTP proxies require connections to be built by reassembling the constituent packets—which requires substantial resources, thereby making this option expensive. Hybrid HTTP proxies are more complex to set up, but once this is done, they are only slightly more expensive than IP filtering despite their much higher versatility. This is because the expensive stage—the HTTP proxy—receives only a small proportion of the traffic, and so need not be particularly powerful.

The cost of Denial-of-Service attacks is difficult to quantify as the scale required depends on how capable the target server is and how fast its Internet connection is. Also, it will likely be illegal to mount this attack, at least on the territory of another country. Legality also affects surveillance, domain deregistration, and server takedown; while easy to do, these mechanisms require adequate legal or extra-legal provisions before ISPs will perform them.

Insertion of False Information

If access to a prohibited Web site is blocked, depending on the mechanism, the user experience will vary. For TCP/IP header and content filtering and Denial of Service it will appear as if there has been an error, which may be desirable if the filtering is intended to be covert. The other options, DNS tampering, proxy and hybrid proxy, domain deregistration, and server takedown all give the option of displaying replacement content. This could be a notification that the site is blocked, to be open about the filtering regime, or it could be a spoofed error message, to be covert. Also, it could be false information, pretending to be from the authors of the content, but actually from somewhere else.

Strategic and Tactical Considerations

It can be useful to compare filtering for censorship with filtering for other purposes. Wiretapping systems, firewalls, and intrusion detection systems share many of the same attributes and problems. In general, such systems may be strategic or tactical. A country may collect strategic communications intelligence by intercepting all traffic with a hostile country regardless of its type, source, or destination using a mechanism such as a tap into a cable. It may also collect tactical communications intelligence in the context of a criminal investigation by wiretapping the phones of particular suspects or by instructing their ISPs to copy IP traffic to an analysis facility.

Similarly, censorship can be strategic or tactical. Strategic censorship may include permanent blocking of porn sites, or of news sites such as the BBC and CNN; this may be done at the DNS level or by blocking a range of IP addresses. An example of tactical censorship might be interference during an election with the Web servers of an opposition group; this might be done by a service-denial attack or some other relatively deniable technique.

Censorship systems interact in various ways with other types of filtering. Where communications are decentralized, for example, through many blogs and bulletin boards, the censor may use classic communications-intelligence techniques such as traffic analysis and snowball sampling in order to trace sites that are candidates for suppression. (*Snowball sampling* refers to tracking a suspect's contacts and then their contacts recursively, adding suspects as a snowball adds snow when rolling downhill.) Countersurveillance techniques may therefore become part of many censorship resistance strategies.

The interaction between censorship and surveillance is not new. During the early 1980s, the resistance in Poland used radios that operated in bands also used by the BBC and Voice of America; the idea was that the Russians would have to turn off their jammers in order to use radio-direction finding to locate the dissidents. Today, many news sites have blogs or other facilities that third parties can use to communicate with each other; so if a censor is reluctant to jam *The Guardian* newspaper, then its dissidents could use blog posts on *The Guardian* site to talk to each other, using pseudonyms. But many of the novel and interesting interactions have to do with applications.

Discussion

Communication is now a part of more and more applications. Some of these are designed for communication, such as Skype, but bring new capabilities; in Skype's case, it provides encrypted communications and is also widely used. Previously, users of cryptography would be likely to draw attention to themselves, especially in authoritarian countries. Today, Skype and other voice-over IP (VoIP) products are used to save money on telephone bills, and provide voice privacy as a side effect.

Another example is given by Google Docs & Spreadsheets. Google purchased an online word-processor product (Writely) and now makes it available to Internet users in many countries as Google Docs & Spreadsheets. People keep their private documents on Google's servers and edit them online via a Web-based interface. Such a document can be shared instantly with other users; this provides a convenient channel for communications. In this case, the communications are the side effect; the reason people use Google Docs & Spreadsheets is to avoid spending money on Microsoft Office.

The general picture is that censors—and wiretappers—perpetually lag behind the wave of innovation. In the 1960s, computer companies fought with telephone companies for less restricted access to the network, and the telephone companies called government agencies to their aid so as to protect their business models (which involved owning all network-attached devices). By the mid-1980s only a few authoritarian states banned the private ownership of modems, and the security agencies of developed countries had acquired the capability to intercept data communications. The explosion in popularity of fax machines in the mid-1980s put the agencies on the back foot again; a handwritten fax still gives reasonable protection against automated surveillance. When e-mail and the Web took off in the mid-1990s, the agencies scrambled to catch up, with proposals for laws restricting cryptography, which turned out to be irrelevant to the real problems that emerged, and more recent proposals for the retention of communications data. Modern Google users may be largely unaffected by all this—their searches, e-mails, word processing, and group communications may all be cohosted.

It will be interesting, to say the least, to see how states deal with the move to edge-based computing. Developed countries tend to observe a distinction between wiretapping that gives access to content, and traffic analysis that gives access merely to traffic data. Most countries require a higher level of warrantry for access to the former. However, the move to the edge blurs the distinction between traffic and content, and there must eventually be a question as to whether this might undermine the existing controls on state interference with communications. Other countries may be less limited by legal scruples than by technical capability and by access. Application service providers such as Google and Yahoo! have to cooperate with the authorities in countries like China where they maintain a physical presence, but may not make all applications available. Small authoritarian states, that enjoy neither the physical presence of the main service providers nor the technological capability, may find their ability to exert technical control over information flows seriously compromised, and may have to rely largely on legal and social mechanisms.

To sum up, the Internet has borders—just like meatspace—and the quality of the borders depends on the situation of the country that erects them.

Conclusion

Ten years ago, Internet utopians like John Perry Barlow held out the prospect that state-sponsored barriers to communication would be swept away, leading to a significant improvement in the human condition. Some less-developed countries denounced this as "U.S. Information Imperialism."

The Internet is more complex than previous mechanisms (such as the postal system and telephones). Control is not impossible, but it requires more sophistication, and the censors are continually playing catch-up as technological innovation changes the game. The migration of communications into the application domain will increase complexity further, and raise interesting policy questions—but there may be different questions in developed and less-developed countries.

The utopians are sometimes seen as having lost; the Internet does have borders now. However, information is much more free than it was ten years ago, and the real question is whether one sees the glass as being half empty or half full. There is no doubt that modern communications technologies—including the mobile phone as well as the Internet—have greatly facilitated the dissemination of news, cultural exchanges, and political activism. Even in developed countries, new technologies from blogs to videophones have increased the potential for surveillance, but have also helped people hold officials to account.

Notes

1. Richard Clayton, Steven J. Murdoch, and Robert N. M. Watson, "Ignoring the Great Firewall of China" in 6th Workshop on Privacy Enhancing Technologies (Cambridge, England, 28–30 June 2006).
2. Maximillian Dornseif, "Government Mandated Blocking of Foreign Web Content," in Security, E-Learning, E-Services: Proceedings of the 17, ed. J. von Knop, W. Haverkamp, and E. Jessen (DFN-Arbeitstagung über Kommunikationsnetze, 2003).
3. Richard Clayton, "Failures in a Hybrid Content Blocking System," in Fifth Workshop on Privacy Enhancing Technologies (Dubrovnik Cavtat, Croatia, 30 May–1 June 2005).
4. Xiao Qiang, "Image of Internet Police: Jingjing and Chacha Online," China Digital Times, http://chinadigitaltimes.net/2006/01/image_of_internet_police_jingjing_and_chacha_online_hon.php (accessed February 19, 2007).
5. Ben Edelman, "Web Sites Sharing IP Addresses: Prevalence and Significance," Berkman Center for Internet and Society (September 2003).

4

Filtering and the International System: A Question of Commitment

Mary Rundle and Malcolm Birdling

Introduction

This book reflects a certain skepticism about filtering trends. Behind this skepticism is both an acceptance that freedom of expression (including the right to seek, receive, and impart information and ideas) is a basic human right under international law, and a sense that many governments' filtering practices represent an obstruction of this right. To ground these assumptions, this chapter seeks first to set out generally what constitutes international law, to whom it applies, and in what contexts, and second to consider how these concepts relate to filtering. At the heart of the matter is the question of if and how legal means can be used to regulate Internet filtering at an international level to protect freedom of expression.

This chapter introduces several elements in considering filtering from a human rights perspective—including international law as commitments among state actors, the setting out of human rights in international law, and filtering as a potential obstruction of the human right to freedom of expression. This chapter finds that international human rights agreements provide a valuable framework for determining what constitutes permissible and impermissible filtering, but that these instruments fall short on the enforcement end due to widespread filtering and states' apparent reluctance to take action against one another. The chapter then turns to consider domestic approaches for holding private actors accountable internationally, but notes that these approaches are inadequate on their own. Finally, the chapter points to the promise of international standards for enabling nonstate actors to prevent broadscale filtering and thereby facilitate the exercise of freedom of expression.

The Backdrop

The modern international system dates from the Peace of Westphalia (1648),[1] which established the principles of 1) state sovereignty[2] and the right of self-determination; 2) legal equality among states; and 3) nonintervention of states in one another's internal affairs. In this system, states are the actors, giving life to international law as they create it together and agree to be mutually bound by it. As such, international law rests on the consent of sovereign states.

States as Intermediaries

Since nonstate actors in current international law are understood to fall under the jurisdiction of states, it is states that have the authority to spell out rules and to bind both themselves and these subordinate actors. If states want international law to apply directly to nonstate actors such as citizens or businesses, they may commit to creating common obligations within their respective jurisdictions; they may also establish international rules and designate bodies to deal directly with nonstate actors.

For the most part[3] states have not created obligations that bind nonstate actors at the international level. Instead, states have been intermediaries between citizens and the international system.

Trend toward Disintermediation

A certain disintermediation may be taking place as international bodies are increasingly dealing directly with citizens. As discussed in some depth here, the primary international treaty addressing civil and political rights carries with it an optional instrument that states may sign onto if they wish to allow private parties to bring complaints to an international body. In addition to that avenue, the Internet may be ushering in a new trend whereby individuals enjoy recognition at the international level. Just as the Internet has reduced the role of middlemen in many areas of e-commerce, so it may be allowing citizens of the world's diverse jurisdictions to interact directly with international institutions. For example, the World Intellectual Property Organization (WIPO) has established an Arbitration and Mediation Center[4] to resolve Internet domain name disputes. Here, individuals are recognized as having *standing*, or the right to bring a case to the tribunal, and so do not have to rely on national governments to do so.[5] By providing a similar type of process that an agency at the national level would, mechanisms for Internet governance are spurring international integration.

Reflecting changing attitudes toward the role of nonstate actors in the global Information Society, forums have been established under the United Nations to foster dialogue among a full range of "stakeholders" on issues relating to the Information Society. The World Summit on the Information Society represented an extensive effort along this line, bringing together thousands of stakeholders for meetings in Geneva (2003) and Tunis (2005).[6] As a result of the Geneva meeting, the U.N.'s Secretary-General convened a Working Group on Internet Governance to feed analysis into the Tunis meeting.[7] While the Working Group was composed of a limited number of individuals from government, the private sector, and civil society, it held open consultations to hear views from a full range of stakeholders. Among issues studied by this group were the roles of all actors in the Information Society.

Continuing in this vein, the U.N.'s Internet Governance Forum, a product of the Tunis meeting, now takes submissions from any contributor and offers an open forum for multistakeholder discussion on matters relating to Internet governance. While this body has not been endowed with decision-making power, it nonetheless can be seen as representing new attempts to factor views of nonstate actors directly into international policy-making.

Despite these signs of nonstate actors' gaining recognition at the international level, formally the international system still treats states as the relevant actors, and others enjoy status only to the limited degree to which states choose to confer it upon them.

Empowering International Institutions

When states consider the prospect of empowering an international agency to serve as a forum for setting and administering global rules, they face the danger that they will create an institution that will eventually gain enough credibility that it in effect becomes freestanding. As that new authority amasses influence at the international level, its authority is no longer consciously considered to derive from the agreement of the individual member states that comprise it, and instead this authority is simply presumed to accompany the institution. At this stage, the authority of the member states themselves may even be questioned if their direction deviates from the central institution's course.[8] Indeed, this tendency is apparent in many people's conceptions today, where international law is perceived to have moral authority due to its international quality. It is no wonder, then, that a state may be wary of assigning powers to an international institution in the first place.

Public International Law and Modern Human Rights

Human rights law in large part concerns the relationship of the individual and groups of individuals to the state. At a fundamental level, it carries questions concerning the source of rights. For example, some people contend that human rights are "natural rights" that are universal as part of the world's inherent nature, or that derive from higher, religious authority and do not stem from mere human beliefs or actions; people subscribing to this view tend to believe that natural rights exist regardless of what a government or society might establish and enforce. Others, such as utilitarian thinker Jeremy Bentham, have categorically rejected the notion of natural rights.[9]

Debates on the source of human rights multiply when considering the application of these rights in an international context. International legal instruments relating to civil and political rights were heavily influenced by the West in the midtwentieth century and reflect a Judeo-Christian heritage. As such, human rights were presented as stemming from the fact that all people have been created by God, and hence all should be on equal footing. Because other regions (e.g., Asia) have not historically had this orientation, there has been an ongoing debate as to whether the rights are truly "universal" at all.

In essence, this international twist is a variation of the question of whether human rights stem from natural rights or from positive acknowledgment of them by the state. If human rights are thought to stem only from their recognition by the state, international human rights are just a matter of negotiation among states as to what they deem priorities to be in light of state interests. On the other hand, if human rights are thought to exist independently of the state, they have a place of their own in the international system and therefore should not be subject to horse trading.

Global Citizens

Some people might argue that society is already so integrated internationally that the relationship between a state and citizens is no longer hierarchical; rather, the relationship is seen as transformed to one of overlap, where a state is ascribed with authority over those "global citizens" who happen to fall within its territorial jurisdiction. Given the amorphous boundaries of cyberspace, this territorial distinction begins to appear murky.

Meanwhile, the Internet lends support to newly emerging forms of transnational, "post-sovereign" political communities. Such groups, including diaspora and aboriginal communities, fit poorly within either a state or a global citizen network framework. Demands for increased autonomy and self-determination by such communities defy the old paradigm of state sovereignty, while particularistic claims challenge the paradigm of universal human rights. Although such communities may have existed previously, the Internet has given them new political life as they can more rapidly create transnational polities that exercise relatively substantial influence. How these new forms of political interaction interrelate with human rights in general, and freedom of expression in particular, is a complex matter.[10]

Quasigovernmental Private Action

In the midst of these ambiguities, additional quandaries arise when the behavior of private, nonstate actors resembles state action. Private actors such as corporations may provide services that people usually conceive of as the state's responsibility. For example, a private actor might build infrastructure (providing water, electricity, roads, or, arguably, an Internet infrastructure). When private actors take on governmental functions—either through direct delegation or mandate by government or as a result of government simply allowing them to carry out activities—should they be considered agents of the state, bound by the same obligations to which the state is bound?

Of course, if a private actor were performing governmental functions across jurisdictions, it could prove challenging to assess on whose behalf it was acting as a state agent. For the sake of maintaining accountability to the public, the international system may need to find a way to hold private actors to a similar standard as states when they act internationally in governmental capacities.

Practical Implications

The questions presented here are not merely esoteric. Rather, their answers very well may determine the kind of regulatory regimes that the Information Society puts in place. More fundamentally, the questions go to the heart of relationships among individuals, states, overlapping polities, multinational enterprises, the international system, and the Information Society as a whole.

The subject of filtering demonstrates some very practical implications of these theoretical issues. For example, filtering poses problems in that a state may claim a sovereign right to determine what constitutes acceptable content that people within its jurisdiction may seek, receive, or impart, whereas the international system may assert a role in overseeing the exercise of human rights, including freedom of expression. Similarly, filtering exemplifies the definitional challenge that presents itself when private action amounts to state action. If a corporation has an effective monopoly on the supply of an Internet service, is it assuming a governmental function if it controls access to information according to what it determines to be acceptable content? Does it matter whether the corporation is doing so of its own accord or whether it is doing so in response to a government mandate? Should such corporations be considered agents of the state, bound by the same freedom of expression obligations to which the state is bound? What responsibilities does a state have for filtering by private actors operating within its jurisdiction? What rights does a person or a group of people have in this mix? How should jurisdiction for filtering be determined in cyberspace?

Before such questions can be approached, it is helpful first to consider the current international legal landscape.

Key International Legal Instruments

Since the end of World War II, "human rights" in the international arena have moved from being largely a tool of political rhetoric to a substantive set of concrete legal obligations among states. The most obvious evidence of this development is the enshrining of rights in a number of binding international documents. At the regional level, countries within several geographical areas have grouped together to form human rights institutions and to create human rights obligations applicable within these areas. Alongside these formations has been the development of a truly international set of human rights, established under the United Nations framework. These rights find form in a set of treaties creating legal obligations on states to do, or to refrain from doing, certain activities. Because these international instruments offer a global approach and enjoy wide ratification in a way that maps well to the Internet's international nature, they are the basis of discussion in this chapter.

The applicability of pre-existing legal instruments to the realm of the Internet has been affirmed by international bodies. The World Summit on the Information Society (referenced earlier) endorsed a Declaration of Principles that, among other things, proclaims that freedom of expression in an Internet context is indeed protected by pre-existing instruments. The question then becomes precisely what do these instruments provide, and are they appropriate for the regulation of filtering in this "new" medium?

Universal Declaration of Human Rights

The starting point for this consideration is the Universal Declaration of Human Rights (the UDHR), which was adopted by the United Nations General Assembly in 1948. Passed in the shadow of World War II, the Declaration is not a treaty, but rather an authoritative statement by the international community of certain values that are said to be so universal in character as to qualify as "human rights"—rights all humans, irrespective of their geographical locations, are said to possess. The preamble entreats all individuals and organs of society to "strive by teaching and education to promote respect for these rights and freedoms and by progressive measures, national and international, to secure their universal and effective recognition and observance."[11]

Article 19 of this seminal document contains the broadly worded statement that "everyone has the right to freedom of opinion and expression; this right includes freedom to hold opinions without interference and to seek, receive and impart information and ideas through any media and regardless of frontiers."[12]

This statement's significance should be understood in the wider context of the importance that the UDHR itself has been accorded across all spheres of human activity. While the UDHR has not been without controversy, today, almost sixty years after its adoption, it is still cited and relied upon on a daily basis by individuals, organizations, and governments across the globe. At the very least it has been used as a firm touchstone by which to measure the morality of individual and governmental action.

The inclusion of a broad, unfettered guarantee of freedom of expression in such a weighty document is a clear statement of international acknowledgment of such a right. The U.S. Restatement (Third) of Foreign Relations Law goes as far as to say that a breach of the UDHR may actually amount to a breach of the United Nations Charter, meaning that the protection of the right to freedom of expression may be a legal obligation on *all states*, irrespective of whether they have ratified any of the international human rights treaties described below.

The International Covenant on Civil and Political Rights

Most of the rights enumerated in the UDHR have now received concrete legal form in a series of treaties created, monitored, and enforced under the auspices of the United Nations. Preeminent among these instruments is the International Covenant on Civil and Political Rights (ICCPR), which provides in part the following:

> Article 19
>
> 1. Everyone shall have the right to hold opinions without interference.
> 2. Everyone shall have the right to freedom of expression; this right shall include freedom to seek, receive and impart information and ideas of all kinds, regardless of frontiers, either orally, in writing or in print, in the form of art, or through any other media of his choice.
> 3. The exercise of the rights provided for in paragraph 2 of this article carries with it special duties and responsibilities. It may therefore be subject to certain restrictions, but these shall only be such as are provided by law and are necessary:
>
>   a. For respect of the rights or reputations of others;
>   b. For the protection of national security or of public order (ordre public), or of public health or morals.

Article 20

1. Any propaganda for war shall be prohibited by law.
2. Any advocacy of national, racial or religious hatred that constitutes incitement to discrimination, hostility or violence shall be prohibited by law.[13]

The ICCPR was adopted by the General Assembly of the United Nations in 1966 and entered into force a decade later. As a treaty, its provisions have direct legal application only in those countries that have voluntarily opted to become parties. This ratification has been extensive. According to the Office of the United Nations High Commissioner for Human Rights, 160 states are party to the ICCPR.[14] Among them are the following countries whose filtering practices are covered in studies by the OpenNet Initiative: Afghanistan, Algeria, Azerbaijan, Bahrain, Belarus, Egypt, Eritrea, Ethiopia, India, Iran, Iraq, Israel, Jordan, Kazakhstan, Kyrgyzstan, Libya, Moldova, Morocco, Nepal, North Korea, the Russian Federation, South Korea, Sudan, Syria, Tajikistan, Thailand, Tunisia, Turkmenistan, Ukraine, Uzbekistan, Venezuela, Vietnam, Yemen, and Zimbabwe.[15]

As with the declaration, the ICCPR is significant as a statement of a fundamental, minimum set of conditions for the observance of human rights. The legitimacy of the ICCPR in this regard can be seen not only in its widespread ratification, but also in the myriad bodies that refer to it. A number of domestic courts, legislatures, nongovernmental organizations (NGOs), and international bodies frequently refer to the ICCPR directly when making decisions in which the rights are implicated.

It is important, then, that the ICCPR also contains a broad, unquibbling guarantee of freedom of expression. Its provisions guarantee, subject to certain limits (discussed later), the "freedom to seek, receive and impart information and ideas of all kinds."

The breadth of this conception is best appreciated by making comparisons to the way similar rights are framed in other documents and interpretations. Many domestic constitutions draw distinctions, for example, between different forms of speech, and afford varying levels of protection depending on the nature of the content. The U.S. Supreme Court, for example, once considered that advertising was outside the scope of constitutional protection accorded to freedom of speech. While the Court has now softened that absolutist position, advertising is still not entitled to the same protection under the U.S. Constitution as other forms of expression. A similar stance has been articulated with regard to "obscene" speech. This tapered rendition of freedom of expression differs from the conception in the ICCPR; indeed, the very words with which the ICCPR right is expressed precludes such a narrow interpretation and demands an expansive understanding of the right.

This broad reading has been confirmed by the United Nations Human Rights Committee (UNHRC), a body of experts established under the ICCPR to scrutinize state compliance with the ICCPR. In considering a challenge to laws restricting commercial advertising, the UNHRC held that the right "must be interpreted as encompassing every form of subjective ideas and opinions capable of transmission to others . . . [including] news and information, of commercial expression, of works of art, etc.; it should not be confined to means of political, cultural or artistic expression."[16] Further, the Committee did not agree that different kinds of expression can be subjected to greater restrictions than others.

However, the rights elaborated in the ICCPR are expressed as being held by natural persons—that is, they pertain to individual human beings rather than explicitly extending as well to legal, or juridical, persons (e.g., corporations). The UNHRC has avoided any difficulties in the freedom of expression context by stating that the right is by its nature "inalienably linked to the person," and that individuals enjoy freedom of expression with respect to their businesses, for example, having a right to use the language of their choice. As such, an individual person's right to freedom of expression should hold even if the primary purpose of the expression is to promote a company.

Article 19 also provides that this right is to apply "regardless of frontiers and through any media." This express lack of qualification is particularly important as it underscores the fact that the right extends across a wide variety of media.[17] As such, arguments that the Internet is somehow different in nature, and immune from scrutiny, should fail.

As a document of some decades' standing, the ICCPR has seen many changes in the structure and organization of the mass media, and its machinery has responded accordingly. The UNHRC has noted that a completely state-controlled media is inconsistent with the right, as are restrictive licensing regimes for television and radio stations.[18] Given the medium-neutral nature of the right, the ICCPR would also be likely to prohibit a similarly restrictive system of state registration for Internet publishers—for example, a system requiring video bloggers to submit to an unduly rigorous licensing regime.

The right to freedom of expression as articulated in these international documents is extremely broad and was intended to be applicable to all types of media—existing now or in the future. Hence, any state restrictions on the distribution of information via the Internet would seem to constitute a restriction (although not necessarily a breach) of the right to freedom of expression under the ICCPR.

Limitations on the Right

The right to freedom of expression as set out in the ICCPR is not absolute, however. The text of Article 19 states that the exercise of the right carries with it "special duties and responsibilities" and that it "may therefore be subject to certain restrictions." While critics of the ICCPR may argue that the exceptions to the right are so broadly drawn as to render the right meaningless, this characterization is not accurate. The permissible scope of such restrictions is in fact narrow.

Article 20 of the ICCPR spells out the most straightforward cases in which restrictions are appropriate; indeed, the language even creates a positive obligation on states to restrict expression in relation to war propaganda and advocacy of national, racial, or religious hatred that constitutes incitement to discrimination, hostility, or violence. States are obliged to prohibit these in their domestic legal systems. In other words, filtering of this form of information would not only be permitted but arguably required by the ICCPR.

Another positive obligation to restrict expression appears in Article 17(2), which obliges states to protect individuals from intentional interference with their honor and reputation.[19]

These Articles cover affirmative requirements to restrict freedom of expression. Outside such cases, it falls to individual states to determine which restrictions they wish to place on the right. The ICCPR does curb the exercise of this power by states by providing in Article 19(3) that such restrictions must be 1) provided by law and 2) necessary for ensuring the respect of the rights or reputations of others, or for the protection of national security, public order, public health, or public morals. It should be emphasized that this is an exhaustive list of the situations in which restrictions are allowed—there are no other grounds on which limitations on freedom of expression are permissible, and states are not permitted to invent further grounds. Similarly, a state cannot cite inconsistent domestic laws as a reason for noncompliance with the human rights provisions of the ICCPR.

The requirement that any limitation must have its basis in law means that there must be some affirmative lawful basis for filtering (whether it be a clearly worded statute, or a similarly clear judicial decision or series of decisions). Vaguely worded statutes will not suffice, nor will the vague exercise of administrative discretion. This precision is important as it allows individuals to understand the restrictions to which their expression may be subject.

The requirement that restrictions must be shown as necessary for a legitimate purpose triggers an inquiry into the proportionality between the extent of the interference with freedom of expression and the importance of the purpose of the restriction. It is not sufficient for a state to make a bare assertion that its actions are necessary to achieve the purpose.

A review of the situations under which the UNHRC has upheld restrictions on freedom of expression, as well as general guidance issued by U.N. bodies, reveal a number of principles that can guide states in determining whether a proposed action meets the ICCPR necessity test.

First, the application of restrictions is to be narrow. This narrowness requirement is particularly important where justifications for restrictions are offered on the basis of alleged national security or public order imperatives. The UNHRC has noted that justifications on these grounds are the most frequently abused by invocation to protect the position of the government of the day, rather than truly to protect citizens' rights.[20] In the filtering context, if a state were to block all political Web sites during an election in the name of public order, it is dubious whether the restrictions would meet the standard of necessity.

Limitations on the freedom of expression in the name of public morals raise similar concerns. The UNHRC initially suggested that states possessed a certain "margin of appreciation" with respect to what was necessary to protect "public morals" in any given jurisdiction.[21] However, the concept of such a margin was expressly rejected by the Committee in a subsequent case concerning other rights.[22] This would tend to suggest that states cannot rely on such a margin when considering their obligations under the ICCPR.

Second, the necessity of restrictions must be convincingly established by the state. In addition to narrowly tailoring exceptions, a state must provide adequate justification for restrictions it imposes. This second principle in showing necessity is generally applicable to all instances where states seek to justify limitations on rights. A state limiting the freedom of expression has a duty to demonstrate convincingly that the measures taken are necessary and proportionate in pursuing legitimate aims.[23] In this regard, the UNHRC has pronounced that "the legitimate objective of safeguarding and indeed strengthening national unity under difficult political circumstances cannot be achieved by attempting to muzzle advocacy of multi-party democracy, democratic tenets and human rights."[24]

The UNHRC has clearly stated that any restrictions must "not put in jeopardy the right itself."[25] In other words, a total clampdown on freedom of expression—even if imposed in the name of ensuring the respect of the rights or reputations of others, or protecting national security, public order, public health, or public morals—would never be deemed justifiable.

Reservations

A final word is warranted before leaving the general subject of limitations. As international law is based on the consent of sovereign states, it is possible for a state to place *reservations* on international treaties it ratifies in some circumstances. These reservations will limit the extent of the reserving state's obligations under the relevant treaty. It is beyond the scope of this chapter to undertake a review of such reservations in the context of the ICCPR, but one salient point is worth noting: A number of states specifically made a reservation to Article 19 to the effect that they retained the power to regulate radio and television broadcasts. These states became parties to the ICCPR before mass communication via the Internet emerged. According to the UNHRC, states will not be permitted to extend a specific reservation to provide a more general exception from the ICCPR rights.[26] Thus, it would be highly unlikely that the UNHRC would accept that a state's reservation with regard to radio and television broadcasts permitted it, by analogy, to regulate the Internet in a similar fashion.

Applying International Law to Filtering

In light of the provisions spelled out above, the vast majority of current filtering practices would seem to fall short of the requirements of international law since 1) most filtering measures are not specifically provided by law, and 2) it is unlikely that these measures would meet the ICCPR necessity test.

Nevertheless, to give some concrete examples of how filtering practices might comply with certain ICCPR provisions procedurally or substantively, this section refers to some specific practices by states.

Measures Provided in Law

With respect to ICCPR requirements that any limitations on the freedom of expression be expressly "provided in law," a state might establish procedures for making its filtering practices open and transparent. Disclosing that such filtering practices are in effect, according to a specific law or court order, is a step in the right direction. An example in this regard is the way that Iran has created a Committee in Charge of Determining Unauthorized Sites, which is legally empowered to identify sites containing prohibited content. To meet the ICCPR standard, the law under which the filtering is carried out should be clear and nonarbitrary.

Ironically, a state can use procedures to impose content restrictions, and these procedures (again, if sufficiently clear and nonarbitrary) can help that state comport with ICCPR obligations to specify policies in precise law. A state may impose licensing requirements—for example, the way Uzbekistan requires cybercafes to comply with a "standardization procedure" carried out by a government agency before starting operation. A state also may enact registration requirements—for example, the way South Korea requires bloggers and Web content developers, or even cybercafes and end users, to associate their online activities with their real-world identities. In addition, a state may assign liability to Internet service providers (ISPs) for content that is delivered to users—for example, the way Iran holds ISPs criminally liable for content. Self-monitoring requirements are another form of procedures that a state can use to restrict activity—for example, the way China drills the message that "the Internet is a public space" to warn people to check their own behavior. In each of these procedural moves, if specific laws are set out, states may in fact be complying with one of the ICCPR's conditions for limitations—even as they erect filtering mechanisms.

Complying with the requirement that restrictions be provided in law does not guarantee that the processes as a whole are compatible with the ICCPR; rather, in imposing the restrictions, states still must comply with the requirement that restrictions be necessary.

Measures Necessary

With respect to the requirement that limitations on the freedom of expression be "necessary," states also can use procedures to target filtering for specific objectives, so that the scope of the filtering is not too broad. Procedures that allow public oversight and accountability act in this vein. For example, Pakistan has established a Deregulation Facilitation Unit to redress grievances in the event of errors or overblocking.

Seeking to comply with the ICCPR requirement that limitations on freedom of expression be justified as necessary, states naturally emphasize the substance of filtering measures, or what they are targeting. States sometimes assert that measures are undertaken for the purpose of respecting the rights or reputations of others (one of the permitted grounds for limitation). For example, China partially justifies its use of rights management tools by saying this filtering helps to enforce intellectual property rights. Similarly, Malaysia's Internet regulatory authority explicitly targets abusive or harassing content.

So, too, states justify limitations as being necessary for the purpose of protecting national security, public order, public health, or public morals. For example, United Arab Emirates cites these goals in justifying its legislation on hacking, the accessing of illegal sites, and the use of digital signatures.

Text box 4.1 breaks this process down into basic elements so as to offer guidance for how a filtering state might avoid violating international law even as it limits freedom of expression. It is important to remember, however, that a state's compliance with these requirements is necessary, but not in itself sufficient, to satisfy the state's obligations. To be compliant, the measures must be *actually* necessary for the purposes the state asserts that they are necessary for—the state's bare assertion that this is the case will not be sufficient.

None of the filtering regimes covered in studies by the OpenNet Initiative appear to have been crafted to meet international commitments on freedom of expression. As states grow more aware of their obligations, it will be interesting to see whether they modify their filtering practices to honor these commitments. In the meantime, it seems the international system is struggling with extensive filtering habits that are out of proportion with legitimate objectives.

When considered in light of technology's tendency to act as a sort of "law" that can govern society,[27] requirements that filtering be provided in law and be necessary are marked with an extra nuance. Surely the idea behind these requirements is to promote precision, to allow people to know what measures apply, and to promote government accountability to the public. Does it not follow that the technologies used in filtering should be precise, transparent, and justifiable as well?

Problem of Enforcement

Having examined obligations that states have agreed upon at the international level, and having briefly explored how these obligations mesh with filtering practices, it is logical next to examine the machinery by which these obligations can be enforced. It is here that the weaknesses of the international system become apparent.

The international human rights instruments rely largely on states themselves to implement their commitments at a domestic level. International enforcement also falls on states themselves. To this end, the ICCPR contains express obligations on states to ensure that this occurs (Article 2).

However, due to political realities, such guarantees are of little use unless they are accompanied by sanctions for violations. It is here that the UNHRC has the potential to play a critical role.

Monitoring under the ICCPR

The ICCPR requires states, upon request by the UNHRC, to provide a report on their compliance with obligations under the treaty. According to the rules of the Committee, a state must prepare a written report, which the UNHRC then examines. State representatives are usually present to answer questions, and the UNHRC also hears from relevant NGOs and other civil society organizations. At the conclusion of the process, the UNHRC issues a report containing

Box 4.1
An unofficial guide to filtering legally

To filter in a way that honors international human rights commitments on freedom of expression, a government can use the following as an unofficial guide:

1. PURPOSE
   The state believes restrictions on freedom of expression are necessary to ____.
   [*Example*: . . . prevent people from using the Internet to stir up violence against a particular ethnic group.]
2. STATEMENT OF WHAT NEEDS TO BE DONE
   Therefore, the government decides to pass a law to ____.
   [*Example*: . . . limit hate speech.]
3. SPECIFIC EXPLANATION OF HOW FILTERING WILL BE CARRIED OUT
   To ensure that people can understand the law and can check to see that its application is not arbitrary, the government spells out ____.
   [*Example*: . . . what exactly is beyond the limits of acceptable speech and how it will be filtered.]
4. PERMITTED LIMITATION AS LISTED IN ICCPR ARTICLE 19 OR 20
   In grounding this action in a justification acceptable by international law, the government indicates that this restriction is necessary ____.
   [*Check all that apply:*]
   - ☐ for respect of the rights or reputations of others;
   - ☐ for the protection of national security;
   - ☐ for the protection of public order;
   - ☐ for the protection of public health;
   - ☐ for the protection of morals;
   - ☐ for the prohibition of propaganda for war;
   - ☐ for the prohibition of advocacy of national, racial, or religious hatred that constitutes incitement to discrimination, hostility, or violence.
5. PROCESS TO TELL PEOPLE WHAT IS HAPPENING AND CORRECT ANY PROBLEMS
   To help ensure that the law is not implemented in an arbitrary or overly broad manner, the state provides a mechanism whereby ____.
   [*Example*: . . . if a Web site is blocked, Internet users receive a message 1) indicating why this filtering has occurred, according to what specific law, and 2) telling them how they can report a problem and receive a response.]

"concluding observations" on the state's compliance including areas of concern and recommendations for action.

The effectiveness of this process is contingent on cooperation by states. Noncooperation has been a frequent problem with the system, and one which the UNHRC is taking an increasingly active role in monitoring. However, the presence of NGOs provides a very real opportunity for the human rights issues experienced in a given jurisdiction to be identified, thereby reducing the ability of a state to subvert the process by providing inaccurate information.

That said, the UNHRC's recommendations under this procedure are simply that—recommendations—and are not binding. Additionally, the institutional constraints and chronic underresourcing endemic within the U.N. system limit the ability of the Committee to conduct searching and comprehensive analysis of the situations within states.[28]

This reporting process is the only supervisory mechanism that applies automatically under the ICCPR. Article 41 of the treaty provides that states may take complaints against other states to the UNHRC if both states have previously agreed that the UNHRC has jurisdiction to do so. Perhaps unsurprisingly, this procedure has never been utilized.

Complaints from Individuals

A potentially more effective procedure is one that allows individuals to make complaints to the UNHRC about a state's failure to secure their rights under the ICCPR. The process is significant because it gives direct enforcement rights to affected people. This standing is in marked contrast to the traditional model of international law, which recognizes only states as actors. Of course, this process is only available if the state concerned has previously become a party to the First Optional Protocol to the ICCPR (Optional Protocol), a separate treaty that provides jurisdiction for this process.

In this Optional Protocol process, the UNHRC begins by determining if the complaint is admissible. This essentially involves a determination of whether the complaint is from a victim of an alleged violation of rights in the ICCPR, whether the individual has exhausted all available domestic remedies, and whether the state concerned is a party to the Optional Protocol.

If a complaint is admissible, the merits are then considered, and the Committee subsequently issues its "views." The use of the term *views* is significant: the UNHRC's role in adjudicating such complaints is to ensure consistency with the ICCPR, and the body is not intended to function as an international court. As a consequence of this arrangement, its decisions are not binding and have normative status only. History has shown that in many cases a state party against whom there has been a ruling *will* comply with the Committee's recommendations—whether that compliance entails offering recompense to an affected individual or repealing an inconsistent piece of legislation.[29]

A starting point when examining the effectiveness of the Optional Protocol mechanism is to examine which states are even party to this supplemental instrument. To date, there are some 109 state parties.[30] Among countries whose filtering practices are studied by the OpenNet Initiative, the following are party to the Optional Protocol: Algeria, Azerbaijan, Belarus, Kyrgyzstan, Libya, Moldova, Nepal, the Russian Federation, South Korea, Tajikistan, Turkmenistan, Ukraine, Uzbekistan, and Venezuela.

While the number of state parties may give the impression of a large degree of support for the Optional Protocol, and while in many cases state parties comply with recommendations, Committee views that are issued under this instrument are often outright ignored by errant states. Two states (Jamaica and Trinidad and Tobago) that have frequently found themselves on the receiving end of adverse views from the UNHRC have denounced the Optional Protocol altogether.[31] In the absence of stronger enforcement powers, a decision to flout the views issued by the UNHRC may simply be a political calculation.

Moreover, the Optional Protocol expressly requires that a complaint come from an individual victim. This limits the ability of NGOs or other representative groups to challenge state practices in the abstract. It would not be possible, therefore, for a group such as Amnesty International to challenge a state's filtering practices before the Committee—the challenge would have to come from an affected individual. This requirement poses problems, especially in light of the fact that in several documented cases individual petitioners faced further persecution from their governments for having exercised their right to petition.[32]

Finally, the limitations of the U.N. system already noted have a constraining effect on the ability of the Committee to conduct thorough analyses of claims brought under the Optional Protocol.

Overall, then, the Optional Protocol mechanism provides a good way for individuals to hold *some* states to account for incursions on the right to freedom of expression. For Internet filtering policies, it is theoretically possible for provisions of the ICCPR and the Optional Protocol to have significant effect. However, given the practical difficulties mentioned here, it is doubtful that this treaty represents an adequate means for deterring and punishing states that oppressively filter Internet content.

The Overall Ineffectiveness of International Law

To summarize: States cannot claim that their obligations under international law surrounding Internet filtering are unclear. To the contrary, the obligations are quite clear. Comprehensive filtering of Internet content amounts to a violation of the broadly conceived right to freedom of expression. For filtering to be permissible under the ICCPR, measures must be grounded in specific law and necessary. However, state compliance remains difficult to secure. The UNHRC affords some possibility for redress, but correction relies to a large extent on the goodwill and political situation of the state that has violated its commitments. While many states may refrain from filtering in order to honor freedom of expression (either because they value this right or because they wish to avoid domestic and international pressure), for errant states, there is little incentive to comply with international law in this area. In short, the weak enforcement capabilities of international human-rights institutions send a message that the international system will tolerate flagrant filtering abuses and fail to defend freedom of expression.

Filtering Curbs through Trade Policy

Taking as a given the notion that freedom of expression is desirable and deserving of protection, but questioning the ability of the international system to enforce commitments under the ICCPR in a meaningful way, one might look to other avenues for enforcement. Because agreements under the World Trade Organization (WTO) include the possibility for dispute settlement backed by economic remedies, it has been suggested that one way to enforce freedom of expression would be to cast it as a market access issue and to seek redress by bringing a case before a WTO panel.[33]

In a nutshell, the theory of such a case would be as follows: If a member had committed to giving market access for the production, distribution, marketing, sale, or delivery of content, but nonetheless was filtering in a way that obstructed this trade, another member whose economy had suffered from the action would request the WTO to establish a panel to hear the case.

The case would not necessarily be clear-cut, however. Similar to the way that the ICCPR allows limitations, Article XIV of the WTO's General Agreement on Trade in Services (GATS) permits members to make exceptions to their market-access commitments if taking measures necessary to protect public morals, health, or safety; to maintain public order; or to bolster consumer protection. Article XIVbis extends these exceptions to include measures in the interest of security.

The WTO case *Measures Affecting the Cross-Border Supply of Gambling and Betting Services*[34] brought by Antigua and Barbuda against the United States demonstrates how these provisions would be understood to interact with market access commitments. In this challenge, the United States relied in part on GATS Article XIV in defending restrictions on the supply of gambling and betting services via the Internet.[35] In determining whether the measures were necessary, the Appellate Body indicated:

> The standard of "necessity" provided for in the general exceptions provision is an objective standard. To be sure, a Member's characterization of a measure's objectives and of the effectiveness of its regulatory approach—as evidenced, for example, by texts of statutes, legislative history, and pronouncements of government agencies or officials—will be relevant in determining whether the measure is, objectively, "necessary."
>
> A panel is not bound by these characterizations, however, and may also find guidance in the structure and operation of the measure and in contrary evidence proffered by the complaining party. In any event, a panel must, on the basis of the evidence in the record, independently and objectively assess the "necessity" of the measure before it.[36]

The Appellate Body then explained how it applies this standard:

> The process begins with an assessment of the "relative importance" of the interests or values furthered by the challenged measure. Having ascertained the importance of the

particular interests at stake, a panel should then turn to the other factors that are to be "weighed and balanced."

A panel then considers two main factors as it continues in its determination of a measure's necessity: "One factor is the contribution of the measure to the realization of the ends pursued by it; the other factor is the restrictive impact of the measure on international commerce."[37]

According to this interpretation, it is the WTO panel itself that is to determine whether a member's exceptions are justified. Although a panel pays deference to a member's decision to invoke Article XIV,[38] the panel makes its own assessment of the importance of the objective and evaluates the measure's effectiveness in accomplishing that objective when balanced against the measure's restrictive effect on trade.[39]

Extrapolated, the implication is that future trade panels could rule illegal a member's filtering practices if the measures conflicted with another member's trade interest. So, for example, China's use of filters to prevent its citizens from accessing Web sites displaying the word *democracy* could be struck down if a panel did not find the purpose of the measure compelling, or if it found the approach too heavy-handed given the negative effects on trade.

In light of these WTO provisions, one could argue that the multilateral trading system supports freedom of expression. While a nice effect, it is important to bear in mind that the WTO's competence is in the area of market access. In this particular international context, the value that governments have embraced and empowered panels to adjudicate concerns open trade, and the effects on freedom of expression are mere offshoots.

To the degree that the institution and its members' acting through it delve into these social questions, they do so reluctantly. For one reason, the WTO Dispute Settlement Body has as its purpose to handle disputes relating to market access; a member is not supposed to bring a claim for the sake of protecting human rights, and indeed government agencies responsible for conducting trade policy are typically focused on economic relationships.

Practically speaking, for a filtering case to come to the WTO, a company would need to lobby its *home* government to bring the case on the basis that another government's measures were hurting the home country's economic interests.[40] However, if a company were hurt economically by the host country's measures, that economic harm might be due to damage suffered from bad public relations in another market. It would be challenging for a home-country government to argue that the host government's measures directly caused these side effects, and it would be difficult to prove the amount of injury in monetary terms.[41] Moreover, the home-country government might not wish to spend its international negotiating capital and dispute settlement resources on such a case.

Although a government might not be inclined to bring such a case before the WTO Dispute Settlement Body, it is feasible that in the future such a hearing might not be so dependent on a government's decision to bring it. For several years now experts have argued that private parties deserve to have standing before tribunals for WTO-related matters.[42] Such an arrangement could result in a deluge of dispute settlement cases, as states would no longer select which disputes to bring according to overall political or economic importance for their economy. Indeed, if states ceased to play this intermediary role, WTO agreements would result in a very extensive regulatory framework for the Information Society.[43]

Viewing this scenario as a matter of using international trade law to enforce human rights, one might ask if the concern were for freedom of expression, or for market access. If it were for market access but the effect were that freedom of expression enjoyed protection, would that be sufficient for those people desiring to see enforcement of human rights by the international system? No doubt privacy advocates would be chagrined at the prospect of the same logic requiring a striking down of limitations designed to protect privacy, with market access in that case hurting the cause of civil liberties.

All in all, dispute settlement in the trade context appears a rather blunt and indirect instrument for enforcing freedom of expression among states. Although the WTO offers an interesting example of enforcement capabilities at the international level, the system has been designed for promoting commerce rather than for protecting human rights. A liberalized trading system may promote the exercise of freedom of expression, but relying on trade policy to protect this fundamental human right could send a message that freedoms are subordinate to trade.

More systemically, integration may eventually bring such issues to a head as value systems are forced to reconcile. By making it possible for people in different places to interact with one another and spurring common institutional approaches, the Internet is causing integration to occur at a pace more rapid than ever experienced. As the distinction between cyberspace and the real world fades with technology's incorporation into nearly all facets of life, this integration arguably will be a fact. In this sense, institutions at the center of interactions over the Internet—including the WTO—may experience a sort of triumph as states become dependent on them instead of granting them piecemeal authority.[44]

The Need for a Different Approach

Reinforcing human rights by targeting states is often unsuccessful because the international system lacks effective enforcement mechanisms.[45] Meanwhile, with respect to freedom of expression in particular, empirical studies by the OpenNet Initiative have shown that the practice of government filtering is on the rise globally, and, as discussed earlier, it is questionable whether such filtering comports with the requirements of the ICCPR. As more and more governments adopt such practices, it seems that countries may be legitimizing these substandard (and arguably unlawful) measures and letting them become part of accepted international practice. Should the international system instead move toward penalizing filtering practices that do not fit within the permissible limitations of the ICCPR?

Fundamentally, states' commitments to enforce protections for human rights are weak because there are still relatively few economic drivers and other factors of state interest. States see little reason to raise state-to-state conflict over the issue of freedom of expression. When it comes to the question of doing so in a neutral, international body designed for this purpose, states do not wish to give up sovereignty by setting up a solid international regime, even if enforcement of human rights is faltering.

Assuming that states consciously are refraining from pushing for stronger international human rights protections, one might ask if there is a tension between the rights of people and the interests of the state.[46]

But is this the end of the story? Might private actors be brought into the equation?

Shifting the Emphasis to Private Actors

Again, in the traditional international system, states have not wanted to negotiate treaty terms to hold companies and other private actors accountable for human rights violations. Generally speaking, states see a sovereign interest in mediating between persons under their jurisdiction and persons elsewhere (including juridical persons). In the filtering context, the home government does not want to pressure its own citizens or companies, even if the state generally favors freedom of expression; meanwhile, the host government often is trying to compel companies to repress freedom of expression (or simply withdraw from its market). Under these conditions, there is little to bring such states to the negotiating table in the name of freedom of expression.

Given the increasingly governmental role played by private actors—for example, providing the means for Internet filtering, or carrying out such filtering themselves—many groups are now seeking ways to hold these private actors accountable. The possibility, in some jurisdictions, of bringing entities before domestic courts for involvement in human rights violations in a third country has received significant attention as a potential tool for protecting the right to freedom of expression in the face of restrictive filtering practices.

The United States' Alien Tort Claims Act

Usually domestic courts will concern themselves only with the application of domestic law and will not consider cases that allege violations of international law. Despite this predominant practice, some countries have adopted legislation to allow domestic courts to consider cases arising under international law. Legal systems that do so to a greater or lesser degree incorporate international law into domestic law. Perhaps the best example of such a process is the United States' Alien Tort Claims Act (ATCA). Passed in 1789, ATCA provides U.S. federal courts with jurisdiction to award damages where an alien sues for a tort (i.e., a civil wrong) committed in violation of "the law of nations" or "a treaty of the United States"—even if the wrong occurred outside the United States.

While the ATCA has been on the statute books for more than two centuries, it is only in the past twenty-five years that it has sprung to life.[47] This vitalization occurred largely as a result of a 1980 Federal Appeals Court decision that held that the "law of nations" included "established norms of the international law of human rights," and that such *norms* could therefore form the basis of an ATCA claim.[48] Since then, the ATCA has led to some sizeable awards against perpetrators of human rights abuses. Awards typically have been in the millions of dollars.[49]

Whereas international law treats *states* as actors, a development in the ATCA has been the extension of liability to *private actors* who have been responsible for assisting with violations.[50] In the domestic context, states themselves are immune from liability under the ATCA.[51]

These developments—targeting nonstate actors in the enforcement of international norms—have prompted academic discussion of the possibility of using the ATCA as a tool for punishing corporations who assist states with Internet filtering; attention has focused in particular on U.S. corporations' involvement with Internet filtering in China.[52] While this prospect is interesting theoretically, it should be noted that any such claims would face several significant hurdles.

At the outset, it would first be necessary to convince a federal court that the right to freedom of expression is actionable under the ATCA. Making this argument would be complicated given the conclusions of the U.S. Supreme Court in its first judgment concerning the ATCA in 2004: the Court concluded that while caution was necessary, claims for breaches of rights were possible, provided that they were defined with specificity as were the limited number of international law rules in the late eighteenth century (when the ATCA was passed), and that they were based "on a norm of international character accepted by the civilized world."[53]

This double hurdle need not be insurmountable. Regarding specificity, freedom of expression in an international context is clearly defined and admits only limited exceptions. While there is room for debate about some borderline cases, the existence of a breach should be clear where a state has a legal culture of wholesale filtering. However, the Supreme Court was skeptical as to whether the UDHR and the ICCPR had achieved sufficient acceptance to allow actionable claims under the ATCA. Such a precedent would inform the deliberations of a court considering a claim that a nonstate actor who had engaged in filtering violated the right to freedom of expression. The court would have the responsibility of determining whether the requisite standard of clarity and acceptance was met in the freedom of expression provisions of the UDHR and the ICCPR. It would seem that a convincing argument could be made that freedom of expression is indeed actionable under the ATCA.[54]

Next the defendant would have to establish the connection between the activities of the corporation and the breach of the right. There has been considerable debate over what standard of involvement is appropriate, and it is not entirely clear what test would be applied by a court adjudicating a potential claim.[55] However, the present leading authority is a 2002 federal Court of Appeals decision, which rejected an argument that it was necessary to show that the company in question was an active participant in the abuse for liability to occur under the ATCA, and which instead held that it was only necessary for the company to give "knowing practical assistance or encouragement" that had "a substantial effect" on the perpetration of the abuse.[56] One leading commentator has suggested that this looser standard is only appropriate where the conduct amounts to a violation of international *criminal* law (such as torture), rather than international *human rights* law (such as a violation of the right to freedom of expression).[57]

The standard that is ultimately applied by a court will have a significant impact on the scope of behavior that is potentially captured by the ATCA. It has been suggested that corporations that facilitate state Internet filtering by providing the required software or hardware may be liable,[58] or that liability may occur where an Internet content provider transfers to a repressive regime information that allows the regime to punish individuals for statements they have made on the Internet.[59] In the latter situation, the connection between the company's actions and the repressive act by the state is clear. However, if the company's actions were more passive—say, agreeing to filter results according to certain government criteria—meeting the test of a connection between the company's activities and the breach may be more difficult.

All in all, there is a very real possibility that this process could be used to enforce the right to freedom of expression by giving individuals standing, and holding companies liable, under the ATCA for their involvement in Internet filtering.

In addition to the ATCA in the United States, there are signs that similar enforcement techniques are being developed in other major jurisdictions—notably within the European Union (EU). In this regard, Professor Dinah Shelton has noted a 1999 resolution of the European Parliament "on EU standards for European enterprises operating in developing countries," which refers to a European Community law that provides that "a corporate decision or policy causing harm abroad may permit tort suits in EU courts against the parent company or branch of the company responsible for the decision."[60] This resolution is significant in that it raises the possibility of ATCA-style claims within the EU system.

In terms of what impact such suits may have, the prospects for successful claims may not be as important as the existence of a formal venue for laying bare the extent of corporate cooperation in filtering activities. It has been suggested that the value of these processes lies not so much in the way the suits award vast damages, but rather in the way they generate sufficient adverse publicity so as to force corporations to cease the impugned activities.[61] As with state actors under the UNHRC process, some companies will be more susceptible to this pressure than others.

These examples may point to a new trend of countries creating mechanisms whereby international law can be enforced domestically, thereby enabling private actors to be subject to claims or to bring them. These approaches may be the most immediate way of accounting for private actions and giving persons a mechanism for seeking redress. Moreover, given the reluctance of states to hold each other to agreed-upon standards, the best hope of reinforcing international human rights may be to make private actors accountable. Nonetheless, these domestic approaches still leave gaps in that they are limited jurisdictionally and cannot afford equal treatment to all people around the world.

Indeed, the ATCA approach is far from the ideal of human rights standards applying equally to all persons around the world. After all, why should today's global citizens suffer disparate enforcement of their rights, with redress available only in limited jurisdictions that in any event are applying a variant of law originally designed to address actions of different (i.e., state) actors?

Corporations could ask similar questions: Why should competing companies be held to different standards, with those having ties to jurisdictions that value freedoms confronting costs that others do not, and with a state applying standards to them that it has failed to require the intended subjects (i.e., its treaty partners) to follow?

International Antibribery Conventions

Of course, the idea of holding corporations to account in one jurisdiction for actions done elsewhere is not new, and valuable lessons for the filtering context can be learned in particular from past attempts to promote ethical behavior among corporations acting internationally. In particular, a hybrid process involving both domestic and international enforcement has developed in recent years in another area pertaining to ethical behavior of private actors, namely, in the area of bribery. Antibribery conventions represent the one area where *binding* rules have been put in place by states acting jointly to regulate responsibilities of transnational corporations and related business enterprises with regard to human rights.[62]

The fight against bribery stands out for its lessons on the futility of single-country attempts to hold companies accountable at the domestic level for their international activities, on the one hand, and the success of broader-based efforts to do so in multiple jurisdictions acting in concert, on the other hand. In 1977 the U.S. Congress passed the Foreign Corrupt Practices Act (FCPA) to make it a crime for U.S. corporations to offer bribes for international contracts. While the FCPA may have given a company a credible reason to refuse to comply with a foreign official's demand for a bribe, that company ended up losing contracts to foreign competitors who not only were permitted to pay this extra expense but also were allowed to take a tax deduction for it. Simply stated, the FCPA put U.S. companies at a tremendous disadvantage vis-à-vis others in their global activities involving foreign direct investment.

At the time, most foreign direct investment was flowing from countries that were members of the Organisation for Economic Co-operation and Development (OECD). As corporations began to be plagued by international corruption scandals and increasingly large bribery demands in the 1980s and early 1990s, there was a political willingness in the OECD to join the United States in standing against corruption. The OECD and five nonmember countries[63] adopted the Convention on Combating Bribery of Foreign Public Officials in International Business Transactions (Anti-Bribery Convention) in 1997.[64] By doing so together, these countries agreed to hold their companies to a common standard and so helped to level the playing field for more ethical conduct. They also adopted the Revised Recommendations of the Council on

Combating Bribery in International Business Transactions to flesh out details in the following areas: international cooperation; the non-tax-deductibility of bribes; accounting, auditing, and public procurement; and measures to deter, prevent, and combat bribery.[65]

Once this critical mass was met in the foreign direct investment community, introducing the idea into an even wider, multilateral setting became quite feasible, and proponents were able to achieve the adoption of the United Nations Convention against Corruption (U.N. Convention) in 2005. By July 2006, the U.N. Convention had 140 signatories and 60 ratifications or accessions.[66]

Beyond addressing the problem of questionable corporate conduct in foreign jurisdictions, the antibribery conventions also suggest a shifting identity of the state in the international system. The arrangements under both the OECD and the U.N. entail similar components including the following:

- Harmonizing domestic law in states that are party to the convention.
- Tailoring domestic law to criminalize undesirable activities on the part of private actors operating abroad.
- Involving civil society to help bring violations to the attention of state parties.
- Establishing transparency and accountability mechanisms for questionable activities of private actors operating abroad.
- Giving international processes central oversight over convention implementation (i.e., prevention, investigation, and prosecution of crimes), with a monitoring of state parties' enforcement of the convention in their respective jurisdictions to ensure rigor.
- Setting out a process whereby state parties may sort out disputes among themselves and bring them before an international body should they not be able to settle the matter.
- Allowing additional mechanisms to be created for further international cooperation under the convention.

Through this international cooperation, states are more able to govern entities under their jurisdiction by holding them to ethical standards while not disadvantaging them vis-à-vis competitors in markets around the world; however, states do so at a price—that is, they are pooling power in a joint body to avoid a race to the bottom. Arguably they are upholding their societies' ethical standards for the sake of their own citizens, but at the same time they may be diluting the relative political power of citizens within their polity as degrees of sovereignty are conceded. As such, perhaps states are giving credit to the concept of global citizenship in the Information Society.

The experience with antibribery conventions suggests international cooperation can help overcome the difficulty that a state faces in holding companies to ethical standards when other markets are governed by different rules. Companies had a tough time under one country's law requiring higher ethical standards until their counterparts elsewhere in the world—that is, the main companies they had to compete against—became subject to similar standards. Once a critical mass of states agreed to a common approach, companies were able to refuse to give in to corruption pressure elsewhere, and they saw their public images and profit margins improve.

This model could provide a viable avenue forward for the area of filtering. In particular, it offers hope that states can cooperate in developing international standards for private actors in the area of freedom of expression, especially as private actors feel pressured to submit to host-country government demands to carry out filtering programs.

Might states cooperate in this way to hold private actors accountable for operations affecting freedom of expression?

The Promise of International Standards

This chapter suggests that international legal instruments designed to protect human rights by holding states accountable have been norm-setting but toothless. The prospects for change in these instruments are not strong because it is difficult for states acting collectively through the international system to establish effective remedies for violations by states. Given this lack of enforceability, the cause of international human rights suffers from a chronic legal deficiency.

Because international law generally does not directly bind private actors, companies today can violate international human rights standards with relative assurance that they will not face charges in an international tribunal. Nonetheless, this apparent impunity may work against those that wish to comply with international human rights standards when governments try to compel companies to restrict freedom of expression through techniques like filtering.

Companies complain they are stuck between Scylla and Charybdis in cases where a host-country government requires a breach of international law by imposing broad filtering mandates that contradict international standards for freedom of expression. Naturally, a company must comply with the laws of the different jurisdictions where it operates, and it is not for the company to decide what the law should be or to straighten out the failures of international law. Rather, the decision for the company to make is whether or not to do business in a given market. However, given the competitive economic pressures brought by globalization, a company may in fact need to do business in certain markets if it is to survive.

Guiding a company's decisions on whether to do business in a market are factors such as the company's charter or management and the potential for profits, though these factors are not rigid. If the company's charter or management calls for certain ethical conduct, and if the jurisdiction where it would like to operate has lower standards, the company might nevertheless choose to do business there in hopes of making a positive difference. If the company's charter or management does not itself call for certain ethical conduct, the company might nonetheless choose to follow higher standards in response to loud calls issued by groups trying to affect company behavior, even if those calls hail from another market altogether. (For example, outcries by loud individuals in the west in 2006 affected the course of western

Internet-related companies operating in China.) In this sense, a company's commitment to higher standards might be displayed for public relations purposes with a view to preserving the company's image (even to stave off negative public relations in other markets), or it could be shown as a manifestation of that company's sincere desire to protect human rights.[67]

While international law as agreed among sovereigns may protect human rights by setting norms only, there may be an additional route to bolstering freedom of expression; that is, states may be willing to draw up a new treaty to apply standards to private actors, and private actors meanwhile could proactively pledge themselves through commitments that they take on voluntarily.

Drawing on the ATCA and antibribery examples, an effective enforcement mechanism could prompt companies to follow international legal standards for the sake of limiting their own liability and exposure to adverse publicity; companies could cite the threat of liability as an excuse when they wished to refuse to comply with mandates to repress freedom of expression. If such an approach were applied on a global level (as in the case of the U.N. Convention against Corruption), it could help avoid the clash of conflicting legal regimes and instead provide companies with a global standard they could say they were obliged to follow.

In this regard, states could begin negotiating a binding treaty complete with domestic harmonization requirements and international cooperation in prevention, investigation, and enforcement. While they do so (a process that will take considerable time), corporations could develop their own codes of ethical conduct for freedom of expression. Such voluntary commitments would allow companies to align themselves in support of human rights and equip themselves with a valid response when asked by repressive regimes to suppress communications; the force of a treaty reinforcing these obligations through legal harmonization and international cooperation would send an added signal to those regimes.

In this sense, then, international law could provide a set of internationally recognized minimum standards that would help reconcile tensions. Since international human rights principles already have been agreed upon and have enjoyed a transnational stamp of legitimacy over the years, these same principles could provide a minimum standard for corporate responsibility. Because additional commitments to follow these standards would be voluntary, they would allow companies to choose to bind themselves in taking an even stronger stand against repressive practices.

Given the tendency of the Internet to push global rules, and given the expectation that the distinction between the real and virtual worlds will fade, a good starting point perhaps would be to pare down the ambition to Internet-related practices. Efforts are already underway in this regard. For example, one of the outcomes of the World Summit on the Information Society was the tasking of the United Nations Educational, Scientific, and Cultural Organization (UNESCO) with facilitating work on "ethical dimensions of the Information Society." This mandate was spelled out in the *Geneva Declaration of Principles* and *Plan of Action* and elaborated in the *Tunis Agenda*. Under the framework of this mandate, UNESCO has begun developing a code of ethics. Interestingly, the first draft of what they are calling a "Code of Ethics for the Information Society" envisions a reporting mechanism, similar to the OECD's Anti-Bribery Convention and the U.N. Convention against Corruption. In addition, the draft instrument affords a mechanism whereby additional, voluntary "Specific Ethical Commitments" may be offered by private actors, who may join states in signing onto the general document.

While this effort is going on in that forum, another process stemming out of the World Summit on the Information Society—that is, the Internet Governance Forum—affords the opportunity for all stakeholders to consider freedom of expression in the Information Society and possibly to articulate shared values. A "Dynamic Coalition on Freedom of Expression" has spontaneously formed following the first meeting of the Internet Governance Forum (Athens, autumn 2006).

By working through state-established intergovernmental organizations, the approaches would avoid chipping away at the institutional groundwork already laid for the international protection of human rights, and instead would enable future human rights endeavors to build upon this foundation. Meanwhile, by paving avenues for nonstate actors to have a meaningful voice in the development and implementation of these protections, the approaches would help operationalize the *Geneva Declaration of Principles*, which called for technical and public policy issues of Internet management to "involve all stakeholders and relevant intergovernmental and international organizations" and to be handled in a way that is "multilateral, transparent and democratic."[68]

Such simultaneous approaches offer the hope of allowing citizens of the world to experience equal human rights in the global Information Society.

Notes

1. The Peace of Westphalia entailed a set of treaties ending the Thirty Years War and the Eighty Years War in Europe.
2. As elaborated in the Montevideo Convention on the Rights and Duties of States (1933), qualities of a sovereign state include a permanent population; a select territory; government (e.g., legislative, judicial, and executive); and the ability to conduct relations with other states.
3. The International Criminal Court, established in 2002, may be viewed as an example where states have allowed prosecution of individuals for genocide, crimes against humanity, war crimes, and the crime of aggression. Article 17 of its founding treaty, the Rome Statute of the International Criminal Court, indicates how the court is designed to complement national judicial systems.
4. This process was established in accordance with the Uniform Domain Name Dispute Resolution Policy (UDRP) of the Internet Corporation of Assigned Names and Numbers (ICANN). See http://www.wipo.int/amc/en/index.html.
5. See "WIPO's Domain Name Dispute Resolution Service," Net Dialogue, http://www.netdialogue.org/initiatives/wipodndrs/. Still, as Net Dialogue notes, the WIPO process is linked to domestic processes since disputing parties may submit a dispute to a traditional court for resolution.
6. Pursuant to United Nations General Assembly Resolution 56/183, the World Summit on the Information Society took place in two phases, the first being in Geneva, Switzerland (2003), and the second in Tunis, Tunisia (2005).
7. The parameters for the Working Group were set out in the WSIS Declaration of Principles (WSIS-03/GENEVA/DOC/0004) and the WSIS Plan of Action (WSIS-03/GENEVA/DOC/0005).

8. This dynamic is common among federations—for example, with the "Commerce Clause" in the United States Constitution being interpreted during the twentieth century to give the U.S. federal government much greater authority over states than originally anticipated by the early American states forming the union, or, more recently, with European Union member states finding themselves much more integrated than the original members of the European Economic Community would have agreed to just a few decades ago.
9. In this regard, Bentham is often quoted as having said that natural rights are "nonsense upon stilts."
10. See Ronald Deibert, *Parchment, Printing, and Hypermedia* (New York: Columbia University Press: 1997) and Ronald Deibert, "Network Power," in Richard Stubbs and Geoffrey Underhill, (eds.) *Political Economy and the Changing Global Order*, 2nd edition (New York: Oxford University Press, 1999).
11. *Universal Declaration of Human Rights*, adopted and proclaimed by U.N. General Assembly resolution 217A(III) of 10 December 1948.
12. Ibid.
13. *International Covenant on Civil and Political Rights*, adopted and opened for signature, ratification, and accession by U.N. General Assembly resolution 2200A(XXI) of 16 December 1966 (entry into force 23 March 1976, in accordance with Article 49).
14. As reflected on the Web site of the High Commissioner for Human Rights, updated December 6, 2006, http://www.ohchr.org/english/countries/ratification/4.htm.
15. Among other countries studied, Cuba, Malaysia, Myanmar, Oman, Pakistan, Saudi Arabia, Singapore, and United Arab Emirates have not signed the ICCPR. The government of China contends: "The signature that the Taiwain authorities affixed, by usurping the name of 'China', to the [Convention] on 5 October 1967, is illegal and null and void" (ICCPR Declarations and Reservations, available at http://www.ohchr.org/english/countries/ratification/4_1.htm).
16. *McIntyre v. Canada*, U.N. Doc. CCPR/C/47/D/359/1989 (1993).
17. In other contexts, this principle is sometimes referred to as that of "technological neutrality."
18. Concluding observations: Guyana, 2000; Concluding observations: Lebanon, 1997. The UNHRC has also taken a strong stand against state media monopolies, stating that "because of the development of the modern mass media, effective measures are necessary to prevent such control of the media as would interfere with the right of everyone to freedom of expression" (General Comment 10, 1983).
19. See United Nations, Economic and Social Council, U.N. Sub-Commission on Prevention of Discrimination and Protection of Minorities, *Siracusa Principles on the Limitation and Derogation of Provisions in the International Covenant on Civil and Political Rights*, Annex, U.N. Doc E/CN.4/1984/4 (1984).
20. See Joseph, Schultz, Castan, *The International Covenant on Civil and Political Rights*, 2nd ed. (New York: Oxford University Press, 2004); see also the Siracusa Principles, supra, which note that these grounds "cannot be used as a pretext for imposing vague or arbitrary limitations and may only be invoked when there exist adequate safeguards and effective remedies against abuse."
21. *Hertzberg v. Finland*, U.N. Doc CCPR/C/15/D/61/1979 (1982). This concept, which is prominent in the jurisprudence of the European Court of Human Rights, has not subsequently been referred to by the Human Rights Committee (see Joseph, Schultz, Castan, *The International Covenant on Civil and Political Rights*, 2nd ed.).
22. See *Länsman et al. v. Finland*, U.N. Doc CCPR/C/58/D/671/1995 (1996), which refers to the right of a member of a minority to enjoy his or her culture; see generally Eyal Benvenisti "Margin of Appreciation, Consensus, and Universal Standards" 31 NYU J. Int'l L. & Pol. 843 (1999).
23. United Nations Human Rights Committee, General Comment No. 31: Nature of the General Legal Obligation Imposed on States Parties to the Covenant, U.N. Doc CCPR/C/21/Rev.1/Add.13 (2004).
24. *Mukong v. Cameroon*, U.N. Doc CCPR/C/51/D/458/1991 (1994).
25. United Nations Human Rights Committee, General Comment No. 10: Freedom of Expression (29 June 1983).
26. See the Human Rights Committee's General Comment 24, where the limited scope of reservations are discussed.
27. Lawrence Lessig, *Code and Other Laws of Cyberspace* (New York: Basic Books, 1999).
28. See P. Alston and J. Crawford, eds., *The Future of UN Human Rights Treaty Monitoring* (Cambridge: Cambridge University Press, 2000).

29. See P. R. Ghandhi, *The Human Rights Committee and the Right of Individual. Communication: Law and Practice* (Dartmouth: Ashgate, 1998).
30. As reflected on the Web site of the High Commissioner for Human Rights (updated December 6, 2006), http://www.ohchr.org/english/countries/ratification/5.htm.
31. See "Jamaica Withdraws the Right of Individual Petition under the International Covenant on Civil and Political Rights" 92 Am. J. Int'l L. 563 (1998); Trinidad and Tobago subsequently re-acceded with a reservation limiting the UNHRC's power to determine complaints regarding death-row inmates, the subject matter of most communications it faced—see http://www.ohchr.org/english/countries/ratification/5.htm.
32. See P. R. Ghandhi, *The Human Rights Committee and the Right of Individual. Communication: Law and Practice* (Dartmouth: Ashgate, 1998).
33. For a more thorough description of these issues, see Tim Wu, "The World Trade Law of Internet Filtering" (May 3, 2006), available at SSRN: http://ssrn.com/abstract=882459. For an additional perspective on how countries have inadvertently opened the floodgates of trade, see Mary C. Rundle, "Beyond Internet Governance: The Emerging International Framework for Governing the Networked World", Berkman Center Research Publication No. 2005-16 (December 13, 2005), available at SSRN: http://ssrn.com/abstract=870059.
34. WTO, *United States—Measures Affecting the Cross-Border Supply of Gambling and Betting Services*, Appellate Body Report (WT/DS285/AB/R) of 7 April 2005.
35. With respect to this particular issue, the Appellate Body held that the measures imposed by the United States fell within the scope of GATS Article XIV.
36. Ibid., para. 304.
37. Ibid., para. 306.
38. This approach loosely embraces the principle of subsidiarity (i.e., governance at the most local level practicable) as the panel checks that the member applying an exception has itself used a process to determine that the interests or values that the measure is protecting are important.
39. This test bears some resemblance to that of necessity under the ICCPR.
40. GATS categorizes the different means by which services are supplied according to different *modes*. Because GATS binds WTO members only where they have specifically agreed to grant market access, and because they may specify these commitments according to the different modes of supply, a panel would need first to determine if the member against whom a claim was brought had even agreed to guarantee market access for that type of service in that particular mode. Of course, for the panel to be applying analysis under GATS, the claimant would need to have cast the activity as a "service."
41. If a case were brought, a panel might defer to ICCPR provisions regarding permissible limitations, but when it would then turn to analyze whether the filtering were done in the least trade-restrictive manner, most likely the answer would be no. In that case, the remedy would be to tell the host government to change its measures or suffer economic sanctions roughly equivalent to the damage the home country had faced.
42. See, e.g., Frieder Roessler, "The Constitutional Function of the Multilateral Trade Order," in Frieder Roessler, ed., The Legal Structure, Functions and Limits of the World Trade Order: A Collection of Essays (London: Cameron May, 2000), 109.
43. Wu asserts that the WTO already constitutes a sweeping regulatory regime, despite the more limited intentions of the signatories to its agreements. Rundle sets out a similar argument that countries have established a loose framework for international Net governance through numerous initiatives in intergovernmental organizations—with this Net regulation then driving integration further and ultimately leading to international federalism. (See note 33.)
44. Ibid.
45. This "failure" of international law stands in contrast to relatively strong regional systems, such as the European Convention on Human Rights, which affords enforceability and, in practice, is binding against the state.
46. Of course, it should be remembered that the state may also be mediating between competing liberties that people have. By way of example, a state may be reconciling freedom of expression and freedom from the discrimination that is brought on by hate speech. This mediating role is particularly important in societies that think in terms of group rights as well as individual rights.

47. See Beth Stevens and Michael Ratner, *International Human Rights Litigation in U.S. Courts* (New York: Transnational Publishers, 1996).
48. *Filartiga v. Pena-Irala*, 630 F.2d 876 (2d Cir. 1980).
49. Professor Dinah Shelton of the George Washington University Law School has surveyed the cases taken to date under the ATCA and concluded that, where successful, awards in the $1–10-million range were usually made. Dinah Shelton, *Remedies in International Human Rights Law*, 2nd ed. (New York: Oxford University Press, 2005).
50. See *Kadic v. Karadzic*, 70 F. 3rd 232 (2nd Cir. 1995); *John Doe v. Unocal Corporation*, 395 F.3d 932 (9th Cir. 2002).
51. This is due to the doctrine of sovereign immunity—see *Argentine Republic v. Amerada Hess Shipping Co*, 488 U.S. 428 (1989).
52. See Neil J. Conley, "The Chinese Communist Party's New Comrade," 111 Penn St. L. Rev. 171 (2006); Jill R. Newbold, "Aiding the Enemy: Imposing Liability on U.S. Corporations for Selling China Internet Tools to Restrict Human Rights," U. Ill. J.L. Tech. & Pol'y 503 (2003).
53. *Sosa v. Alvarez-Machain*, 542 U.S. 692, 724 (2004).
54. Conley, "The Chinese Communist Party's New Comrade," 200–201.
55. See Olivia De Schutter, ed., *Transnational Corporations and Human Rights* (Oxford: Hart, 2006), 59ff.
56. *John Doe v. Unocal Corporation*, 395 F.3d 932 (9th Cir. 2002).
57. Andrew Clapham, *Human Rights Obligations of Non-State Actors* (New York: Oxford University Press, 2006), 266.
58. Jill R. Newbold, "Aiding the Enemy: Imposing Liability on U.S. Corporations for Selling China Internet Tools to Restrict Human Rights," U. Ill. J.L. Tech. & Pol'y 503 (2003).
59. Conley, "The Chinese Communist Party's New Comrade."
60. Shelton, *Remedies in International Human Rights Law*, 172–173.
61. See Olivia De Schutter, ed., *Transnational Corporations and Human Rights* (Oxford: Hart, 2006).
62. See UN Economic and Social Council, Report of the Sub-Commission on the Promotion and Protection of Human Rights: Report of the United Nations High Commissioner on Human Rights on the Responsibilities of Transnational Corporations and Related Business Enterprises with regard to Human Rights (E/CN.4/2005/91), February 15, 2005, available at http://www.globalpolicy.org/socecon/tncs/2005/0215untncreport.pdf.
63. Argentina, Brazil, Bulgaria, Chile, and the Slovak Republic joined the thirty OECD members in this effort. (For a list of OECD members, see http://www.oecd.org/document/58/0,2340,en_2649_201185_1889402_1_1_1_1,00.html.)
64. See "OECD Convention on Combating Bribery of Foreign Public Officials in International Business Transactions," Organisation for Economic Co-operation and Development," http://www.oecd.org/document/21/0,2340,en_2649_201185_2017813_1_1_1_1,00.html.
65. See "1997 Reused Recommendation of the Council on Combating Bribery in International Business Transactions," Organisation for Economic Co-operation and Development," http://www.oecd.org/document/32/0,2340,en_2649_201185_2048160_1_1_1_1,00.html.
66. See "United Nations Convention against Corruption," United Nations Office on Drugs and Crime, http://www.unodc.org/unodc/crime_convention_corruption.html.
67. Do companies' motivations matter when the community opposed to illegal filtering practices is considering what approach to take? It could be that the distinction in motivations is not so important, and that what matters is simply the result—that is, that decision-makers for the company ensure that company policies support freedom of expression. Still, true convictions would seem to matter for the sake of maintaining social values in favor of freedom of expression.
68. Geneva Declaration of Principles, paras. 48 and 49.

5

Reluctant Gatekeepers: Corporate Ethics on a Filtered Internet

Jonathan Zittrain and John Palfrey

Introduction

Picture a corporate boardroom in the headquarters of a large information technology company in the north of Europe. The chief business development executive has just made a pitch to the board: the company should offer its Internet-based service, delivered over a variety of devices, in east Asia. Her plan is that the firm should start with the white-hot Chinese market and then turn to Vietnam, Thailand, and Singapore. Each of these new markets promises enormous growth.

In each case, the plan calls for a strategy of first entering into joint ventures with local Internet companies, then seeking local investors to set up a stand-alone subsidiary once each trial is successful. Competitors, she argues, will not be far behind. The company might well find itself in the posture of the follower if it does not move quickly. Several board members, each of them outside investors, sound a note of approval.

The general counsel, though, has a few words of warning before the board takes a vote on the proposal. He is concerned about the regulatory requirements that the corporation will face in these new markets. The company needs to be prepared to censor the content it is offering, to disallow users to publish certain information through the service, and to turn over information about the identities of its subscribers upon demand. These are typical requirements when operating almost anywhere—even liberal democracies identify information to be removed, such as that which infringes copyright, or meets some test of obscenity. They require help identifying users at times, and some impose blanket data retention requirements for these purposes.

But in more authoritarian places like China the practices have extra bite. The information the government seeks to censor can relate to civic dialogue and freedom, and the people they seek to identify might be political dissidents or religious practitioners. Often, the requirements to redact or block will be stated or implied only generally without specific requests for individual cases, which means that the company must be prepared to operate in something of a gray zone, trying to divine what the regulators have in mind—and act to censor without explicit orders to do so.

To support his case, the general counsel notes that some of America's most prominent Internet companies have found trouble trying to follow local law against a backdrop of international criticism. Yahoo! has been faulted for turning over information about a journalist that allegedly led to his arrest and imprisonment—for no crime that a court in Yahoo!'s home jurisdiction of California could recognize. Cisco has been attacked for selling the routers and switches that make censorship and surveillance possible. So, too, has Microsoft, for offering a blog service that generates an error rejecting "profanity" when a user includes the word *democracy* in the title of a blog. Google has come under fire for offering a search product in China that omits certain search results compared to what its other offerings provide. Side-by-side comparisons of a Google image search for *Tiananmen Square* in http://google.com and http://google.cn starkly show the results of censorship; for anyone who can see both sets of images, the latter lacking any shots of a person staring down a tank in 1989, is forced to consider what it would be like to live under an authoritarian regime. There is no reason why we should be any different, he concludes.

Successful technology companies must now focus on more than simply implementing great ideas that people will pay for. In the earliest days of the Internet, the relevant markets were modest in size and close to home. A local Internet Service Provider once could profit by offering a dialup Internet access service over plain old telephone lines to people who lived near the corporate headquarters. Few of the big players involved were large, publicly traded entities. Revenue projections commonly looked like hockey sticks pointing toward bright blue skies. And, most important for the purposes of this chapter, states left alone the Internet and the companies that built it and its many services. The prevailing orthodoxy was that a regulator that required too much of companies doing business on the Internet would unduly restrict the early growth of online activity, and might find associated high-tech jobs going elsewhere. Few states placed any kind of liability or responsibility on intermediaries for troubles arising from the activities and transactions they facilitated.

More than ten years into the Internet revolution, these are no longer the facts on the ground. The Internet is big business in which entrenched players—and not just what were once called dot-coms—with colossal market capitalizations compete with one another over multi-billion-dollar revenue streams. Their markets span much of the globe. Most important, some states have increasingly forced companies that provide Internet services to do more to regulate activity in the Internet space. This approach applies a new kind of pressure on nearly every corporation whose business involves information and communications technologies (ICTs), especially when the pressure is piecemeal or downright contradictory from one jurisdiction to another, and when the desired regulation contravenes the values of the company's owners or customers. While liberal democracies have so far remained remarkably hands-off as the Internet has matured, the desire of more closed regimes to tap the Internet's economic potential while retaining control of the information space confines the options for these firms.

As this book makes plain, over the past five years there has been a steady rise of Internet filtering practices from a handful of states in 2002 to over three dozen states in 2007. The most extensive of these filtering regimes are found in states in the Middle East and North Africa, Asia and the Pacific, and the Commonwealth of Independent States. The job of online censorship and surveillance is difficult for the state to manage itself, if not altogether impossible.

To carry out these practices, states turn to private firms to provide the tools and services necessary to effect the censorship and surveillance. Most of the high-profile incidents of this type have involved well-known technology companies based in the United States and their efforts to enter the Chinese markets. But this issue is about more than a few companies and about more than one emerging market. Almost any business in the information technologies or telecommunications space might find itself in this position. These private firms include hardware manufacturers, software firms, online service providers, and local access providers, among others.

The shareholders in large technology companies reasonably expect continued growth of market volume or share, and improved profit margins. The pull of markets farther from home is powerful. The shares in these firms are often publicly traded by investors in the state in which they are chartered. In many instances, the social norms and conceptions of civil liberties in the new target market are dissonant with the norms and liberties enjoyed where the senior executives and most powerful shareholders of the corporation live. An everyday act of law enforcement in an authoritarian market looks like a human rights violation to a more liberal one. That act may in fact contravene international human rights standards—and some shareholders, concerned about matters beyond growth and profits, are starting to ask hard questions of corporations about their involvement in such practices.

The ethical problem arises when the corporation is asked to do something at odds with the ethical framework of the corporation's home state. Should a search engine agree to censor its search results as a condition of doing business in a new place? Should an e-mail service provider turn over the names of its subscribers to the government of a foreign state without knowing what the person is said to have done wrong? Should a blog service provider code its application so as to disallow someone from typing a banned term into a subject line?

These questions—prompted by the hard cases that lie between simple acts of law enforcement and clear violations of international norms—are not easily answered through legislation or international treaty. Laws fashioned in this fast-moving environment to lay out what orders corporations must resist in authoritarian states—really, laws about laws—may function as a hopelessly trailing indicator. The firms involved in this quandary should not be seen as a single bloc. They represent a range of levels and types of involvement in censorship and surveillance regimes.

In the context of the cyberlaw literature, these questions ask us to assess "second-order" regulation of the cyberenvironment. From a public policy angle, the question is not whether to impose any control over private actors online, but rather what constraints might be placed on those private actors with respect to the first-order regulation. When states disagree with each other, private actors chartered in one state and operating in the other can become proxies in the fight.

The most efficient and thorough way to address this conundrum is for the corporations themselves to take the lead. The corporations, as an industry, are best placed to work together to resolve this tension by adopting a code of conduct to govern their activities in these increasingly common situations. This approach could, at a minimum, clarify to end-users what they need to know about what companies will and will not do in response to demands from the state. At best, the industry might be able to resist the most excessive first-order demands of the state with a corresponding benefit for civil liberties online. The corporations should call upon the knowledge and goodwill of NGOs, academics, public officials, and others to help frame this code of conduct. The drafters of the code should consider neither the firms nor the markets to be singular in terms of their respective ethical obligations, but rather consider them to be disaggregated. The goal of drafting and putting in place a code should be to establish a meaningful, flexible, and lasting solution to the problem of corporate ethics on a filtered Internet, a solution that may be as much process as substance, creating mechanisms for the resolution of questions as they arise that earn the acquiescence of their first-order regulators, and the respect of their customers and their second-order regulators.

First-Order Regulation of the Online Environment

The initial debate over the regulation of the online environment, as we describe in chapter 2, was whether or not states could regulate online activity. Cyberlibertarians—often derided as cyberutopians—took the provocative view that cyberspace was so different that states could not reach it. That debate is now settled. The answer is that they *can*, more or less in the ways that they have regulated offline activity. Whether or not states *should* regulate the online environment in comparable manner to how they have regulated in the past is a more complicated matter.

We refer here to "first-order regulation" of the Internet as this first generation of questions. The large issues covered in Lawrence Lessig's *Code and Other Laws of Cyberspace*, the definitive text in this area, comprise a reliable list.[1] Should the state regulate speech online—whether hate speech, political speech, or otherwise? Should the state step in to protect user privacy? Or listen in on the conversations of citizens in the service of law enforcement? What is the proper role of the state in granting and enforcing intellectual property rights in ideas and expression, or brand and trade secrets, in the online environment? In each instance, virtually every state with a significant population online has exerted some control of this ordinary sort.

The story of this chapter, though, is about whether regulation should come into play in response to this first-order regulation of private actors doing business in other jurisdictions. The relevant first-order regulation is the extent to which states have required corporations to censor search results, to configure software in such a manner as to block certain expression, to collect and turn over data, and so forth.

Second-Order Regulation of the Online Environment: More State Control, Greater Pressure on Private Parties

As more states place pressure on intermediaries to help control the online space, other states may try to prevent such control, often by imposing their own regulations—a form of second-order regulation of the online environment. This notion of second-order regulation presents a new issue: how to evaluate the regulation that some states place on some firms, based in other jurisdictions, when it comes to activity in the online environment.

This issue has arisen most prominently in the context of the United States Congress inquiring into the activities of several of its most prominent technology firms in the Chinese markets, though as we argue in this chapter, the issue is much broader than such a precise frame would suggest. The first-order regulation is China's requirement that a search engine censor the results that are presented to users in response to a search query, as part of broader practices prohibiting online service providers from disseminating information that may "jeopardize state security and disrupt social stability."[2] At issue is not simply whether the first-order regulation is warranted, but rather whether the United States should regulate the activity of the firm chartered in its jurisdiction when it competes in the Chinese markets. In some cases, no first-order regulation has yet been applied, but regulation of the second-order type—of the export control variety—has been proposed.

One reason for focusing on this ethical problem at an early stage of its development is that in a global technology marketplace, such second-order regulatory issues are likely to continue to arise. A mode of responding to these issues in the context of online censorship and surveillance may pay dividends over time as structurally similar quandaries come to the fore.

New Markets, New Modes of Control, New Challenges

As Faris and Villeneuve's review of the data in chapter 1 indicates, Internet filtering occurs primarily in three regions of the world: the Middle East and North Africa, Asia and the Pacific, and the Commonwealth of Independent States. China continues to be the case that garners the most public attention, given the size of its market and the extent to which the state has set in motion the world's most sophisticated filtering regime. But China is far from alone, as more than two dozen states carry out some form of Internet censorship and surveillance online. Further, large, regionally powerful states—China, the Russian Federation, and India, for instance—that provide downstream Internet service to smaller states are poised to pass along their filtering as well.

To add to the complexity of the matter, the mode and extent of censorship and surveillance varies substantially from one state to another as the data in this book make plain. There are several ways for states to filter and monitor. The most direct means is through the use of technology. In its simplest form, the state requires the reprogramming of the routers that lie between the individual end-user and the broader network. The job of the new code is to block certain packets from reaching their destination or simply to learn and record the contents of those packets and who is sending or receiving them. Sometimes it is apparent to the end-users that their requests for certain Web pages have been blocked by the state thanks to special messages substituted for the destinations the users have sought; more often, it is not so apparent. The manner and extent to which censorship takes place online is easier to prove, while surveillance is more elusive.

Online censorship and, potentially, surveillance, is carried out through nontechnical means as well. These controls are sometimes imposed by law: end-users might be prohibited from accessing or publishing certain information that is deemed to undermine public order or other state interests. Such laws are typically very broad, hard to understand, and even harder to follow with any degree of precision. These controls are also imposed most effectively as part of a package of *soft controls*, whereby cultural norms drive censorship or surveillance into the home or local community, often resulting in extensive self-censorship.

Integrated Modes of Online Control: Combining the Technical and the Legal

The most salient form of filtering is direct technical control implemented by legal controls trained on private actors who lie between an end-user and the network at large.[3] The state, unable to carry out filtering effectively on its own, requires private actors to carry out the censorship and surveillance for them. This requirement comes as a formal or informal condition of holding a license to provide Internet-related services in that state.

So, for a large search engine like Google, the mandate from the state may be to ensure that search results provided to citizens of that state do not include links to online content banned in that jurisdiction. In some cases, like insistence by the German and French governments that search results to Nazi propaganda be excised, Google's censorship of search results is controversial only to die-hard civil libertarians, especially when the ways to circumvent such filtering are open secrets. (Germans wishing to search for Nazi propaganda can simply use google.com instead of google.de.) In other cases, like China, where a much broader range of politically and culturally sensitive results are excluded, the public response is one of broader concern.

Likewise, the provider of a blog publishing tool may be prompted to include controls that disallow an individual publisher from including certain words in the title of a blog post. Microsoft found itself in this quandary in 2005. After a successful launch of its MSN Spaces product in the United States market, Microsoft rolled out a Chinese version of the service. MSN

Spaces operated differently in China than it did in the United States, however. If a blogger using the U.S.-branded version of the service decided to type *democracy* into the title of a blog post, there is no problem. In China, that same blogger is presented with an error message: "You must enter a title for your space. The title must not contain prohibited language, such as profanity. Please type a different title." Automated screening of content is also coupled with specific interventions: in 2006, MSN abruptly pulled the blog of a Chinese-based journalist using the pseudonym Michael Anti, apparently at the behest of Chinese authorities.[4] Corporations hosting blogs told ONI researchers in interviews of the persistent fear of being asked to perform one-off censorship tasks of this sort. It is plain that these firms do not relish this job, but fear retribution if they do not comply with the local mandates.

An Internet service provider might be required to keep records of the online activity of all or some of its subscribers, or to monitor who seeks to access certain kinds of content. The provider of a Web-based e-mail service might be required to turn over the e-mail messages of a user identified by the government. Yahoo! has been faced with this dilemma several times. In the United States, Yahoo!'s lawyers routinely respond to law enforcement requests for information about subscribers, pairing an IP address or an e-mail name with other subscriber information. But in China, the stakes are different for the same activity: in at least two instances, Yahoo!'s local affiliate, now Alibaba, has turned over information about users of its e-mail service that allegedly has landed journalists in jail. The crime involved, related apparently to political dissent, would be no crime at all if committed in the state where Yahoo! is chartered.

Though less of a concern to multinational firms, cybercafés can be required to maintain logs of who uses their computers. The cybercafé owner can be called upon to report on the identity of a certain Web surfer who used a given PC during a given time interval. Some are asked to call a special number on the fly if the online activity of a customer sets off certain alarms bells.[5] As Internet connectivity increases, often through broad access at shared terminals, this mode of control continues to become more effective over time.

Two Taxonomies of Private Actors Facing This Quandary

Different technology firms are called upon by states to carry out quite different online censorship and surveillance tasks. In seeking to fashion a policy response, it helps to disaggregate the firms implicated in this matter. Two taxonomies, one more helpful than the other, offer ways to disaggregate these firms and those firms that may soon join them in this awkward position.

The first approach is to consider the nature of the firms' business, which is most useful for determining the firms that might get drawn into an ethical controversy of this sort. We include this taxonomy primarily as it is the orientation that casual observers ordinarily bring to the issue. While useful for the purpose of determining to whom this issue is relevant, this taxonomy is far less helpful in terms of informing what to do about it.

The more useful taxonomy considers the nature and level of involvement of the firms in the online censorship and surveillance regimes. The second taxonomy points the way forward more clearly toward a solution by identifying the various ways in which firms are implicated in these regimes and offering means of distinguishing the different types of ethical obligations they may bear.

Types of Firms

Several types of corporations might find themselves called upon to act as gatekeepers. The first corporations to find themselves involved in the censorship and surveillance controversy were technology hardware providers that sold the switches and routers involved in these regimes. In many parts of the world, Internet security firms sell the services and products used in the censorship and surveillance regimes. More recently, content and online service providers, whose customers are typically end-users, have been implicated. Looking ahead, other telecommunications service providers may well find themselves in a similar position as technologies and forms of digital content converge.

Hardware Providers First, technology hardware manufacturers face scrutiny for their sales of routers, switches, and related services to the regimes that carry out online censorship and surveillance practices. According to the critique of human rights activists, companies like Cisco and Nortel that profit from the sale of the hardware that blocks the flow of packets online or enables states to trap and trace online communications are acting unethically. The problem, the critique goes, is akin to the Oppenheimer problem in the context of nuclear technologies. While nuclear technologies can provide energy efficiently to those who need it, the same means can also power weapons of mass destruction of unprecedented power. The hardware manufacturers respond that the technologies sold to regimes that censor and practice surveillance are precisely the same as those technologies sold to firms and governments in states that do not carry out such regimes. This issue is not new, these firms respond. Dual-use technologies present this issue in an untold number of contexts. And the blame should be placed on those who implement the dual-use technologies in the suspect manner, not on those who produce the "neutral" technologies.

Software Providers The second class of firms implicated in this matter includes those corporations that sell the software and services that determine what gets blocked, recorded, or otherwise impeded. Internet security firms—such as Secure Computing, Websense, Fortinet, and others—often serve states, corporations, and other institutions that seek to impede the free flow of packets for one reason or another. A library, for instance, might wish to block underage patrons from accessing pornography online. A similar software package could enable a state to configure a proxy server between a citizen and the wider Internet to block or track certain packets. Many of the states in the Middle East and North Africa that have filtering regimes in place rely upon software packages, and corresponding lists of banned sites, developed and compiled in the United States. These firms make similar arguments to those of the hardware providers: their technologies and services are dual-use in nature. The tool that can protect a child from seeing a harmful image can also keep a citizenry away from politically or culturally sensitive information online. The human rights critique, the firms argue, should be trained on the regimes that apply the services in a manner that violates laws and norms, not at the service providers who make the tools and update the lists. But the lists of banned sites include some nongovernmental organizations that observers suggest have no place there, if in fact, for instance, the notion is just to protect children.

Online Service Providers Most recently, the providers of Internet-based applications have found themselves facing hard questions about their activities in such regimes. A wide range of firms fall in this category: ISPs, e-mail service providers, blog-hosting firms, search engines, and others. ISPs are asked to route traffic in certain ways to prevent citizens from accessing or publishing certain content; likewise, ISP data retention policies are a hot topic of debate in many jurisdictions, as the personal data they keep about citizens is at once sensitive and potentially useful in the context of law enforcement activities. E-mail service providers, such as Yahoo!'s local partner in the Chinese context, are routinely asked to turn over information related to subscribers. The makers of Weblog software and hosting services, such as Microsoft's MSN unit, are asked to block certain information from being published and told to take down the postings or entire blogs of subscribers. Search engines, including Google, are required to limit the results that appear in response to certain queries entered by citizens. The nature of the ethical questions each of these types of firms face varies with the nature of the service they provide and the type of participation the state asks of them. In most instances, corporations respond that they have an obligation to obey local law with respect to services they offer in all jurisdictions.

Corporations often perceive that they do not have the option of resisting the demands of law enforcement officials, for fear that the corporation or their local employees will face sanctions or that their license to operate will be revoked. Some corporations, recognizing the risks inherent in doing business in certain regimes, have limited the types of services that they offer in those contexts to avoid being placed in an uncomfortable role. Google, for instance, decided not to introduce its popular blogging and e-mail tools in the Chinese markets to avoid the possibility of being forced to turn over much information about subscribers, other than possibly basic search query data. In an ironic twist, in Iran, Google has been accused locally of "censorship" for failing to bring all of its services into the Iranian market.

Online Publishers Corporations that publish information online are also caught up in this issue, though their situation is somewhat more straightforward. As a general matter, online publishers are treated as other publishers in the states in which they operate, so the ordinary media restrictions that attach to newspapers and other traditional media also attach in the online space. The notion of providing a single news or information service from one place in the world that is accessible at any other place, so long as it is not censored, remains a viable model. Large media companies, such as the BBC or CNN, tend to adopt this posture. The BBC pays a price for this approach: everything it publishes is blocked in China. When their content is filtered at the destination by the state, they are not complicit.[6] The ethical issue arises only for those firms with local offices and offerings specifically targeting a state that censors online material.

Telecommunications and Other Content Delivery Providers Additional classes of corporations soon could be recruited as gatekeepers. For instance, as mobile telecommunications providers continue to thrive and begin to function as digital content providers, it is only a matter of time before these intermediaries will be pressed into service by states as a requirement of their licenses to operate. Providers of Voice-over Internet Protocol services have already found that their services are sometimes blocked, as in United Arab Emirates. Filtering and surveillance, though posing new technical challenges, may follow. Firms that serve other businesses in delivering online content—including rich media, such as streaming audio and video, in addition to traditional Web pages—also may be subject to such restrictions. Any large-scale intermediary that plays a role in delivering digital information to an end-user might find itself an arm of the state in the online environment—and will have to answer to the same questions as their peers in the hardware, software, and Internet services industries.

Types of Involvement

Another way to categorize the firms that face increasingly difficult ethical questions in this context is to assess not the type of firm, but the type of involvement that a given firm has in the censorship or surveillance regime in question. Though the first taxonomy is simpler, this second taxonomy draws the ethical questions into greater relief. This second taxonomy provides a basis for the different types of ethical obligations that might apply to various firms.

Direct Sales to States of Software or Services to Filter Online Content This category includes those firms that seek to profit from the sale of software or online services, including constantly updated block lists, that states use to implement their online censorship regime. Since these services typically require updates related to the lists used for blocking and since the revenues track directly to the censorship service itself, these firms are the most intertwined with online censorship. An important further distinction emerges between those firms that provide software and those that provide software plus the service of an updated list of sites to block.

Direct Sales to States of Software or Services for Surveillance This category includes those firms that seek to profit from the sale of software or online services, including suites of Internet security systems, that states use to implement their online surveillance regime.

Direct Sales of Dual-use Technology Used in Filtering Online Content This category includes those firms that seek to profit from the sale of Internet-related hardware, including related software and services, that states use to implement their online censorship regime.

Direct Sales of Dual-use Technology Used in Online Surveillance This category includes those firms that seek to profit from the sale of Internet-related hardware, including related software and services, that states use to implement their online surveillance regime. Often, this hardware is sold with related software and services, such as training and support. The more the hardware provider is aware of the usage of the equipment and the more the revenues from services are recurring (rather than a one-time sale of hardware), the more complex the ethical posture the company faces.

Offering a Service that Is Subject to Censorship This category includes those firms that seek to profit from the provision of online services that result in a citizen of a state accessing information in a manner that is censored, such as through a search engine with results omitted or an ISP that refuses access to certain parts of the Internet.

Offering a Service that Censors Publication This category includes those firms that seek to profit from the provision of online services that disallow a citizen of a state from publishing certain information online or that takes down published information at the behest of a state.

Offering a Service with Personally Identifiable Information, Subject to Surveillance This category includes those firms that seek to profit from the provision of online services that capture personally identifiable information about a citizen of a state and where that information may be monitored, searched, or turned over to state authorities upon request.

In certain contexts, the executives of a firm in any of these categories might argue that they do not face a hard ethical question. For instance, in the case of an e-mail service provider that turns over information to a law enforcement officer about a subscriber in a manner that prevents commission of a crime—or, in the most extreme example, an act of terrorism—the corporation may not only have no qualms about its actions, but in fact be proud of its role. By contrast, when the information sought by the state is related to a political dissident whose every action is lawful, or protected by international norms, the ethical landscape is transformed. The same is true with respect to censorship: the blocking or taking down of hate speech, in the context of Germany and France, may well be viewed differently than the blocking or taking down of the expression of certain religious beliefs, for instance. The ethical question in any given instance may ultimately turn less on the precise role of the corporation in the digital ecosystem and more on the nature of the information or the manner in which it is requested of the corporation.

Potential Responses

Reasonable people disagree as to the best means of resolving these emerging ethical concerns. One might thus contend that there is no ethical problem here—or, at least, that the ethical problem is nothing new. If an Internet censorship and surveillance regime is entirely legitimate from the perspective of international law and norms, the argument goes, then a private party required to participate in that regime has a fairly easy choice. If the executives of our hypothetical corporation based in Europe disagree on a personal level with a censorship and surveillance regime, then they should simply exercise their business judgment and refuse to compete in those markets. Alternatively, those executives could decide to refuse to comply with the demands that they believe put their firm in a position in which their ethics are compromised—and then accept the consequences, including possibly being forced to leave the market, that befall them as a consequence of their resistance.

One option from a public policy angle, then, is to do nothing—to accept the status quo, and to let the trend play itself out. In the unlikely event that online censorship and surveillance were to cease across the globe, or if states were to stop calling upon private actors to get the job done, or if corporations were to stop expanding into other markets, the problem might be most cleanly resolved. But absent such changes in the facts as they stand, the stakeholders in these issues have a series of possible ways to move forward to resolve the conflicts.

Industry Self-Regulation

The most likely—and most desirable—means of resolving this problem in the near-term would be for the relevant corporations themselves to come up with a sustainable manner of ensuring that they operate ethically in these charged contexts. It is surprising that no major firm has gone public with such an ethical code before entering a market, such as China, where such problems are sure to present themselves. With firms now competing in those markets, the need to do so is no less acute, whether or not legislative or other action follows.

In the simplest form, individual firms could each develop their own principles, much like a privacy policy on today's Internet; statements could clarify to users, shareholders, and others how the firm will handle these situations. Microsoft set forth a partial version of such a policy at a speech by General Counsel Brad Smith in 2005, in which he pledged the company to follow a "broad policy framework" for responding to restrictions on the posting of blog content.[7] The policy included three specific commitments:

> Explicit standards for protecting content access: Microsoft will remove access to blog content only when it receives a legally binding notice from the government indicating that the material violates local laws, or if the content violates MSN's terms of use.
>
> Maintaining global access: Microsoft will remove access to content only in the country issuing the order. When blog content is blocked due to restrictions based on local laws, the rest of the world will continue to have access. This is a new capability Microsoft is implementing in the MSN Spaces infrastructure.

> Transparent user notification: When local laws require the company to block access to certain content, Microsoft will ensure that users know why that content was blocked, by notifying them that access has been limited due to a government restriction.[8]

Microsoft's step to set forth these three commitments is laudable. And despite putting in place these commitments, which sets the firm apart from most competitors, Microsoft's executives have continued to exercise leadership in the industry in the effort to come up with a common set of principles.

But as a policy matter, a firm-by-firm model of this sort, though potentially an expeditious way forward, would suffer from the variation among approaches bound to ensue. Users would be forced to sort through legalese, much as privacy policies and terms of use force the curious to do on today's Internet, and to compare policies of the relevant firms—a task few people are prepared to invest the time to undertake, and which would disadvantage those who cannot easily parse fine print. And by not standing together, the firms would only have as much leverage as each firm has to begin with.

The more promising route would be for one or more groups of industry members to come up with a common, voluntary code of conduct that would govern the activities of individual firms in regimes that carry out online censorship and surveillance. Such a process is underway, coordinated by the Center for Democracy and Technology and by Business for Social Responsibility. Google, Microsoft, Vodafone, and Yahoo! are actively working together on a code. This process profitably includes additional nonstate actors such as NGOs and academics, including the Berkman Center for Internet & Society at Harvard Law School and the University of St. Gallen in Switzerland. Regulators with relevant expertise and authority have been actively involved in the drafting process. The code is intended to set out common principles with enough detail to inform users about what to expect, but without being so prescribed as to make the code impossible to implement from firm to firm and from state to state. The code might also provide a roadmap for when a firm might refuse to engage in regimes that put them in a position where they cannot comply with both the code and with local laws.

If the industry itself does not succeed through such an approach, the likelihood increases that an outside group will come up with a set of principles that will gain traction and place pressure on the companies to act. The Paris-based Reporters Sans Frontières have drafted such a set of principles, as have a group of academics with their base at the University of California, Berkeley. An outsider's code might be something to which firms could be encouraged to subscribe, on the model of the Sullivan Principles and the Apartheid-era South Africa. An institution might emerge to support the principles and the companies that subscribe to them.

Whether drafted by industry members, outsiders, or a combination thereof, the elements of such a code might either be general in nature—a set of core commitments such as transparency, rule of law, the rights of free expression and individual privacy, and so forth—or more specific, according to a taxonomy of the second sort described earlier. The more specific the code, the more useful, almost certainly, though the reality of getting competing businesses to agree to detailed business practices of this sort is daunting.

As a substantive matter, the code might address censorship and surveillance together, or might disaggregate these topics. Sales to governments of technologies that enable censorship and surveillance presents yet another set of problems that might be taken up in such a code.

By way of example, the framework for such principles on censorship and surveillance might take the following form:

Censorship: Commitment and Guiding Principles At the core of the censorship framework is a company's commitment to the right to free expression. Specific elements of such a commitment might include the following:

1. Formalization. A commitment to establish and carry out formal internal processes for responding to all requests for censorship, whereby the company will respond to requests to censor online information only when presented with formal, written requests from state officials at the appropriate level of authority to make such a demand.
2. Limitation of Scope. Where one state requires that certain online content must be censored, a company will make its best efforts to publish that content in all other markets that the company serves where the content is permitted to be accessed or published online.
3. Reduction of Collateral Censorship. A commitment to make an active effort to uncover instances in which online content that is censored does not fit local legal definitions of what is meant to be censored. A company will work with local authorities to remove from lists of sites to be censored, or otherwise ensure that customers and employees can access, inadvertently blocked online sites or information. A company will maintain a policy for processing complaints about overcensored sites and will take action where complaints are determined to be meritorious.
4. Awareness. The net result of a company's activities in a given country is greater awareness of censorship and filtering by users and lawmakers than if the company were not offering its services in that country. A company seeks to indicate when information that otherwise would have been available is not made available to a user. When one of the company's users is a source of information censored online, the company will seek to inform that user that information they published has been censored. The company will publish, or work with others to publish, information about how censorship works in practice in countries where the company does business and will share data with researchers who study these matters. The company is also committed to supporting the efforts of the international community to uphold universal human rights.

State Demands for User Information: Commitment and Guiding Principles At the core of the framework related to state demands for user information is a company's commitment to the rights of its users to privacy. Specific elements of such a commitment might include the following:

1. Protection from Forced Disclosure. A commitment to establish and uphold rigorous procedural protections to ensure that the company only discloses user information to foreign governments when absolutely necessary under local law.
2. User Notification and Education. A commitment to providing general information about the risks to the company's users of using the company's services on a worldwide basis, as well as specific information about the risks of specific activities in certain settings where those risks are particularly high.
3. Consciousness of Data Location. A commitment to locating servers in places that are unlikely to result in the unethical, forced disclosure of user information. Server location will be based, where possible, in countries with a demonstrated commitment to due process of law and to reliable and consistent rule by legitimate governments. A company will disclose to users the location of its servers hosting their personally identifiable information where possible.

A critical part of such a voluntary code, regardless of its substantive terms, would be to develop an institution that would be charged with monitoring adherence to the code and enforcing violations. One way to accomplish this goal would be for states to adopt the code as law, by passing ordinary legislation and then bringing to bear the full law enforcement capabilities of the state to back it up. Another way could be to imagine an institution—perhaps not a new institution, but a pre-existing entity charged with this duty—that would include among its participants representatives of NGOs or other stakeholders without a direct financial stake in the outcome of the proceedings. This institution may or may not have state regulators involved as partners to ensure compliance. The institution would play an essential role in ensuring that the voluntary code of conduct not only has force over time, but also that it continues to address the ethical issues as they change.

The development of the code itself solves only a small part of the problem; it is in the successful application of the code that a long-term solution lies. In the context of other instances of corporate codes of ethics implicating human rights, such as the sweatshops issue, getting to the code was the easy part.

Law

The legal system might provide one or more ways to resolve the ethical dilemmas facing corporations in the context of states that censor or carry out surveillance online. That said, classic state-based regulation—of the second-order variety—is unlikely to be the most effective means of addressing this particular problem over time. Individual states might require corporations chartered in their jurisdiction to refrain from certain activities when operating in other states.

The analogy in the United States context runs to the Foreign Corrupt Practices Act, which disallows corporations chartered in the United States from bribing foreign officials and other business dealings that would violate U.S. law if carried out in the home market. A "hands-tying" regulation of this sort might be combined with other approaches—including the voluntary code, whether or not embodied in formal law—that might attack parts of the problem, but would unlikely resolve the conflict outright. Such approaches might include funding for pro-democracy activities in the online context, banning the sale of certain technologies, banning the location of servers in certain places, or applying pressure in the context of trade negotiations on those states that are placing the corporations in the hard position.

A member of the United States Congress, Rep. Chris Smith of New Jersey, introduced the Global Online Freedom Act (GOFA) in 2006[9] and again in 2007.[10] This proposed legislation would establish "minimum corporate standards for online freedom" and would impose export controls on the sales of any item "to an end user in an Internet-restricting country for the purpose, in whole or in part, of facilitating Internet censorship." The legislation's intent is laudable: to limit the extent to which United States–based corporations participate in censorship and surveillance in other states.

The shortcomings of GOFA point to the difficulty of enacting second-order regulation on this topic at this moment in history. Some of the provisions, such as the requirement that no servers are to be located within the borders of a state deemed to be "a designated Internet-restricting country," might well achieve the statute's aims by simply disallowing most companies from competing in the foreign market in question; providing a service from abroad will often be too slow or too limited by state-level firewalls or filtering to provide a compelling service to the targeted customers, as Google learned in China.[11]

GOFA would also require a United States–based corporation to check with the State Department before providing "to any foreign official of an Internet-restricting country information that personally identifies a particular user of such content hosting service." When combined with a private right of action for any citizen aggrieved by a violation of that section, these provisions are likely to be such an administrative burden on both private and public parties as to be unworkable. The export controls are impossible to evaluate on the merits, since the legislation simply calls upon the secretary of state and the secretary of commerce to work out regulations within ninety days of enactment of the Act. Given the fact that no specifics are provided on the export controls in the proposed Act after months of formal and informal hearings, drafting, and discussion, one is led to believe that coming up with such regulations in ninety days will be a substantial challenge.

Another reason not to rely upon traditional legal mechanisms in this context is that a globally coordinated set of standards will almost certainly take so long to put in place that the contours of the problem will have changed beyond recognition by the time of enactment. Changes to the relevant statutes or treaty may be equally hard-won. The challenge of coordinating adjustments over time across multiple regimes would be enormous. Laws fashioned in this fast-moving environment will function as a hopelessly trailing indicator, especially if an industry-led process does not precede the legislative approach to the problem. The GOFA drafting experience in the United States suggests that law should be seen as a component of a solution, and perhaps the way to memorialize what the relevant industry members adopt, but not the initial approach.

One possibility for a viable long-term solution would be for the industry consensus to be given the status of law over time. This approach would help to address three of the primary shortcomings of the industry self-regulation model. First, self-regulation can amount to the fox guarding the chicken coop. Second, self-regulation permits some actors to opt out of the system and to gain an unfair competitive advantage as a result. Last, the self-regulatory system could collapse or be amended, for the worse, at any time—and may or may not persist in an optimal form, even if such an optimal form could be reached initially.

This mode of ratifying an industry self-regulatory scheme has instructive antecedents. Most immediately relevant, the Sullivan Principles—proposed initially by one man—eventually became incorporated into U.S. law: the Anti-Apartheid Act in 1986 that embodied the Sullivan Principles passed over President Reagan's veto.[12] In the technology context, a series of proposed laws in the United States—some more advisable as public policy than others—have had a similar history. In the case of the Security Systems Standards and Certification Act of 2001 (SSSCA), the Consumer Broadband and Digital Television Promotion Act of 2002 (CBPTPA), and the Audio Home Recording Act of 1992 (AHRA), the industry came to consensus as to a feasible solution to a common problem, which the Congress then took up as possible legislation. The analogy here is not to the merits of each proposal, each of which suffered from deep flaws. The analogy runs instead to the process of the industry working through the details of a common problem, with lawmakers coming along thereafter to ratify the agreement.

The advantages of such a process are several. This approach would lead to a more stable regulatory regime, bringing with it the benefits of administrative, enforcement, and appellate mechanisms. Depending on what emerges from the process, the Congress or their colleagues in other jurisdictions could decline to ratify the agreement if the industry had not moved the bar high enough. This approach would also solve possibly the toughest problem of industry self-regulations, whereby industry outliers who do not opt in may enjoy an unfair advantage, especially in a context like this one where the behavior is hard to codify as good or bad. The function of ratifying the industry-led agreement ex post facto would be to level the playing field for all relevant firms. Local firms might retain their advantage—they would have only the first-order regulation to contend with, not the second-order—but that is another problem of globalization altogether.

International Governance

Problems in cyberspace rarely have been solved through coordinated international action, though there is no inherent reason to believe that international cooperation or governance could not play a meaningful role in resolving these ethical dilemmas. The United Nations has not been involved in extensive regulation of the online space. The primary U.N.-related entity to play a regulatory role in anything related to the Internet is the International Telecommunication Union (ITU), which has a long history in the coordination between states and private parties in the telecommunications sector. The ITU's role has included the coordination of country codes to facilitate international telephone dialing, which parallels the port allocation process in Internet governance generally handled by the nonprofit Internet Corporation for Assigned Names and Numbers (ICANN). Put in the ITU's own expansive terms, its role ranges "from setting standards that facilitate seamless interworking of equipment and systems on a global basis to adopting operational procedures for the vast and growing array of wireless services and designing programmes to improve telecommunication infrastructure in the developing world."[13] But these activities have generally focused on interoperability within the telecommunications sector broadly, and have not extended far into the Internet governance realm.

Other than the ITU, the U.N.'s work relevant to this problem has been handled through the Internet Governance Forum (IGF), chaired by Nitin Desai and under the secretariat of Markus Kummer. The IGF has the authority to conduct an international dialogue on issues related to the Information Society, which has provided a forum for broaching issues but is neither chartered, nor likely, to accomplish any degree of change. An international treaty process, though cumbersome, could emerge as the way ahead. Some activists have considered litigation under existing human rights agreements. More likely than a treaty process, though, the IGF could be called upon to raise this issue squarely with the global community to determine the most promising course of action. Unlike the analogous Sullivan Principles process, though, the technical aspects of the Internet filtering and surveillance issue make it unlikely that a true global community conversation would ensue. Rundle and Birdling have taken up related issues in much greater detail in chapter 4 of this volume.

Other Modes of Pressure

Human rights activists, academics, and shareholder advocates have played an important role to date in the public discourse related to this issue. The United States Congress has held hearings on this matter to draw attention to the actions of large technology firms. The New York City Comptroller has recently filed shareholder actions with certain technology firms to prompt action on these topics. Human rights organizations and investor groups around the world have hosted forums related to corporate involvement in such regimes. While the involvement of NGOs and other outsiders in the process of addressing these ethical issues is not a solution in itself, it is clear that these stakeholders play an important role in any next steps.

Conclusion

The most promising approach to addressing the ethical dilemma facing multinational corporations doing business in states that carry out online censorship and surveillance is for the relevant community to develop a voluntary code of conduct, with the possibility that such a code be redacted into formal law at some later stage. The code can emphasize procedural safeguards so that Internet users will know the extent to which their communications have been restricted, altered, or censored due to the contributions of a signatory. The code can bind firms to act only where required, and only where the demands placed upon it are specific and formal. That code must be coupled with the establishment of a reliable mechanism for monitoring and compliance assurance. This approach could, at once, be responsive to the nuanced issues involved, flexible over time as the technologies and politics shift, and sustainable over the long-term. Such a process ought to include at the table the NGO community in a supportive, nonadversarial, mode. State regulators might also be drawn into the process in constructive ways. The affected industry need not—and ought not—go it alone.

Though the environment is too complex and unstable for the standard modes of lawmaking to work in the near-term, states do have a role to play in helping to resolve this tension. A patchwork of competing state laws that restrict corporations chartered in one locale in how they do business in this regard could be counterproductive in others. The challenges inherent in framing the Global Online Freedom Act of 2006 and 2007, in the United States context, point to some of the many hazards of this approach.

The proper role of the state in the context of addressing this problem is twofold. First, those states that are more concerned with what their corporations are doing elsewhere should support and encourage the corporations as they seek to work together to raise the bar for themselves and their competitors. That support might come in the form of involvement and encouragement as the industry works with the NGO and academic communities to derive a set of ethical guidelines. Support might also mean using leverage in trade negotiations to lessen the extent that corporations are placed in this position in the first place—in other words, true state-to-state battles against first-order regulation so that second-order regulation is not necessary. Where constructive, states might consider rule-making that ties the hands of their corporations to provide support for their refusal to operate outside of the bounds of these ethical constraints. But states are unlikely to be able to lead constructively and quickly enough to address this problem alone. States may in fact play their role best as "fast-followers" to ensure that the industry-led process results in meaningful and effective second-order regulation of corporate action in these contexts.

On a fundamental level, the states that are increasing Internet filtering and surveillance themselves are best positioned to resolve this tension. In some instances, the primary driver for change might be a careful review of the human rights obligations, whether through treaty or otherwise, that place limits on state sovereignty to act in this manner. Human rights activists may prompt this review through litigation if states do not undertake it themselves. In other instances, the driver might be economic; there is little argument that the development of a competitive environment for businesses using ICTs is a positive factor in economic growth, particularly of developing economies. In either event, states that place restrictions on Internet usage and seek to leverage network usage for purposes of surveillance outside the bounds of human rights guarantees do so at some political and economic peril.

Multinational corporations have every incentive to work hard toward an industry-led, collaborative approach to resolving the tension, regardless of how states act. An industry-led approach could have, at a minimum, the benefit of improved clarity. If the code is well-drafted and well-implemented, users of Internet-based services would know what to expect in terms of what their service provider would do when faced with a censorship or surveillance demand. The benefit of such an approach could well extend further. By working together on a common code, and harnessing the support of their home states, the NGO community, investors, academics, and others, the ICT industry might well be able to present a united front that would enable individual firms to resist excessive state demands without having to leave the market as a result of noncompliance. The ICT industry should strive to provide the best possible services without compromising civil liberties, the generativity of the network, and its democratizing potential.

Notes

1. Lawrence Lessig, *Code and Other Laws of Cyberspace* (New York: Basic Books, 1999, 2006).
2. See Internet Society of China, "Public Pledge of Self-Regulation and Professional Ethics for China Internet Industry," http://www.isc.org.cn/20020417/ca102762.htm.
3. See Joel R. Reidenberg, *States and Internet Enforcement*, 1 U. OTTAWA L. & TECH. J. 213 (2003–04); see also John Palfrey and Robert Rogoyski, "The Move to the Middle: The Enduring Threat of 'Harmful' Speech to Network Neutrality," 21 *Washington University Journal of Law and Policy* 31 (2006).
4. Robert Scoble, Scobleizer, http://scobleizer.com/2006/01/03/microsoft-takes-down-chinese-blogger-my-opinions-on-that/ (accessed January 3, 2007) (for a prominent Microsoft employee's discussion of his company's takedown of the Michael Anti blog). See also Rebecca MacKinnon's contemporaneous account, http://rconversation.blogs.com/rconversation/2006/01/microsoft_takes.html (accessed January 3, 2007).
5. Confirmed in multiple interviews by ONI researchers with representatives of U.S. companies doing business in China.
6. See Jonathan Zittrain, "Internet Points of Control," 43 B.C. L. Rev 653 (2003) for a taxonomy of Internet points of control, including conceptions of source and destination filtering.
7. Microsoft Press Pass—Information for Journalists, "Microsoft Outlines Policy Framework for Dealing with Government Restrictions on Blog Content," January 31, 2006, http://www.microsoft.com/presspass/press/2006/jan06/01-31BloggingPR.mspx (accessed January 4, 2007).
8. Ibid.
9. "H.R. 4780 [109th]: Global Online Freedom Act of 2006," GovTrack.us, http://www.govtrack.us/congress/billtext.xpd?bill=h109-4780 (accessed January 3, 2007).
10. "Smith Reintroduces the Global Online Freedom Act," PR Newswire, http://www.prnewswire.com/cgi-bin/stories.pl?ACCT=104&STORY=/www/story/01-08-2007/0004502076&EDATE= (accessed February 15, 2007).
11. Andrew McLaughlin, "Google in China," the Official Google Blog, http://googleblog.blogspot.com/2006/01/google-in-china.html (accessed January 3, 2007).
12. Winston P. Nagan, "An Appraisal of the Comprehensive Anti-Apartheid Act of 1986," *Journal of Law and Religion*, vol. 5, no. 2 (1987): 327–365.
13. "Role and Work of the Union," International Telecommunication Union, http://www.itu.int/aboutitu/overview/role-work.html (accessed January 3, 2007).

6

Good for Liberty, Bad for Security? Global Civil Society and the Securitization of the Internet

Ronald Deibert and Rafal Rohozinski

Introduction

The spectacular rise and spread of NGOs and other civil society actors over the past two decades is attributable in part to the emergence and rapid spread of the Internet, which has made networking among like-minded individuals and groups possible on a global scale. Powerful search technologies like Google, "me-media" tools such as blogs and MySpaces, and communicative systems like Skype make it easy to form virtual communities, mobilize support, and effect political change. Widespread access to inexpensive digital cameras, editing systems, and distributional channels allows anyone with desire and a few hundred dollars to become a potential Spielberg or Riefenstahl. Causes of all shapes and sizes seek and find moral and financial support on a global basis and, consequently, local politics now plays itself out on a planetary scale.

But the technological explosion of global civil society has not emerged without unintended and even negative consequences, particularly for nondemocratic and authoritarian states. The Internet has raised new, nimble, and distributed challenges to these regimes, manifest in vigorous opposition movements, mass protests, and in some cases even revolutionary changes to long-established political authority. Even among democratic states, the explosion of global civil society has presented serious challenges, though of a slightly different nature. Just as progressive and social justice groups have made use of the Internet to advance global norms, so too have a wide variety of resistance networks, militant groups, extremists, criminal organizations, and terrorists. Whereas once the promotion of new information communications technologies (ICTs) was widely considered benign public policy, today states of all stripes have been pressed to find ways to limit and control them as a way to check their unintended and perceived negative public policy and national security consequences.

In this chapter we examine the ways in which states have targeted the Internet and have begun to assert their power in cyberspace as a means to control and limit global civil society. While *global civil society* is used often and widely today, there is no consensus as to its meaning or significance, particularly among social scientists.[1] Typically, the concept is used to describe those collective associations that citizens have formed to influence public policy, whether domestic or international, such as Amnesty International, the World Wildlife Fund, or the International Campaign to Ban Landmines. Most, though not all, ascribe to global civil society a positive association, and see networks of global civil society reigning in sovereign states while simultaneously pushing for rights, justice, and environmental rescue. Critics, on the other hand, often question the significance, ideological bias, and/or inherently democratic nature of global civil society.[2]

Although important, it is not the intent of our chapter to engage fully these conceptual debates around global civil society, other than to agree with John Keane when he says that "[l]ike all other vocabularies with a political edge, its meaning is neither self-evident nor automatically free of prejudice."[3] We take, therefore, as a definitional starting point that the London School of Economics' (LSE) Civil Society Project has developed:

> Civil society refers to the arena of uncoerced collective action around shared interests, purposes and values. In theory, its institutional forms are distinct from those of the state, family and market, though in practice, the boundaries between state, civil society, family and market are often complex, blurred and negotiated. Civil society commonly embraces a diversity of spaces, actors and institutional forms, varying in their degree of formality, autonomy and power. Civil societies are often populated by organisations such as registered charities, development non-governmental organisations, community groups, women's organisations, faith-based organisations, professional associations, trades unions, self-help groups, social movements, business associations, coalitions and advocacy groups.[4]

The LSE definition captures some of the most important elements of civil society, particularly the theoretical (though not always practical) distinction from the state and market, the recognition of collective associations around shared political purposes, and the diversity of its many manifestations. Adding the term *global* to *civil society* simply acknowledges those associations whose political activities take them beyond the confines of their own sovereign state.

Following Keane, it is important to note that, in spite of the "progressive" bias that influences the organizational examples given in the LSE definition, global civil society itself can comprise a wide range of contrary ideological positions. What is less than clear is whether the concept has normative content that excludes groups, like Al Qaeda, for example, that employ violence to further their ends. Historically, civil society has been strongly associated with minimizing violence, furthering dialogue, and expanding spheres of peace. Whether and where "uncivil" society groups fit into this equation is a debate that falls outside of the purview of this chapter; yet it has been precisely those militant and extremist actors that have been both significant beneficiaries and employers of the Internet, and have in turn been identified by authorities as a putative justification for policies aimed at reigning in and securing cyberspace.

To address these issues, we breakdown *global civil society* into three spheres of agency: *civic networks*, *resistance networks*, and *dark nets*. *Civic networks* refer to progressive environmental, peace, and social justice movements that are most typically associated with the term *civil society*. *Resistance networks* include those more radical groups who are opposed to the status quo and whose activities can be considered illegal in some jurisdictions, making them the target of law enforcement and intelligence agencies. Groups advocating electronic civil disobedience, for instance, are examples of resistance networks. A third form of agency is the least well known, and has tended to fall outside of the scope of most scholarship on global civil society. This category, which we call *dark nets*, includes armed social movements, criminal networks, and the underground economy linking diaspora communities worldwide. Although it is the latter two forms of agency that are most often used as justifications for assertions of state power in cyberspace and are the primary targets of filtering, one of the main contentions of our chapter is that such targeting can have collateral impacts on civic networks as well. In other words, state filtering policies and practices are altering the dynamic ICT environment not just for resistance networks and dark nets, but for civic networks as well.

Global Civic Networks in the Internet Environment

The past several decades have witnessed a rapid expansion of global civic networks. The source of this expansion is undoubtedly complex, and reflects a variety of independent factors having to do with the end of the Cold War, the rise of a set of global values and political causes, the decline of civic participation in traditional structures of political participation, the increase of development initiatives, and no doubt other factors as well. However, there should be no doubt that this expansion also has been the result of the enabling role played by the new media environment, and in particular the growing use of the Internet by civic networks beginning in the 1990s.

Global civic networks were among the earliest adopters of Internet technologies for their collective activities, and have been at the forefront of innovative uses of new media, like SMS, VoIP, and blogs. The medium's constitutive architecture—distributed, decentralized, and relatively cheaply and easily employed—fits with the organizational and political logics of global civic networks. As John Naughton observes, by facilitating access to published data, information, and knowledge; lowering the barrier to information dissemination and overcoming traditional gatekeepers in media; facilitating rapid communication and information sharing on a global scale; and helping to form virtual communities of people with shared interests, the Internet's material properties (how they were constituted in the 1990s) fueled a remarkable and unprecedented expansion of global civic networks.[5] In short, global civic networks both contributed to and were empowered by the evolving environment of Internet communications.

The origins of civic networks' use of the Internet can be traced back to the early 1980s when social change and activist groups began to employ computer networks as a mode of information dissemination and organization.[6] These early networks were largely "basement operations" with individuals donating their time and computing equipment to assist in the activities of their NGOs. By the late 1980s, more formal links had been established among some of these networks in England (GreenNet) and the United States (PeaceNet and EcoNet), and then later Sweden (NordNet), Canada (Web), Brazil (IBASE), Nicaragua (Nicarao), and Australia (Pegasus). In 1990, the networks jointly founded the Association for Progressive Communications (APC), a global umbrella network that still exists as a coordinating and advocacy NGO with a significant presence in Internet governance forums throughout the world today.

Perhaps the earliest demonstration of the Internet's facilitation of civic networks' organizational and networking capacities can be found at the 1992 U.N. Earth Summit in Rio de Janeiro. The Earth Summit was a unique event involving the extensive and official participation of numerous NGOs from around the world. Leading up to the summit, the U.N. and the APC established a network to facilitate communications among NGOs and disseminate official summit information. As Rory O'Brien and Andrew Clement note: "Backgrounders to the issues, draft policies, country briefings, and logistical bulletins were posted by the UN to a set of computer conferences shared internationally on all APC networks. This allowed several thousand civil society groups around the world to be kept informed at very little cost to the UN."[7] The global, distributed nature of the NGO participation—in other words, the fact that groups not physically present at Rio nonetheless were able to participate—was instrumental in the formulation of the several "alternative treaties" that were put forth from the parallel NGO summit, called "the Global Forum," held simultaneously with the Earth Summit.

The type of civic networking demonstrated at the Earth Summit recurred throughout the 1990s, having a tangible impact on local, regional, and international rule-making forums. For example, according to scholarly observations and those of the participants themselves, the Internet played a critical role in the International Campaign to Ban Landmines (ICBL).[8] Although the campaign did not employ computer networks in any substantial sense until about 1995, from that point on the Internet was vital to collecting and disseminating information and forming strategy in the ICBL across its membership in more than seventy countries. The networks were, according to participants, crucial in lowering organizational costs and integrating into decision-making structures members from poorer, developing countries. More importantly, it dramatically augmented the intellectual capacity of the ICBL-member NGOs, who were able to bring analytically and empirically informed analyses to the table when meeting with states on the landmines issue. It also knit the diverse participants together into a relatively coherent unit, particularly with regard to the ICBL strategy.[9]

Although perhaps the most prominent, the ICBL was not the only instance of global civic networks being empowered by the Internet, nor was the model it supplied—working within legitimate processes of political participation, albeit in very novel, challenging ways—generalized elsewhere. Indeed, the 1990s also saw the Internet employed by a growing number of resistance networks—anarchist, antiglobalization, environmental justice, and political opposition movements.[10] In the landmark case of the opposition to the Multilateral Agreement on Investment (MAI) negotiations, for example, civic networks organized a multipronged campaign of resistance and protest across numerous countries and involving hundreds of loosely linked autonomous groups and individuals.[11] Their activities broadsided state policy-makers involved in the negotiation process and by some accounts led to the eventual cessation of the MAI negotiations. The MAI campaign, in turn, morphed into a broader platform of civic networking around antiglobalization, most notably characterized by the street demonstrations of Seattle, Quebec City, Genoa, and elsewhere.[12] Today, this broadly distributed network of individuals concerned with economic and social justice continues to bristle with Internet-based political activity, although street demonstrations have been mitigated in the post–9/11 security environment.

Another celebrated networking campaign of resistance occurred in support of the Zapatista liberation movement.[13] The Zapatistas are a revolutionary national liberation group based in the Mexican province of Chiapas. Beginning in the early 1980s, the Zapatistas formed an armed independence movement that attracted an international web of support among anti-globalization and other activists. *Hacktivists* in support of the Zapatista cause developed distributed Denial-of-Service tools that were employed as mechanisms of protest against the Mexican government, and brought forth one of the first instances of online civil disobedience.[14] The methods of the Zapatista electronic civil disobedience campaign were, in turn, duplicated in other similar acts directed against perceived injustice and corporate power throughout the 1990s, with ambiguous but always controversial results.[15] The rise of resistance networks, in turn, drew the attention of state intelligence and law enforcement worldwide.[16]

Over the past several decades, the Internet has provided a technological foundation and material support for the massive flourishing of advocacy, rights, and justice movements worldwide. These movements have pressed upon traditional structures of political participation in ways that many believe are contributing to a fundamental change in world order. At the very least, these network-enabled transnational social movements have altered the operational environment for states, international organizations, and corporations who have been forced to address civil society stakeholders in all policy arenas. In some cases, such as the so-called color revolutions of the former Soviet Union, civic and resistance networks have actually been responsible for the overthrow of long-established authority structures.

Internet Protection and Hacktivism

The importance of the Internet as a material foundation and explosive engine for civic and resistance networks has not gone unnoticed. Within a dynamic, technologically savvy sector of civil society, a transnational social movement has emerged around what might be called *Internet protection*—that is, collective securitization whose aim is to uphold the Internet as a forum of free expression and access to information through advocacy, training, policy development, and technological research and development.[17] Though coming at the problem from different backgrounds, Internet protection advocates are beginning to network around a shared agenda of communications security and privacy, freedom of expression, equal access, the protection of an open public domain of knowledge, and the preservation of cultural diversity. The participants in this social movement include local, regional, and global nongovernmental organizations, activists, and policy networks including major international rights organizations such as Human Rights Watch and Amnesty International, as well as the OpenNet Initiative (and its partner institutions) itself.

Critical to the constitution of this social movement has been the support provided by major nonprofit research and advocacy foundations such as the Ford Foundation, Markle Foundation, Open Society Institute, and the MacArthur Foundation. The support of these nonprofit foundations has included not only financial resources but also networking opportunities, venues for collaboration, and research and development coordination. To be sure, this type of support has had an important impact. However, the resources provided by these donor agencies do not rival the collective financial capacities that can be marshaled by states. Nor do they always come without unintended consequences. Scholars have noticed funding of this sort can promote the emergence of patron-client ties between donors and recipients, rather than horizontal links among civic networks.[18] They may also create a hostile environment for civic networks due to the impression of outside "interference" and "meddling"—particularly if the NGOs are perceived as a thin vehicle for one state's foreign policy within the jurisdiction of another state. One recent study found that nineteen countries, concentrated mostly in Africa, the Middle East, and the former Commonwealth of Independent States (CIS), have enacted or proposed laws over the past five years that restrict the activities of civil society.[19]

One area where the asymmetric capacities of civic networks may be most tangibly felt is in building code, software, and other tools explicitly designed with an Internet protection paradigm in mind. From the outset, the Internet's character has been shaped not only by states and corporations but also by the distributed base of users themselves. Skilled computer geeks, hacktivists, and other individuals have been responsible for some of the most innovative Internet technologies, from open source/free software platforms to P2P networks and encryption systems. Although "Internet protection" technologies go back decades, in recent years there has been a more concerted and organized research and development effort, working in tandem with the policy, governance, and awareness efforts described earlier. These efforts include tools to support anonymous communications online, such as the Tor system; tools that circumvent Internet censorship, such as psiphon or Peacefire; and tools that support privacy online, such as PGP, ScatterChat, and others. These tools are, in turn, increasingly localized to different country contexts, distributed via nongovernmental organizations and human rights networks, and built into training and advocacy workshops organized by the Internet protection civic networks described above.[20] As we show in a later section, however, state filtering efforts have deliberately targeted these Internet protection tools as a way to control and limit networking activities of civic and resistance networks.

Toward Uncivil Society Networks and the Rise of Dark Nets

The bulk of scholarship on the rise and spread of civic networks has tended to focus on those elements of civil society that are explicitly nonviolent in nature and liberal in outlook and spirit. These characteristics reflect, in large part, the "peace dividend" of the immediate post–Cold War era that provided a hiatus from decades of interstate conflict, and where the bulk of visible or public sphere transnational networks tended to center on issues of social justice, environment, and universal rights and values. Of these elements, even the most extreme forms, such as the anticorporate resistance networks described earlier, were still largely characterized by nonviolent methods and centered in and around Western industrialized activist circles.

However, outside the focus of mainstream scholarship, other social movements and resistance groups discovered and began to appropriate the Internet, recognizing the unprecedented capabilities it offered for organization, communication, mobilization, and action. These actors, ranging from militants, insurgents, criminal elements, and diaspora and migrant communities, expanded exponentially, aided by the largely unfettered and unregulated growth of the Internet throughout the developed and developing world. Much less was said or known about these networks—whose activities and aims were facilitated by the Internet in much the same way as were global civic and resistance networks, but whose aims were often criminal, covert, and sometimes violent. These *dark nets* can be divided roughly into three categories.

The first and most well known of the dark nets are the armed social movements. Armed social movements can represent a multiplicity of local causes, but their ability to share tactics, contacts, and at times drink from the same ideological well make them appear as a unified global network. In the post–9/11 era, Al Qaeda and the jihad movements represent perhaps the most visible manifestation of this kind of armed-social-movement dark net. However, they are by no means the first and only networks of this kind. In the 1990s, the old paradigm of wars among nation states was displaced by a new form of warfare—what Mary Kaldor calls "new wars."[21] What sets "new wars" apart from the previous generation of Cold War–era armed struggles is the participants' ability to leverage the emerging global networked economy—in particular the illicit global economy—to become self-reliant for the arms, money, and political support required to pursue armed struggle against state and nonstate actors. Many of the "new wars" that occurred during the 1990s, particularly those in the developing world where First World militaries were neither involved as supporters or peacekeepers, were fought essentially as transnational civil wars where armed formations pursued both guerilla and conventional warfare against government and rival groups. In conflicts that included Sri Lanka, Somalia, the former Yugoslavia, western Africa, and Chechnya, "new wars" demonstrated that globalization had made armed social movements capable of challenging and at times defeating state actors without the need of state-based patrons or backers.

More importantly, this new generation of armed social actors also increasingly embraced the Internet, recognizing the capacity afforded to affect both their supporters and opponents.

Significantly, it was these groups, rather than First World militaries, that were the first to leverage the Internet as a means to wage information operations that redefined the main field of battle away from the military and toward the political sphere.[22] Beginning with the first Chechen war, the videotaping of attacks on the Russian military became more important than the military significance of the attacks themselves. When shown to supporters, as well as the Russian public (via rebroadcast on Russian television, and later on the Internet), their shock value was enough to convey the impression that the Russian military was being defeated. Similar tactics were adopted and further refined by Hezbollah in its resistance against Israeli occupation of southern Lebanon prior to their withdrawal in 2001. Hezbollah produced reports in the form of music videos that were both broadcast across Hezbollah's terrestrial television station (al Manar), as well as made available for download from Web sites the movement had established as part of its strategic communications and information warfare strategy.

These video shorts proved highly effective, and have since undergone several significant evolutions, paralleling the spread and popularity of such online resources as YouTube and Indymedia that are used regularly by global civic networks and resistance groups. They are now one of the key instruments used by these movements to attract interest in their causes and are a significant feature of the more than 4,500 active jihad Web sites, chat rooms, and forums. As the resources necessary for producing multimedia technologies continue to fall, and access to inexpensive digital cameras and editing software increases, the threshold and number of video and other multimedia products in circulation has grown exponentially. Meanwhile, the age of the producers has sharply declined. During the early months of the second Intifada, for example, several of the more compelling PowerPoint slides circulating on the Internet depicting the brutality of the Israeli reoccupation of the West Bank were produced by a fourteen-year-old living in a refugee camp in Lebanon.[23]

In addition to changing the nature of the conflicts, the video clips have also served to change the nature of the movements themselves. They have eliminated the need for strict command and control, especially for smaller and more marginal movements which can now claim legitimacy for their actions by virtually piggybacking on the perceived effectiveness and success of others. They also give the impression of a unity and scale among groups that in reality simply does not exist. As a result, much as the discourse of human rights and other universal values provides a moral center that binds many of the civic networks together, the depictions of resistance, wrapped in religious undertones, provide a means for smaller, more local struggles to identify with and benefit from a broader ideological pool. When networked in this way, this ideological pool serves to demonstrate that resistance is not only possible, but also positively effective.

The Internet is only one of the tools used by armed social movements in the pursuit of their cause, but it is certainly the one that, because of its largely unregulated character and relative freedom of access, causes the greatest concern for states under threat from such actors. It is seen, at least in part, as the sea in which global militants find sanctuary of the kind that Mao postulated in his classical treatise on People's War. The difficulty, then as it is now, is how to effectively separate the insurgents from the people, or armed social movements from the Internet, in a manner that does not destroy the latter.

Transnational criminal networks are a second form of dark nets. These actors, who can be large or small, local or transnational, exploit the relative anonymity offered by the Internet, as well as the absence of harmonized national laws defining cybercrime, to circumvent or avoid prosecution. Much of the activities of these actors involve old crimes, such as fraud and theft, which have been adapted to the new possibilities offered by the emergence of the e-economy. In other cases, jurisdictions with poorly functioning or nonexistent laws are used to hide otherwise criminal activities, such as distribution of child pornography, out of the reach of authorities in jurisdictions where such activities are clearly criminalized.

Globally, the incidences of reported cybercime is increasing in both developed and developing economies. In the Russian Federation, for example, acknowledged as a source of some of the most imaginative forms of cybercrime, incidences reportedly grew by almost 300 percent between 2003 and 2006.[24] Yet accurate comparative statistics makes measuring global cybercime difficult. For example, in the United States—an economy in which losses caused by cybercrime were cited by one Treasury Board official as exceeding $105 billion—only in 2006 did the Department of Justice belatedly begin the process of establishing a baseline for measuring cybercrime. In part, the absence of reliable statistics reflects the difficulty faced by local police and justice institutions who have to police activities that may not be defined or considered criminal in their jurisdiction (or against which they have few tools). Quite simply the globalization of criminality has far exceeded the capacity of states to define or harmonize an effective global mechanism to contain or police it. Consequently, despite notable efforts such as the Council of Europe's Convention on Cybercrime, criminal activity and networks continue to multiply and expand into new regions and activities. Russian hackers are implicated with identity theft and credit card fraud in the United States and Europe. Nigerian gangs have become omnipresent in a variety of scams and wire fraud, while Chinese and Israeli gangs preside over a global distribution network of pirated DVDs and software. The result of this criminal use of the Internet is that in local jurisdictions the first real awareness of Internet use in their local community comes accompanied with a request for prosecution. In one particularly egregious case that occurred in the late 1990s, the entire ".tj" domain was registered by a U.S.-based entity that used it to host child pornography. Local Tajik authorities were forced to pursue legal action to claim the domain, a fact that did little to portray the social benefits of an unregulated Internet to the morally conservative Tajik society.

A third dark net, and perhaps the hardest to define, consists of the multitude of private social networks that exist among migrant and diaspora communities and that play an important function in supporting the economic and social ties that bind these communities to their kin and communities of origin. These "private interest" networks are the least well known and analyzed, as penetrating them requires gaining the trust of the communities. Often these networks serve specific social functions, circumventing cultural or social taboos, or serving highly specific economic interests. As a consequence, they are often deliberately "closed," and thus may even be denied or downplayed by the communities they serve.[25]

These networks are, nonetheless, among the most active of the dark nets, and hence tend to get labeled with the same negative image as armed social movements or criminal networks because they appear to support or even appropriate the same means used by the latter. For example, diaspora communities are often used to facilitate the movement of funds outside of the formal banking system, especially among migrant communities in the Gulf, South Asia, and the Horn of Africa. So-called *hawala* networks, underground banking networks that in some counties carry a volume of funds equal to or larger than the official banking system, have fallen under suspicion as having been the source of funds ending up in the hands of local militant groups.[26] While this may be the case, both the number of hawala transfers used by terrorist groups as well as the amounts needed for carrying out terrorist attacks are relatively small given the overall volume of hawala transfers, and could just as easily have been hidden in regular banking or other online transactions (e.g., PayPal).

Although not often analyzed together by scholars with the civic networks described earlier, the "uncivil" dark nets are as much a part of global civil society as are the former. Following the LSE definition employed in our chapter, they constitute an arena of uncoerced collective action around shared interests, purposes, and values. Their institutional forms are distinct from those of the state, family, and market, though in practice, the boundaries between them are often complex, blurred, and negotiated. What differentiates them from the civic networks is, of course, their perceived illegitimacy, making them the target of state security and law enforcement. In the following section we turn to the ways in which states have attempted to control and contain the challenges presented by civic networks, resistance networks, and dark nets through Internet filtering, surveillance, and control.

Assertions of State Power Over Civic, Resistance, and Dark Nets in Cyberspace

As the other chapters and the country summaries of this volume make clear, the problem of Internet filtering and censorship is growing in scope, scale, and sophistication worldwide. What began as a practice confined to a small handful of nondemocratic regimes has expanded to countries throughout every region of the globe, and includes nondemocratic, transition, and democratic countries. How much of this filtering can be attributed to attempts by states to control the challenges presented by civic, resistance, and dark networks? Here there is no plain answer, as the motivations for Internet filtering and censorship vary among states and are often shrouded in secrecy and deceit. We can, however, identify several areas from our research where states are asserting control over the Internet as a means to limit the threats posed by the varied elements of global civil society. As we show, even in those cases where the targets are clearly dark nets, there can be collateral impacts on the communications environment for civic networks.

Filtering Data Analysis

The results of the ONI's testing as outlined in this volume present a wide array of categories targeted for filtering, in both English and local languages, across numerous countries and several regions. In this section, we highlight several categories of filtering where it can be imputed that states are deliberately targeting content or communication channels of civic, resistance, and/or dark nets. We also note those instances of collateral filtering, where filtering of content or communications channels of dark and resistance networks impact civic networks as well.

Human Rights

One area of importance to civic networks, both as a normative underpinning and a source of content produced by those networks, is human rights. As many civic networks are critical of states' records in the areas of human rights, many of the affected states have been targeting the sources of that content for filtering. Pakistan, Myanmar, India, Iran, Uzbekistan, Algeria, Ethiopia, Tunisia, Vietnam, China, Syria, Saudi Arabia, and Thailand all block access to at least one Web site categorized by the ONI as "human rights." Among those countries, China, Vietnam, Tunisia, Uzbekistan, Pakistan, Myanmar, and Iran are all pervasive filterers of human rights–categorized content. Among the forty-eight sites that the Chinese government blocks in this category are the Web sites of Chinese Rights Defenders, Human Rights in China, the Asian Human Rights Commission, Amnesty International, Human Rights Watch, and Olympic Watch—essentially the full panoply of international and country-specific organizations with an interest in China's human rights record. Iran's coverage is similar to China's, although it also singles out prominent individuals for filtering, such as the infamous blogger, Hoder. Vietnam tends to focus on country-specific human rights NGOs operating in the Vietnamese language, Tunisia strikes a balance between "international" and "country-specific," as does Myanmar, while Uzbekistan targets mostly independent media, television, and radio Web sites related to Uzbek human rights. For its part, Pakistan targets almost exclusively those human rights sites related to the Balochistan liberation movement. Furthermore, eight states (Algeria, Ethiopia, Iran, Myanmar, Saudi Arabia, Sudan Tunisia, and Yemen) block at least one women's rights site on the ONI's testing lists, with one state, Iran, blocking the highest amount (seventeen of seventy tested). Additionally, seven states (China, Ethiopia, Iran, Myanmar, Pakistan, Syria, and Tunisia) all have at least one blocked in the category "minority rights," with China blocking all sites tested related to Tibet. Overall, the ONI's testing results show a concerted effort among many states to target human rights–related content.

Independent Media and Free Expression

In many states, control over major media, like television and radio, is seen as an important lever of power that can be tightly regulated and controlled, leaving independent media as one of the only sources of news and free expression for civic and resistance networks. A total of seventeen states (Algeria, Bahrain, China, Ethiopia, Iran, Kyrgyzstan, Myanmar, Oman, Pakistan, Saudia Arabia, Sudan, Syria, Thailand, Tunisia, Uzbekistan, Vietnam, and Yemen) blocked at least one Web site categorized by the ONI as a content provider in the "independent media" category. Not surprisingly, there is a strong degree of overlap among those countries that block a high amount of human rights content and independent media content. Other notable instances of filtering of independent media occurred in countries during election periods, a point discussed in more depth later in this chapter. Nineteen states (Algeria, Azerbaijan, Bahrain, China, Ethiopia, India, Iran, Kyrgyzstan, Myanmar, Oman, Pakistan, Saudi Arabia, Sudan, Syria, Thailand, Tunisia, Uzbekistan, Vietnam, and Yemen) blocked at least one site in the "free expression" category. Of those countries, Syria, China, Iran, Myanmar, Tunisia, and Vietnam block a high amount, with Uzbekistan, Saudi Arabia, and Ethiopia blocking a moderate amount.

Internet Protection and Hacktivist Tools

As mentioned earlier, civic and resistance networks have been actively developing software tools to protect and preserve freedom of speech and access to information online. The ONI ran tests to capture filtering targeted against these tools and found several significant country cases. China, Iran, Yemen, Sudan, Tunisia, Oman, and Saudi Arabia all block access to a high amount of URLs in the ONI's "anonymizers and circumvention" category. Uzbekistan blocks access to a relatively few though significant number of anonymizer and proxy sites, as does Vietnam, Myanmar, and Syria. China blocks access not only to known circumvention sites, but sites that are known to provide information and tutorials about censorship circumvention. In the cases of other states, we conclude that some of this filtering is the result of the use of categories built into commercial filtering products used by these regimes. Sudan, Tunisia, Oman, and Saudi Arabia all use SmartFilter, which has an "anonymizer" blocking category, while Yemen uses Websense, which has a "proxy avoidance" category. The filtering system used in Iran varies between ISPs and the specific product used is currently unknown. The targeting of anonymizers and circumvention tools used by civil society (civic, resistance, and dark) suggests states are moving to counter the Internet protection efforts described in this chapter.

Tools of Communication

States have also blocked some of the major media of communication used by all spheres of civil society, including free e-mail services and VoIP. In both cases, the filtering may be motivated by concerns over economic protection and monopoly preservation. However, the collateral impact of the filtering is felt strongly by civil society networks that rely on such low-cost means of communicating. Iran, Syrian, Yemen, and Myanmar all block access to a small but significant number of popular free e-mail services. United Arab Emirates (UAE) highly targets VoIP Web sites for filtering. UAE and Myanmar both block the popular VoIP tool Skype and UAE is joined by Syria in blocking www.dialpad.com and www.iconnecthere.com. Vietnam also blocks two sites in the ONI's VoIP category (www.evoiz.com and www.mediaring.com). Jordan blocked access to Skype in 2006, citing national security concerns.

Hacking and WAREZ

Both resistance and dark nets (and to a lesser degree civic networks) can occasionally make use of Web sites found in the ONI's "hacking" and "WAREZ category." For civic and resistance networks, some of these Web sites and resources provide tools, information, and strategies associated with hacktivism that can be useful to their networking, social mobilization, and political activism. For dark nets, the Web sites of most interest are found in the "WAREZ" category and relate to illicit trade in pirated software and other material, although other dark nets make use of hacker tools as well. Iran, Yemen, United Arab Emirates, Saudi Arabia, Sudan, and China all block a high amount of ONI's "WAREZ" and "hacking" categories, Tunisia blocks a somewhat lesser amount, while South Korea, Oman, Myanmar, and Azerbaijan all target a minimal amount. Hacktivist groups, such as cultdeadcow.com (Algeria, China, Iran, and Yemen), hacktivismo.com (Tunisia and Yemen), nmrc.org (Algeria and Yemen), and thehacktivist.com (Iran and Yemen) are also caught in the "hacking" and "WAREZ" net and blocked in some of these countries. In ONI's 2005 tests, Yemen did not block any sites in our "hacking" category. Now, one of the two ISPs tested, YNET, blocked access to twenty of forty-six URLs we tested, most likely as a result of enabling the "hacking" category on their Websense filtering system.

Miltancy, Extremists, and Armed Separatist Movements

Many states justify their filtering practices as a way to target those members of dark nets that are armed social movements—that is, either militants, extremists, or armed separatist movements. Eleven states, (Algeria, Azerbaijan, China, Ethiopia, Iran, Myanmar, Pakistan, Saudi Arabia, Syria, Tunisia, and Yemen) all block at least one URL that the ONI categorized as a content provider in the "militancy, military group" category. Most of these Web sites relate to country-specific security issues involving extremist groups or organizations advocating or being associated with violent change. As mentioned earlier, Pakistan blocks all Web sites related to the Balochistan insurgency.

The results of the ONI's testing strongly suggest a concerted effort among some states to target the content and communicative infrastructure of global civil society, including civic, resistance, and dark nets. Perhaps not surprisingly, most of the states that do so tend to be democratically challenged or nondemocratic regimes, as these states face the stiffest challenges from these networks. In the following two sections, we turn to a more detailed examination of evidence from blogging and blocking efforts around key periods such as elections.

Securing and Filtering Blogs

As tools of individual self-expression, blogs and blogging have important implications for civic networks, resistance networks, and even some dark nets.[27] First, blogs can provide a source of independent and alternative news from traditional mainstream media. This is especially important in light of the fact that many countries in the world strictly control traditional print and broadcast media, but it is also relevant to areas of the world where such controls are absent. In the United States, for example, the relationship between blogging and traditional forms of journalism has been prominently debated. In nondemocratic and repressive countries, blogging can provide a window into events and issues not covered by the mainstream or government-controlled media. Not surprisingly, many dissidents and activists have been attracted to blogging.

Second, blogging can provide an easy tool for individuals and organizations to disseminate information to a global constituency, that is, for coordination and organizational purposes. In this respect, blogging does not differ fundamentally from traditional Web sites, which also provide the means to publicize information worldwide. Rather, what makes blogs unique is the ease of posting and syndication. Individuals who have no expertise in computer programming and HTML editing can very easily update their blogs, opening Web publication to a wide audience. Additionally, those living in regions of the world with low-bandwidth connections to the Internet can more easily edit blogs than Web sites. Indeed, new technologies allow people to update their blogs using only cell phone text messages.

Third, blogs can provide NGOs and other groups with a new means to attract support for their organizations, particularly in the area of fund-raising and recruiting. Civil society networks continually struggle to get their message out, and often have a difficult time penetrating the mainstream media, particularly regarding their successes. Potential donors and supporters can acquire, through the window of blogging, a sense of immediacy and detailed understanding of the nature and operations of collective activities direct from the source and the field.

However, blogs and blogging do not come without potential negative implications. Because blogging can threaten state control of media, and have become a popular tool of dissidents, militants, and activists, bloggers can find themselves the object of threats, physical violence, and arrests. As shown below, blogging has become a focus of attention by authorities in nondemocratic and repressive regimes, with many bloggers being silenced through arrest or intimidation. Additionally, states that filter Internet communications are beginning to target blogs with increasingly refined forms of censorship, including parsing through entries and removing objectionable keywords and phrases.

The intention of the research in this section is to examine the global effort to silence bloggers. We assess where bloggers are targeted by authorities, for what reasons, and using what mechanisms of silencing (e.g., arrest). As part of the collection of information from Web sites, news articles, and blogs, a number of assumptions have emerged that inform our analysis.[28] First, the actions of states create fundamental challenges to a thorough and quantifiable examination into the issue of silencing bloggers. This is largely due to states' unwillingness to publicize the actions taken against bloggers.[29] Second, because bloggers at risk tend to be located in nondemocratic regimes, accused bloggers are often burdened by limited or no access to legitimate justice systems. Those detained may be held without charge for an indefinite period of time, without knowledge of the charges against them or access to legal representation. When faced with a trial, it is often illegitimate. Third, repressive regimes, such as China and Iran, are the primary perpetrators of blogger targeting. Bloggers in these two states are the least likely to be informed of the charges against them, and the most likely to face lengthy detention, although other countries are following suit.

The targeting of bloggers by authorities has increased on an annual basis over the past five years. We examined many cases of blogger arrests dating back to 1995, although as blogging is a new and growing medium we assume that few cases would have arisen before 2002. The first case of a blogger facing charges by a state was an Egyptian blogger in 2003.[30] The blogger was charged with violating Egypt's religious laws. He was sentenced for an undisclosed prison term and remains incarcerated. Since that time, the number of bloggers arrested has increased on a yearly basis, with a large jump in the rate of arrests beginning toward the end of 2004 (see figure 6.1).

The cause for the increase in arrests of bloggers is likely due to blogging being an increasingly popular medium, particularly for dissidents and activists. As the rate of blogging has increased, so has the threat by blogs to state authorities. Most nondemocratic regimes place stringent controls on media and freedom of expression, including the Internet. Over the past three years, bloggers and the practice of blogging have created an alternative, independent source of news and media. Quite apart from the content of what is published by bloggers (which itself can be threatening), the very independence of blogging undermines state control over media. Hence it is not surprising to find an increasing amount of attention paid to bloggers and blogging by nondemocratic and repressive regimes.

While one might assume that states target bloggers for challenging the legitimacy and authority of the state itself, we found that there are a number of key declared causes for the arrest and detention of bloggers. The most common declared cause of blogger arrests is "antistate activity." This tends to include bloggers who challenge or insult the leadership of the state or incite antigovernment activity, such as protests or violence. Inciting racial hatred and espionage are the second and third highest stated causes of blogger arrests. (See figure 6.2.)

The limitation of this analysis is the reliance on information provided by state authorities themselves on the cause of a blogger's arrest. In numerous cases, those closely linked to

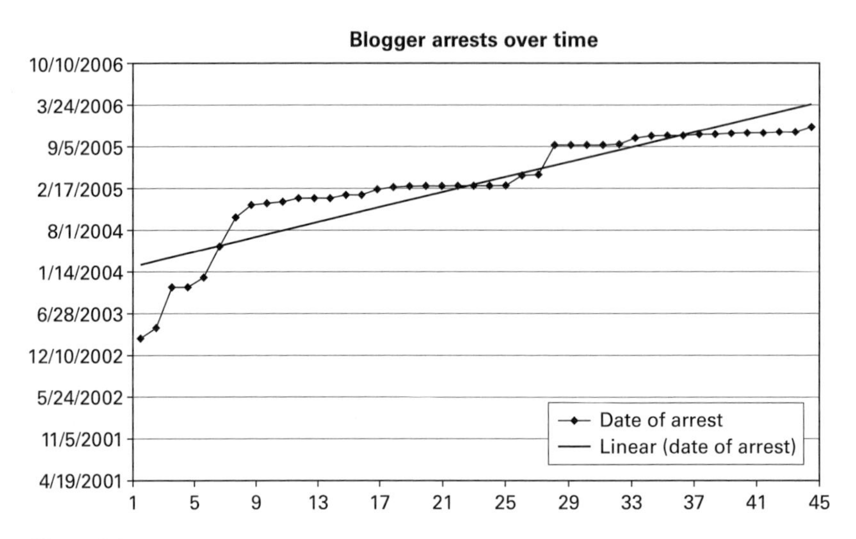

Figure 6.1
Blogger arrests over time.

the blogger have argued that the state falsified claims or made their case on trumped-up charges, and did so for the sole purpose of silencing unwanted views. It is not easy to assess the validity of these claims. It should be noted that on a number of occasions those making the claims were also charged and imprisoned soon afterward.[31]

While Iran and China are key perpetrators of threats against bloggers, the effort to silence bloggers is widespread. Bloggers are facing sentences throughout Asia, Europe, North Africa, and Europe (see figure 6.3). As figure 6.3 illustrates, bloggers are facing charges throughout the world. While Iran remains by far the largest contributor to worldwide blogger arrests, Bahrain has prosecuted the second largest number of bloggers. China, Malaysia, and the United States are not far behind.

We also found that a preponderance of sentences handed down to bloggers found guilty of a crime is for undisclosed periods of time. While the increase in the number of bloggers facing charges and arrest is in and of itself a worrying trend, perhaps more alarmingly is the finding that 40 percent of those arrested face charges and sentences that are not made public. Most bloggers arrested will be detained without access to legal recourse until the state, of its own volition, chooses to release information on the blogger's sentence and occasionally release the blogger either outright or pending a trial or retrial.

Of those bloggers whose sentences are public, the majority spend a maximum of one year incarcerated. Often among this group, a longer sentence was set and then shortened during appeals through the justice system. (See figure 6.4.)

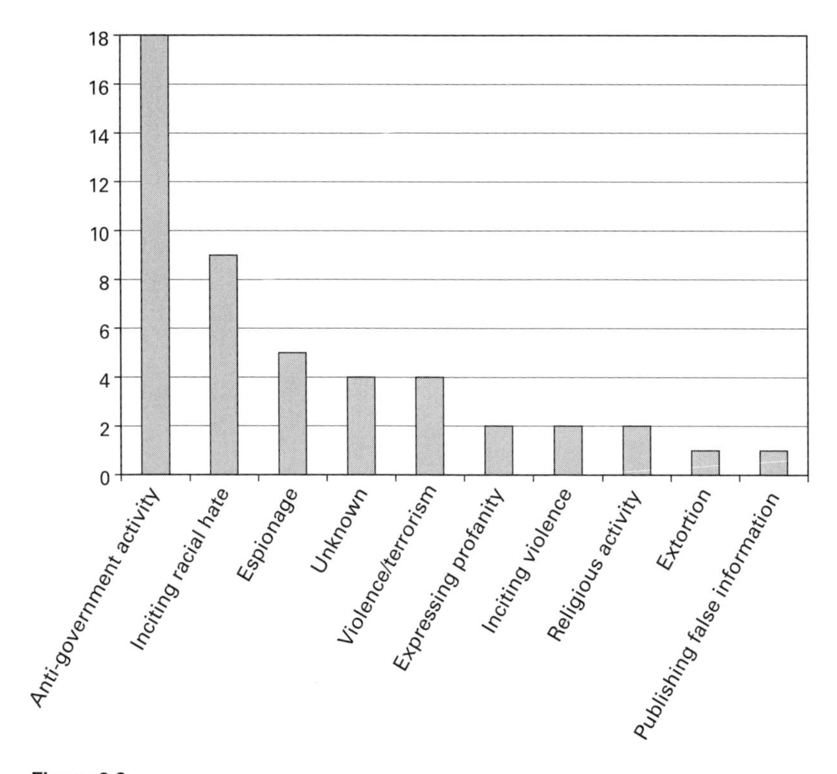

Figure 6.2
Stated causes of blogger arrests.

Blog Filtering

One of the best indications of the political nature of blogging can be found in the extent to which blogs are subject to unlawful Internet filtering. Among two of the states that monitor and control Internet communications that the ONI has investigated, namely China and Iran, blogging has become one of their major focuses of attention, although a number of other countries filter blogs and blogging services as well.

Blogging has become especially popular in China. Although it is difficult to determine with precision the number of blogs hosted in the country, one source indicates that Chinese servers host more than twenty million bloggers.[32] These bloggers post content on topics ranging from daily diaries to political commentary, both critical and supportive of the Chinese state. However, dissidents and human rights activists have been particularly drawn to the medium of blogging. Given the nature of China's Internet filtering and surveillance regime, and the way in which blogging's instantaneous publication threatens the restrictions placed on freedom of speech in that country, it is no surprise that Chinese authorities have intensified

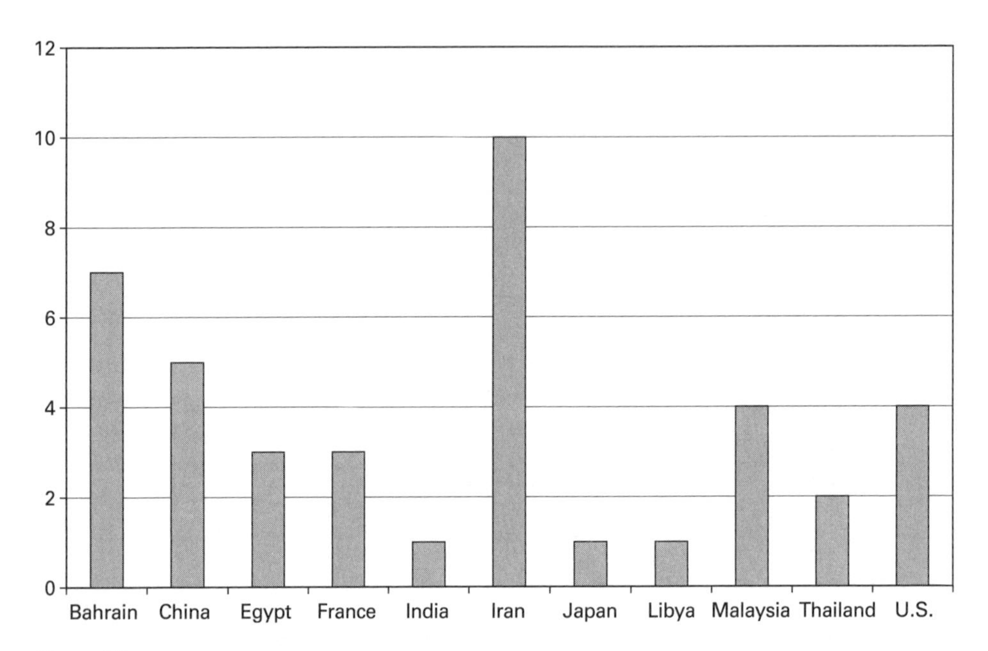

Figure 6.3
Numbers of bloggers arrested by country.

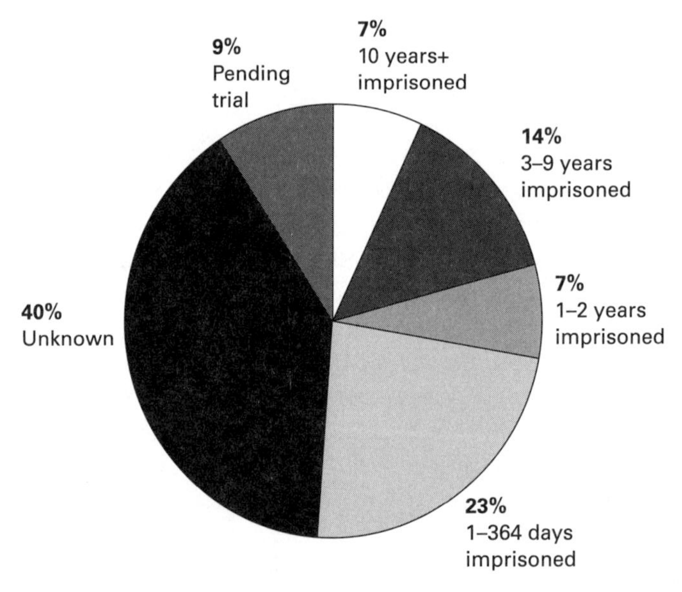

Figure 6.4
Lengths of blogger sentences.

their efforts to monitor and control blog content. These efforts have ranged from shutting down blogging services entirely to filtering blogging services for objectionable content.

In March 2004, the state closed three popular domestic blog Web sites—www.blogcn.com, www.blogbus.com, and www.blogdriver.com—reportedly because a blogger posted a controversial letter regarding the Tiananmen Square incident and the SARS outbreak. All three providers were eventually allowed to reopen, but were required to implement filtering mechanisms. These filtering systems search for sensitive keywords when users attempt to post material. The ONI tested these filtering systems from computers based within China using a list of banned keywords that Chinese hackers discovered and published to a Chinese bulletin board system in August 2004. The list includes terms in categories such as national minorities' independence movements, the Tiananmen Square incident, Falun Gong, proper names of Chinese Communist Party leaders, and sensitive nonproper nouns (such as generic words relating to uprisings or oppression), and were said to be employed by authorities on popular instant messaging services. Using this list of keywords as a basis for testing, the ONI found that Blogbus and BlogCN filtered only 18 and 19 of the keywords, respectively, while Blogdriver filtered 350 of the terms.

An analysis of the filtered content that the ONI tested shows the areas of content about which the Chinese government is especially sensitive. The filtered keywords generally fall into five categories:

1. National minorities' independence movements: the well known Tibetan cause is represented as well as Xinjiang and Inner Mongolia. The inclusion of some Taiwanese politicians' names also fall into this category as they are all people who are known to support Taiwan independence.
2. The Tiananmen Square incident in 1989: it is referenced by the full name, "Tiananmen massacre," the Chinese custom of referencing important events by the number of the month and the day (in this case, 6-4), and also by reference to people involved—a mother of one of the victims who has been campaigning for human rights. The name of Zhao Ziyang, former Chinese Communist Party (CCP) general, is also included in this category.
3. Chinese communist leaders: a list of the top leaders, past and present, are included along with a particularly creative rewriting of Jiang Zemin by replacing one of the characters of his name by the character for *thief*.
4. Falun Gong: a list of different names for Falun Gong including various spellings with characters that sound the same, often used to circumvent filtering.
5. Keywords relating to uprisings or suppression: a list of words referring to uprisings or suppression.

Blogs are clearly seen within China as being threatening to the state because of the ways in which they facilitate rapid and easy freedom of expression. As blogs have grown in popularity, the Chinese authorities have focused their attention on blogs, bloggers, and blogging services accordingly.

As with China, Iran has seen blogging become a popular activity for dissidents and human rights activists both within and outside the state. By some counts, there are 65,000 individual blogs written in Persian, and numerous others written in English on Iranian issues by Iran expatriots located around the world. A list of Iranian blogs is archived at Blogs X Iranians.[33] One Iranian blogger, Hossein Derakshan (a.k.a. Hoder), has become a very prominent blogger in the Internet activist community, and has been profiled in major media around the world.

However, blogging and bloggers in Iran have become the target of censorship by a filtering regime that ranks with China as one of the world's most extensive. The 2005 and 2006 ONI's tests on Internet filtering in Iran included checks on blogs and popular blogging services. The 2005 tests were performed using sets of lists in two different categories: the ONI's general global list of blogging sites applied to all countries (of which one was blocked) and a high-impact list of 533 Iranian-related blogs. Of the 533 blogging sites that were checked, 86 were found to be blocked. Moreover, there was a dramatic increase in the number of blogs blocked during the time frame in which the ONI conducted its tests; 35 of the 86 sites were accessible one year earlier. The ONI tested a large number of blogs on several of the large blogging domains and found that, while Iran blocks a significant number of individual blogs, the state has not taken the (technically) easier step of preventing access to entire blogging domains. Most likely this is because the Iranian government wants to allow access to most blogging services (the exceptions being Moveable Type and Live Journal, which are blocked in their entirety) while focusing on individual blogs that threaten the regime.

The 2006 tests show a continuation of these trends. For example, all seven ISPs tested in Iran block access to the highly popular blog www.boingboing.net. Iran (four of seven ISPs tested) also blocks access to the popular blogger tool www.technorati.com and the photo-sharing site Flickr, while all Iranian ISPs block access to the video posting and distribution site YouTube.

Although China and Iran are the most aggressive in terms of targeting blogs and bloggers, the ONI found evidence other countries are following suit. In Syria, for example, we tested 159 blog URLs in the "Free Expression and Media Freedom" category and found 117 blocked. However, all the 117 blocked blogs were hosted on Google's Blogspot and are blocked as a result of the Blogspot service in its entirety being blocked. Ethiopia and Pakistan also block all of Blogspot. In the case of the latter, the motivation to block Blogspot comes from a desire to block access to blogs hosted on the service containing imagery offensive to Islam. However, the Pakistan Telecommunications Authority has chosen to block access to these blogs by blocking all of Blogspot, thus collaterally filtering even those Web sites critical of the blogs containing the imagery. United Arab Emirates, Ethiopia, India, and Tunisia all block at least one Blogspot blog. Saudi Arabia, Sudan, and Tunisia block access to BoingBoing, most likely as a result of those countries' use of SmartFilter, which categorizes BoingBoing as "nudity." Like Iran, Saudi Arabia also blocks access to the video file–sharing blog service, YouTube.

Just as blogging is becoming a popular form of self-expression and communication generally, activists, dissidents, NGOs, and other global civil society actors are also increasingly blogging. Blogs are even more efficient than typical Web sites in providing a quick and easy means for individuals and organizations to communicate and distribute information, especially in regions of the world with low bandwidth. Blogging is becoming increasingly politicized, particularly among nondemocratic and repressive regimes, but also in "free" and "partly free" parts of the world. The number of bloggers targeted for silencing has grown in proportion roughly to the spread and increasing popularity of blogging. In some parts of the world, blogging can be an attractive alternative to state-controlled media, but one which presents significant security risks for individuals undertaking the blogging. Blogging content is increasingly subject to Internet censorship and surveillance. Among the countries that the ONI has studied, China and Iran have the most refined systems of blog filtering in place, although there is blog filtering in other countries as well. As the tendency worldwide is toward an increase in the scope and scale of Internet censorship and surveillance, we should anticipate the number of countries targeting blogs for filtering to increase as well.

Evolving Techniques: Just-in-Time Blocking, DoS, and Computer Network Attack

Since 2003, research collected by the ONI indicates an increase in the number of countries applying filtering to an expanding number of categories, many of which affect civic networks, resistance networks, as well as dark nets. However, increasing awareness of filtering practices has also provoked a degree of blowback, evident in both the negative publicity in the global media targeting the worse offenders, such as China, Iran, and Uzbekistan, and calls for adjusting U.S. foreign policy to label countries following such practices as pariahs. There is, of course, a question mark over the degree to which establishing the global norm of a free and open Internet is possible, given that such a stance would contradict the concerns shared by many security agencies, in democratic and nondemocratic states alike, as to the degree to which the Internet can and does serve as a sanctuary for armed social movements, and hence is in need of enhanced rather than decreased policing. Likewise, criminal exploitation of cyberspace and particularly efforts aimed at stopping sexual exploitation of children means that calls for the complete removal of filtering are unlikely to meet with success.

It is equally true, however, that not all countries have the political will, economic clout, or natural resource base of a China or Iran. Many Third World countries are dependent on different forms of foreign assistance, or are sensitive to sanctions that may disrupt trade or the movement of migrant workers. Consequently, being labeled as a pariah, with any of the attendant negative publicity and possibility of sanctions, is of consequence. Yet controlling unwanted political agency, whether it comes in the form of prodemocracy groups, independent media channels, or armed social movements is increasingly critical, particularly in authoritarian states, or countries with less institutionalized and more fragile systems for managing political change (such as elections). Among these states, the perceived costs of maintaining a national filtering policy may be seen as either too high, too difficult to maintain, or simply undesirable for other reasons.

However, the costs of no control may be even higher. In this respect, the "color revolutions" that occurred in the former Soviet republics of Ukraine, Georgia, and Kyrgyzstan between 2003 and 2005—which leveraged the Internet and other forms of communications as a means to force political change by way of mass civil action—may be seen as milestones toward the evolution of *just-in-time blocking* identified by the ONI. Just-in-time blocking differs from the first-generation national filtering practices of countries like China and Iran in several significant ways. First, and most importantly, just-in-time blocking is temporally fixed. Unlike the evolving block lists used by national firewalls, just-in-time blocking occurs only at times when the information being sought has a specific value or importance. Usually, this will mean that blocking is imposed at times of political change, such as elections, or other potential social flashpoints (important anniversaries or times of social unrest). In the CIS, this kind of filtering was documented by the ONI during the March 2005 Kyrgyz parliamentary elections,[34] the March 2006 Belarus presidential elections,[35] and the October 2006 Tajik presidential elections.[36] It has also been alleged in other regions, including Bahrain, Uganda, and Yemen during the run-up to their 2006 presidential and parliamentary elections.[37]

Second, the exact techniques by which just-in-time blocking is occurring differs greatly from traditional national firewalls. In some cases, such as in the Tajik and Ugandan elections, existing public order laws are used that require ISPs to filter out sites detrimental to national security. In Tajikistan, ISPs received orders to block two opposition Web sites "in compliance with the national concept of information security developed in year 2003" as they were deemed to "aim to undermine the state's policies in the sphere of information."[38] Similarly the Uganda Communications Commission (UCC) ordered the two national Internet providers, MTN and Uganda Telecom, to block radiokatwe.com, a Web site critical of the government citing "serious concerns."[39] The Uganda case came to light as the technique used by the ISP resulted in a further 657 completely unrelated Web sites that shared the same IP address being blocked. Bahrain blocked several Web sites in the run-up to the country's parliamentary elections in 2006, and Yemen banned access to several media and local politics sites ahead of the country's 2006 presidential elections. Likewise, Bahrain also briefly blocked access to Google Earth in 2006, citing national security concerns, as did Jordan in the same year with respect to Skype.

In other cases, blocking has been accomplished by covert or special technical means. During the Belarus elections, a variety of techniques were observed, ranging from apparent errors in the propagation of domain name information, causing Web sites to be inaccessible from ISPs within Belarus, through to technical failures that disconnected all Internet access in Minsk during the period of street demonstrations that followed the election. One of the most often seen techniques is the use of Denial-of-Service attacks (DoS) against ISPs hosting targeted Web sites or services. This form of blocking is particularly effective as it can occur anonymously, with no demands being made, and presents investigators with the difficult task of pinpointing the source of the attack, which in at least one case was purchased from rogue hackers on the open market (in the CIS). During the Kyrgyz 2005 elections, a sophisticated DoS attack was carried out against a national ISP (El Cat) that hosted several independent (and pro-opposition) media sites. "Extortion notes" requiring that the ISPs remove the opposition sites accompanied the attacks. El Cat was particularly vulnerable as it was dependent on a few relatively "narrow" connections to the Internet and, as a result, the DoS attacks on the opposition sites threatened to disrupt access to all its other commercial Internet operations, which included a large number of commercial clients. In Belarus, DoS attacks were used against several opposition Web sites (hosted outside of Belarus). In this latter case, the attacks were not accompanied by any claims of responsibility or demands. The attacks did end shortly after the elections, as it was clear that the opposition was defeated and its street protest would not prevail.

This later form of just-in-time blocking, which takes an offensive rather than defensive character (as in most traditional forms of filtering), is likely to gain in popularity. The expansion of broadband access, particularly in less-developed countries with lower levels of knowledge of "bot nets" and other "crack attacks," will almost certainly lead to an increase in these kind of disruptions as "bot herders" exploit unprotected computers and broadband connections. Other factors also make this form of offensive blocking particularly appealing. The first is that such attacks are difficult to trace to an exact source (particularly as they can be bought) and thus allow for "plausible deniability." It is also difficult for individuals or nonstate groups to get assistance in tracking down the source of such attacks, as they do not have access to the necessary legal instruments to do so. For example, in the Kyrgyz election case, the extortion notes sent by the attackers originated from a computer located in the United States. However, to enlist U.S. authorities' assistance, the means to do so—Multilateral Legal Assistance Treaties (MLATs)—need to be initiated by states. In this case, the Kyrgyz ISP affected by the attacks was told that in order to get help from the FBI (or other U.S. law enforcement agencies) they would either have to launch a request through the Kyrgyz Ministry of Justice (or Interior), or file a civil case directly in a U.S. state court in which the computer allegedly responsible for sending the letter was located. In both cases, bureaucratic realities and costs prevented the Kyrgyz ISP from taking any further action. These barriers, combined with the relative ease in which such attacks can be "plausibly denied" by their perpetrators, make them a potentially effective tool for preemptive attacks against information resources. Indeed, use of these kinds of attacks against "terrorist" sites is currently under active consideration by a number of states, including perhaps most importantly the United States.[40]

Indirect filtering by way of DoS or other computer network attacks (CNAs) also requires much less in the way of infrastructure, and is thus less costly and less difficult to maintain than national firewalls. As a result, it opens the door for substate actors to engage in their own denial of access campaigns using CNAs. In the Russian Federation, and the CIS, for example, winning elections is as much about mobilizing your supporters as it is about preventing the mobilization of opposition groups and parties. The relative ease and low cost of conducting DoS or other CNAs makes it probable that such tactics will become part of the normal way in which elections campaigns are run. It is also likely that activist groups from across the political spectrum will employ these means as a way to raise awareness or cause lasting damage to their opponents, indirectly causing damage to the openness of the Internet by likely leading to further calls for regulation and policing. Indirectly, it may induce yet further blowback against unfettered use of the Internet by civic networks.

Conclusions

As the evidence presented in this chapter makes clear, a simple correlation between the Internet environment and the expansion of global civil society can no longer be taken for granted. While it is certain that civic networks, resistance networks, and dark nets exploded in the 1990s and early 2000s, the material and political conditions of the communications environment of the time were favorably structured for such an outcome. Largely oblivious to the unintended consequences of the Internet environment, policy-makers were actively encouraging the growth and penetration of information and communications technologies worldwide through FDI and development projects. Not until the appearance and impact of civic networks, resistance networks, and dark nets—human rights advocates, antiglobalization activists, militants, extremists, and jihadists—did state military, intelligence, and law enforcement take active measures to secure the Internet through filtering and surveillance and begin to rethink the encouragement of open and uncontrolled global communications networks. As we show in this chapter, the scope, scale, and sophistication of Internet-content filtering and surveillance and other methods of Internet control are growing rapidly and spreading globally. Although these security practices are aimed primarily at "uncivil" society, dark nets, and those actors considered to be a national security threat, the measures affect the operational environment for civic networks as well.

It is important to underline, however, that, notwithstanding these assertions of state power and control, the Internet and civic networks likely will never fully be reigned in. A sprawling, distributed, and highly potent sphere of global civic networks has been unleashed that moves in and around sovereign states. These networks of autonomous agents are highly creative and can be technologically sophisticated. Most noteworthy has been the growing solidification and international presence of a formidable transnational social movement around Internet protection. This movement has put the filtering and surveillance activities of states and corporations under an intense "sous-veillance" grid, exposing unaccountable and nontransparent practices while pushing for access to information and freedom of speech worldwide. Their efforts include grassroots research and development initiatives to build software and advance knowledge and capacity that helps secure human rights online. Although the pendulum presently has swung in the direction of state control worldwide, hacktivists occasionally are able to puncture through.

Alongside state efforts at control, as well as the emerging militarization of cyberspace, the Internet has become an object of geopolitical contestation among states and nonstate actors alike across each of its layers: infrastructure, code, law, and ideas. The outcome of this competing securitization process is not clear—for state sovereignty, for human rights, or for openness on the Internet. While states have more power and legal means to directly influence the Internet, and together are creating mutually constitutive (if not explicitly defined) norms of control, civil society actors are able to create tools and publish information that expose and occasionally even undermine these measures. For the foreseeable future, then, we believe the Internet will have no "natural" tendency; it will be a media environment that morphs in continuous tension, creating new forms of agency that in turn produce effects that shape the Internet itself. Given the multilayered complexity of this environment, it seems apparent that no one agent will be able to dominate cyberspace entirely, but many will be able to push technologies, regulations, and norms that affect it.

Notes

1. See Marlies Glasius, Mary Kaldor, and Helmut Anheier, eds. *Global Civil Society 2006/7* (London: Sage, 2005).
2. See, for example, David Rieff, "Civil Society and the Future of the Nation-State: Two Views," *The Nation*, vol. 268, no. 7 (1999): 11–16.
3. John Keane, *Global Civil Society?* (New York: Cambridge University Press, 2003), xv.
4. "What Is Civil Society?" The London School of Economics and Political Science, http://www.lse.ac.uk/collections/CCS/what_is_civil_society.htm.
5. John Naughton, "Contested Space: The Internet and Global Civil Society," in *Global Civil Society*, ed. Helmut Anheier Marlies Glasius and Mary Kaldor (London: Sage, 2001).
6. This section draws upon Ronald Deibert, "Deep Probe: The Evolution of Network Intelligence," *Intelligence and National Security*, vol. 18, no. 4 (2004): 175–200.
7. Rory O'Brien and Andrew Clement, "The Association for Progressive Communications and the Networking of Global Civil Society: APC at the 1992 Earth Summit," The Association for Progressive Communications, www.apc.org/english/about/history/rio_92.htm.
8. See Richard Price, "Reversing the Gun Sights: Transnational Civil Society Targets Land Mines," *International Organization*, vol. 52, no. 3 (1998): 613–644.
9. Kenneth Rutherford, "Internet Activism: NGOs and the Mine Ban Treaty," *The International Journal on Grey Literature*, vol. 1, no. 3 (2000), 99–106.
10. Porta, Donatelladella, "Global-net for Global Movements? A Network of Networks for a Movement of Movements," *Journal of Public Policy*, vol. 25, no. 1 (2005): 165–190.
11. See Ronald J. Deibert, "International Plug 'n Play? Citizen Activism, the Internet, and Global Public Policy," *International Studies Perspectives*, vol. 1 no. 3 (2000): 255–272.
12. A recent comprehensive survey of participants at five globalization protests found that 80 percent of those surveyed found out about the protests via the Internet. See Dana Fisher, "How Do Organizations Matter? Mobilization and Support for Participants at Five Globalization Protests," *Social Problems*, vol. 52, no. 1 (2005): 102–121.
13. For a nuanced analysis that compares that MAI and Zapatista cases with a discussion of the role of the Internet, see Josee Johnston, "Solidarity in the Age of Globalization: Lessons from the Anti-MAI and Zapatista Struggles,"

Theory and Society, vol. 32, no. 1 (2003): 39–91; Harry Cleaver, Jr., "The Zapatista Effect: The Internet and the Rise of an Alternative Political Fabric," *Journal of International Affairs*, vol. 51, no. 2 (1998): 621–640.

14. Hacktivism combines the notion of "hacking" in its original positive sense, as someone interested in exploring technology, with social and political activism.

15. Jenny Pickerill, "Radical Politics on the Net," *Parliamentary Affairs*, vol. 59, no. 2 (2006): 266–282; and Brian Huschle, "Cyber Disobedience: When Is Hacktivism Civil Disobedience?" *The International Journal of Applied Philosophy*, vol. 16, no. 1 (2002): 69–83.

16. See, for example, Canadian Security and Intelligence Services, Report No. 2000/08: Anti-Globalization—A Spreading Phenomenon (August 2000), http://www.csis-scrs.gc.ca/en/publications/perspectives/200008.asp.

17. For an overview, see Ronald Deibert, "Black Code: Censorship, Surveillance, and the Militarisation of Cyberspace," *Millennium*, vol. 32, no. 3 (2003): 501–530.

18. See Sarah Henderson, "Selling Civil Society: Western Aid and the Nongovernmental Organization Sector in Russia," *Comparative Political Studies*, vol. 35, no. 2 (2002): 139–168.

19. See "Recent Laws and Legislative Proposals to Restrict Civil Society and Civil Society Organizations," *International Journal for Not-for-Profit Law*, vol. 8, no. 4 (August 2006), http://www.icnl.org/knowledge/ijnl/vol8iss4/art_1.htm; and Peter Ackerman, "The Right to Rise Up: People Power and the Virtues of Civic Disruption," *The Fletcher Forum of World Affairs*, vol. 30, no. 2 (2006).

20. See, for example, the NGO-in-a-Box Security Edition, http://ngoinabox.org/boxes/security/.

21. Mary Kaldor, New and Old Wars: Organised Violence in a Global Era. Cambridge, Polity Press, 1999.

22. See Rafal Rohozinski, "Bullets to Bytes: Reflections on ICT and 'Local Conflict,'" in *Bombs and Bandwidth: The Emerging Relationship Between IT and Security*, ed. R. Latham (New York: New Press, 2003), 288, 24cm.

23. Ibid.

24. Russian federal law captures cybercrime under 111 separate statutes ranging from "unlawful access to computer information," and "creation, use and distribution of malware," through to fraud and illegal distribution of porn materials and items containing child porn.

25. See Rafal Rohozinski, "How the Internet Did Not Transform Russia," *Current History*, vol. 18, no. 4 (2004): 175–200.

26. See Rafal Rohozinski, "'Secret Agents' and 'Undercover Brothers': The Hidden Information Revolution in the Arab World" (unpublished paper), http://www.ssrc.org/programs/itic/publications/ITST_materials/rohozinskibrief3_4.pdf.

27. Thanks to Julian Wolfson for assistance in the research for this section.

28. The silence associated with the arrests of bloggers provides little hard data or resources available to quantify the occurrences of attacks on bloggers by hostile state actors. To undertake this survey two methods have been used: Google searches and use of the LexisNexis legal, news, public records, and business information database. Google searches used the following keywords: *blogger arrest*, *blogger arrested*, *cyber dissidents*, *blogger detention*, *blogger trial*, *blogger detained*, *blogger crime*, *weblog* and *arrest*, *weblog* and *arrested*, *web log* and *trial*. The Google search provided a broad spectrum of information only some of which was useful for the research. To separate the data into relevant categories, an assessment was made as to whether the action taken against bloggers was associated with their online activities. Crimes that were not associated with their blogging directly were not included in the analysis. For example, the blogger arrested for robbing a bank was excluded from the data set. For searches on the LexisNexis database a similar method was undertaken. The following keywords were used: *blogger arrest*, *blogger arrested*, *cyber dissidents*, *blogger detention*, *blogger trial*, *blogger detained*, *blogger crime*, *weblog* and *arrest*, *weblog* and *arrested*, *web log* and *trial*. The research for this section was carried out in October 2005.

29. Phil Deans, "The Internet in the People's Republic of China," in ed. J. D. Abbot, The Political Economy of the Internet in Asia and the Pacific. Westport: Praeger, 2004: 123–139. Shanth; Kalathil and Taylor C. Boas, Open Networks, Closed Regimes: The Impact of the Internet on Authoritarian Rules. Washington, D.C.: Carnegie Endowment for International Peace, 2003.

30. See "The Arabist for November 3rd, 2005," http://arabist.net/archives/2005/11/03/.

31. For more information see the case of Najmeh Oumidparvar of Iran, charged and imprisoned for twenty-four days on March 2, 2005, for posting on her blog a defense of her husband, Mohamad Reza Nasab Abdolahi, previously detained for his posts to his blog. And the case of Mojtaba Saminejad, arrested in Iran on February 12, 2005, for posting articles to his blog on the arrest of three other bloggers.

32. Li Qian, "Chinese Weblogs Thriving," China Daily, http://www.chinadaily.com.cn/china/2007-01/11/content_781602.htm.
33. Blogs by Iranians, http://blogsbyiranians.com/.
34. "Election Monitoring in Kyrgyzstan," Special Report: Kyrgyzstan, http://www.opennetinitiative.net/special/kg/.
35. Ibid.
36. "Tajikistan Blocks Access to Web Sites in the Run-Up to Presidential Election," New Eurasia, http://tajikistan.neweurasia.net/?p=116.
37. "Internet Filtering in Uganda," Internet Censorship Explorer, http://ice.citizenlab.org/?p=190, infra p. 209.
38. http://tajikistan.neweurasia.net/?p=116.
39. http://ice.citizenlab.org/?p=190.
40. In mid-2006, the U.S. Department of Defense was well underway in preparing the country's first National Military Strategy for Operations in Cyberspace. While the details of the strategy are expected to remain classified, the identification of cyberspace as distinct "domains of operations" equal to land, air, sea, and space, mark an acknowledgment of its importance to national military capabilities and national security. The strategy is expected to unify and expand the Computer Network Operations that are presently distributed among several separate commands (Joint Task Force—Computer Network Operations, 67th Network Warfare Wing, as well as dedicated resources of the National Security Agency and elsewhere). In December 2006, the U.S. Air Force announced the establishment of the U.S. Cyberspace Command (formally becoming the 8th Air Force), which is expected to become the global force provider for all U.S. cyberspace operations and will include both offensive and defensive Computer Network Operations. The formal announcement of this capability is expected to accelerate the emergence of similar capabilities among other military powers. Already, both China and the Russian Federation have declared doctrines for pursuing cyberspace operations. China's doctrine of "Integrated Network Electronic Warfare," for example, considers computer network attacks as essential to developing a first strike capability. Special units consisting of reservists drawn from among China's research and computational elite have been formed within the PLA, and since 2006, these units have reportedly participated in large-scale exercises. The entry of the United States and major regional superpowers into cyberspace operations is likely to spur an arms race as military establishments worldwide seek to develop both offensive and defensive capabilities. The militarization of cyberspace will create further means for states to regulate and control national cyberspace and will likely lead to further restriction on both civil and dark networks. For an extended discussion, see Ronald Deibert and Rafal Rohozinski, *The Global Politics of Internet Securitization* (forthcoming, 2008).

Regional Overviews

Introduction to the **Regional Overviews**

Every country wishes to share in the prospective benefits of the Internet. However, there are no countries that are completely comfortable with the newfound freedoms of expression and access to information the Internet brings. As a result, there are few countries left in the world today that have not debated, planned, or implemented Internet filtering. In the following eight regional overviews, we provide broad summaries that exhibit the ways in which the countries within each region are grappling with the implications of Internet freedom and the challenges of regulating online content.

Three of the eight regional overviews—Asia, the Middle East and North Africa (MENA), and the Commonwealth of Independent States (CIS)—synthesize the findings of the technical tests and the background research carried out in these regions. These regional overviews present the results of the forty country studies in a greater context.

The other five regional overviews—United States/Canada, Europe, Latin America, sub-Saharan Africa, and Australia/New Zealand—are written to extend the coverage of the study beyond the forty countries in which we were able to test in the first year of this global filtering study. These overviews are based solely on background research and secondary sources; the OpenNet Initiative (ONI) did not carry out technical filtering tests in these regions with the exception of two countries—Ethiopia and Zimbabwe—in sub-Saharan Africa and one country—Venezuela—in Latin America. Although these overviews fall short of a truly comprehensive global view of Internet filtering, we believe that they cover the major issues and trends as of spring 2007.

In general, the regional overviews are structured to cover the targets of and approaches to Internet content regulation, though the individual composition of the eight overviews varies in accordance with the quantity, focus, and strategies of regulation and filtering employed by the countries within a given region. As ONI continues to investigate and document Internet filtering in future years, we expect to expand our regional coverage to include more countries.

The overviews for Asia, MENA, and CIS exhibit considerable variation in filtering practices between and within those regions. This variation is seen not only in the depth, breadth, and foci of filtering, but also in the legal, technical, and administrative tools used to enact filtering. For example, the overview of Asia presents a region with a range of filtering targets and strategies as wide and diverse as its political and cultural landscape. The CIS overview displays a more narrow range of activity, reflecting perhaps the common history of the region. The MENA report evinces a region with extensive social filtering regimes and a growing penchant for targeting political speech.

By contrast, the general picture that emerges from Europe, the United States, Canada, Australia, and New Zealand is one of more narrowly focused targeting of online content and a more diverse mix of strategies for restricting access to that content. Filtering plays an important part in these regions and countries, though it tends to be voluntary and focused on a much narrower set of issues—primarily child pornography and, in a few cases, hate speech. The primary content regulation strategies in these countries tend to rely more heavily on taking down domestically hosted Web sites and in removing Web sites from search results than on the technical filtering of foreign-hosted Web sites. This is not surprising given the large proportion of total Internet content hosted on local servers in these regions. The targets of content restrictions vary by country. Within this set, Australia is the most aggressive toward combating obscene content, while the United States goes to the greatest effort to remove Web sites that are suspected of breaching copyright law. Germany and France are the most vigorous in addressing online hate speech.

Latin America generally shares the same complement of targets and strategies as documented in Europe, the United States, Canada, and Australia. However, the legal and administrative means for restricting access to content are not as advanced in Latin America as they are in these other countries and, therefore, the policy and practice of Internet blocking and content restrictions have not been applied as widely. As the legal structures and technical tools are further developed in the next several years, we may see a marked change in content regulation in Latin America.

Finally, sub-Saharan Africa has implemented the lowest level of regulatory restrictions on content of any region to date. One country, Ethiopia, has a systematic filtering regime, while Uganda has one reported incidence of filtering. In Africa the obstacles to viewing and posting content online are based on infrastructure and economics—few people have access to the Internet. This region is another in which we expect to see increased content regulation activity in the future, particularly as Internet access expands.

In the regional overviews that follow, ONI presents information on the current ways that regions approach Internet filtering and content restrictions. These summaries in turn provide a context for the forty specific country summaries addressed by ONI in this first report.

Internet Filtering in **Asia**

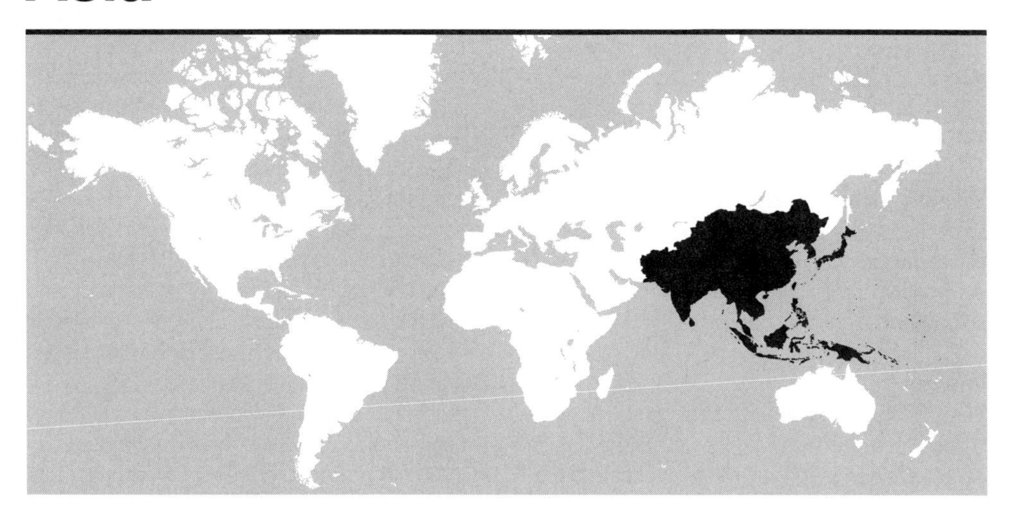

Overview

It is not surprising that Asia, a region with extraordinary cultural, social, and political diversity, is home to a broad range of approaches, policies, and practices toward Internet censorship.

ONI conducted in-country testing in Afghanistan, China, India, Malaysia, Myanmar (Burma), Nepal, Pakistan, Singapore, South Korea, Thailand, and Vietnam. Afghanistan, Malaysia, and Nepal do not use technical filtering to implement their policies on information control, but China, Myanmar, and Vietnam rely heavily on pervasive filtering as a central platform for shaping public knowledge, participation, and expression. The filtering practices of Thailand and Pakistan are more targeted, as ONI testing indicated that they blocked a substantial number of sites across categories of content considered sensitive or illicit. The remaining countries in Asia tested by ONI filtered on a selective basis and on targeted topics, including India (ethnic and religious conflict), South Korea (sites containing North Korean propaganda or promoting the reunification of North and South Korea), and Singapore (pornography).

Of countries filtering political content, China, Myanmar, and Vietnam blocked with the greatest breadth and depth, spanning human rights issues, reform and opposition activities, independent media and news, and discrimination against ethnic and religious minorities. Thailand and Pakistan blocked political content to a much more limited degree than China, Myanmar, or Vietnam.

A narrower range of social content was blocked in Asian countries. Many countries, including Vietnam, cited obscene content as a major justification for engaging in technical filtering. Singapore, Thailand, China, Pakistan, and Myanmar actually blocked pornographic content to varying degrees. Pakistan filtered a number of sites posting Danish cartoon images of the Prophet Muhammad widely condemned as blasphemous, while India also blocked a limited number of sites providing extreme viewpoints on religion. South Korea and Thailand filtered a small selection of gambling sites.

Conflict and security blocking was carried out by Myanmar, China, South Korea, India, Pakistan, and Thailand most frequently in regard to groups or movements implicated in "secessionist" or pro-independence activities, or in regard to disputed territories and border conflicts.

Myanmar, China, Vietnam, Thailand, and Singapore filtered Internet tools, including free Web-based e-mail providers, blog hosting services, and more frequently proxies and other circumvention tools. South Korea blocked pirated software on a nominal basis.

Internet in Asia

Some of the most and least connected countries in the world are located in Asia: Japan, South Korea, and Singapore all have Internet penetration rates of over 65 percent, while Afghanistan, Myanmar, and Nepal remain three of thirty countries with less than 1 percent of its citizens online.[1]

Among the countries in the world with the most restricted access, North Korea allows only a small community of elites and foreigners online. Most users must rely on Chinese service providers for connectivity, while the limited number of North Korean–sponsored Web sites are hosted abroad.

Even with an Internet penetration rate of only 10 percent, China was host to 137 million Internet users at the end of 2006.[2] The Chinese government predicted that within two years China would overtake the United States in becoming the country with the greatest number of Internet users worldwide.[3] Similarly, though India's Internet community is the fifth largest in the world, users amounted to only about 4 percent of the country's population in 2005.[4]

Afghanistan, Myanmar, and Nepal are among the world's least-developed countries. Despite the constraints on resources and serious developmental and political challenges, however, citizens are showing steadily increasing demand for Internet services such as Voice-over Internet Protocol (VoIP), blogging, and chat. The Internet market in Nepal is growing rapidly as a result of a competitive Internet service provider (ISP) market and low Internet access prices.[5] In Afghanistan, the Internet and information communications technology (ICT) have been identified as important sources of growth and development, with the potential to create opportunities for disadvantaged groups such as women, whose literacy rate stands at least 10 percent lower than the overall adult literacy rate of 28 percent. [6]

The range in access to broadband and high-speed Internet in Asia on a national basis is also staggering. South Korea has the highest rate of Internet penetration in the world: more than 89 percent of South Korean households had Internet access, and 75 percent of them used broadband in 2005.[7] As a result of heavy investment in its broadband infrastructure following the Asian financial crisis in the late 1990s, South Korea provides its citizens with a national network that carries data at speeds up to 50 Mb/s.[8] In 2005 the number of Internet users in Singapore reached 2.42 million, or 67 percent of the population,[9] one of the highest Internet penetration rates in the world. Singapore became the "first fully connected country in the world" by acting on a commonly held belief that the integration of technology is essential to achieving economic growth.[10] Home access is commonplace, with residential dialup and broadband subscriptions totaling more than 2.1 million.[11] Although Thailand has a penetration rate of 19 percent, homes and businesses in Bangkok and other major cities account for most of the connectivity;[12] only around 15 percent of schools in 2004 had access the Internet, and broadband access for households is at less than 2 percent penetration.[13] In Pakistan, broadband and high-speed Internet is accessible only to wealthier individuals or businesses: the majority of home Internet users are connected by modem, while cybercafés tend to split one modem or DSL connection over many computers, reducing connection speed.

Regional, language, and ethnic differences also impact access to Internet services and ICT infrastructure, and frequently reflect other disparities in national development priorities and resource allocation. China's longstanding policy of extracting taxes and other resources from rural areas to fund coastal development has resulted not only in alarming rural-urban income disparities, but also has contributed to a growing digital divide: while a quarter or more of residents in major cities such as Tianjin are online, in poorer and western provinces the rate is usually less than 10 percent.[14] Access in India is gradually expanding from the eight most heavily populated urban centers, where 41 percent of users are concentrated, to small cities and towns.[15] Since 71 percent of the population lives in rural areas, and since the gap between rural and urban teledensity is increasing, the majority of Indians are shut out of the Internet.[16] In Thailand and Vietnam it is believed that Internet use will increase as content (including search engines) becomes available in local languages rather than English.

User-generated content and media, which has ballooned in the scale and the scope of its influence, continued to shape—and in many cases redefine—the dissemination and generation of information in many Asian countries. In media climates where news publications are frequently owned by the state or controlled by business interests with close ties to ruling parties, bloggers and other independent content providers are becoming an increasingly trusted source of news, and in many cases have broken stories that are picked up by mainstream media. At the same time, the popularity of blogs and portals discussing political issues and reform in countries such as Malaysia and China indicate that citizen-generated content is filling an important information deficit in highly controlled media environments.

For example, despite the government's requirement that "persistently" political blogs and Web sites would be required to register and then abstain from engaging in election campaigning in the run-up to the 2006 general elections in Singapore, "citizen media" uploaded footage of opposition party rallies taken with handheld video cameras and cell phones to media-sharing sites such as YouTube and Google Video. This participation marks a departure from perceptions that the vast majority of Singaporeans "do not consider the Internet to [sic] useful for political engagement and civic participation."[17] Although very few Nepalis have access to the Internet, it has nevertheless become an important source of independent news in Nepal.[18] When King Gyanendra assumed authoritarian control in 2005, Nepali bloggers became an important political voice and source of information to the world about the situation unfolding inside the country, as traditional media were either shut down or heavily censored.[19] In a study of MSN users in Hong Kong, India, Malaysia, Singapore, South Korea, Taiwan, and Thailand, 41 percent of bloggers were "active" (spending three hours or more each week blogging), and with the exceptions of India and South Korea, a majority of bloggers in these countries were women.[20] Online citizens' media has played an important role in South Korean politics and Internet culture in recent years, led by ohmynews.com, a Seoul-based online newspaper that publishes articles mostly written by 50,000 citizen journalists and is considered the most influential news source in South Korea.[21] OhmyNews has been widely acknowledged as strongly influencing the 2002 election of Korean President Roh Moo-hyun.[22]

Age is an essential demographic factor to consider when tracking trends in Internet use and growth in Asia. In Vietnam—which has an Internet penetration rate of 17 percent, where more than half of the population is under thirty, and where a significant portion of individual users use cybercafés for online gaming and access to the Internet—control over these venues is an important priority for the state.[23] In China, eighteen- to twenty-four-year-olds comprise over 35 percent of all Internet users.[24] Over 70 percent of the 20.8 million bloggers (of which around 3.15 million were active) are under thirty.[25] In Thailand those under twenty-five years account for over half of Thai users.[26] A Windows Live Spaces report on a thriving blogging community in India, estimated at 14 percent of Internet users, found that a vast majority of bloggers are men under the age of thirty-five, which conforms to the demographic snapshot of Indian Internet users as predominantly male, middle class, and young.[27]

Legal and regulatory frameworks

Each of the countries that practice pervasive filtering in the region have issued ambitious regulations that aim to bring Internet users under government supervision and control, even if the feasibility of such oversight remains in doubt. Myanmar, China, and Vietnam engage in constant, unremitting supervision of and interference with other forms of media. Well-established strategies include the shuttering of reformist newspapers and Web sites, the institutionalized supervision over content, and the intimidation and harassment of dissidents, journalists, and human rights activists.

In the regulation of cyberspace, the corresponding phenomenon is the delegation of policing and monitoring responsibilities to ISPs, content providers, private corporations, and users themselves. These frameworks are not structured to accommodate only voluntary self-regulation along industry lines, but rather they exact compliance with state-imposed requirements through the looming threat of shutdowns, loss of license, fines, job dismissals, and even criminal liability. Vietnam and China apply a more direct form of censorship through the detention of cyberdissidents, while in Pakistan the Supreme Court authorized the police to register criminal cases against publishers of content blaspheming the Prophet Muhammad, even though no one was apprehended. Although there are no known cases of individuals detained for merely viewing prohibited online content, there are scores of journalists, writers, and activists who have been imprisoned on the basis of publishing criticism of government policies on the Internet, even in the form of song lyrics or discussion of political reform over VoIP.[28] The hidden, cumulative cost of these tiered and overlapping controls is self-censorship and a chilling effect that pervades all speech.

China can point to dozens of regulations that systematically proscribe nine to eleven types of illegal content, and the number of regulations is growing. The government has imposed new regulations to keep pace with, and even anticipate, the explosion of online video sharing, blogging, and other Web 2.0 platforms, proposing real-name registration requirements for bloggers and national regulations for online video and short film content.[29] With the support of a legal framework where even unemployment rates and certain family planning statistics are state secrets, the central propaganda organ issues instructions throughout the government hierarchy to media organizations, hosts such as BBS and blog platforms, and other content providers to suppress discussion of an ever-expanding list of proscribed topics.

Whether or not a legal basis for filtering is implicit in content regulations, in many Asian countries filtering has proceeded despite the lack of clear authority to do so. This includes countries with established democratic systems and protections for the press and other forms of speech. For example, while India is in the process of centralizing its filtering at the international gateway level and therefore improving its efficacy, many still question whether its primary legal authorization for filtering, the 2000 IT Act, is valid in light of constitutional requirements for limits to freedom of expression.

In Thailand, where human rights protections and press freedom have deteriorated in recent years, the legal authority for filtering is not clear. Indeed, the practice of filtering may in fact contradict protections in the 1997 Constitution that guarantee Thai citizens the rights to express opinions, to communicate by "lawful" means, and to access information.[30] The first military coup in fifteen years, in September 2006, amplified the uncertainty over the legitimacy of government policy, particularly through the declaration of martial law that precipitated claims of increased filtering.[31] The new military government took controversial and unilateral measures such as abrogating the Thai Constitution and banning new political parties, but ONI testing revealed that the post-coup content targeted for filtering was generally continuous with the filtering regime established by former Prime Minister Thaksin Shinawatra's government.

Defamation laws

A popular tool for silencing critics in countries such as Singapore, Malaysia and China, defamation laws and other forms of civil and criminal liability have begun to be applied to compel independent news sources, bloggers, and others to remove or retract online content.

In Singapore, defamation suits levy civil liability and heavy damages on independent and critical voices, from opposition party politicians to regional publications with domestic circulation.[32] Thai journalists and other critics of former Prime Minister Thaksin Shinawatra's ruling party had been similarly targeted, in line with a well-established precedent for using defamation suits to silence those fighting corruption.[33] Individuals perceived as criticizing the King, an act of lèse majesté, can be found liable under both defamation laws and the criminal code. In Malaysia, the first defamation suits against bloggers were inaugurated in January 2007, where the *New Straits Times* paper and several of its executives sued Jeff Ooi (www.jeffooi.com) and Ahirudin Attan (www.rockybru.blogspot.com) simultaneously for both blog posts and reader comments critical of their coverage.[34]

Implementation of filtering

However, even where legal authority for technical filtering and other forms of Internet censorship has been clearly established, filtering remains a contested practice.

At times an important source of conflict between users and government is the clumsy execution of imprecise methods, leading to a much broader scope of filtering than what was authorized. This was the case in Pakistan in February 2006, where a strong public outcry to "blasphemous" Danish cartoons depicting the Prophet Muhammad contributed to the blocking of twelve sites posting the images. The initial blockage quickly mushroomed into a mandate to filter all blasphemous content and resulted in the collateral blocking of the Blogspot domain for most of 2006, a consequence of the use of IP blocking. In India, the collateral blocking of Web sites occurred in response to CERT-IN orders in August 2003 and July 2006,[35] where ISPs in both incidents cut off access to parent Web sites including Google's blogspot.com, typepad.com, and Yahoo!'s geocities.com. One exception to the elastic filtering frequently encountered in Asian countries is North Korea, where access to online content is limited to the few dozen Web sites in Kwangmyong, the nation's domestic intranet.

In the implementation of technical filtering, the content blocked also frequently departs from pre-established or publicly acknowledged targets. For example, despite its putative focus on cleansing the Web of "harmful" social content such as obscenity,[36] the South Korean government uses its authority to define "harmful" content to focus on pro-North Korean or pro-reunification material. ONI testing found very little blocking of sensitive social content. The variations in filtering, if not the type of content blocked, between the two state-owned ISPs in Myanmar were surprising given the government's lockdown on information and all forms of media. India's IT Act, cited as the authority for the creation of the filtering certification body CERT-IN, prohibits only the publication of obscene content; however, CERT-IN has used its authority to issue blocking instructions against religious and ethnic inflammatory content.

Nonstate actors—especially ISPs—are deputized not only to shoulder monitoring duties and legal responsibility for unauthorized online behavior, but in many Asian countries they are relied upon to implement technical filtering. Countries that filter effectively at the international gateway were the exception in Asia. The two main state-owned ISPs in China each control a backbone network, and filtering was remarkably consistent between them. Blocking between South Korean ISPs was extremely consistent, even though both Korea Telecom (KorNet) and Hanaro Telecom (HanaNet) are publicly held corporations. In most other countries, the implementation of filtering primarily at the "margins" of state action has led to significant disparities across ISPs (especially in a crowded market), potential overblocking, and other inconsistencies such that users in the same country experience their "right" to access information differently and are able ultimately to view and interact with different portions of the Internet.

Additionally, the scope of a state's legal regulation of online activity often belies an implementation of state policy that is not entirely monolithic. Myanmar, China, Vietnam, India, and Pakistan all have regulations requiring cybercafés to monitor the online activities of their users and demand personal identity information. Generally these policies are difficult to enforce. For example, though the Myanmar government requires that users be registered and that screenshots of their activity be taken every five minutes, cafés do not always comply and CDs of screenshots are requested only sporadically.[37]

Impact of economic and social factors

Economic incentives and social factors have a definite impact on filtering practices in Asia. Global industry players such as Google and Microsoft have engaged in censorship in order to benefit from state investment in providing improved speed and quality of access to approved content while strengthening technical filtering, particularly in China.

On June 29, 2006, India's Department of Telecommunications (DOT) reportedly instructed around 150 ISPs to block the Web site of the People's War Group (PWG), a Maoist paramilitary group that was hosted on Geocities. A month later, the DOT informed ISPs that Yahoo! had removed the PWG's site, apparently the first time a service provider had voluntarily removed a Web site to avoid being blocked.[38]

The Chinese and Vietnamese governments must contend with the challenge of maintaining control over specified corridors of information as the space for approved or harmless topics grows increasingly vast. Myanmar has taken a blunter approach than its authoritarian neighbors, and in the case of China, its stalwart ally and aid provider. Internet access in Myanmar is structured so that broadband costs are prohibitive for most of its citizens, and dialup access comes bundled with state-monitored, fee-based e-mail service and a small collection of pre-approved sites on the country's intranet. In a reported attempt to not only censor communications but also preserve its monopoly over telephone and e-mail services as MPT's revenues dipped, the government blocked free e-mail services at points in 2006.[39] ONI testing confirmed that Yahoo! Mail, Gmail, Hushmail, and mail2web were blocked, with the ISP Myanmar Posts and Telecom taking the additional precaution of blocking thirteen additional e-mail sites, including Hotmail and Fastmail. Similar concerns about loss of state revenue have factored into similar tightening of VoIP services in Pakistan and China.

Economic motivations may also work to achieve an opposite effect, where governments explicitly refrain from Internet censorship in order to encourage growth. A number of Asian countries known for effective and sophisticated systems of information control, such as Singapore and Malaysia, demonstrated surprisingly low levels of filtering. For these governments, the strength of their historical interventions in freedom of the press and free speech may partially obviate the need for rigorous filtering.

In contrast to its approach toward other forms of media, the official policy of Singapore's Media Development Authority (MDA) has been to apply a "light-touch" regulatory framework to the Internet, promoting responsible use while giving industry players "maximum flexibility." Thus Singapore filters content on a symbolic scale, but also relies on established controls over print and broadcast media to set a precedent for citizens' online behavior, leading one scholar to call media regulation a "dual regulatory regime."[40] Greater control of cyberspace may be obviated by existing restrictive laws, political ties to the judiciary, and ownership and intimidation of the media that are already used to suppress dissenting opinion and opposition to the ruling People's Action Party (PAP).[41] Taken together, these economic and legal controls contribute to a climate of pervasive self-censorship of political commentary.

In Malaysia the state pledges not to censor Internet content in its "Bill of Guarantees" to companies approved for its Multimedia Super Corridor (MSC), a high-tech business center and communications infrastructure designed to help the country become an international information technology leader.[42] However, rather than filter content, Malaysia's Communications and Multimedia Act (CMA) targets "indecent" and "offensive" online content by subjecting publishers and authors to civil and/or criminal liability. Internet content publishers in Malaysia operate under the constant risk that the CMA and numerous other laws regulating speech and content on traditional media will be interpreted or amended to extend to Internet publications.[43] Notably, the bloggers Jeff Ooi and Ahirudin Attan were targeted under defamation laws and not the regulatory framework for online speech, which delineates fines and criminal penalties for persons using a content applications service to provide content that is "indecent, obscene, false, menacing, or offensive in character."[44]

In Afghanistan and Nepal serious political and economic challenges have perhaps made technical filtering impracticable, but this does not mean their citizens have unfettered access to information and communication via the Internet. Political instability has affected not merely the quality of access, but also the question of access altogether. The Internet in Afghanistan was banned altogether by the Taliban in July 2001—primarily because it was thought to broadcast obscene, immoral, and anti-Islam material, and the few Internet users at the time could not be easily monitored because they obtained their phone lines from Pakistan.[45] In 2005, citing deteriorating security conditions in Nepal due to Maoist violence, the Nepali king imposed authoritarian rule and a week-long media blackout and cut off all Internet access in the country.

In countries whose governments consider free access to information and unrestricted freedom of expression to be threats to social stability and public order, filtering is overwhelmingly targeted at local language content and country-specific issues.

China, Myanmar, and Vietnam filter a significant portion of content addressing their own human rights record and practices. Both China Netcom and China Telecom chose to block only one of the major international news organizations tested—the BBC—but they denied their users access to a significant number of overseas Chinese-language media representing different positions on the political spectrum. News in languages spoken by ethnic minorities in contested regions was also blocked in China. In Vietnam, sites only in English or French were rarely blocked, but sites in Vietnamese only tangentially or indirectly critical of the government—such as those with content focusing on local communities, world news, or voicing strong anti-Communist sentiments—were inaccessible. In Pakistan, though Balochi and Sindhi independence and human rights sites have been filtered, other Web sites pertaining to Pashtun secessionism were fully accessible. In this case, filtering may be seen as unnecessary, as the majority of Pashtuns are illiterate in their local language.

Transparency

All countries in Asia engaged in technical filtering exhibited a lack of transparency in the legal authorization, technical processes, or implementation of filtering. Governments chose to remain silent on their source of authority to filter content from its citizens, or relied on indirect or implicit authority found in existing laws and regulations. Citizens in many countries were not put on notice of filtering as it occurred, and instead the cause of content inaccessibility was identified to be the result of inadvertent or unintentional error. Virtually all governments in Asia have yet to develop procedures for official notification of blocking to Web site owners, or appeal mechanisms for individuals to challenge blocking decisions in independent tribunals.

Governments often fail to disclose the extent of filtering to the general public. One notable exception is Singapore: before most Asian countries even had infrastructure in place to begin engaging in technical censorship of the Internet, Singapore announced in 1999 that a list of 100 pornographic Web sites would be blocked by the three major ISPs. As a "gesture of concern" to demonstrate the government's commitment to "Asian" values,[46] this figure has been continually cited in coverage of Internet censorship in Singapore, though the extent of actual filtering has remained symbolic.

In South Korea, state regulation (through the Internet Content Filtering Ordinance in 2001)[47] reportedly required ISPs to block as many as 120,000 Web sites on a state-compiled list, as well as mandating that Internet access facilities accessible to minors, such as public libraries and schools, install filtering software.[48] However, ONI testing indicated that Internet filtering in South Korea is not as extensive as reports have suggested. In Thailand as well, the distinct lack of transparency in the filtering process has persisted through the change in Thai governments. Adding to the uncertainty, a range of figures for the number of sites blocked by the Thai government continues to be circulated but not confirmed. A Thai police Web site citing the number of blocked sites at over 34,000 sites since 2002 has been taken down.

However, both South Korean and Thai ISPs do employ a blockpage, the most transparent notification of filtering. For example, sites blocked by KorNet through DNS tampering in South Korea resolve to a blockpage hosted by the police at 211.253.9.250. This blockpage not only states that the page has been lawfully blocked but also displays the user's IP address, which suggests the possibility of tracking the viewers that have visited the blocked site. Only a few countries provide clear notice that access is being denied because of proscribed content. In contrast, China's filtering methods are set up so that users in China who cannot access content due to IP address blocking, DNS tampering, or keyword search string filtering receive a network timeout or error page. When a keyword block is triggered, further requests made to the target site (IP address) are blocked (including attempts to access otherwise permissible sites) for a variable period ranging from five to thirty minutes.

Civil society mobilization

In Asia, civil society groups (as characterized/defined by Deibert and Rohozinski in Chapter 6) have carved out a prominent role in monitoring Internet censorship, advocating for greater access to information and freedom of expression and creating a space for issues shut out or marginalized in mainstream discourse. In response to the collateral blocking of the entire Blogspot domain in India and Pakistan, bloggers and other loosely based coalitions mobilized quickly to generate media attention to the blocking and called for government transparency in the process.

Vigilant monitoring and advocacy by civil society activists in India, rather than government disclosure, has contributed to a greater understanding of technical filtering processes there. For example, the July 13, 2006, notice to block seventeen Web sites in the wake of the Mumbai train bombings issued by CERT-IN was not reproduced on any official government Web site but scanned and posted on an individual's blog. A number of individuals have filed requests seeking greater disclosure about the criteria and authorization for filtering under the 2005 Right to Information Act, but information has not been forthcoming.

Similarly, in Pakistan, the government never provided an official declaration confirming the blanket block on Blogspot.com since March 2006 or the rationale for it. Rather, the investigation and awareness-building around the controversial overblocking was initiated by two individuals through their "Don't Block The Blog" campaign. In the months after two Malaysians became the first bloggers to be sued for defamation in January 2007, bloggers in Malaysia and around the region formed the protest campaign "Bloggers United, No Fear," organized a legal defense fund trusteed by Datin Paduka Marina Mahathir, the daughter of former Prime Minister Mahathir Mohamad, and initiated a boycott of the plaintiff New Straits Times.

Many countries in Asia have achieved highly restricted media environments, and the Internet has become a tool for savvy civil society activists who must operate in them. Commonly the organization and resources required to shut down the Internet as an alternative medium of communication are far more expensive than the requirements of transmitting information online. For example, for many months in 2005 Radio Free Asia had been reporting about villagers in Shanwei in China's Guangdong province. These villagers had been protesting the construction of a wind power plant that would threaten their livelihoods and provide them with inadequate compensation for their expropriated land. After news about the police shooting and killing several Shanwei villagers broke in December 2005, the government attempted to suppress information about the incident by shutting down cybercafés in neighboring areas, cutting off Internet access to residents, stopping queries for the town's name on search engines, and erasing blogs mentioning the incident as soon as they were posted.[49] In spite of a lockdown in the area, a rights defense group was able to conduct an investigation into the incident and post it online.[50]

Conclusion

Notwithstanding a diverse range of approaches to Internet censorship, most of the governments in Asia where ONI conducted in-country testing are expanding their mandate to filter sensitive content, both technically and through "soft controls" such as legal regulation and delegated liability. Technical filtering is far from refined in most Asian countries, but is becoming an increasingly important tool in an armament of possible controls on free expression and the flow of information. It is also most clearly demarcated along national lines rather than using any regional or categorical formula. Accordingly many Asian governments focus overwhelmingly on content relating to sensitive political information and in local languages. Although filtering has been adopted as state policy for many Asian countries, the practices and implications of filtering continue to be contested.

Author: Stephanie Wang

NOTES

1. ICT Statistics, International Telecommunications Union, http://www.itu.int/ITU-D/ict/statistics/ict/index.html.
2. China Internet Network Information Center Report, Nineteenth Statistical Report on the Development of the Internet in China, (*di shijiu zhongguo hulian wangluo fazhan zhuangkuang tongji baogao*), issued January 23, 2007.
3. Ibid.
4. Paul Budde Communication Pty Ltd., India-Key Statistics and Telecommunications Market Overview, July 30, 2006, p. 2.
5. Paul Budde Communication Pty Ltd., Nepal: Telecoms Market Overview and Statistics, July 30, 2006, p. 11.
6. World Bank, World Development Indicators (2006), http://devdata.worldbank.org/external/CPProfile.asp?PTYPE=CP&CCODE=AFG; *Vision 2020: Islamic Republic of Afghanistan Millennium Development Goals Report 2005,* p. 123, p. 21
7. International Telecommunication Union, *World Telecommunication Indicators 2006.*
8. See Kristin Kalning, "Forget reality TV. In Korea, online gaming is it," MSNBC.com, February 21, 2007, http://www.msnbc.msn.com/id/17175353/.
9. Internet World Stats, Singapore, http://www.internetworldstats.com/asia.htm.
10. Terence Lee, "Internet control and auto-regulation in Singapore," Surveillance & Society 3(1): 74–95, http://www.surveillance-and-society.org/articles3(1)/singapore.pdf.
11. InfoComm Development Authority (IDA), Statistics on Telecom Services for 2006 (July–December), http://www.ida.gov.sg/Publications/20061205181639.aspx (listing 1,489,500 residential dialup subscriptions and 657,900 residential broadband subscriptions as of October 2006).
12. 2005 Information and Communication Technology Survey. National Statistical Office: Thailand, http://web.nso.go.th/eng/en/stat/ict/ict05_rep.pdf. See also Paul Budde Communication Pty Ltd., Thailand-Internet, 2006, p. 1.
13. Paul Budde Communication Pty Ltd., Telecommunication Sector Snapshot: Thailand, 2006.
14. China Internet Network Information Center Report, Nineteenth Statistical Report on the Development of the Internet in China, (*di shijiu zhongguo hulian wangluo fazhan zhuangkuang tongji baogao*),issued January 23, 2007.
15. Internet and Mobile Association of India, Internet in India:2006, http://www.iamai.in/research_index.php3.
16. Telecom Regulatory Authority of India, Indian Telecom Services Performance Indicators April–June 2006, October 2006, p. 40, http://www.trai.gov.in/Reports_content.asp?id=29;
17. Terence Lee, "Internet control and auto-regulation in Singapore," Surveillance & Society 3(1): 74–95, http://www.surveillance-and-society.org/articles3(1)/singapore.pdf.
18. See Vincent Lim, "Blogging for Democracy in Nepal," AsiaMedia, April 13, 2006, http://www.asiamedia.ucla.edu/article.asp?parentid=43000.
19. See Mark Glaser, "Nepalese Bloggers, Journalists Defy Media Clampdown by King," Online Journalism Review, February 23, 2005, http://www.ojr.org/ojr/stories/050223glaser/.
20. http://advertising.microsoft.com/asia/NewsAndEvents/PressRelease.aspx?Adv_PressReleaseID=296.
21. See http://ohmynews.com; Jun Kwanwoo, "'Citizen journalism' wins hearts and minds," Dawn, March 30, 2007, http://www.asiamedia.ucla.edu/article.asp?parentid=66921.
22. See Christopher M. Schroeder, "Is this the Future of Journalism?" Newsweek, June 18, 2004, http://www.msnbc.msn.com/id/5240584/site/newsweek/.
23. Paul Budde Communication Pty Ltd., 2006, Vietnam:Internet, p. 10, July 30, 2006. See also John Boudreau, "Bay Area Entrepreneur Leads Way in Online Gaming in Vietnam," The Mercury News, January 17, 2007, http://www.mercurynews.com/mld/mercurynews/business/16475618.htm.
24. China Internet Network Information Center Report, Nineteenth Statistical Report on the Development of the Internet in China, (*di shijiu zhongguo hulian wangluo fazhan zhuangkuang tongji baogao*),issued January 23, 2007.
25. Xinhua News Agency, "China has 20.8 million bloggers," January 10, 2007.
26. National Electronics and Computer Technology Center (NECTEC), Thailand MICT Indicators 2005 (February 2005), http://iir.ngi.nectec.or.th/download/indicator2005.pdf.
27. The Press Trust of India, "Indians prefer good-old diary to blogs," November 27, 2006.
28. Deutsche Presse-Agentur, "Vietnam youths arrested over internet chats released after 9 months," August 16, 2006. See also Human Rights in China, Press Release, "Dissident writer Zhang Lin to be tried next week," June 15, 2005, http://www.hrichina.org/public/contents/press?revision_id=22932&item_id=22931.
29. Xinhua News Agency, "China to issue new regulations to censor online video programs," August 16, 2006.

30. See Article 19, Freedom of Expression and the Media in Thailand, December 2005, at 38. The 1997 Constitution has since been abrogated. Council for Democratic Reform, Announcement no. 3, September 17, 2006.
31. See Freedom Against Censorship Thailand Web site, http://facthai.wordpress.com, (accessed April 4, 2007).
32. U.S. Department of State, Country Reports on Human Rights Practices 2006: Singapore, at 1.e., 2.a., 2.d., 3, http://www.state.gov/g/drl/rls/hrrpt/2006/78790.htm.
33. Bangkok Post, "Defamation law being misused, seminar told," January 10, 2005, http://www.asiamedia.ucla.edu/article.asp?parentid=19333.
34. South China Morning Post, "Newspaper sues Internet bloggers for defamation," January 19, 2007, reprinted at http://www.asiamedia.ucla.edu/article.asp?parentid=61629.
35. In August 2003, CERT-IN issued an order to ISPs to block the mailing list kynhun on Yahoo! Groups belonging to the militant outfit Hynniewtrep National Liberation Council. See http://pib.nic.in/archieve/lreleng/lyr2003/rsep2003/22092003/r2209200314.html.
36. See Korea Internet Safety Commission, http://www.icec.or.kr/.
37. The Irrawaddy Online, "Something is better than nothing," April 2, 2004, http://www.irrawaddy.org/art/2004/april01.html.
38, See Shivam Vij, "The discreet charms of the nanny state," National Highway, October 2006, http://www.shivamvij.com/2006/10/the-discreet-charms-of-the-nanny-state.html.
39. Reporters Without Borders, "Internet increasingly resembles an Intranet as foreign services blocked," July 4, 2006, http://www.rsf.org/article.php3?id_article=18202; *The Irrawaddy,* "Junta blocks Google and Gmail," June 30, 2006, http://www.irrawaddy.org/aviewer.asp?a=5924.
40. See Cherian George, "One country, two systems; for how long?" http://singaporemedia.blogspot.com/2007/03/one-country-two-systems-for-how-long.html
41. U.S. Department of State, Country Reports on Human Rights Practices 2006: Singapore, at 2.a., http://www.state.gov/g/drl/rls/hrrpt/2006/78790.htm.
42. See Malaysia Multimedia Super Corridor Web site, http://www.msc.com.my/msc/rollout_status.asp.
43. See, for example, Star Online, "Government looking at gaps in Printing Act," July 27, 2006, http://www.thestar.com.my/news/story.asp?file=/2006/7/27/nation/14961817&sec=nation ("The Government will study if the Printing Presses and Publications Act should be amended to include the electronic media and the Internet media.")
44. Malaysian Communications Multimedia Act of 1998, § 211(1). See Reme Ahmad, "Case revives debate of freedom of speech versus the right to redress," *The Straits Times,* January 20, 2007.
45. BBC News, "Taleban outlaw Internet," July 13, 2001, http://news.bbc.co.uk/2/hi/south_asia/1437852.stm.
46. Terence Lee, "Internet control and auto-regulation in Singapore," Surveillance & Society 3(1): 74–95. http://www.surveillance-and-society.org/articles3(1)/singapore.pdf.
47. Electronic Frontiers Australia, Internet Censorship: Law & Policy Around the World (2002), http://www.efa.org.au/Issues/Censor/cens3.html#sk.
48. See, for example, Han Chae-yun and Yi Huso, "On-again and off-again: Korean on/off-line LGBTQ/Iban community blocked," The Sungkyun Times, September 2002, http://web.skku.edu/~sktimes/251/society.html.
49. Esther Pan, "China's angry peasants," Foreign Council on Foreign Relations, http://www.cfr.org/publication/9425/#4; International Freedom of Expression eXchange, "Beijing imposes new blackout on village shootings," http://www.ifex.org/alerts/layout/set/print/content/view/full/71219/.
50. See EastSouthWestNorth blog, at http://www.zonaeuropa.com/20060112_1.htm.

Internet Filtering in

Australia and New Zealand

Introduction

Australia maintains some of the most restrictive Internet policies of any Western nation, while its neighbor, New Zealand, is less rigorous in its Internet regulation. Without any explicit protection of free speech in its constitution,[1] the Australian government has used its "communications power" delineated in the constitution to regulate the availability of offensive content online,[2] endowing a government entity with the power to issue take-down notices for Internet content hosted within the country. A number of state and territorial governments in Australia have also passed legislation making the distribution of offensive material a criminal offense, as the constitution does not afford that power to the national government.[3]

The Australian government also promotes and finances an "opt-in" filtering program, in which Internet users voluntarily accept filtering software that blocks offensive content hosted outside of the country. At present there are no plans for a countrywide Internet service provider (ISP)-level filtering regime, though Australia's handling of hate speech, copyright, defamation, and security signals the government's desire to increase the scope of its Internet regulation.

New Zealand by contrast is less strict in its Internet regulation. The government maintains a more limited definition of offensive content that can be investigated by a designated government entity, although—unlike in Australia—the definition includes hate speech (despite it being illegal in both countries). Furthermore, the government has not passed legislation to allow issuance of takedown notices for such content and its enforcement of Internet content regulation by prosecution almost solely focuses on child pornography. Although New Zealand Internet copyright policies have not yet been formalized, its defamation and security policies are fairly similar to Australia's.

Overall, however, Australia maintains a stricter regime of Internet censorship and regulation than New Zealand and much of the Western world, though not at the level of the more repressive governments that ONI has studied.

Offensive content

Australian and New Zealand approaches to offensive content on the Internet are somewhat similar in structure, in that they both rely on classifications systems and entities with the power to investigate online content. But their approaches are very different in terms of what is considered offensive and what is done about the offending content.

Australian laws relating to the censorship of offensive content are based on the powers delineated in and protections omitted from the Australian constitution. Section 51(v) of the document gives the Parliament power to "make laws for the peace, order, and good government of the Commonwealth with respect to: (v) postal, telegraphic, telephonic, and other like services."[4] With no explicit constitutional protection of free speech, the Australian government has invoked its "communications power" to institute a restrictive regime of Internet content regulation.

The Broadcasting Services Amendment (Online Services) Bill 1999, an amendment to the Broadcasting Services Act 1992, establishes the authority of the Australian Communications and Media Authority (ACMA)[5] to regulate Internet content. The ACMA is empowered to look into complaints from Australians about offensive content on the Internet and issue takedown notices. The ACMA is not mandated to scour the Internet for potentially prohibited content, but it is allowed to begin investigations without an outside complaint.[6]

Web content that is hosted in Australia may be removed by the ACMA if the Office of Film and Literature Classification finds that it falls within certain categories as defined by the Commonwealth Classification (Publications, Films and Computer Games) Act 1995, a cooperative classification system agreed to by the national, state, and territorial governments.

The levels and definitions of prohibited content are as follows:

- R18—Contains content that is likely to be disturbing to those under eighteen. This content is not prohibited on domestic hosting sites if there is an age-verification system certified by the ACMA in place.
- X18—Contains nonviolent sexually explicit content between consenting adults. This content may be subject to ACMA takedown provisions if hosted on domestic servers.
- RC—Contains content that is Refused Classification (child pornography, fetish, detailed instruction on crime, and so on)[7] and is prohibited on Australian-hosted sites.

The classification system chosen for Internet content is not the publications classification system but the more restrictive standard used for films. As a result, some content allowable offline is banned when brought online.[8]

Once the determination has been made that content hosted within Australia is prohibited, the ACMA issues a takedown notice to the Internet Content Host (ICH). It is not illegal for the ICH to host prohibited content, but legal action could be taken against it by the government if it does not comply with the take-down notice.

For offensive content hosted outside of Australia, the ACMA itself determines whether content is prohibited and notifies a list of certified Web-filter manufacturers to include the prohibited sites in their filters.[9] To obtain certification, these certified "Family Friendly Filters" must agree to keep lists of prohibited sites confidential.[10] ISPs are then required to offer a Family Friendly Filter to all of their customers, though customers are not required to accept them.[11] As a result, content taken down in Australia could be posted outside of the country and still be accessible to the majority of Australian Internet users. Electronic Frontiers Australia, a nonprofit group dedicated to protecting online freedoms, reports that at least one site taken down has moved to the United States, even keeping its URL and ".au" domain. It is not known how many sites have moved overseas in this fashion.[12]

States and territories have instituted a variety of laws that criminalize the downloading of illegal content and the distribution of content that is "objectionable" or "unsuitable for minors."[13] The state of Victoria, for example, in §57 of its Classification (Publication, Films and Computer Games) (Enforcement) Act 1995, makes it illegal to "use an on-line information service to publish or transmit, or make available for transmission, objectionable material."[14] There is not complete uniformity between the states, however. In Western Australia, for example, it is *not* illegal to distribute R18 and X18 to adults online (though the ACMA can still issue takedown notices), but the possession of *any* RC content (not just child pornography as is the case in other states) is illegal.[15]

Beyond its regulation of online content, the Commonwealth is implementing new Internet filtering initiatives. In June 2006 the Australian government announced an AU$116.6 million initiative called "Protecting Australian Families Online." Of this, AU$93.3 million will be spent over three years to provide all families with free Web filters, though they will still be optional. Further, the National Library of Australia is now required to use Web filters on all of its computers. All other libraries are to be provided with free Web filters and encouraged to use them on their computers as well.[16] Finally, perhaps in a nod to elements of the government—especially members of the Labor Party—pushing for a system like Cleanfeed in the United Kingdom, the government will be testing an ISP-level blocking system in Tasmania. The Minister for Communications, Information Technology and the Arts, Helen Coonan, however, remains opposed to implementing this system on a countrywide basis.[17]

In a related development, all mainland states in Australia recently banned access to YouTube over school networks because of a video uploaded depicting a seventeen-year-old Australian girl being abused, beaten up, and humiliated by a group of young people. Eight youths have been charged in connection with the assault.[18] The blocking has continued to worsen rifts between state schools and some nonstate schools, such as Melbourne Grammar School, which have chosen to protect free speech and allow unfiltered access to the Internet.[19]

New Zealand, on the other hand, does not have any government legislation directly regulating Internet content.[20] Officials have claimed, however, that the Films, Videos, and Publications Classification Act of 1993, which defines "objectionable" material, covers Internet materials as well.[21] Under the Act, any material that "describes, depicts, expresses, or otherwise deals with matters such as sex, horror, crime, cruelty, or violence in such a manner that the availability of the publication is likely to be injurious to the public good" is considered objectionable and is illegal to distribute or possess.[22] Specifically listed is any material that promotes or supports "the exploitation of children, or young persons, or both, for sexual purposes; or the use of violence or coercion to compel any person to participate in, or submit to, sexual conduct; or sexual conduct with or upon the body of a dead person; or the use of urine or excrement in association with degrading or dehumanising conduct or sexual conduct; or bestiality; or acts of torture or the infliction of extreme violence or extreme cruelty."[23] There is also a decision-making procedure described in the Act for any content that might be objectionable but does not fall within this specific list, including discriminatory and hateful material.[24] This law has formed the basis of the Department of Internal Affairs' (DIA) enforcement of Internet censorship in the country.

Like Australia's ACMA, the DIA "proactively" investigates potentially banned material[25] and submits any such material not already classified to the Office of Film and Literature Classification for a ruling.[26] This office then classifies the material as "unrestricted," "objectionable," or "objectionable" except in certain circumstances of restricted access or for "educational, professional, scientific, literary, artistic, or technical purposes."[27]

Unlike in Australia, however, there is no explicit legal mechanism for the take-down of objectionable material. Instead, the nonprofit InternetNZ is in the process of establishing an industrywide code of conduct that would require its signers to agree not to host illegal content.[28] As a result, the government focuses its efforts on prosecuting the distributors or possessors. The Films, Videos, and Publications Classifications Amendment Act 2005 sets the penalty for distributing objectionable material at a maximum of ten years in prison (up from a maximum of one year) and for knowingly possessing objectionable materials at a maximum of five years in prison or a NZ$50,000 fine.[29] According to various sources, the DIA has almost completely focused its enforcement of Internet censorship on child pornography.[30]

Hate speech

Both Australia and New Zealand have legislation addressing hate speech generally, and both have applied this legislation to the Internet through different means. New Zealand, however, has an institutionalized investigation system, while Australia does not.

Australia addresses hate speech through the Racial Discrimination Act 1975, which makes it "unlawful for a person to do an act, otherwise than in private, if: the act is reasonably likely, in all the circumstances, to offend, insult, humiliate or intimidate another person or a group of people; and the act is done because of the race, colour or national or ethnic origin of the other person, or of some or all of the people in the group."[31]

Australian courts applied this law to the Internet for the first time in October 2002 in the *Jones v. Toben* case. Jeremy Jones and the Executive Council of Australian Jewry brought a lawsuit against Frederick Toben, the director of the Adelaide Institute, because of material on Toben's Web site (www.adelaideinstitute.org) that denied the Holocaust. The Federal Court, ruling that publication on the Internet without password protection is a "public act," found that posting this material online was in direct violation of §18C of the Racial Discimination Act 1975 (quoted above) and called for the material to be removed from the Internet.[32]

Australia does not, however, give the ACMA authority to investigate complaints or issue takedown notices for hateful or racist materials online, even if they would be illegal under the Racial Discrimination Act 1975.[33] Schedule 5 of the Broadcast Services Act 1992 gives the ACMA authority only over materials deemed "offensive" within the classification scheme described earlier. As a result, there appears to be no venue other than the courts in which to pursue complaints about hateful or racist materials online. However, Chilling Effects reports that Google received notice on May 5, 2006, of a site in its search results that "allegedly violates section 18C of the *Racial Discrimination Act 1975*" and removed it from the Google Australia site (www.google.com.au).[34] This may be indicative of a new notice-based system taking form.

New Zealand, on the other hand, has both explicit prohibition of discrimination based on race, religion, age, disability, sexual orientation, and so on in §21(1) of the Human Rights Act 1993,[35] as well as explicit prohibition of the publication of material that "represents (whether directly or by implication) that members of any particular class of the public are inherently inferior to other members of the public by reason of any characteristic of members of that class, being a characteristic that is a prohibited ground of discrimination specified in §21(1) of the

Human Rights Act 1993"[36] in §3e of the Films, Videos and Publication Classifications Act 1993. The DIA uses these statutes to pursue investigations into potentially discriminatory material.

Copyright

Australia is applying copyright law to the Internet in a vigorous attempt to expand its role in limiting copyright infringement. New Zealand, on the other hand, is more slow-moving and has yet to enact legislation directly relevant to Internet copyright.

Australia's copyright laws underwent significant overhaul following the acceptance of the Australian-United States Free Trade Agreement in 2004. Pursuant to that agreement, Australia was required to bring its copyright laws closer in line with those of the United States.[37] Some of the relevant requirements included:

1. agreeing to World Intellectual Property Organization (WIPO) Internet treaties,
2. implementing an "expeditious" takedown system of copyright infringing materials,
3. strengthening control over copyright protection technology circumvention,
4. agreeing to copyright protection standards, and
5. increasing the length of copyright to life + seventy years from its previous level of life + fifty years.[38]

Most of these provisions were implemented in the US Free Trade Agreement Implementation Act 2004,[39] though new regulations in response to requirement (3) were recently implemented in the Copyrights Amendment Act 2006.[40]

After implementing a system of copyright more consistent with that of the United States, the Australian government decided to pursue another overhaul of its copyright laws in 2006 to, as ABC Science Online reports, "keep up with the rapidly changing digital landscape."[41] The proposed amendments to the Copyright Act 1968 were worrisome to many. Google argued that certain provisions would allow copyright owners to pursue legal action against it and other search engines for caching material without obtaining express permission from each site. This would "condemn the Australian public to the pre-Internet era," Google argued.[42] Other critics contended that the proposed amendments would make possession of an iPod or other music-listening device designed to play MP3s illegal, and uploading a video of yourself singing along to a pop song a crime.[43]

Although these two final concerns have been remedied in the resulting Copyrights Amendment Act 2006 (it is still legal to own an iPod and it is allowable to post a lip-synching video),[44] the caching issue still appears to be unresolved. There is an exception in the act that allows computer networks of educational institutions to cache copyright-protected online material "to facilitate efficient later access to the works and other subject-matter by users of the system."[45] However, this does not appear to offer the exception that Google sought.

Overall, though, the amendments allow for increased exceptions to the copyright laws to establish more realistic fair use of copyrighted material, such as "time-shifting, format-shifting and space-shifting" (e.g., recording a television show to watch later, scanning a book to view it electronically, and transferring material from CDs to iPods, respectively), and greater protection of parody and satire.[46]

The Australian judiciary has been active in setting precedents in copyright enforcement online. In a landmark decision in December 2006, the Federal Court upheld a lower court ruling that found the Web site operator of mp3s4free.net, Stephen Cooper, and the hosting ISP, E-Talk, liable for copyright infringement. Cooper's site did not itself host any copyright-protected material, but rather served as a search engine through which users could find and download copyright-protected music for free. In its ruling, the court found that merely linking to copyright-protected material was grounds for infringement. In addition, the court found that ISP E-Talk was also liable for copyright infringement because it posted advertisements on the site and was unwilling to take the site down.[47] Interestingly, Dale Clapperton of Electronic Frontiers Australia has argued that this decision could be used against search engines such as Google. In an article in the *Sydney Morning Herald,* he stated that "what Cooper was doing is basically the exact same thing that Google does, except Google acts as a search engine for every type of file, while this site only acts as a search engine for MP3 files."[48]

In New Zealand, there is no legislation in effect that explicitly relates copyright law to the Internet. Current New Zealand copyright law is contained within the Copyright Act 1994, which makes exceptions for time-shifting of television programs but none for format- or space-shifting of content. In addition, copyright is set at life + fifty years.[49]

The Copyright (New Technologies and Performers' Rights) Amendment Bill currently being considered in New Zealand, however, would dramatically change the digital copyright landscape into one that more closely mirrors the Digital Millennium Copyright Act (DMCA) of the United States. If passed, the bill would allow for format-shifting and space-shifting of music,[50] criminalize the distribution of the means to subvert technological protection measures protecting copyrighted content, and establish a system in which ISPs are required to remove copyright-infringing content and notify the poster if "[the ISP] obtains knowledge or becomes aware that the material is infringing."[51] This removal system is somewhat different from the U.S. system of notice-and-takedown in that it requires knowledge of infringement and not simply notification.[52]

Defamation

Through a variety of court cases, both Australia and New Zealand have applied their respective defamation laws to the Internet, and both countries, with New Zealand courts following the Australia courts' example, have controversially expanded their jurisdiction in defamation suits to online materials hosted outside of their borders.

Defamation in Australia, except for a small range of cases, is handled through state and territorial law.[53] And until December 2005, states and territories maintained largely nonuniform codes of defamation.[54] After what amounted to a threat that the Commonwealth would act if states and territories did not, the states and territories finally decided to enact uniform laws in December 2005.[55] Since defamation laws in Australia are applied where material is seen, read, or experienced, nonuniform laws meant that writers and publishers had to be wary of different sets of laws all over the country under which they might be sued under various definitions of defamation.[56] Now the laws are uniform, so this liability risk has been mitigated. No legislation specifically targets defamation on the Internet and, therefore, its regulation is essentially the same as that for all other publications.[57]

The judiciary has played an important role in setting online defamation policy because of jurisdictional issues. In a major decision in December 2002, the Australian High Court ruled that a party within Australia can sue a foreign party in Australian court for defamation resulting from an online article hosted on a foreign server. The specific case involved a lawsuit pitting Joseph Gutnick, an Australian businessman, against Dow Jones over a defamatory article written about him in Barron's Online in October 2000. Dow Jones argued that since its servers (and therefore the article) are in the United States, the defamation case should have been tried in the United States. A decision allowing the case to be tried in Australia, they argued, would restrict free speech around the world because it would require authors and publishers to take into account the laws of foreign countries under which they could be sued when publishing material online.[58]

The court countered, however, that the "spectre of 'global liability' should not be exaggerated. Apart from anything else, the costs and practicalities of bringing proceedings against a foreign publisher will usually be a sufficient impediment to discourage even the most intrepid of litigants. Further, in many cases of this kind, where the publisher is said to have no presence or assets in the jurisdiction, it may choose simply to ignore the proceedings. It may save its contest to the courts of its own jurisdiction until an attempt is later made to enforce there the judgment obtained in the foreign trial. It may do this especially if that judgment was secured by the application of laws, the enforcement of which would be regarded as unconstitutional or otherwise offensive to a different legal culture."[59] The parties eventually settled for AU$180,000 in damages and AU$400,000 in legal fees."[60]

New Zealand defamation law was first found to apply to online material in a District Court decision, *O'Brien v. Brown,* in late 2001. In the case, Patrick O'Brien, CEO of the New Zealand domain manager Domainz, sued Alan Brown, the head of a Manawatu ISP, for Brown's posting of harsh criticisms and calls for fraud investigation into Domainz on a publicly available Internet Society of New Zealand bulletin board.[61] The judge in the case found that the Internet afforded no additional freedom of expression to the defendant than any other medium and, further, that publication on the Internet required a *greater* award of damages than through another medium because of the ease with which Domainz's potential customers and clients could access the defamatory material.[62]

In addition the New Zealand courts have followed in Australia's example in determining the jurisdiction for defamation suits over online content hosted in a foreign country. Ironically enough, the relevant suit involved an Australian defendant. In 2004 the Wellington High Court found that the University of Newlands (based in New Zealand) could sue Nationwide News Ltd. (based in Australia) in New Zealand court for Nationwide's inclusion of the plaintiff in a list of "Wannabe Unis" and "degree mills" in its online newspaper, *The Australian.* This essentially eschewed the United States' rule of "single publication" and more closely aligned New Zealand defamation policy with Australia.[63]

Security

Both Australia and New Zealand have taken steps toward greater Internet security in their countries, passing laws to give government agencies greater authority to investigate illegal activities online.

Australia's Internet surveillance regime is primarily based on two laws. The first is the Telecommunications (Interception and Access) Act 1979. This act, amended in June 2006, prohibits intercepting telecommunications or accessing, without first notifying both the sender and the receiver, stored telecommunications by any person or entity, except in cases such as the installation or maintenance of telecommunications equipment.[64] It also establishes two warrant systems, controlled by the Attorney General, by which law enforcement may gain access to these communications: "telecommunications service warrants" (for real-time interception) and "stored communications warrants" (for access to stored communications without a requirement to notify the communicants).[65]

The second relevant law is the Surveillance Devices Act 2004, which significantly increases the authority of law enforcement to install surveillance devices such as key-stroke recorders under newly created "surveillance device warrants."[66] Electronic Frontiers Australia has expressed worry that these warrants will be used by law enforcement to avoid applying for a telecommunications service warrant, essentially allowing them to intercept communications where a telecommunications service warrant would not have been authorized.[67]

Further, in 2003 the Australian Internet Industry Association (IIA) attempted to establish a code of practice requiring ISP signatories to retain user information for six or twelve months and provide it to law enforcement upon official request. Specifically, personal data—such as name, address, and credit card details—were to be retained by ISPs for six months after a customer ends service with that ISP or twelve months after the record is created, whichever is longer. Operational data, such as proxy logs and email information, were to be kept for six months after creation of the data.[68] Law enforcement could request this information using the certificate system set up in the Telecommunications Act 1997,[69] which allows private information to be disclosed if "an authorised officer of a criminal law-enforcement agency has certified that the disclosure is reasonably necessary for the enforcement of the criminal law."[70] The code was skewered by privacy advocates,[71] and it is still listed as "not yet ratified" and "in public consultation" on the IIA's Web site, even though it was released four years ago.[72]

In New Zealand, the most relevant piece of legislation to Internet security is Supplemental Order Paper 85 to the Crimes Amendment Bill No. 6, passed in 2003. The act essentially makes it illegal to hack or intercept electronic communications, but exempts the Police, Security Intelligence Service, and the Government Communications Security Bureau acting under interception warrants as described by the Crimes Act 1961. As Keith Locke of the Green Party points out, however, these warrants "can be quite broad in their application and cover a class of people."[73]

Conclusion

Australian laws and policies toward the Internet are restrictive relative to similar Western countries, while New Zealand is less stringent. The Australian government has instituted a strict takedown regime for offensive content, and various states and territories have made distribution of said content a criminal offense. The government is pursuing voluntary programs to increase home filtration of the Internet, and Australia's evolving hate speech, copyright, defamation, and security policies offer further justification for restricting Internet content. So far, the government has resisted calls to implement ISP-level blocking of offensive content on a countrywide basis, though there is significant political backing to implement one.

New Zealand, on the other hand, has instituted a more limited classification system—though it does include hate speech—with no takedown notices and has not even formally adopted copyright legislation that applies to the Internet. Its broad defamation and security policies, however, are more reminiscent of Australia.

Overall, though, Australia's Internet censorship regime is strikingly severe relative to both its neighbor and similar Western states. It is not, however, at the level of the most repressive regimes that ONI has examined.

Author: Evan Croen

NOTES

1. Roy Jordan, "Free Speech and the Constitution," Parliamentary Library, June 4, 2002, http://www.aph.gov.au/LIBRARY/Pubs/RN/2001-02/02rn42.htm.
2. See Australian Constitution, §51(v), http://scaleplus.law.gov.au/html/pasteact/1/641/0/PA000700.htm.
3. Electronic Frontiers Australia, "Internet censorship laws in Australia," March 31, 2006, http://www.efa.org.au/Issues/Censor/cens1.html.
4. Australian Constitution, §51(v), http://scaleplus.law.gov.au/html/pasteact/1/641/0/PA000700.htm.

5. The Australian Communications and Media Authority was formed in July 2005, merging the Australian Broadcasting Authority and the Australian Communications Authority. See ACMA Overview, http://www.acma.gov.au/WEB/STANDARD//pc=ACMA_ORG_OVIEW.
6. Electronic Frontiers Australia, "Internet censorship laws in Australia," March 2006, http://www.efa.org.au/Issues/Censor/cens1.html.
7. Ibid.; Office of Film and Literature Classification, Guidelines for the Classification of Films and Computer Games, 2005, http://www.oflc.gov.au/resource.html?resource=62&filename=62.pdf.
8. Electronic Frontiers Australia, "Internet censorship laws in Australia," March 2006, http://www.efa.org.au/Issues/Censor/cens1.html.
9. Australian Communications and Media Authority, "Internet regulation," February 2007, http://www.acma.gov.au/web/STANDARD//pc%3DPC_90169.
10. Schedule 1, Codes for Industry Co-Regulation in Areas of Internet and Mobile Content (Pursuant to the Requirements of the Broadcasting Services Act 1992), May 2005, http://www.acma.gov.au/acmainterwr/aba/contentreg/codes/internet/documents/iia_code.pdf.
11. IIA Guide for ISPs, March 2006, http://www.iia.net.au/index.php?option=com_content&task=view&id=121&Itemid=33.
12. Electronic Frontiers Australia, "Internet censorship laws in Australia," March 31, 2006, http://www.efa.org.au/Issues/Censor/cens1.html.
13. Ibid.
14. Classification (Publications, Films and Computer Games) (Enforcement) Act 1995, §57, http://www.austlii.edu.au/au/legis/vic/consol_act/cfacga1995596/s57.html.
15. Electronic Frontiers Australia, "Internet censorship laws in Australia," March 31, 2006, http://www.efa.org.au/Issues/Censor/cens1.html.
16. Senator Helen Coonan, "$116.6 million to protect Australian families," June 2006, http://www.minister.dcita.gov.au/media/media_releases/$116.6_million_to_protect_australian_families_online.
17. Stephen Deare, "ISP level porn filtering won't work, says Coonan," June 2006, http://www.cnet.com.au/broadband/0,239036008,240063710,00.htm.
18. Stephen Hutcheon, "YouTube bans don't work: Internet founder," March 2007, http://www.stuff.co.nz/3986172a11275.html.
19. Ibid.
20. InternetNZ, "Internet governance in NZ," http://www.internetnz.net.nz/net-in-nz/governance.
21. Electronic Frontiers Australia, "Internet censorship: Law and policy around the world," March 28, 2002, http://www.efa.org.au/Issues/Censor/cens3.html#nz.
22. Films, Videos, and Publications Act 1993, §3, http://rangi.knowledge-basket.co.nz/gpacts/reprint/text/2005/se/042se3.html.
23. Ibid.
24. Ibid.
25. Department of Internal Affairs, Censorship and the Internet, November 2006, http://www.censorship.dia.govt.nz/diawebsite.nsf/wpg_URL/Services-Censorship-Compliance-Censorship-and-the-Internet?OpenDocument.
26. Department of Internal Affairs, Censorship Compliance, December 2006, http://www.dia.govt.nz/diawebsite.nsf/wpg_URL/Services-Censorship-Compliance-Index?OpenDocument.
27. Films, Videos, and Publications Act 1993, §23, http://rangi.knowledge-basket.co.nz/gpacts/public/text/1993/se/094se23.html.
28. InternetNZ, "ICOP," May 2006, http://www.internetnz.net.nz/issues/current-issues/ICOP.
29. Department of Internal Affairs, Amendment Act of 2005, April 2005, available at http://www.dia.govt.nz/diawebsite.nsf/wpg_URL/Services-Censorship-Compliance-Amendment-Act-2005?OpenDocument.
30. Keith Manch and David Wilson, "Objectionable material on the Internet: Developments in enforcement," 2003, www.netsafe.org.nz/Doc_Library/netsafepapers_manchwilson_objectionable.pdf; Electronic Frontiers Australia, "Internet censorship: Law and policy around the world," March 28, 2002, http://www.efa.org.au/Issues/Censor/cens3.html#nz.
31. Racial Discrimination Act 1975, §18C, http://austlii.law.uts.edu.au/au/legis/cth/consol_act/rda1975202/s18c.html.
32. Galexia, "Article: Jones v Toben: Racial discrimination on the Internet," Oct 2002, http://www.galexia.com/public/research/articles/research_articles-art22.html#fn357.
33. Australian Department of Communications, Information Technology and the Arts, "Racism and the Internet," November 2002, www.dcita.gov.au/__data/assets/word_doc/10892/Racism_and_the_Internet.doc.
34. Chilling Effects, "Google removal complaint: §18C of Australia's Racial Discrimination Act of 1975," May 2006, http://www.chillingeffects.org/international/notice.cgi?NoticeID=4266.
35. Human Rights Act 1993, §21(1), http://www.legislation.govt.nz/browse_vw.asp?content-set=pal_statutes.
36. Films, Videos and Publication Act 1993, §3e, http://rangi.knowledge-basket.co.nz/gpacts/reprint/text/2005/se/042se3.html.
37. Austrade, "The Australian-United States Free Trade Agreement in brief," http://www.fta.gov.au/ArticleDocuments/AUSFTA Client Brochure_Final_200606.pdf.aspx.

38. Australian Department of Foreign Affairs and Trade, Intellectual Property, http://www.dfat.gov.au/trade/negotiations/us_fta/outcomes/08_intellectual_property.html; Australian Department of Foreign Affairs and Trade, A Guide to the Agreement: Intellectual Property, http://www.dfat.gov.au/trade/negotiations/us_fta/guide/17.html.
39. Australian Copyright Council, Free Trade Agreement Amendments, January 2006, http://www.copyright.org.au/publications/G085.pdf.
40. Australian Department of Foreign Affairs and Trade, Intellectual Property, http://www.dfat.gov.au/trade/negotiations/us_fta/outcomes/08_intellectual_property.html.
41. Judy Skatssoon, "Google warns Aust copyright laws could cripple the Internet," ABC Science Online, November 7, 2006, http://www.abc.net.au/news/newsitems/200611/s1782921.htm.
42. Ibid.
43. Associated Press, "Proposed changes to Australian copyright laws could make iPod user into criminals," November 21, 2006, http://www.iht.com/articles/ap/2006/11/21/technology/AS_TEC_Australia_Copyright_Crime.php.
44. The Attorney-General, Copyright Amendment Bill 2006: Frequently Asked Questions, December 2006, http://www.ag.gov.au/agd/WWW/MinisterRuddockHome.nsf/Page/RWPC7B0742318EF6A58CA25723B008145FC.
45. Copyright Amendment Act 2006, http://www.comlaw.gov.au/ComLaw/Legislation/Act1.nsf/0/5A32BBA137EC7020CA2572440000E793/$file/1582006.pdf.
46. Australian Copyright Council, Copyright Amendment Act 2006, January 2007, http://www.copyright.org.au/g096.pdf.
47. Asher Moses, "Copyright ruling puts hyperlinking on notice," *Sydney Morning Herald,* December 19, 2006, http://www.smh.com.au/news/web/copyright-ruling-puts-linking-on-notice/2006/12/19/1166290520771.html.
48. Ibid.
49. Ministry of Economic Development, Copyright Protection in New Zealand, November 29, 2005, http://www.med.govt.nz/templates/Page____7290.aspx.
50. Copyright (New Technologies and Performers' Rights) Amendment, §44, http://www.parliament.nz/NR/rdonlyres/5A88D15B-C4A1-42C2-AE75-9200DD87F738/51071/DBHOH_BILL_7735_40199.pdf.
51. Judith Tizard, "Digital copyright bill: Questions & answers," Official Web site of New Zealand government, December 21, 2006, http://www.beehive.govt.nz/ViewDocument.aspx?DocumentID=28179.
52. Ibid.
53. Electronic Frontiers Australia, "Defamation laws and the Internet," January 14, 2006, http://www.efa.org.au/Issues/Censor/defamation.html#2006.
54. The Attorney-General, "Belated state defamation laws," December 15, 2005, http://www.ag.gov.au/agd/WWW/MinisterRuddockHome.nsf/Page/Media_Releases_2005_Fourth_Quarter_15_December_2005_-_Belated_State_defamation_laws_-_2372005.
55. Ibid.
56. Rhonda Breit, "Uniform defamation laws: A fresh start or the same chilling problems?" The Brisbane Institute, May 11, 2006, http://www.brisinst.org.au/resources/brisbane_institute_defamation.html.
57. Electronic Frontiers Australia, "Defamation laws and the Internet," January 14, 2006, http://www.efa.org.au/Issues/Censor/defamation.html.
58. OUT-LAW News, "Australia rules on where to sue for internet defamation," December 10, 2002, http://www.out-law.com/page-3184.
59. *Dow Jones & Company Inc. v. Gutnick,* December 10, 2002, http://www.austlii.edu.au/au/cases/cth/high_ct/2002/56.html.
60. Jack Goldsmith and Tim Wu, *Who Controls the Internet?: Illusions of a Borderless World.* New York: Oxford University Press, 2006, p. 148.
61. Caslon Analytics, "Brown, O'Brien, and Domainz," http://www.caslon.com.au/defamationprofile10.htm#brown, (accessed March 16, 2007); FindLaw, "Say No Evil: Defamation in Cyberspace," http://www.findlaw.com/12international/countries/nz/articles/852.html, (accessed March 16, 2007).
62. FindLaw, "Say No Evil: Defamation in Cyberspace," http://www.findlaw.com/12international/countries/nz/articles/852.html,(accessed March 16, 2007)
63. Sarah Harrison, "Overseas website caught by New Zealand defamation laws," AJ Park, November 1, 2004, http://www.ajpark.co.nz/library/2005/03/oseas_website_defamation_laws.php.
64. Telecommunications (Interceptions and Access) Act 1979, §7 and §108, http://www.austlii.edu.au/au/legis/cth/consol_act/taaa1979410/.
65. Electronic Frontiers Australia, "Telecommunications privacy laws," October 19, 2006, http://www.efa.org.au/Issues/Privacy/privacy-telec.html.
66. Ibid.
67. Electronic Frontiers Australia, "EFA comments on Surveillance Devices Bill 2004," May 18, 2004, http://www.aph.gov.au/senate/committee/legcon_ctte/completed_inquiries/2002-04/surveillance/submissions/sub8.pdf.
68. Internet Industry Association, Cybercrime Code of Practice, §7, September 2003, www.iia.net.au/cybercrime_code_v2.doc, (accessed March 16, 2007).
69. Ibid, §8.

70. Telecommunications Act 1997, §282 (3), http://www.comlaw.gov.au/comlaw/legislation/act-compilation1.nsf/0/D22A9.DC08193D6D8CA25726D007B99E4/$file/Tele1997_Version2_WD02.pdf
71. Steven Dreare, "Privacy advocates rip into ISP cyber-crime code," *PCWorld,* August 21, 2003, http://www.crime-research.org/news/2003/08/Mess2102.html.
72. Internet Industry Association, Cybercrime Code of Practice, http://www.iia.net.au/index.php?option=com_content&task=category§ion-id=3&id=22&Itemid=33, (accessed March 16, 2007).
73. Keith Locke, Fact Sheet on Government Plans for E-mail Snooping and Computer Hacking on the Public, Green Party of Aotearoa New Zealand, March 31, 2001, http://www.votegreen.org.nz/searchdocs/other4819.html.

Internet Filtering in the **Commonwealth of Independent States**

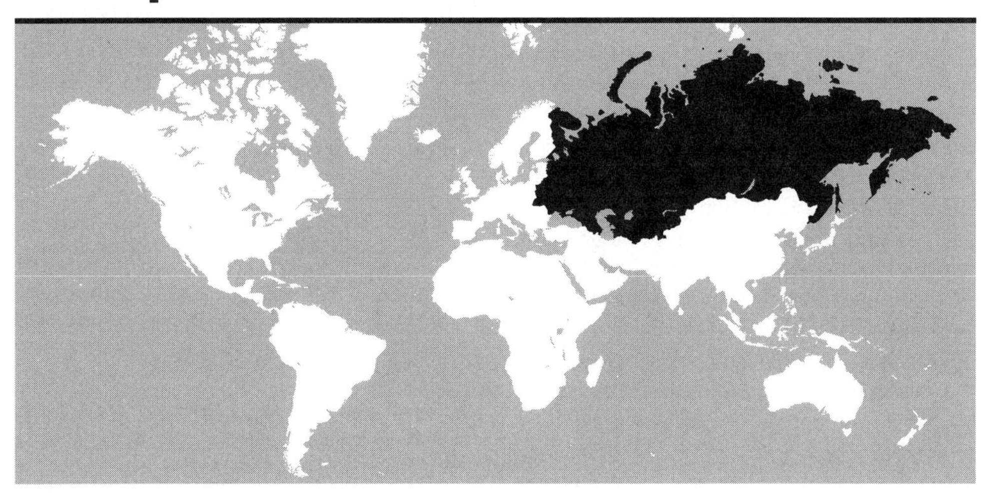

Overview

As a former superpower—with a tradition of authoritarianism, poorly developed independent media, and lack of private rights—the Commonwealth of Independent States (CIS) would seem to be an ideal setting for substantial and pervasive Internet controls.[1] The reality, however, is variegated and complex. While the CIS region is home to some of the world's most repressive measures and advanced techniques for subtly "shaping" Internet access, it also showcases examples of just how profoundly the Internet can affect social and political life.

States within this region have a conflicted relationship with the Internet. Most have adopted national development strategies that emphasize information technology (IT) as a means for economic growth, with some even declaring their intent to become regional "IT powerhouses." IT development is favored because it is seen to leverage the comparative advantage of the ex-Soviet educational system with its emphasis on mathematics and engineering, and the strong tradition of innovation in the computing and technology sector. Until its demise in 1991, the Union of Soviet Socialist Republics (USSR) was one of the few countries with a "homegrown" capacity in supercomputing, cryptography/crypto-analysis, and worldwide signals intelligence gathering. Currently many former Soviet citizens are among the leaders of the global IT industry.

At the same time, CIS governments are wary of the civil networking and resistance activities that these technologies make possible. In recent years, Ukraine, Georgia, and Kyrgyzstan have experienced "color revolutions," where networked opposition movements (albeit movements that are more reliant on cell phones than on the Internet) have effectively challenged and overturned the results of unpopular (or allegedly fraudulent) elections. Neighboring governments fear that these challenges were made possible by opposition groups leveraging IT to organize domestic protest (often with the help of foreign-funded NGOs) and are therefore wary of leaving the sector unregulated and without control. Many now see the Internet and other communications channels in national strategic terms, and these countries have increasingly turned to security-based arguments—such as the need to secure "national informational space"—to justify regulation of the sector.

In 2006 ONI tested for the presence of filtering in eight of the eleven CIS countries: Azerbaijan, Belarus, Kazakhstan, Kyrgyzstan, Moldova, Tajikistan, Ukraine, and Uzbekistan. Background and baseline testing was also carried out in a further two countries: the Russian Federation and Turkmenistan, although in these two cases limitations on the testing methodology do not allow us to claim comprehensive results.

Of the eight countries in which ONI tested, our results did not yield significant patterns of substantial or pervasive filtering. Only Uzbekistan pursued pervasive filtering of the kind found in China, Iran, and some parts of the Middle East.[2] In almost all countries some degree of filtering was present, but this filtering occurred mostly on corporate networks (such as educational and research networks) where accepted usage policies (AUPs) dictated that inappropriate content was not permitted or in "edge locations," such as Internet cafés, where the reasons for filtering were more benign (conserving bandwidth) or left to the discretion of the Internet café owners themselves.

At the same time, in all eight countries authorities had taken steps of one kind or another to restrict or regulate their national informational space. These measures include:

- expanded use of defamation and slander laws to selectively prosecute and deter bloggers and independent media from posting material critical of the government or specific government officials (however benignly, including, as was the case in Belarus, through the use of humor);
- strict criteria pertaining to what is "acceptable" within the national media space, leading to the deregistration of sites that did not comply (Kazakhstan);
- moves to compel Internet sites to register as mass media, with noncompliance then being used as grounds for filtering "illegal" content;
- national security concerns (Ukraine); and,
- formal or informal "requests" of ISPs.

The net effect of these sanctions (legal and quasi-legal) is to create overall environments that encourage varying degrees of self-censorship among ISPs, who are fearful of jeopardizing their licenses, and among individuals for whom prosecution or imprisonment is too high a price to pay for voicing criticism, which at times amounts to little more than a form of digital graffiti.

The CIS region: Ethno-cultural diversity and a shared historical space

To define the CIS as a region understates the sheer diversity of the countries and peoples that fall within the former Soviet Union's historical boundaries. Straddling a swath of Eurasia from the Pacific to the doorsteps of Europe, the Arctic Circle, and the deserts of Central Asia, this vast landmass takes in twelve time zones, some 350 million people, and more than a hundred distinct ethnic groups encompassing all the world's major religions and at least three major linguistic communities (Slavic, Turkic, Farsi). At the ethno-cultural level, diversity is a defining commonality of this region.

At the same time, the CIS forms an historical community that for seventy years constituted the world's second major economic, military, and political superpower of the twentieth century, rooted in the same traditions of modernism as the West but oriented around a different set of ideological and organizational principles. These principles emphasized a centralized and administered form of governance where the state rather than the market decided issues of economic and social production and where overarching leadership was vested in the Communist Party, whose rule was substantiated by ideological precepts that did not allow for dissention or opposition.

Despite this complex multinationalism, the former Soviet Union was dominated by Russia, which endowed the region with a common language (Russian) and popular culture, as well as defined trade, political, and even social ties (including the creation of substantial Russian minorities in some states, which persist to the present day). Even following the USSR's dissolution and the newly independent states' adoption of national languages and scripts (in Azerbaijan, Ukraine, Uzbekistan, and others), CIS countries retained strong ties with Russia. Transportation, communications, and energy routes continue to bind the region together. Russia is currently a major energy supplier to many CIS states, giving it considerable political muscle in the region (which it has not been shy to flex, when needed).

The region's shared political heritage, and the fact that many present-day leaders in the CIS governments and economies were also in positions of authority during the Soviet era, means that a great deal of formal and informal coordination exists among and between member states, despite political differences that are at times difficult. Furthermore, the loose, informal coordination among officials is helped along by the fact that most countries share the same legal codes and procedures, as well as similar organizational characteristics of the security forces and the distribution of powers among the judiciary, executive, and legislative branches of government.

The Internet in the CIS: Access and political significance

Internet penetration rates in the CIS region are relatively low and clustered among the urban youth—both male and female, perhaps reflecting the "equality" between sexes of the Soviet period.[3] Income levels in the CIS are generally low, while the costs of computers and connectivity are relatively high. This means that Internet use is lower than would be expected. Overall, Internet penetration in Russia lags behind that of other industrialized nations (15 percent as of 2005),[4] and is relatively high only in large cities (particularly Moscow and St. Petersburg). Among the CIS countries, Belarus has the highest Internet penetration rate of 30 percent; Ukraine and Moldova lag behind with less than 10 percent penetration rate, while the states of Central Asia have the lowest Internet penetration rates. Azerbaijan and Kazakhstan lead this latter subgroup with around 8 percent, followed by Kyrgyzstan. The least connected countries are Uzbekistan (3 percent for 2004) and Tajikistan, where only 1 percent of the population has access to the Internet.

However, in all cases these figures may be misleading. Most Internet users rely on shared Internet access, through their places of work or study, as well as via Internet cafés, whose use is very high in some countries, (for example, Internet cafés users account for over 50 percent of all users in Kyrgyzstan).[5] This shared use—and in some cases the creative use of networks such as Fidonet to route traffic to and from the Internet—may result in considerable underestimation of the actual number of users.[6]

The importance of the Internet to political life varies from low in Tajikistan to high in Uzbekistan. In Russia the relevance of the Internet as a source of news is reported as low; however, this estimation is changing as the Internet remains one of the few outlets for direct criticism of the government. Moreover, an important aspect of the Internet's political significance—as a person-to-person backchannel for communications and social networking essential to daily life in Russia (where personal contacts and an "informal economy of favors" remains a key to "getting ahead")—remains understudied.[7] In this sense, it is interesting to note that in Uzbekistan information obtained from the Internet is accepted as being more accurate than from other sources, reflecting the culture's strong social networking aspect.

Legal and normative environment

In general, the tendency in all CIS countries has been toward greater government regulation of the Internet to bring it in line with existing regulations that control the mass media (in Russia, Uzbekistan, and Belarus, for example). To date, government actions to enforce more restrictive Internet environments have rarely been challenged—perhaps a reflection overall of the weakness of "opposition" parties in most countries, as well as poorly defined or tested laws governing the role of independent media. Nonetheless, some exceptions exist. For example, in Tajikistan and Azerbaijan concerted (if quiet) action by "civic" actors led to the reversal of policies aimed at removing politically sensitive content from cyberspace. In Tajikistan political Web sites that were banned during the December 2006 election were restored. In Azerbaijan a banned Web site that was critical of the government's policy of raising prices was restored and its author released from police detention. Both cases are significant because the initial order to "ban" the Web sites was opaque from a legal perspective.

The constitutions of nearly all CIS countries enshrine principles of freedom of expression and prohibit censorship. Nevertheless, often these provisions are interpreted "flexibly" when it comes to implementation. In Kazakhstan authorities often resort to various quasi-legal or "administrative" mechanisms to suppress "inappropriate" information or shut down oppositional domain names. In Uzbekistan the law on mass media holds journalists and editors responsible for the "veracity" of published materials, which has caused independent media and bloggers to practice self-censorship. The "objectivity" test is applied also in Belarus, where independent journalists, editors, and opposition leaders are frequently subject to prosecutions and arrests.

In legislation and regulation, Russia remains a leader in the region, and increasingly has been proactive in seeking influence and extending assistance to other CIS states. Since late 2000 Russia's "Doctrine of Information Security" has been adapted (in various forms and guises) as the basic precept defining the national strategic value of the Internet and the "national informational space" in most CIS countries. [8] Likewise, Russia's legal approach to Internet surveillance for law enforcement (that is, the System for Operational-Investigative Activities or SORM-II, which allows security services unfettered physical access to ISP networks) has influenced the way in which other CIS countries have approached the problem (see the next section). Some, including Kazakhstan, have adopted the Russian system, while others have mirrored its approach. In Russia, Belarus, Moldova, and Ukraine, specialized units under the Ministry of Internal Affairs (Department "K") have been established to combat "computer crime" with specialized technical units also established in other security services.

Surveillance

Obtaining a telecommunications license in Russia and other CIS states requires close cooperation with state security agencies. Since the mid-1990s a key requirement has been that providers allow law enforcement and other security agencies full monitoring access to the communications systems. In Russia the enabling acts and system used to monitor telecommunications, including the Internet, comes under the rubric of SORM-II, which came into effect in 2000.[9]

At the regulatory and technical level, SORM-II requires ISPs to provide the Federal Security Service (FSB) with statistics about all Internet traffic that goes through the ISP servers (including the time of an online session, the IP address of the user, and the data that were transmitted).[10] ISPs themselves are responsible for the cost and maintenance of the hardware and connections. ISP objections to SORM-II, which raised concerns about individual privacy, resulted in the providers being stripped of their licenses.[11]

In many respects, SORM is not unlike a combination of the Unites States' Communications Assistance to Law Enforcement Act (CALEA)[12] and the recent "warrantless" provisions for wiretapping, including the USA PATRIOT Act[13] passed after the attacks of 9/11. Russian legislation formally protects individual privacy, prohibiting wiretapping of any kind without a court order.[14] As a consequence, SORM requires government personnel to obtain a court order to intercept telephone conversations, electronic communications, or postal correspondence.[15] In reality, however, the FSB will not bother to seek a warrant. Recently a senior FSB official sought to apply similar registration requirements for all mobile phones with Internet capabilities. However, despite this formidable surveillance potential, there is doubt about the actual capacity of the FSB to analyze the data collected.[16]

At present, several CIS countries have followed Russia's lead in implementing Internet surveillance.

- Kazakhstan followed the Russian example requiring ISPs to install special software in order to register and maintain electronic records of customers' Internet activities.
- Azerbaijan made an unsuccessful attempt to employ technologies similar to the SORM-II. At present surveillance does occur, but mainly by way of visits to ISPs and Internet cafés by officials from the State Security Service.
- In Uzbekistan the principal intelligence agency, the National Security Service (SNB), monitors the Uzbek segment of the Internet and works with the main regulatory body to impose censorship. As all ISPs must rent channels from the state monopoly providers. Credible anecdotal evidence strongly suggests that Internet traffic is recorded and monitored via a centralized system purchased from an Israeli vendor.
- In Ukraine, the security services have developed a capacity to monitor Internet traffic and legislation has been proposed to limit access to "questionable" content for reasons of national security. The security services are also empowered to initiate criminal investigations and use wiretapping devices.
- In Belarus, special services conduct active and warrantless surveillance of Internet activities under the pretext of national security using a system similar to SORM-II.

Transparency

Former British Prime Minister Winston Churchill once said when asked about the Soviet Union, "It is a riddle, wrapped in a mystery, inside an enigma; but perhaps there is a key. That key is... national interest." Transparency with regard to filtering practices varies across the region, but in all cases it is defined by the interest of the state (or the group that holds the reins of power). Protection of state interests (usually cast in terms of national security or the protection of public or cultural values) generally trump the written rules for regulation of Internet content, although often the laws themselves are ambiguous and open to interpretation. In addition, the restrictive practices of states are often fairly subtle. As an example, Uzbekistan—which was until recently the most egregious Internet censor in the region—denied that it was engaged in censorship practices. The plausibility of this claim was increased because filtering was neither uniform nor universal across all ISPs, which left open the possible, although highly improbable, chance that observed filtering practices were self-imposed by ISPs rather than proscribed by higher ups. Such subtle approaches allow the state "plausible deniability" of any wrongdoing and require a great deal of contextual research to uncover the sources of the practice.

Overall a general lack of transparency affects most political/legal issues in the CIS, not only the issue of Internet filtering. Often official laws are breached in subtle but effective ways. For example, in Azerbaijan the author of a Web site critical of the government was detained without formal arrest; this was never followed up by any formal legal sanctions. In other cases, such as the pervasive filtering policies of Internet cafés throughout the region, the decision to limit content is formally controlled by the café owners, so it is difficult to argue whether their filtering results from a fear of sanction for allowing politically sensitive material to be accessed, or from personal choice. Certainly for most Internet café owners, the objective is to make a living, not to run for office. If certain content stands in the way of business, then it is not a difficult choice to decide what measures to take. In Tajikistan, for example, research suggests that filtering is really based on economic choices rather than any overt fear of political sanction from the security forces.

Emergent forms of information control

Overt Internet filtering, such as that undertaken by China or Iran, is unlikely to occur in the CIS. First, only in a very few cases (Uzbekistan, Turkmenistan) is the government willing to create an informational blockade of the country that could, in turn, jeopardize economic prospects and stifle the "scientific potential" of these technologies. Second, as noted above, governments generally have more subtle legal and quasi-legal methods for putting pressure on content and access providers to remove or otherwise eliminate "undesirable" content, so there is little need to resort to overt technical means such as filtering. Third, many CIS states are dependent on development aid and trade and have oriented themselves toward integration with Europe and the broader global economic system. Engaging in widespread filtering of the kind conducted by China or Iran would present the risk of being labeled as an "international human rights pariah," an eventuality that most CIS countries would rather avoid. Fourth, and perhaps most important, those CIS states that are concerned by the Internet's empowering potential—that is, their potential to make possible further "color revolutions"—have found more subtle technical means for ensuring that these capacities are curtailed, if and when necessary.

Event-based interventions

The CIS is the first region in which ONI research documented the presence of "event-based" filtering. This form of filtering differs in technical execution from more conventional filtering forms (such as those that rely on bloc lists) and is more difficult to track and definitively ascertain. For example, during Kyrgyzstan's 2005 parliamentary elections, two ISPs were disrupted by distributed denial of service attacks (DOS), and then a "hacker for hire" posted threats to the affected ISPs' visitor logs, stating that unless these sites stayed offline the attacks would continue.[17] The DOS attacks effectively disrupted the ISPs' services because the hacker exploited the ISPs' narrow bandwidths and dependence on a single satellite-based connection. To this day is it unclear who hired the hackers responsible for the attack, although an investigation by ONI found that they were based in Ukraine (and were also responsible for an attack on a U.S. site using the same "bot" network). The opposition accused the government of ordering the attacks as a means of undermining the opposition. The government responded by ordering the affected ISPs to keep their resources online, but this was impossible because the DOS attack had degraded their ability to provide any services. In the end, the attack was stopped as a result of U.S. legal action against the originating "bot net," which had also been attacking a U.S. site. When the "bot net" was taken down, the attacks against the Kyrgyz sites also stopped.

During the March 2006 presidential elections in Belarus, several opposition Web sites became suddenly inaccessible, ostensibly by innocuous network faults and DNS failures. Likewise, at the peak of protests against the election results, a major Minsk-based ISP ceased to provide dialup services owing to "technical problems." These occurrences meant that important independent media and opposition political Web sites were not accessible at periods when the information they were conveying could have had political significance or acted as a catalyst for further political action. Although nothing transpired that could be identified as extralegal filtering, *de facto* access was not available when and where needed, with some evidence suggesting that tampering may have been afoot.[18]

This form of "event-based" information control, which temporally "shapes" Internet access, can be said to represent the emerging "2.0 version of Internet controls." Not unlike the shorter supply line chains that boasted manufacturing efficiencies under just-in-time production, event-based filtering can also be considered to be "just in time" as it offers greater efficiencies in denying access to information when and where it is needed. At the same time this form of targeted and time-limited filtering is much harder to prove, which also removes the potential liabilities of being "caught" undertaking more deliberative filtering.

Upstream filtering

For its size, the CIS region has a relatively underdeveloped telecommunications system, much of which remains centered on Russia. At the same time, the region itself is contiguous with (or borders) Europe, Asia, and—via the circumpolar route—North America. This centrality means that most countries in the region obtain connectivity from several different sources beyond Russia. This situation has created some interesting patterns in filtering behavior, such as similar content becoming inaccessible across several different countries, but with different filtering patterns amongst content providers within any single country. ONI research into this phenomenon is still preliminary, and thus we are not yet in a position to provide conclusive evidence or observations on its implications.

However, preliminary indications suggest that providers reselling connectivity to CIS countries may be providing pre-filtered access, passing on filtered content either as part of their service offering or as a consequence of the policies they use to manage traffic on their own networks. This form of blocking, which we have dubbed "upstream" filtering (indicating that the filtering is happening in a jurisdiction other than that of the state in question), was first observed during ONI testing in Uzbekistan in 2004. At that time the traffic of one Uzbek ISP was clearly filtered using a pattern similar to that employed by Chinese ISPs. Further investigation revealed that the Uzbek ISP was buying connectivity from China Telecom, which in this case may have sold access to its network as it would to a regular Chinese client. Our 2006 testing suggested similar patterns of prepackaged filtering affecting Internet services within several other CIS states where ISPs had purchased their connectivity from a Russian provider.

Conclusion

The CIS region is experiencing a general trend toward greater regulation and control of the national information space, which includes the Internet. Although most CIS countries do not practice the substantive or pervasive filtering—Uzbekistan and Turkmenistan excepted—Internet content control through regulation or intimidation is growing throughout the region. In most cases, the legislative and judicial framework for filtering (or other restrictions) is ambiguous and open to interpretation. Moreover the laws are often unevenly applied, with "flexible" implementation often paired with other more subtle (but effective) measures designed to promote self-restraint (or self–censorship) of both ISP providers as well as content producers. Information control—in particular the protection of national informational space—is clearly an issue of concern throughout the CIS, and has encouraged more stringent attention to telecommunications surveillance (as has been happening in other parts of the world, most notably the United States). In addition, measures to protect regimes in power and stifle opposition are often couched in the language of "national security" and have resulted in the development of new measures and techniques aimed at temporally "shaping" access to information at strategic moments, such as "event-based filtering." Another innovation that merits further investigation is "upstream filtering." Although these new measures are not present in all CIS countries, they are indicative of a new seriousness with which strategies for information control are being developed.

In 2007 a number of critical elections will take place in Russia and several other CIS countries. In Russia, exiled billionaire Boris Berezovsky has expressed his intent to overturn the existing regime. The Internet and other forms of communications technologies are expected to play an important role in the electoral process, and as such they will no doubt be the object of many actors' attention.

Last, the re-emergence of stronger states in the region following more than a decade of transition and general unhappiness concerning U.S. policies in the region (which have, over the past ten years, promoted media freedom and an active if foreign-funded civil society), is also sparking a degree of "blow-back" and renewed competition between East and West. For example, ONI research found that many ".mil" sites are not reachable in the CIS, suggesting that these may be subject to "supply-side" filtering by U.S. authorities.[19] Between greater assertiveness on the part of CIS states and the stimulus of renewed interstate competition, the CIS is a region to watch as a global actor shaping norms that will govern the Internet into the future.

Authors: Rafal Rohozinski, Vesselina Haralampieva

NOTES

1. The CIS consists of eleven countries: Armenia, Azerbaijan, Belarus, Georgia, Kazakhstan, Kyrgyzstan, Moldova, Russia, Tajikistan, Ukraine, and Uzbekistan. Turkmenistan has been an associated member since 2005. With a strong political and economic influence over its neighboring countries, Russia remains the predominant political actor and strategic economic power in the group.
2. Turkmenistan's Internet is even more tightly restricted, with access available only via a single government provider. While our lack of test results do not allow us to conclusively map the extent of filtered content, preliminary analysis indicates that the Turkmen authorities employ a "white list" that allows only permitted sites to be visited.
3. Internet users in the CIS are predominantly young, aged between fifteen and twenty-five. Around 55 percent of all users in Azerbaijan belong to this age group, compared with 60 percent in Kyrgyzstan and similar percentages in Uzbekistan. The number of women using Internet in Uzbekistan and Kazakhstan is equal to or larger than the number of their male counterparts. The proportion is slightly in favor of men in Ukraine, while in Tajikistan only 22.5 percent of the Internet users are women.
4. See International Telecommunication Union, *World Telecommunication Indicators 2006.*
5. In Kazakhstan 28.4 percent of users access Internet at home, and 27.5 percent in Azerbaijan. The workplace is also a critical access point in Kazakhstan (27.2 percent), Moldova, Belarus, and Uzbekistan. In contrast, cybercafés in Kyrgyzstan are the main Internet access point in the country (for approximately 57 percent of users).
6. Rafal Rohozinski, "Mapping Russian cyberspace: Perspectives on democracy and the Net," United Nations Research Institute for Social Development (UNRISD) Discussion Paper 115, October 1999. Available at unpan1.un.org/intradoc/groups/public/documents/UNTC/UNPAN016092.pdf.
7. Alena Ledeneva, *How Russia Really Works: The Informal Practices That Shaped Post-Soviet Politics and Business,* Ithica: Cornell University Press, 2006.
8. Doctrine of the Information Security of the Russian Federation, September 9, 2000, No. Pr-1895, http://www.medialaw.ru/e_pages/laws/project/d2-4.htm.

9. See http://www.libertarium.ru/libertarium/37988.
10. See http://www.iworld.ru/magazine/index.phtml?fnct=page&p=93433812, (last accessed April 10, 2007).
11. See http://www.libertarium.ru/libertarium/14424/def_article_t?PRINT_VIEW=YES and http://www.techweb.com/wire/story/TWB19990726S0003 (last accessed April 1, 2007).
12. The Communications Assistance for Law Enforcement Act (CALEA), Pub. L. No. 103-414, 108 Stat. 4279 (1994).
13. Uniting and Strengthening America by Providing Appropriate Tools Required to Intercept and Obstruct Terrorism (USA PATRIOT ACT) Act of 2001, (H.R.3162), http://thomas.loc.gov/cgi-bin/query/z?c107:H.R.3162.ENR.
14. Article 23 of the Constitution of the Russian Federation, http://www.constitution.ru/en/10003000-01.htm.
15. See http://www.worldpoliticswatch.com/article.aspx?id=416.
16. Interview with Andrei Richter, Director, Media Law and Policy Institute, Moscow State University, in Moscow, Russia, March 28, 2006; Interview with Alexey Simonov, President, Glasnost Defense Foundation, in Moscow, Russia, March 27, 2006.
17. See "Election monitoring in Kyrgyzstan," ONI Special Report, February 15, 2005, http://www.opennetinitiative.net/special/kg/.
18. See "The Internet and elections: The 2006 presidential election in Belarus," ONI Internet Watch 001, http:// www.opennetinitiative.net/belarus/.
19. The inaccessibility of U.S.military Web sites was not limited to the CIS region but was also observed in numerous countries around the world. Future research will focus on this issue of filtering that is carried out by Web site hosts based on geolocation.

Internet Filtering in
Europe

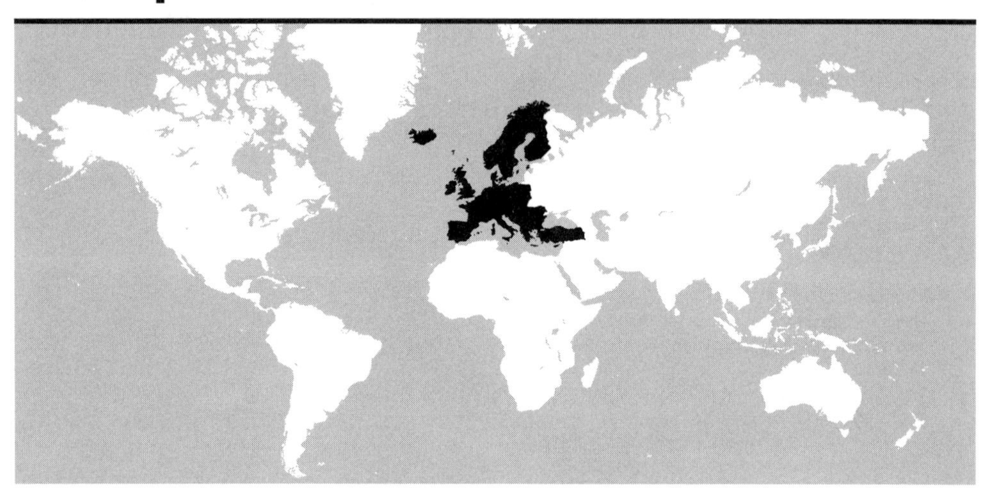

Introduction

In less than a decade, the Internet in Europe has evolved from a virtually unfettered environment to one in which filtering in most countries, particularly within the European Union (EU), is the norm rather than the exception. Compared with many of the countries in other regions that block Internet content, the rise of filtering in Europe is notable because of its departure from a strong tradition of democratic processes and a commitment to free expression. Filtering takes place in a variety of forms, including the state-ordered takedown of illegal content on domestically hosted Web sites, the blocking of illegal content hosted abroad, and the filtering of results by search engines pertaining to illegal content. As in most countries around the world that engage in filtering, the distinction between voluntary and state-mandated filtering is somewhat blurred in Europe. In many instances filtering by Internet service providers (ISPs), search engines, and content providers in Europe is termed "voluntary" but is carried out with the implicit understanding that cooperation with state authorities will prevent further legislation on the matter.

The scope of illegal content that is filtered in Europe largely is limited to child pornography, racism, and material that promotes hatred and terrorism, although more recently there have been proposals and revisions of laws in some countries that deal with filtering in other areas such as copyright and gambling. Filtering also takes place on account of defamation laws; this practice has been criticized, particularly in the UK, for curtailing lawful online behavior and promoting an overly aggressive notice-and-takedown policy, where ISPs comply by removing content immediately for fear of legal action. ISPs in Europe do not have any general obligation to monitor Internet use and are protected from liability for illegal content by regulations at the European Union (EU) level, but must filter such content once it is brought to their notice. Therefore the degree of filtering in member states depends on the efforts of governments, police, advocacy groups, and the general public in identifying and reporting illegal content.

Efforts over the past decade have been underway to create a set of common policies and practices at the EU-level on Internet regulation. This is viewed as necessary to promote regional competitiveness and commerce, to counter Internet crime and terrorism, and to serve as a platform to share best practices amongst nations. Notable advancements in regulation at the EU level—although not directly in the area of filtering—include the definition of ISP liability toward illegal content and obligations toward data retention.

Regional regulation

A recurring theme throughout this overview will be the overlapping nature of individual country-level law and regionwide regulation. Countering criminal activity on the Internet and promoting the overall competitiveness of the Internet industry have been the primary reasons cited to develop a regional regulatory framework.[1] A regional approach in Europe has its beginnings with a request by the European Council to the European Commission in April 1996 to produce "a summary of problems posed by the rapid development of the Internet" and to assess the need for regulation. The Commission produced a report titled "Illegal and Harmful Content on the Internet" and a Green Paper on "The Protection of Minors and Human Dignity in Audiovisual Services" in response. Based on these documents, "a common framework for self-regulation (of the Internet) at the European level" was drafted, which culminated in an Action Plan on Promoting Safe Use of the Internet. The plan, adopted on January 25, 1999 and operational up to 2002, outlines the basic principles underlying Internet content regulation at the European level.[2] Broadly, undesirable content on the Internet is classified either as "illegal" or "harmful."

The scope of "illegal" content tends to vary between countries, although there are certain issues where there is a greater amount of consensus, such as child pornography, trafficking in human beings, racist material, material promoting terrorism, and all forms of Internet fraud (such as credit card fraud).[3,4] "Harmful" material, as defined in the plan, is that which might offend the values and sentiments of others and could pertain to politics, religion, or racial matters, and could also vary significantly between cultures.

The plan emphasizes the need for action in five broad areas in order to curb illegal and harmful content on the Internet:[5]

1. promoting voluntary industry self-regulation and content monitoring schemes, including the use of hotlines for the public to report illegal or harmful content;
2. providing filtering tools and rating systems that enable parents or teachers to regulate the access of Internet content by children in their care, while allowing adults access to legal content;
3. raising awareness about services offered by industry among users to allow them to leverage the Internet more fully;
4. exploring the legal implications of promoting the safer use of the Internet; and
5. encouraging international cooperation in the area of regulation.

Europe also maintains a regional policy that is generous in limiting ISP liability under the Electronic Commerce Directive, 2000/31/EC. Article 12, the "mere conduit" exception provision, absolves ISPs from liability for information transmitted over their networks as long as they did not initiate the message, select or modify the information, or select the intended recipients. The exemption also extends to the "automatic, intermediate and transient" storage of information, provided it is for a "reasonable period." The latter is left to be specified by member states. Article 13 deals with caching—granting exemption from liability for the "automatic, intermediate and temporary storage of information" that is carried for the exclusive purpose of making onward transmission more efficient. Article 14 addresses the liability associated with hosting content, stating that ISPs "will not be liable for hosting information, provided they do not have actual knowledge that the activity is illegal and, upon obtaining such knowledge, act quickly to remove it."[6] Finally, Article 15 precludes ISPs from any general obligation to monitor content or data transmitted or stored through their services. Further, ISPs are not required to actively seek facts that might indicate illegal activity.[7] These provisions granting ISPs substantial immunity from liability over illegal content are consistent with the law and practice of many other countries around the world that seek to expand Internet use and promote freedom of expression.

Social filtering

Action to regulate obscene content started with individual countries and the implementation of voluntary ISP-level filtering programs. The landmark model of large-scale voluntary ISP filtering in Europe originated in the UK.[8] BT, Britain's largest ISP, serving about a third of the country's home Internet users, launched Project Cleanfeed in June 2004[9] in consultation with the British Home Office. Under the auspices of this project, BT filters Internet content based on a blacklist of Web sites hosted anywhere in the world that contain images of child abuse as defined by the amended Protection of Children Act, 1978.[10] The list is compiled by the Internet Watch Foundation (IWF), a not-for-profit organization, in consultation with government, industry, the police, and the public. IWF provides the list to its members, which today include ISPs, mobile network operators, content providers, and search engines such as Google and Yahoo![11] Those attempting to access the illegal content hosted abroad receive an error message as if the particular page were unavailable as a result of other connectivity problems.[12] Illegal content that is hosted within the UK, including child abuse images and content that is criminally obscene or incites racial hatred, is required to be taken down by ISPs and content providers under a notice-and-takedown regime.[13] Although this form of filtering is termed "voluntary," by the end of 2007 all broadband consumer ISPs in Britain are expected to have implemented a similar system, failing which, regulatory enforcement might be considered.[14,15] Other countries, such as Norway, Sweden, Denmark, and Italy, have implemented similar programs, while Finland is currently considering doing so.[16]

Filtering also takes place through "voluntary self-regulation" by search engines. As of early 2005 all major search engines in Germany — Google, Lycos Europe, MSN Deutschland, AOL Deutschland, Yahoo, T-Online, and t-info — have formed an organization that coordinates filtering of search results that are harmful to minors, based on a list provided by a government agency in charge of media classification. The move is seen as a response to pressure for voluntary self-regulation by industry at the EU level, and arguably to the fear among industry that a failure to comply will result in increased legislation. The system has been criticized, however, for a lack of transparency,[17] since the search engines cannot disclose the list of Web sites to the public, as per a codex signed by them.[18] In addition, disclosure would defeat the purpose of filtering search results, as the sites are removed only from the search results, not from the Internet.

Internet content is also monitored through online surveillance by authorities in the UK. The Child Exploitation and Online Protection Centre (launched in April 2006) made thirteen arrests in July 2006 after beginning investigations into pay-per-view Internet services.[19] The police in Britain have also been vested with the power to pass on to banks the personal details of those who access illegal content online using credit cards, based on an amendment to the Data Protection Act (1998).[20] Banks will then cancel the cards as a breach of their terms of service.

The public in nineteen European countries assists in identifying and reporting illegal content —particularly in the area of child pornography — through a network of hotlines that have been implemented on the basis of a recommendation at the EU level.[21] In Austria authorities were able to uncover a "child-pornography ring" involving seventy-seven countries in February 2007, based on a report by a man working for a Vienna-based Internet file-hosting service.[22] Recent reports show that the Save the Children Denmark Hotline, financed jointly by Denmark and the European Commission's Safer Internet Plus Programme, had nearly 9,000 reports of child abuse images in 2006 alone.[23] The police in Spain were able to arrest ninety people in 2004 in the country's largest operation against the distribution of child pornography, facilitated by the hotlines. The INHOPE Association acts as the coordinator of the network of hotlines, including in countries outside Europe such as Australia, Brazil, Canada, South Korea, Taiwan, and the United States.[24]

Although early filtering efforts had fairly limited agendas, proposals and laws are emerging in many nations toward filtering in other social realms, such as gambling and betting. A proposal was drafted in 2002 to revise Swiss federal laws on lotteries and betting, such that those providing access to games that are considered illegal face fines up to 1 million Swiss francs or up to a year of imprisonment. This effort was suspended in 2004, and no further action has been taken since. As of February 2006 ISPs in Italy are required to block access to Web sites that offer online gambling. The list of Web sites to be blocked is compiled by the Autonomous Administration of State Monopolies (AAMS, a part of the Ministry of Economy and Finances), which issued the decree.[25] The most broad-based proposal yet for filtering comes from Norway, where the government is considering blocking access to foreign gambling sites, Web sites that "desecrate the Flag or Coat of Arms of a foreign nation," sites that promote hatred toward public authorities, contain hate speech or promote racism, offensive pornography sites, and peer-to-peer sites that offer illegal downloads of music, movies, or television shows.[26]

Nationalistic filtering

There are no examples in Europe of filtering carried out to silence political opposition such as those that the ONI has documented in other regions. There are, however, examples of filtering that seeks to maintain the legitimacy of government institutions and preserve national identity. In December 2002 a local Swiss magistrate, Françoise Dessaux, ordered several Swiss ISPs to block access to three Web sites hosted in the United States that were strongly critical of Swiss courts,[27] and to modify their DNS servers to block the domain appel-au-people.org.[28] The Swiss Internet User Group and the Swiss Network Operators Group protested that the blocks could easily be bypassed and that the move was contrary to the Swiss constitution, which guarantees "the right to receive information freely, to gather it from generally accessible sources and to disseminate it" to every person. However, there was strong enforcement, as the directors of noncompliant ISPs were asked to appear personally in court, failing which they faced charges of disobedience.

On March 7, 2007, the video-sharing Web site YouTube was blocked in Turkey as per a court order, following the posting of certain videos on the site that were found to be derogatory toward Turkey's founding father, Mustafa Kemal Ataturk, the Turkish people in general, and the Turkish flag. The blocking invoked Article 301 of the Turkish Penal Code, known as the main obstacle to freedom of speech, which defines insults toward Ataturk as well as "Turkishness" as a crime. Turkey's leading ISP, Turk Telecom, complied with the order but petitioned to the court to allow access to the site to be restored. The court agreed on the condition that the particular videos were removed. The two-day blocking was heavily criticized both within Turkey and abroad and likened to "closing a library because of a single book that was found to be improper."[29]

Hate speech

European states are also increasingly taking action against online hate speech, applying their offline policies to the Internet. Some efforts raise important issues such as the jurisdiction over material on the Internet. For example, a French court in 2000 ruled that U.S.-based Yahoo! Inc. is liable under French law for allowing the people of France access to auction sites that include Nazi memorabilia and demanded that Yahoo! must ensure that this content is impossible to access from France or face fines.[30] The case was brought by two French not-for-profit organizations[31] dedicated to fighting anti-Semitism.[32] Yahoo! brought suit in a U.S. District Court in San Francisco, claiming that the French court's ruling was unenforceable in the United States. The U.S. court ruled in Yahoo!'s favor in November 2001, but in 2004 a panel of the 9th U.S. Circuit Court of Appeals overturned the ruling by the lower court on the grounds that it "did not have sufficient jurisdiction over the French parties."[33] After reconsidering the decision, the 9th U.S. Circuit Court of Appeals dismissed Yahoo!'s case in January 2006 despite claiming jurisdiction over the matter because Yahoo! had already removed the materials and, therefore, the requirement to block would not have done any actual First Amendment harm.[34]

Similarly, the German Federal Court of Justice ruled in December 2000 that material glorifying the Nazis and denying the Holocaust must be censored as per German law, regardless of where it is hosted, based on a case involving an Australian-based Holocaust revisionist who was using the Internet to spread his message denying the atrocities of World War II.[35] In another case, seventy-eight ISPs in Nordrhein-Westfalen were ordered to block access to two foreign Web sites in 2002 that contained neo-Nazi content.[36] The same regional government of Düsseldorf also took an anti-censorship activist to court for posting hyperlinks on his Web site to radical rightwing content that had been censored.[37]

Other European countries also have laws against Holocaust denial and ban material that promotes racial hatred. These have been "harmonized" in a protocol to the Council of Europe's cybercrime treaty, which requires that "any written material, any image, or any other representation of ideas or theories, which advocates, promotes or incites hatred, discrimination or violence, against any individual or group of individuals, based on race, color, descent or national or ethnic origin, as well as religion if used as pretext for any of these factors" and "material which denies, minimizes, approves of or justifies crimes of genocide or crimes against humanity" must be made illegal by the signatories.[38] As with all illegal content, once brought to their attention, ISPs must either take down or block the relevant Web sites depending on whether the sites are hosted within the country or abroad.

Defamation

Member states of the EU have expressed the need for a simplified framework to be applied with respect to rules concerning defamation by media or publications via the Internet and other electronic networks. The general principle in cases of defamation concerning the media—that the law of the country where the defamed person lives is applicable—implies that media organizations must know the privacy and defamation laws of each European country, which is criticized as impractical. In Italy, for example, in 2000, a man in "a trans-border custodial battle" claimed that his ex-wife, now resident in Israel, was responsible for posting statements and images on the Internet that were defamatory of him and derogatory of his ability to care for their two daughters. The Italian Supreme Court, or Suprema Corte di Cassazione, overturned a prior verdict from a lower court, affirming that Italy's laws of libel apply to content on foreign Web sites accessible by Internet users in the country.[39] The Court held that while the offending statements were posted outside of Italy, the effects were felt within the country and were therefore subject to the national laws.

The issue of the need for a unified framework was brought to the fore once more in February 2007 as a part of the European Parliament's second reading of the Rome II Regulation, which seeks to establish rules on the applicable law to noncontractual obligations relevant to publications via the Internet and other electronic networks. The Parliament's proposed amendment is that the law applicable should be that of the country to which "the publication or broadcast is most directed," which is to be determined "by the language of the publication or broadcast, or by sales or audience size in a given country as a proportion of total sales or audience size, or by a combination of these factors." Further, the amendment suggests that if these are not easy to determine, "the relevant law will be the one of the country where editorial control is exercised." With regard to the right to reply, it is suggested that the applicable law should be that of the country in which the publisher or broadcaster has its "habitual residence." The text, which has been adopted by the Parliament, is not expected to find easy favor with the European Council and must undergo a standard conciliatory procedure where member states and Members of European Pariliament, in equal representation, debate the proposal, and it will be approved as a regulation if an acceptable compromise is reached.[40]

In their current form, defamation laws at the country level, particularly in the UK, have been criticized for leading to a "Web takedown" culture where ISPs immediately remove content that is allegedly defamatory when brought to their notice, for fear of facing law suits. The concern in the UK, as in other nations, is that this can have a "chilling effect" on lawful online content and behavior.[41]

A landmark precedent in the UK led the way for the establishment of a notice-and-takedown system. In *Laurence Godfrey v. Demon Internet Limited,* a defamatory statement was made on a posting to a newsgroup called "soc.culture.thai," available on a server at the provider Demon Internet Limited. The message was found to be forged and only appeared to come from Godfrey. Despite a request by Godfrey to take down the content, as it was defamatory of him, the ISP did not comply. As a result, he claimed damages for libel under §1 of the Defamation Act, 1996, and settled with Demon out of court.[42]

Libel law in the UK has been known to be particularly sympathetic to libel plaintiffs—and is often contrasted with the law in the United States in this context—such that many individuals from outside countries have sued publications in the UK, despite a relatively small circulation there, for a better chance of winning. However, the *Jameel v. Wall Street Journal Europe* case significantly increased press protections against libel claims in October 2006.[43] There has also been debate over whether the protection of the reputation of individuals is in conflict with the Human Rights Act of 1998, insofar as it might infringe upon the right to free speech.[44]

Copyright

A few countries in Europe have begun to employ Internet filtering to combat copyright infringement, evolving toward the notice-and-takedown approach used in the United States. In Denmark, as per a ruling of the Copenhagen City Court on October 2006, TDC, the country's largest ISP, blocked access to a Web site that distributes illegally copied music.[45] In February 2007, as mentioned earlier, Norway proposed filtering on a much larger scale that would include blocking of peer-to-peer sites offering illegal downloads of music, movies, and television shows.[46]

On March 16, 2007, the police arrested the owner of www.arenabg.com, which is one of Bulgaria's largest BitTorrent trackers and one among the country's ten most popular Web sites,[47] providing links to copyrighted music, movies, and software.[48] Although the owner was released within twenty-four hours, the Web site was filtered by police order for the period March 16–19, on the grounds that it was "necessary to prevent foreign interference with the torrent trackers."[49] The order to filter the site was lifted by the General Office for Fighting Organized Crime, but has resulted in considerable citizen protest for what is considered unjust treatment toward the owners and operators of torrent sites.[50] Following the arrest, other tracker Web sites have reportedly closed, some under threat of confiscation of property by the police, or have moved their servers abroad to avoid prosecution under the Bulgarian Copyright Law. The extent of actual filtering of these sites in the country is not known because there are differing reports regarding accessibility by various ISP subscribers. Given that BitTorrent trackers point to content but do not host it, the legal recourse to deal with the copyright violation associated with these Web sites is especially unclear.[51]

Law suits concerning alleged copyright infringement by search engines have been raised in a few countries, with recent rulings in favor of a notice-and-takedown policy that could arguably serve as a precedent for other countries in the region. In February 2007 the Brussels Tribunal found Google Inc. to be in violation of national copyright laws in a case raised by Copiepresse of Belgium, a trade group representing seventeen of Belgium's French- and German-language newspapers, and the company was fined 2.4 million pounds for the breach.[52] As per a translation of the ruling, "the reproduction and publication of headlines as well as short extracts, and the use of Google's cache, the publicly available data storage of articles and documents, violate the law on authors' rights."[53] The former refers to the Google News service,[54] while the latter to Google Web Search. The outcome is that Google cannot include references to articles, pictures, or drawings of Copiepress members through its Google News service without prior agreements, and must remove Belgian newspaper content from its search results. Failure to comply will result in fines of 25,000 euros a day.

Google intends to appeal against the judgment, stating that Web search results and the news service in fact drive more traffic toward the newspaper Web sites, and that Google News does not earn any advertising revenue from this. Copiepress, however, holds that by allowing users to bypass the front pages of newspapers and link directly to articles, newspapers lose advertising revenue. In addition, by making old newspaper material available through its cache, newspapers effectively lose the ability to charge customers for access to their archives, while Google Web Search does in fact earn advertising revenue for this service. The court ruling also states that all copyright holders can notify Google in case of infringement, and the search engine will have to remove content within a twenty-four-hour period or pay a 1,000 euro daily fine.[55] This could lead to an attitude of risk aversion and immediate compliance on the part of ISPs, content providers, and search engines—similar to instances of alleged defamation—in the face of potential law suits.

Google had run into similar difficulty in France with respect to its news service when Paris-based Agence France Presse (AFP) had sued the company for USD 17.5 million in 2005. The suit was dropped in April 2007, following a licensing agreement where Google would be allowed to use stories and photographs from AFP for its news aggregator and for other Google services, including products that Google is expected to launch in the future. The financial terms of this arrangement have not been publicly disclosed.[56] Out-of-court settlements in Europe for copyright infringement should not be surprising, because the legal defenses available in the region for alleged infringers are relatively weak.[57]

At the regional level, Intellectual Property Rights pertaining to Internet content are addressed by two directives: the Copyright and Related Rights in the Information Society adopted on April 9, 2001, and the Electronic Commerce Directive 2000/31/EC, which came into force on June 8, 2000. Article 5(1) of the Copyright Directive exempts ISPs from liability for copyright infringement where "reproduction is transient or incidental" or where copies are an integral part of a technological process "whose sole purpose is to enable onward transmission in a network between third parties by an intermediary or a lawful use of a work or other subject-matter to be made." The Copyright Directive also exempts ISPs from liability where the copies have "no independent economic significance"; this is left to be adjudged independently by courts in the respective member states. As per the first condition, ISPs and telecommunications operators do not need to request permission to transmit transient copies across their networks. However, the second condition implies that ISPs still face a situation of differing degrees of liability across the member states of the EU, and the directive has been criticized in this regard.[58] The Electronic Commerce Directive deals with the liability of ISPs toward content more generally, but with important implications for copyright. In particular, the directive provides a "mere conduit" exception, limits liability for content associated with the caching and hosting functions, and exempts ISPs from any general obligation to monitor.

Security

Security concerns in Europe have resulted in legislation concerning the surveillance and monitoring of Internet use. Although distinct from filtering, these have many parallels in their potential impact upon online freedom of speech. A recent and controversial area of legislation at the EU level in this regard pertains to the surveillance of traffic data and its retention. As per the European Data Retention Directive, which was passed in March 2006 and must be put into effect for Internet traffic by March 2009,[59] ISPs in the various nations are required to retain specific data pertaining to communications—in particular, with regard to Internet access, e-mail and telephony—for a period of at least six months but not exceeding two years. The data to be retained do not concern the content of communications. The aim is to bring about a "common code" of data retention in order to facilitate the tracing of illegal content and the source of attacks against information systems, and to identify those who use the electronic communications networks for terrorist activities and organized crime.[60] As the directive is implemented across the member states, privacy groups are concerned about the ability of ISPs, search engines,[61] and Web companies to retain data and monitor people's online habits. Moreover, the retention period of up to twenty-four months has been argued to be an unjustifiable length of time.[62]

An example of security legislation at the country level is a proposed law drafted in March 2007 in Sweden, which would give the national defense intelligence agency power to monitor all cross-border phone calls and e-mail traffic without court order. This will be carried out by the National Defence Radio Establishment in the form of searches for sensitive key words through the use of computer software. With some suggested amendments, the Swedish Legislative Council has approved the proposal to go forward. Concerns for privacy have been raised, including for communications within the country, which are often routed via servers hosted abroad.[63] Critics include the country's national security police agency, SAPO, which considers the proposal to be in violation of "personal integrity."

Conclusion

Filtering of online content takes a variety of forms among the states of Europe. Examples include orders issued by states to ISPs to take down Web sites that contain illegal content if they are hosted within the country, blocking orders by enforcement authorities for illegal content hosted abroad, and search engines that filter results pertaining to illegal content as a form of self-regulation. Although forms of filtering by search engines and ISPs are often referred to as "voluntary self-regulation" in some countries, there appears to be an implicit understanding that cooperation with government orders will forestall further legislation.

Filtering in European countries has also given rise to several legal disputes over the question of jurisdiction involving content that is hosted abroad. While the degree of filtering that takes place tends to vary among states, there is a concern in many countries over an apparent increase in the overall extent of filtering, as manifested in recent proposals and revisions in laws. Filtering in European states has, however, largely been confined to content that is illegal, and the extent has been tempered by public dialogue, adherence to law, and commitment to free speech, although the latter is more constrained than it is in the United States.

At the EU level there have been efforts over the past decade to create a common platform of "harmonized" Internet regulation. With regard to the filtering of online content, the emphasis has been on greater cooperation among industry, the public, and enforcement authorities within states, and increased voluntary industry self-regulation. Although EU level discussions were initially focused on various forms of illegal content online (in particular child pornography and racist and xenophobic content), there is increased attention being paid toward the use of the Internet for terrorism and organized crime in recent years. The latter has spurred legislation in the area of data retention, and much debate on the need for greater security measures versus the associated implications for privacy. There have also been recent advancements in terms of regulation at the EU level in the areas of defamation law, copyright, and defining ISP liability for online content. Creating a common platform for legislation at the regional level is a slow and complex process given the significant differences in the cultures and existing legislations in the countries of the European Union.

Author: Sangamitra Ramachander

NOTES

1. See http://europa.eu.int/ISPO/legal/en/internet/communic.html#f10 (accessed May 11, 2007).
2. This has been followed by the Safer Internet Action Plan (2002–2005) and the Safer Internet Plus Programme (2005–2008).
3. See http://europa.eu.int/ISPO/legal/en/internet/communic.html#f10 (accessed May 11, 2007).
4. Even in the case of child pornography, variations between countries exist pertaining to the definition of child pornography, the range of criminal activities that are subject to legislation (the possession, production, and dissemination of material, and so on), the means of investigation, and the penalties. For an overview of the national-level legislation and initiatives to counter child pornography in various countries, see http://www.inhope.org/en/about/about.html (accessed May 11, 2007).
5. See http://europa.eu.int/ISPO/legal/fr/internet/actplan.html (accessed May 11, 2007).
6. http://www.jisclegal.ac.uk/pdfs/isp_liability.pdf (accessed May 11, 2007).
7. However, member states might impose additional obligations for ISPs to immediately convey information to relevant authorities "of alleged illegal activities undertaken, or information provided by recipients of their service." ISPs might also have to provide, on request, information that enables the "identification of recipients of their service with whom they have storage agreements." See http://www.jisclegal.ac.uk/pdfs/isp_liability.pdf (accessed May 11, 2007).
8. Project Cleanfeed is cited as the "first mass censorship of the web attempted in a Western democracy," http://observer.guardian.co.uk/uk_news/story/0,6903,1232422,00.html.
9. For further information on Project Cleanfeed, see http://www.cl.cam.ac.uk/~rnc1/cleanfeed.pdf (accessed May 11, 2007).

10. As of early 2006, 35,000 illegal images were being blocked daily and four million access attempts were recorded in a period of four months among BT subscribers.
11. http://www.theregister.co.uk/2006/12/29/iwf_feature/ (accessed May 11, 2007).
12. Although BT records the number of access attempts, it does not retain information pertaining to the identity of persons who attempt to access these Web sites. See http://technology.guardian.co.uk/news/story/0,,1704342,00.html (accessed May 11, 2007).
13. Internet Watch Foundation, "Frequently Asked Questions by the Media," (page modified January 15th, 2006), http://www.iwf.org.uk/media/page.70.215.htm (accessed May 11, 2007).
14. http://publicaffairs.linx.net/news/?p=518 (accessed May 11, 2007).
15. Project Cleanfeed was introduced in the aftermath of Operation Ore, an operation in the UK that formed a part of a large-scale international police operation to track down pedophiles on the Internet, under which 6,500 police investigations, 1,200 arrests, and 655 convictions were made in the country. The accused were identified based on credit-card information used to access a pedophile Web site hosted in the United States, passed on to the UK by the FBI. (The U.S. counterpart of the project, which preceded Operation Ore, is known as Operation Avalanche). See http://news.bbc.co.uk/2/hi/uk_news/2445065.stm (accessed May 11, 2007).
16. http://press.telenor.com/PR/200505/994781_5.html; http://www.financialmirror.com/more_news.php?id=2574; http://en.wikipedia.org/wiki/Internet_censorship (accessed March 7, 2007); http://www.edri.org/edrigram/number5.1/italy_blocking (accessed May 11, 2007).
17. http://blogs.law.harvard.edu/ugasser/2005/03/10#a52 (accessed May 11, 2007).
18. http://www.heise.de/english/newsticker/news/56817 (accessed May 11, 2007).
19. http://news.bbc.co.uk/1/hi/uk/5213058.stm (accessed May 11, 2007).
20. http://www.theregister.co.uk/2006/06/27/child_convictions_passed_to_banks/ (accessed May 11, 2007).
21. For the list of countries running hotlines and the organizations involved, see http://ec.europa.eu/information_society/activities/sip/projects/hotlines/index_en.htm.
22. http://www.cnn.com/2007/WORLD/europe/02/07/kids.online.porn.ap/index.html.
23. http://www.redbarnet.dk/Files/Filer/Seksuelt_misbrug/Pressemedfebruar07_eng.doc (accessed May 11, 2007).
24. http://ec.europa.eu/information_society/activities/sip/projects/hotlines/index_en.htm (accessed May 11, 2007).
25. http://www.edri.org/edrigram/number4.12/italybetting (accessed May 11, 2007).
26. Article available in Norwegian, http://www.dagbladet.no/dinside/2007/02/12/491719.html, cited in: http://www.opennetinitiative.net/blog/?p=144 (accessed May 11, 2007).
27. The contested Web sites were www.appel-au-peuple.org, http://de.geocities.com/justicecontrol, and www.swiss-corruption.com.
28. http://www.fitug.de/news/newsticker/newsticker120203210053.html.
29. http://www.edri.org/edrigram/number5.5/youtube-turkey (accessed May 11, 2007).
30. http://www.cdt.org/publications/policyposts/2005/5 (accessed May 11, 2007).
31. La Ligue Contre Le Racisme Et l'Antisemitisme (LICRA) and L'Union Des Etudiants Juifs De France.
32. http://www.tomwbell.com/NetLaw/Ch03/YahoovLICRA.html (accessed May 11, 2007).
33. http://www.cdt.org/publications/policyposts/2005/5 (accessed May 11, 2007).
34. BBC News, "The Law, borders, and the Internet," January 24, 2006, http://news.bbc.co.uk/2/hi/technology/4641244.stm (accessed May 11, 2007).
35. Center for Democracy and Technology, "Foreign courts' exercise of jurisdiction over Web content seen in other cases," July 11, 2001, http://www.cdt.org/publications/pp_7.06.shtml. For more information on this case, please refer to the Toben case in the Australia and New Zealand Regional Overview.
36. For further details, see http://md.hudora.de/publications/200306-gi-blocking/200306-gi-blocking.pdf.
37. http://www.edri.org/edrigram/number2.22/filtering (accessed May 11, 2007).
38. I. Brown, "Internet censorship: Be careful what you ask for." *Proc. International Conference on Communication, Mass Media and Culture,* Istanbul, October 2006.
39. http://www.cptech.org/ecom/jurisdiction/defamation2.html (accessed May 11, 2007).
40. http://www.edri.org/edrigram/number5.3/romell (accessed May 11, 2007).
41. Libel law in Britain—known internationally to be particularly strict—was loosened in October 2006. See Sarah Lyall, "High court in Britain loosens strict libel law," *The New York Times,* October 12, 2006, http://www.nytimes.com/2006/10/12/world/europe/12britain.html.

42. Yaman Akdeniz, "Case Analysis of Laurence Godfrey v. Demon Internet Limited," 1999, http://www.cyber-rights.org/reports/demon.htm; Consumer Project on Technology, CPT's Page on Defamation and Libel Cases, http://www.cptech.org/ecom/jurisdiction/defamation2.html.
43. http://www.nytimes.com/2006/10/12/world/europe/12britain.html.
44. http://www.lawcom.gov.uk/docs/defamation(1).pdf (accessed May 11, 2007).
45. http://www.flickr.com/photos/jesper/336756697/ (accessed May 11, 2007).
46. Article at http://www.dagbladet.no/dinside/2007/02/12/491719.html (in Norwegian), cited February 13, 2007, in http://www.opennetinitiative.net/blog/?p=144.
47. BitTorrent is "a peer-to-peer (P2P) communications protocol for file sharing," and is a "method of distributing large amounts of data widely without the original distributor incurring the entire costs of hardware, hosting and bandwidth resources." In this system, "when data is distributed using the BitTorrent protocol, recipients each supply data to newer recipients, reducing the cost and burden on any given individual source, providing redundancy against system problems, and reducing dependence upon the original distributor." A BitTorrent client is any client that implements the BitTorrent protocol, and "each client is capable of preparing, requesting, and transmitting any type of computer file over a network, using the protocol. A peer is any computer running an instance of a client. To share a file or group of files, a peer first creates a 'torrent'. This is a small file which contains metadata about the files to be shared, and about the 'tracker', the computer that coordinates the file distribution. Peers that want to download the file first obtain a torrent file for it, and connect to the specified tracker which tells them from which other peers to download the pieces of the file." See http://en.wikipedia.org/wiki/BitTorrent (accessed May 11, 2007).
48. In May 2006 the Web site administrator and systems operator of www.arenabg.com had been arrested and subsequently released on lack of grounds for arrest.
49. http://www.novinite.bg.
50. http://torrentfreak.com/government-blocks-torrent-site-citizens-protest/ (accessed May 11, 2007).
51. Ibid.
52. http://www.telegraph.co.uk/news/main.jhtml?xml=/news/2007/02/13/wgoogle113.xml (accessed May 11, 2007).
53. http://www.out-law.com/page-7758 (accessed May 11, 2007).
54. Introduced in Belgium in 2006, Google News shows headlines, photos, and the first few lines of news stories with links to the full versions on the Belgian newspaper Web sites.
55. http://www.edri.org/edrigram/number5.3/google-belgium (accessed May 11, 2007).
56. http://news.com.com/2100-1030_3-6174008.html (accessed May 11, 2007).
57. http://wistechnology.com/article.php?id=3548 (accessed May 11, 2007).
58. http://www.jisclegal.ac.uk/ispliability/ispliability.htm (accessed May 11, 2007).
59. http://www.europarl.europa.eu/oeil/file.jsp?id=5275032 (accessed May 11, 2007).
60. http://www.privacyinternational.org/article.shtml?cmd%5B347%5D=x-347-63514 (accessed May 11, 2007).
61. At present, IP addresses, search queries, and cookie details are retained by Google in Europe for eighteen to twenty-four months. After this period, server logs are anonymized and it is no longer possible to identify users.
62. See http://www.edri.org/edrigram/number5.6/google-data-retention (accessed May 11, 2007).
63. http://www.edri.org/edrigram/number5.5/sweden-wiretapping (accessed May 11, 2007).

Internet Filtering in
Latin America

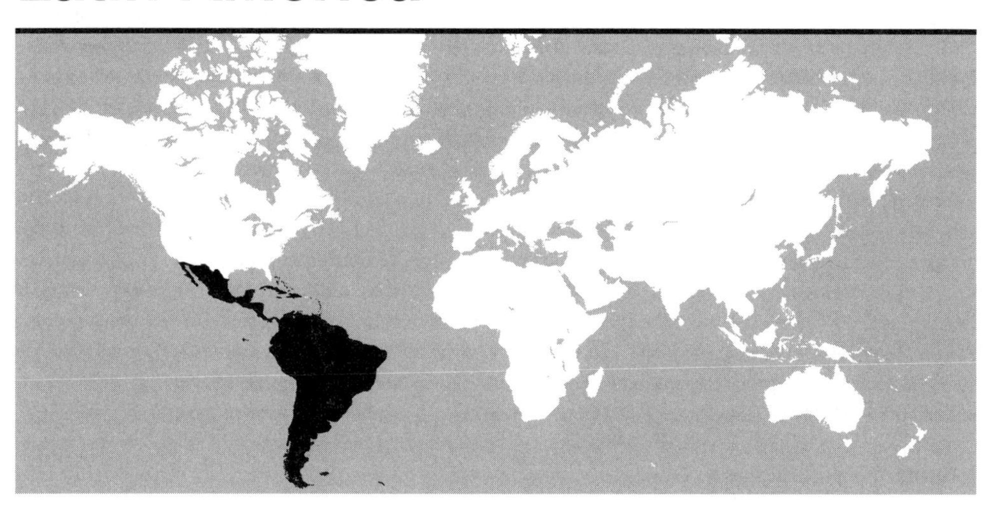

Introduction

With the exception of Cuba, systematic technical filtering of the Internet has yet to take hold in Latin America. The regulation of Internet content addresses largely the same concerns and strategies seen in North America and Europe, focusing on combating the spread of child pornography and restricting children's access to age-inappropriate material. As Internet usage in Latin America increases, so have defamation, hate speech, copyright, and privacy issues.

The judiciary in Latin America has played an important role in shaping and tempering filtering activity, a development common to North America and Europe. At the same time, there has been a wide range of legal and practical responses to regulating Internet activity. Latin American countries have relied primarily upon existing law to craft remedies to these challenges, though a growing number of Internet-specific laws have been debated and implemented in recent years. These issues have been addressed primarily through the application of cease and desist orders in conjunction with requests to have materials removed from search engine results.

Though most Latin American countries have ratified the American Convention on Human Rights,[1] a regional treaty that guarantees the freedom of expression, speech continues to be threatened by government authorities, drug cartels, and others. In particular, journalists have long been targets of a range of attempts to obstruct or limit speech, from government threats to withhold publication licenses to outright intimidation and physical violence. In 2006 and 2007 journalists in Argentina, Bolivia, Brazil, Colombia, Ecuador, Guatemala, Guyana, Haiti, Mexico, Nicaragua, Paraguay, Peru, and Venezuela were threatened, physically attacked, murdered, or simply disappeared.[2] For journalists working in Latin America, death threats were commonplace. In 2006 Mexico surpassed Colombia as Latin America's deadliest country for journalists (second only to Iraq globally), while Cuba has the world's second-biggest prison for journalists.[3]

The level of openness of the media environment in Latin America is reputed to be subject to considerable self-censorship, particularly in Brazil, Colombia, Mexico, and Venezuela.[4] Because of threats from local drug cartels or other gangs and individuals, many journalists practice self-censorship,[5] including many in Colombia who avoid reporting on corruption, drug trafficking, or violence by armed groups. Drug gangs waging a campaign of intimidation in Mexico not only tack notes to corpses and publish newspaper ads, but have also posted a video on YouTube where an alleged Zeta member (a group of cartel operatives) is tortured and decapitated.[6] The few Cubans who gain access to the Internet are limited by extensive monitoring and excessive penalties for political dissent, leading to a climate of self-censorship.

Internet in Latin America

Most countries in Latin America recognize the value of the Internet as an integral part of modern life. For example, numerous groups in Chile have recommended legislation to make access to the Internet a right, alongside access to clean water and shelter.[7] However, the high value placed on Internet access has not in fact resulted in uniformly unfettered access. Although the Cuban government declared Internet access a "fundamental right" of the Cuban people,[8] all Internet access there requires government authorization and oversight by the Cuban Ministry of Computer Technology and Communications.[9]

While estimates vary, the regional penetration rate appears to be approximately 12 percent.[10] More than half of the Internet users in Latin America are in Brazil and Mexico, though Jamaica, Chile, and Argentina have the highest penetration rates (at 44 percent, 34 percent, and 26 percent, respectively).[11] Penetration rates in Argentina, Brazil, and Mexico are clustered close to 17 percent. In Bolivia, only one person in twenty is connected, and in Cuba less than one person in fifty.[12] In 2004 Cuba had the lowest penetration rate in the region, trailing even Nicaragua (2.3 percent) and Paraguay (2.5 percent).[13] Brazil, Mexico, Argentina, and Chile are also the leaders in high-speed Internet access, accounting for 90 percent of all broadband subscribers in 2006 and forming the top four markets for ADSL in the region.[14] Despite the region's low Internet penetration, fixed line and mobile phone subscription continues to grow at an annual rate of 50 percent.[15]

In countries such as Argentina, Chile, and Colombia the process of deregulation has led to a surge in more affordable and increasingly popular services such as voice over Internet protocol (VoIP). Nominally the Cuban Internet service provider (ISP) market was fully competitive by 2000,[16] in contrast to the monopolies in the various telephone, data, and television markets.[17] However, all ISPs remain under government control and oversight; of the ISPs, only CENIAI provides personal internet access to Cuban citizens.[18]

Physical, legal, and economic limits on access to the Internet can constitute the most significant form of governmental control. The Cuban government strongly restricts not only private ownership of computer hardware,[19] but also many public access points to Cuban intranets.[20] In addition to the state prohibition of private computer sales, the Cuban police have also confiscated existing private computers and modems.[21] The lack of private resources forces most Cubans to use public access points, which may allow access only to national e-mail and Cuban intranets.[22] In Venezuela, Internet use is concentrated among young, male, educated city residents, with more than 60 percent of users coming from Caracas and all but the lowest income sector represented.[23] Despite programs promoting Internet use by poor and rural Venezuelans, access for 60 percent of the population remains essentially nonexistent, and basic public education does not incorporate Internet technologies.[24]

At the same time, many governments in Latin America have committed to investing in expanded public access points and creating community telecenters, such as cybercafés, where most users in the region access the Internet.[25] In countries such as Honduras, cybercafés and other public access centers have become the local "telephone booth," providing cheaper and more readily available Internet telephony.[26] Though VoIP is available throughout the region, the regulatory landscape is still evolving, with sometimes contradictory reports on the legality of the service. VoIP is illegal in Cuba, but it is offered with stringent restrictions in countries such as Guyana, Paraguay, and Costa Rica. Licensing requirements also legally restrict which operators can offer VoIP in Bolivia,[27] Mexico, Venezuela, Colombia, Ecuador, Peru, and the Dominican Republic, though these restrictions are not enforced in many countries.[28] In Chile and Brazil, the VoIP markets operate as if unregulated, but they are also evolving.[29] In October 2006, even after deregulation, Telefónica Chile was fined nearly USD1 million for antitrust violations in blocking VoIP calls.[30]

The introduction of Internet services in Latin America has offered citizens opportunities to affect their social and political landscape. For example, bloggers in Mexico inaugurated their coverage of elections in the 2006 presidential campaigns. Social networking sites are also immensely popular. Orkut in Brazil was host to eleven million of Orkut's more than fifteen million users.[31]

Social content

The protection of children is a widely used rationale for filtering the Internet in Latin America. Despite the generally sparse extent of Internet regulation, countries throughout the region have focused on making the access and provision of pornographic material illegal online.

The bulk of the regulatory responsibility for filtering has been delegated to ISPs and public Internet access points such as cybercafés. For example, in 2006 the Venezuelan National Assembly passed a law to safeguard children from illicit content on the Internet, requiring ISPs both to limit content on their servers and to provide free filtering software to users in order to promote self-regulation.[32] Examples of similar mechanisms include the 2002 Argentine Internet Providers Law, which requires all ISPs to provide filtering software to users upon request,[33] and a Colombian law demanding that ISPs monitor their content and report any illegal activity to the government.[34] Colombia's "Internet Sano" (healthy Internet) campaign calls for public education on "decent" ways of using the Internet as well as penalties for improper usage.[35] In Peru it is mandatory for all businesses to have filters installed in all computers designated for use by children.[36] In Buenos Aires businesses offering Internet services that fail to install pornography filters on computers for use by children are subject to fines or temporary closures.[37] Definitions of pornographic content are not always clear; Argentine ISPs expressed concern that the instruction to filter "specific sites" was not adequately precise.[38]

Regulation of child pornography is steadily being expanded to include the Internet. In 2003 Brazil made child pornography illegal in any medium, explicitly including the Internet.[39] Similar laws have been approved and implemented in Buenos Aires[40] and Colombia.[41] In Argentina's proposed draft law on cybercrimes, child pornography is criminalized in "any medium of communication."[42]

In addition to efforts at protecting children from explicit online content, other social content deemed offensive has occasionally come under fire. Since the 1997 presidential declaration regarding "Free Speech on the Internet" that guaranteed Internet content the same constitutional protections for freedom of expression, Argentina has become a haven for neo-Nazi and race-hate groups around the region.[43] In 2000 an Argentine appellate court affirmed a lower court's dismissal of a claim that a Yahoo! site selling Nazi memorabilia violated Argentina's anti-discrimination law (no. 23.592),[44] holding that the equivalent restrictions of non-Internet speech would be unacceptable.

A recent case involving a social networking site illustrates some of the tensions between law enforcement needs and individuals' right to privacy. In 2005 the Brazilian government took issue with Google's social networking site, Orkut, when it became evident that it was being used for the sale of illegal drugs[45] and child pornography, and had also become a domain for racist speech.[46] The National Reporting Center of Cyber Crimes, which operates in partnership with the Ministério Público Federal, brought civil and criminal court lawsuits against Google's Brazilian unit alleging failure to stop the spread of child pornography and hate speech.[47] In 2006 Google agreed to comply with the Brazilian government's request that they track all users and hand over the identities of users involved in these and other illegal activities.[48]

Defamation

The bulk of filtering in the Latin American region arises from court order. Conceptually, defamation covers a broad swath of unlawful acts in the region, primarily distinguished by the status of the person(s) harmed. In addition to defamation of individuals and antidiscrimination laws banning hate speech (group defamation), the majority of countries in Latin America have laws against *desacato* (disrespect, insult against, or comtempt for public figures).[49]

Hate speech is regulated by some Latin American countries. In Brazil, the Criminal Code includes the crime of prejudice on the basis of race, color, religion, ethnic background, or national origin.[50] The Brazilian Constitution, which establishes racism as a crime not entitled to bail or statutes of limitation, has been used as the legal basis for search engine takedowns.[51] In Argentina's antidiscrimination law, a crime is considered more serious if racism is involved.[52]

In Argentina the defendant in the case *Jujuy.com v. Omar Lozano* was found liable for publishing slanderous content on his Web site after imputing adulterous conduct to a couple and failing to remove the content promptly. An injunction was imposed and damages were set at USD40,000.[53] In a "defamation of the public image" case, the Brazilian court ordered the country's seven largest ISPs to block the Web site of a travel company based in the United States called "Tours Gone Wild," which reportedly sells and promotes sexual tourism packages to Rio. A Brazilian citizen had sued the Web site claiming photos were used on the Web site without permission.[54] These judicial strategies may be incorporated into future legislative moves by the Brazilian government or other Latin American countries.

Although many countries have declared *desacato* laws unconstitutional,[55] others—such as Panama[56] and Venezuela[57]—are increasing restrictions on press freedom through such defamation laws. A 1999 *desacato* case in Costa Rica, where the journalist Mauricio Herrera Ulloa published accounts of the illegal acts of a public official, led to a judicial order to remove the name of the plaintiff from a newspaper Web site and to criminal convictions against Herrera Ulloa.[58] However, the Inter-American Court of Human Rights ruled that the conviction of Herrera Ulloa was a violation of his right to freedom of expression under the American Convention on Human Rights.[59]

In 2006 a Brazil court extended the 1967 Press Law[60] to apply to Internet publications and fined a magazine, *Veja Online,* for defaming an ex-official in an article published online.[61] The 2006 elections in Brazil provide a prominent example of *desacato* being brought into cyberspace, as well as the self-regulating stance taken by Brazilian ISPs. Senate candidate (and former President) Jose Sarney sued and won his case in the electoral court (which exists in part to "ensure that all candidates are fairly represented in the media," though this court does "not generally cover defamation")[62] against a blogger who posted a cartoon of Sarney. Even after Alcilene Cavalcante deleted the cartoon as requested by the court, the ISP that hosted her blog (http://www.uol.com.br) proceeded to remove the blog without a court directive to do so. Sarney also filed to sue Cavalcante's sister Alcineia, who divulged details of the case on her blog. Again, without any court order, the ISP also removed Alcineia Cavalcante's blog.[63]

Privacy and confidentiality

The judiciary also continues to play an active role in parsing the scope of privacy rights and confidentiality of data by experimenting with filtering orders. For example, in 2005 a court in Brazil ordered the daily newspaper *Folha Online* to remove from its Web site 165 URLs that detailed how Brasil Telecom allegedly used a Canadian consulting company to spy on its competitor Telecom Italia. On trial for the abuses alleged in these articles, Brasil Telecom requested that the judge issue a writ against *Folha Online.* The articles were published in print a year before the takedown writ was issued, but the Web site was held to have violated the confidentiality of a judicial investigation.[64] However, after protests, the judge reduced the number of pages to be blocked the next day.[65]

The Brazilian judiciary has also engaged with Google over the privacy concerns regarding sexual content that appeared on its video-sharing site YouTube in 2005. After a Brazilian model and her boyfriend sued YouTube for hosting a sexually explicit video they claimed violated their right to privacy, Google agreed to take down the video, but it continued to be put back online by users. In January 2007 a São Paulo judge ordered telecommunications companies to block YouTube until the video was removed from the Web site. Several ISPs, including Brasil Telecom, announced their intention to comply with the court ruling.[66] Days later, the judge revoked his order and lifted the ban on the entire site.[67]

Security and political speech

With the exception of Cuba, there has been no reported technical filtering of content relating to security or political speech. Since it established its first full-time Internet connection in 1996,[68] the Cuban government has combined access restrictions with severe penalties for illegal uses—including violations such as counter-revolutionary writing[69]—to deter free expression online.[70] Regulation outlaws Internet use "in violation of Cuban society's moral principles or the country's laws," as well as e-mail messages that "jeopardize national security."[71] Moreover, the government restricts Internet use by having all legal Cuban Internet traffic pass through state-run ISPs, which use software to detect politically dissident information, and requires ID and registration for Internet use.[72] E-mail messages are monitored prior to being sent or delivered.[73]

Copyright

Many countries in Latin America, including Argentina and Brazil, have attempted to shore up intellectual property rights (IPR) protections by drafting and updating laws and ratifying international agreements such as the World Intellectual Property Organization (WIPO) Copyright Treaty. Other countries, such as Chile and Mexico, have been criticized for having antiquated or weak laws that fail to meet international threshold requirements set by the UN and WIPO.[74] Uneven regulation of IPR is often coupled with a level of enforcement characterized as insufficient or anemic.[75] For example, one of the objectives of the U.S. government in signing a free trade agreement with Chile was to improve protection against piracy for U.S. copyright and trademark holders.[76] As a proposed replacement for the North American Free Trade Agreement, the Free Trade Agreements of the Americas (FTAA) would include every country in Central America, South

America, and the Caribbean, except Cuba. Although the United States has pushed for greater intellectual property protections, negotiations have been stalled since 2005.[77]

At the same time, the drive for enhanced IPR regimes, often led by the United States, has been controversial for a range of reasons, from lack of public support to the nature of the civil law system in many countries.[78] For example, these international "individualistic" and "exclusionary" frameworks have been criticized as alien to many of the unique cultures of the region, indigenous rights, and traditions of collective rights. Panama's IP laws recognize indigenous folklore and knowledge, and in 2000 it became the first country in the world to conceive of a *sui generis* IP system for the protection of indigenous crafts and knowledge.[79]

Other factors

In Latin America economic factors can have a significant impact on citizens' Internet access. In Cuba a combination of Cuban government policy, the U.S. trade embargo, and personal economic limitations prevents the vast majority of Cuban citizens from accessing the Internet. Access is likely restricted even further by the U.S. government's sponsorship of reverse filtering, which encourages Web sites to prevent access from Cuba and other countries.

In Venezuela, President Hugo Chávez's announcement on January 8, 2007, of re-nationalization plans for the telecom CANTV[80] has heightened fears of expanded regulation and content restrictions as the government assumes greater control of Internet media. A recent article notes that CANTV has held 83 percent of the Internet market since the market's privatization,[81] so any changes in filtering through a nationalized CANTV will have a strong impact on Internet users.

Conclusion

Governments and especially courts in Latin America are engaged in an adaptive process of regulating online activity and content. Only Cuba employs systematic technical filtering, with many countries delegating the responsibility for filtering content unsuitable for minors to ISPs. In addition, a wide range of actors—including government officials, telecom companies, individuals, and judges—have attempted to induce or enforce filtering on a case-by-case basis, often with negotiated and shifting results. The ad hoc approaches that have been applied thus far suggest that efforts to control Internet content in Latin America are still unsettled and contested; this promises to be an area of considerable change in the coming years.

Authors: Jehae Kim, Patricio Rojas, Joanna Huey, Kathleen Connors, Stephanie Wang

NOTES

1. Article 13, American Convention on Human Rights, O.A.S.Treaty Series No. 36, 1144 U.N.T.S. 123, entered into force July 18, 1978, reprinted at http://www1.umn.edu/humanrts/oasinstr/zoas3con.htm.
2. Committee to Protect Journalists, Americas, http://www.cpj.org/regions_06/americas_06/americas_06.html and http://www.cpj.org/regions_07/americas_07/americas_07.html; World Association of Newspapers, "Media employees killed in 2006," http://www.wan-press.org/rubrique.php3?id_rubrique=863; and World Association of Newspapers, "Killing the messenger: Report of the global inquiry by the International News Safety Institute into the protection of journalists," March 2007, http://www.wan-press.org/IMG/pdf/REPORT_FINAL.pdf.
3. World Association of Newspapers, "Press freedom, world review, November 2005–May 2006," June 3, 2006, http://www.wan-press.org/print.php3?id_article=12552.
4. See http://www.cpj.org/attacks05/americas05/americas05.html.
5. Inter American Press Association, "IAPA conclusions on Press Freedom in the Americas," March 19, 2007, http://www.sipiapa.org/espanol/pressreleases/chronologicaldetail.cfm?PressReleaseID=1869.

6. Julie Watson, "Mexican drug gangs spread fear through Internet, newspaper ads, messages tacked to dead," Associated Press, April 12, 2007.
7. See the Banda Ancha campaign of Atina Chile (http://www.atinachile.cl/taxonomy/term/150); the Metropolitan Technological University (http://www.utem.cl/cyt/derecho/gobierno.html); and the Center for Informational Rights of the University of Chile (http://www.cedi.uchile.cl/), as well as its book listing (http://www.cedi.uchile.cl/catalogo/derechoinformatico/index.html).
8. Patrick Symmes, "Che is dead," *Wired,* February 1998, http://www.wired.com/wired/archive/6.02/cuba.html.
9. Reporters Without Borders, "Going online in Cuba: Internet under surveillance," October 2006, http://www.rsf.org/IMG/pdf/rapport_gb_md_1.pdf.
10. World Bank Group, *World Development Indicators Database,* April 2006 (data are from 2005), Latin America & Caribbean Data Profile, http://devdata.worldbank.org/external/CPProfile.asp?PTYPE=CP&CCODE=LAC.
11. Paul Budde Communication Pty Ltd., Regional: Internet: The Americas – 2006, p. 27.
12. International Telecommunications Union, ITU ICT Eye, Cuba, at http://www.itu.int/ITU-D/icteye/DisplayCountry.aspx?countryId=63.
13. Paul Budde Communication Pty Ltd., Regional: Internet: The Americas – 2006, p. 27.
14. Ibid. p. 28.
15. World Bank Group, *World Development Indicators Database,* April 2006 (data are from 2005), Latin America & Caribbean Data Profile, http://devdata.worldbank.org/external/CPProfile.asp?PTYPE=CP&CCODE=LAC.
16. International Telecommunication Union, Trends in Telecommunication Reform 2000–2001, p. 193, http://www.ituarabic.org/arabbook/2004/GTTR-2000.pdf; see also http://www.cuba.cu/sitios.php?idrcategoria=8&base=0 (listing Cuban internet providers).
17. International Telecommunication Union, Trends in Telecommunication Reform 2000–2001, p. 193, http://www.ituarabic.org/arabbook/2004/GTTR-2000.pdf.
18. Dana Bomkamp and Maria Soler, Information Technology in Cuba, http://www.american.edu/carmel/ms4917a/Internet%20Diffusion.htm.
19. Patrick Symmes, "Che is dead," *Wired,* February 1998, http://www.wired.com/wired/archive/6.02/cuba.html.
20. Reporters Without Borders, "Going online in Cuba: Internet under surveillance," October 2006, http://www.rsf.org/IMG/pdf/rapport_gb_md_1.pdf.
21. Patrick Symmes, "Che is dead," *Wired,* February 1998, http://www.wired.com/wired/archive/6.02/cuba.htm.
22. Reporters Without Borders, Mexico: Annual Report 2007, http://www.rsf.org/article.php3?id_article=20539; Reporters Without Borders, Cuba:Annual Report 2007, http://www.rsf.org/article.php3?id_article=20534.
23. *Global Competitiveness Report 2001–2002,* Harvard Center for International Development, http://www.cid.harvard.edu/cr/profiles/Venezuela.pdf.
24. Ibid.
25. Paul Budde Communication Pty Ltd., Regional: Internet: The Americas – 2006, p. 27.
26. Paul Budde Communication Pty Ltd., Regional: Infrastructure: VoIP NGNs in the Americas – 2005, December 31, 2006, p. 36.
27. In Bolivia, however, there is virtually no enforcement of this law. There is private use of VoIP, and most Internet cafés provide a VoIP service. E-mail to Open Net Initiative, April 2, 2007.
28. International Telecommunication Union, The Future of Voice: Ruling Voice over IP: Challenges for Regulators in Latin America, June 16, 2006, http://www.itu.int/osg/spu/ni/voice/documents/Background/VoIP_LatinAmerica_Nathaly_Rey.pdf.
29. Paul Budde Communication Pty Ltd., Regional: Infrastructure: VoIP NGNs in the Americas – 2005, December 31, 2006, pp. 32–3.
30. International Telecommunication Union, REGULATORY Newslog, VoIP Regulation due early 2007, January 4, 2007, http://www.itu.int/ituweblogs/treg/default,date,2007-01-09.aspx.
31. *The New York Times,* "A website born in U.S. finds fans in Brazil," April 10, 2006, http://www.nytimes.com/2006/04/10/technology/10orkut.html?ex=1302321600&en=81a68673b731539d&ei=5088.
32. Law no. 38.529, Asamblea Nacional de la Republica Bolivariana de Venezeula, "Ley de Protección de Niños, Niñas y Adolescentes en salas de uso de Internet, Vídeo Juegos y otros Multimedia" November 5, 2006, http://www.asambleanacional.gov.ve/ns2/leyes.asp?id=741&dis=1 (in Spanish).
33. Law no. 25.690. See InfoLeg, Centro de documentacion e informacion del Ministerio de Economia, "Establécese que las empresas ISP (Internet Service Provider) tendrán la obligación de ofrecer software de protección que impida al acceso a sitios específicos," November 28, 2002, http://infoleg.mecon.gov.ar/infolegInternet/anexos/80000-84999/81031/norma.htm (in Spanish).
34. Privacy International, Silenced: Latin America Profile, January 1, 2003, http://www.privacyinternational.org/article.shtml?cmd%5B347%5D=x-347-103798.

35. Ministerio de Comunicaciones, "Que Es Internet Sano: Idea de la Campana," http://www.internet-sano.gov.co/que_es.htm (in Spanish).
36. El Congreso De La Republica, Ha dado la Ley siguiente: Ley que Establece la Obligacion de Filtros Antipornograficos en Instituciones Educativ as y Bibliotecas que Bridnen Acceso a Internet, http://www2.congreso.gob.pe/Sicr/Relat Agenda/proapro.nsf/ProyectosAprobadosPortal/E72D2E5A68F3267E0525715400031C02/$FILE/12756Fil trosantipornograficos.pdf (in Spanish).
37. Law no. 863, La Legislatura de la Ciudad Autonoma de Buenos Aires, "Sanciona con fuerza de Ley," August 15, 2002, http://www.cedom.gov.ar/es/legislacion/normas/leyes/html/ley863.html (in Spanish).
38. The "protection software that impedes access to specific sites" is for the protection of children, but the vague terms of the law make companies confused about the requirements for the filtering software. Lanacion, "Las proveedoras denuncian censura en Internet," January 12, 2003, http://www.lanacion.com.ar/Archivo/nota.asp?nota_id=465416 (in Spanish).
39. Privacy International, Silenced: Latin America Profile, January 21, 2003, http://www.privacyinternational.org/article.shtml?cmd%5B347%5D=x-347-103798.
40. Law no. 863, La Legislatura de la Ciudad Autonoma de Buenos Aires, "Sanciona con fuerza de Ley," August 15, 2002, http://www.cedom.gov.ar/es/legislacion/normas/leyes/html/ley863.html (in Spanish).
41. Ministerio de Comunicaciones, Law no. 679 of 2001, http://www.internetsano.gov.co/ley679.htm.
42. Draft Law on Cybercrimes, Article 15, Argentina Association for Internet Security, http://www.asira.org.ar/07bas_del_proycomision.htm.
43. Colin Barraclough, "Race-hate groups find virtual haven in Argentina," August 23, 2002, http://www.csmonitor.com/2002/0823/p07s02-woam.htm; Decreto 1279/97, Declárase comprendido en la garantía constitucional que ampara la libertad de expresión al servicio de INTERNET, http://mepriv.mecon.gov.ar/Normas/1279-97.htm (in Spanish).
44. Privacy International, Silenced: Argentina, January 21, 2003, http://www.privacyinternational.org/article.shtml?cmd%5B347%5D=x-347-103569; Legal mania, "La Libertad de Expresion y la Difusion de sumbolos Nazis," June 12, 2000, http://www.legal-mania.com/actualidad_general/simbolos_nazis.htm (in Spanish).
45. BBC News, "Google site 'used by drug gang,'" July 22, 2005, http://news.bbc.co.uk/2/hi/technology/4706489.stm.
46. *The New York Times,* "A website born in U.S. finds fans in Brazil," April 10, 2006, http://www.nytimes.com/2006/04/10/technology/10orkut.html?ex=1302321600&en=81a68673b731539d&ei=5088.
47. SaferNet Brasil, "Crimes on Orkut: Understand the chronological evolution of the case that defies the institutions of the democratic judicial state in Brazil," September 14, 2006, http://www.denunciar.org.br/twiki/bin/view/SaferNet/CrimesOrkutEn.
48. Ellen Nakashima, "Google to give data to Brazilian court," *Washington Post,* September 2, 2006; Richard Waters, "Brazil lawyers lean on Google," *Financial Times,* August 2, 2006, http://www.ft.com/cms/s/d2420942-32f9-11db-87ac-0000779e2340.html.
49. Organization of American States, " 'Desacato' Laws and Criminal Defamation," Chapter V, 6, 2004, http://www.cidh.org/Relatoria/showarticle.asp?artID=310&lID=1.
50. Law no. 7716, 1997, http://www.cejamericas.org/doc/proyectos/raz-sistema-jud-racismo2.pdf.
51. See Chilling Effects database, http://chillingeffects.org/international/notice.cgi?NoticeID=6673.
52. Law no. 23.592.
53. The defendant was found liable under Article 1113, 2, of the Civil Code. Alfredo M. O'Farrell, lawyer at Marval, O'Farrell & Mairal, e-mail to Open Net Initiative contact, November 14, 2006.
54. Opovo, "TJ manda Bloquear site por favorecer turismo sexual," February 7, 2007, http://www.opovo.com.br/opovo/brasil/668706.html.
55. Argentina, Costa Rica, Paraguay, Preu, Guatemala, Honduras, Chile, and Mexico have declared *desacato* laws to be unconstitutional. See Organization of American States, "The Office of the Special Rapparteur for Freedom of Expression of the IACHR Expresses its Satisfaction with Decisions in Guatemala and Honduras Declaring Desacato Laws Unconstitutional," July 5, 2005, http://www.oas.org/OASpage/press_releases/press_release.asp?sCodigo=PREN-126E; Committee to Protect Journalists, "Attacks on the press in 2005, Americas," 2005, http://www.cpj.org/attacks05/americas05/snaps_americas_05.html; and International Freedom of Expression Exchange, "Venezuela: Government Tightens '*desacato*' laws," 2005, http://www.ifex.org/en/content/view/full/65788/.

56. This new law "makes it a crime to publish information received ... concerning a third party ... without express permission of that third party, if the individual concerned claims that the publication would cause him or her prejudice." Panama: Proposed Criminal Code Severely Restricts Freedom of Expression and Information, Article 19, February 22, 2007, http://www.article19.org/pdfs/press/panama-criminal-code.pdf; and International Freedom of Expression Exchange, "President signs into law penal code amendments that threaten press freedom," March 27, 2007, http://www.ifex.org/en/content/view/full/82026. Though Panama had made moves in 2005 to improve press freedoms, *desacato* is still legally valid. Committee to Protect Journalists, "Attacks on the press in 2005: Americas," http://www.cpj.org/attacks05/americas05/panama_05.html.
57. Human Rights Watch, "Venezuela: Curbs on free expression tightened," March 24, 2005, http://hrw.org/english/docs/2005/03/24/venezu10368.htm.
58. For more details on the case, see Article 19, Global Campaign for Free Expression, *amicus curiae* brief on Inter-American Court of Human Rights Case No. 12. 367 "La Nacion," Defamation Law as Restriction on Freedom of Expression, http://www.article19.org/pdfs/cases/costa-rica-written-comments-in-ulloa-v.-costa-.pdf.
59. Inter-American Court of Human Rights, Case of *Herrera-Ulloa v. Costa Rica*. Judgment of July 2, 2004, http://www.corteidh.or.cr/docs/casos/articulos/seriec_107_ing.pdf.
60. See Reporters Without Borders, Brazil: Annual Report 2006, http://www.rsf.org/article.php3?id_article=17415, stating that the 1967 press law has never been repealed. In this law, "insults" and "libel" are crimes whose sentences can be increased if a public official has been targeted. See also Press Law no. 5.250/67, Article 75, http://www.sipiapa.com/projects/laws-bra7.cfm.
61. Knight Center for Journalism in the Americas, University of Texas at Austin, http://knightcenter.utexas.edu/newsmonitor_article.php?page=6913; and *Consultor Jurídico,* "Comunicação na rede Para Justiça, Internet está sujeita à Lei de Imprensa," (in Portuguese) November 22, 2006, http://conjur.estadao.com.br/static/text/50369,1.
62. Reporters Without Borders, "Electoral court censors blog that posted cartoon of senatorial candidate," September 8, 2006, at http://www.rsf.org/article.php3?id_article=18801.
63. Global Voices Online, "Election and censorship dialectics in the Brazilian blogosphere," September 1, 2006, http://www.globalvoicesonline.org/2006/09/01/election-and-censorship-dialectics-in-the-brazilian-blogosphere/.
64. Reporters Without Borders, "Judge cuts back censorship order, allowing websites to reinstate banned pages," December 16, 2005, http://www.rsf.org/article.php3?id_article=15908.
65. Ibid.
66. Reuters, "Phone companies in Brazil blocking YouTube," January 9, 2007, http://www.reuters.com/article/oddlyEnoughNews/idUSN0841810920070109.
67. Reporters Without Borders, "Judge lifts blocking order on YouTube," January 9, 2007, http://www.rsf.org/article.php3?id_article=20342.
68. Patrick Symmes, "Che is dead," Wired, February 1998, http://www.wired.com/wired/archive/6.02/cuba.html.
69. The government "concedes that some sites are blocked, but say these are 'terrorist, xenophobic, or pornographic.' Websites based in the US which publish articles by dissidents from within Cuba are generally inaccessible." BBC News, "Web censorship: Correspondent reports: Cuba: Stephen Gibbs, Havana," May 29, 2006, http://news.bbc.co.uk/2/hi/technology/5024874.stm.
70. Reporters Without Borders, "Going online in Cuba: Internet under surveillance," October 2006, http://www.rsf.org/IMG/pdf/rapport_gb_md_1.pdf.
71. 1996 Decree-Law 209; see Reporters Without Borders, "Going online in Cuba: Internet under surveillance," October 2006, http://www.rsf.org/IMG/pdf/rapport_gb_md_1.pdf.
72. Ibid.
73. Reporters Without Borders, "Going online in Cuba: Internet under surveillance," October 2006, http://www.rsf.org/IMG/pdf/rapport_gb_md_1.pdf.
74. See R. Craig Woods, "Comment and Casenote: The United States-Chile free trade agreement: Will it stop intellectual property piracy or will American producers be forced to walk the plank?" *Law and Business Review of the Americas,* Spring 2004 (10): 425.
75. Marcos J. Basso and Adriana C.K. Vianna, "The Internet in Latin America: Barriers to intellectual property protection: Intellectual property rights and the digital era: Argentina and Brazil," Sypmosium article, *The University of Miami Inter-American Law Review,* Spring 2003 (34): 277. See also James F. Smith, Katherine C. Pearson, and Michael L. Rustad, Book Review: "Why Mexico? Why Mexican Law? Why Now? A Review Essay of *Mexican Law*, by Stephen Zamora, Jose Ramon Cossio, Leonel Pereznieto, Jose Roldan-Xopa, and David Lopez," *Penn State International Law Review,* Fall 2005 (24): 412.

76. See R. Craig Woods, "Comment and Casenote: The United States-Chile free trade agreement: Will it stop intellectual property piracy or will American producers be forced to walk the plank?" *Law and Business Review of the Americas,* Spring 2004 (10): 425.
77. Caribbean Media Corporation, "CARICOM still holding out FTAA hope," December 8, 2006, reprinted by BBC Monitoring International Reports.
78. See Guillermo Cabanellas, The Internet in Latin America: Barriers to Intellectual Property Protection: Law of the Internet in Argentina," Symposium article. *The University of Miami Inter-American Law Review,* Spring 2003 (34): 247. See also Irma De Obaldia, "Note: Western intellectual property and indigenous cultures: The case of the Panamanian indigenous intellectual property law," *Boston University International Law Journal,* Fall 2005 (23): 337.
79. See Irma De Obaldia, "Note: Western intellectual property and indigenous cultures: The case of the Panamanian indigenous intellectual property law," *Boston University International Law Journal,* Fall 2005 (23): 337.
80. Alberto Padillo. "Analysis. Chavez living up to radical promise," CNN.com, January 16, 2007, http://www.cnn.com/ROOT/WORLD/americas/01/16/venezuela.chavez/index.html.
81. La Red. "CANTV se convertiría en el carrier que no pudo ser," http://www.lared.com.ve/archivo/telco12-01-07.html (accessed October 2, 2007).

Internet Filtering in the

Middle East and North Africa

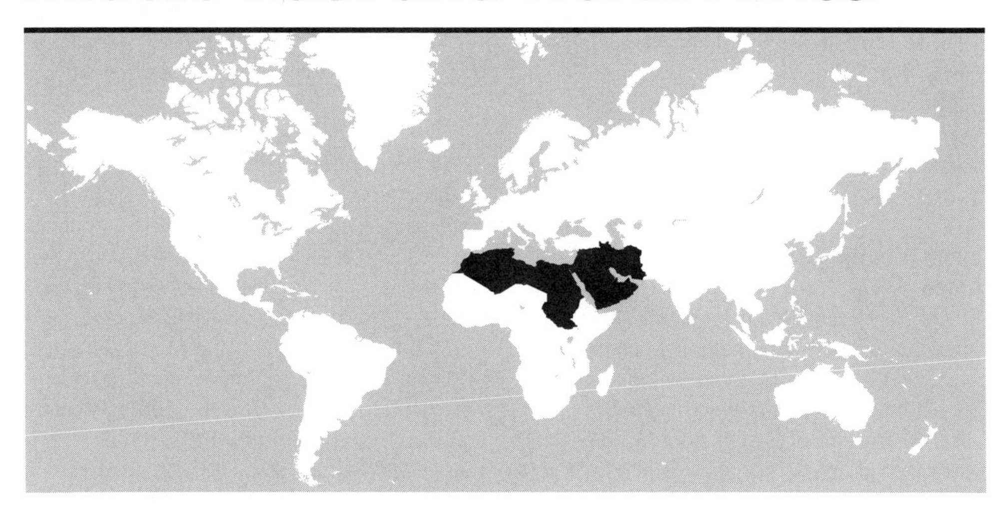

Overview

ONI conducted in-country testing for Internet filtering in sixteen countries in the North Africa and Middle East region. We found that eight of these countries broadly filter online content: Iran, Oman, Saudi Arabia, Sudan, Syria, Tunisia, United Arab Emirates, and Yemen. Another four—Bahrain, Jordan, Libya, and Morocco—carry out selective filtering of a smaller number of Web sites. ONI found no evidence of consistent technical filtering used to deny access to online content in Algiers, Egypt, Iraq, or Israel.

Most of the sites targeted for blocking are selected because of cultural and religious concerns about morality. Political filtering, however, is the common denominator in the region. Bahrain, Jordan, Libya, and Syria focus their filtering efforts primarily on political content. Iran, Oman, Saudi Arabia, Sudan, Tunisia, the United Arab Emirates, and Yemen, on the other hand, not only extensively filter political content but also pervasively block content that is perceived to be religiously, culturally, or socially inappropriate.

Regional and internal political conflicts are also behind content blocking. For example, Syria and the United Arab Emirates block all Web sites within the Israeli domain. Morocco blocks Web sites arguing for the independence of Western Sahara.

Internet censorship in the Middle East and North Africa is multilayered, relying on a number of complementary strategies in addition to technical filtering; arrest, intimidation, and a variety of legal measures are used to regulate the posting and viewing of Internet content.

Introduction

Most of the states in the Middle East and North Africa introduced the Internet in their countries to promote economic development and competitiveness; however, they soon realized that the Internet made it more difficult for them to control the flow of information both within the country and across international borders.

States' power to regulate social, economic, and political activities started to erode as citizens and other nonstate actors, empowered by the

Internet, started to create and disseminate information. The Internet, along with satellite television networks, has effectively broken the monopoly of many Middle Eastern and North African governments. The availability and accessibility of information, as well as the ability to create and disseminate information anonymously, has led to a sense of freedom among many Arab Internet users.

Internet in the Middle East and North Africa

Though some countries in the region enjoy widespread and easy access to the Internet, the region is also home to some of the least-connected countries in the world. While in 2006 some 61 percent of Israelis and 35 percent of Emiratis had regular access to the Internet, Internet penetration still lags behind in most of the region. In fact, according to the International Telecommunication Union, less than 4 percent of people in the Arab world use the Internet regularly.[1] In many countries, poor infrastructure along with economic barriers remain the biggest obstacle to expanding access to the Internet. In Yemen, for example, less than 1 percent of the population uses the Internet (there are 0.87 users per 100 inhabitants), and there are only 300,000 personal computers in the country (1.5 per 100 inhabitants).[2] In Syria fewer than six out of one hundred people regularly use the Internet. In Iraq the Internet penetration rate is 0.1 percent.

Interestingly, broadband Internet access is growing faster in the Middle East and Africa than in any other region in the world. The number of broadband subscribers grew by 38 percent in 2006, while the number of those subscribers using DSL access technology grew by 82 percent, to 4.3 million.[3]

The highest rates of broadband penetration in the region are found in Qatar, the United Arab Emirates, and Lebanon. Half of all households in Qatar, almost one-third in the United Arab Emirates, and one-quarter in Lebanon have a broadband connection. Three countries—Tunisia, Qatar, and Egypt—experienced a remarkable broadband growth rate, doubling in a year. At the same time, DSL subscriptions in the United Arab Emirates increased by almost two-thirds.[4]

State response: Censorship

As the Internet has proven to be a new space for the extension of power and for different nonstate players to compete for influence, states have perceived this as a potential threat. The response has been filtration and surveillance. For the many restrictive governments of the Middle East and North Africa, embracing the Internet meant providing citizens with access to troubling content and ideas, along with new methods to circumvent traditional controls on life and discourse. In reaction, many governments in the region have chosen to restrict online freedom, giving the Middle East and North Africa one of the most repressive Web environments in the world. The Middle East and North Africa is home to five of the thirteen countries listed as enemies of the Internet by Reporters Sans Frontieres.[5]

ONI testing has confirmed that governments and Internet service providers (ISPs) have blocked content they have designated as morally offensive, in violation of public ethics and order, or critical of governments, leaders, or ruling families.

To one degree or another, the Gulf countries, as well as Iran, Sudan, Tunsia, and Yemen, block content related to pornography, homosexuality, dating, and provocative attire. Some of these countries also censor topics considered sensitive or forbidden under Islam, such as gambling, alcohol, and drugs, along with Web sites that feature nudity, even if in a non-erotic context. A few countries, such as Saudi Arabia and the United Arab Emirates, ban access to Web sites that are critical of Islam and those that promote conversion to Christianity.

Many states in the region were found to block political content, or to have blocked such content in the past. For example, Bahrain, Saudi Arabia, Syria, and Tunisia consistently block Web sites of opposition groups. Egypt has intermittently blocked the Web site of the Muslim Brotherhood, an Islamist group critical of the government, as well as the site hosting the online version of the Labor Party's newspaper, which had previously been banned in its hard copy. Yemen temporarily blocked political Web sites in the run-up to the 2006 presidential elections, while Bahrain did the same ahead of parliamentary elections. One political Web site was found to be blocked in Jordan.

Several countries, including Bahrain, Saudi Arabia, and Tunisia, also restrict access to material from human rights organizations, particularly sites that have published reports that are especially critical about those countries.

Legal and regulatory frameworks

Communications services have been liberalized in several countries in the Middle East and North Africa in the past few years, and there are attempts to liberalize more markets in other countries. Some countries have passed legislation that regulates the telecommunications sector and allows the participation of the private sector in the communication industry. The past few years have also witnessed the establishment of telecommunications regulatory authorities.

Most of the countries in the region do not have Internet-specific legislation, though some countries have started to adopt these laws.

In February 2006 the United Arab Emirates issued a federal law designed to combat cybercrime. This law criminalizes certain online activities such as "setting up a website or publishing information for groups calling for facilitating and promoting ideas in breach of the general order and public decency," and "setting up a website or publishing information for a terrorist group under fake names with intent to facilitate contacts with their leadership, or to promote their ideologies and finance their activities, or to publish information on how to make explosives or any other substances to be used in terrorist attacks."[6]

In October 2006 Saudi Arabia also issued a law that criminalizes, among other things, "[e]avesdropping on, tapping or obstructing information sent through the Internet or a computer without legal justification," "[d]efaming others or harming them through the different means of information technology," and "[e]stablishing a Web site for terrorist organizations and/or publishing it in order to aid the leaders of these organizations or any of their members, or promoting their ideas, or financing them, or publishing how to make explosives or other weapons used in terrorist acts."[7]

Journalists and citizen journalists have been detained under emergency laws, vague media laws, or penal codes. Others have faced extralegal harassment and intimidation from security agencies. Governments have further blocked access to new services, citing security concerns. In 2006 the voice-over Internet protocol (VoIP) service Skype and Google Earth were briefly banned in Jordan and Bahrain, respectively. In both cases, the government cited security concerns.

The use of Internet is also regulated by ISPs' terms of use that in some cases mandate that users not carry out activities that contradict the social, cultural, political, religious, or economic values of the state. In some cases, users are asked to sign written agreements to this effect.

Transparency

Some countries in the region openly acknowledge their practice of Internet filtering. Saudi Arabia and Sudan publish details about what they filter, how, and why. They also make available information about their Internet filtering policies, procedures, and other related materials, such as the impact of their filtering systems on connectivity. An Iranian official recently boasted that Iran has censored ten million Web sites, and that they add 1,000 Web sites to the blacklist every month.[8] However, even where a state admits some filtering, it may not admit targeting political opposition, dissidents, or critical human rights reports.

Some ISPs acknowledge filtering by serving blockpages when users try to access banned content. A blockpage usually alerts users that they tried to access illegal Web sites; some invite users to suggest the removal of the block on the Web sites if they think they were erroneously blocked. Some ISPs also ask users to volunteer suggestions for the blacklists.

Countries such as Syria and Tunisia attempt to hide their filtering regimes by returning blockpages disguised to look like error messages. Users in Libya receive time-out messages when they try to access banned content.

Overblocking

Internet filtering will inherently lead to either overblocking or underblocking of targeted content. Many countries in the region are reasonably successful at blocking what they openly declare to be the target of their filtering system, without excessively high rates of overblocking. Others, however—such as Iran, Saudi Arabia, and the United Arab Emirates—not only extensively block targeted content but they also unnecessarily overblock unrelated content. For instance, Iran and the United Arab Emirates block www.flickr.com entirely because they have deemed some of the photographs posted on the site objectionable. Also most of ISPs in countries such as Saudi Arabia, Syria, Tunisia, the United Arab Emirates, and Yemen prevent Internet users from legitimately using privacy and anonymizing tools and online translation services because they can be used to bypass the filtering systems.

Filtering tools

The majority of the ISPs in the region rely on commercial filtering software, primarily applications produced by U.S.-based companies Secure Computing and Websense. This software allows ISPs, often acting on the behest of governments, to filter by category based on lists of pages updated by the company. The categories that ISPs choose to filter can differ widely between countries. In some cases ISPs block individual Web sites' URLs or entire top-level domains, as in the case of Syria and the United Arab Emirates, both of which block access to the Israeli top-level domain. In addition, some ISPs block search strings that contain objectionable keywords. For example, the Yemeni ISP Ynet blocks the use of the word *sex* in search strings, and the Emirati ISP Etisalat bans the use of several keywords that could return erotic images.

In addition, some ISPs block access to cached copies of certain Web sites as an extra measure to prevent access to their content. Most notably, the U.S.-based, Arab-language online newspaper www.arabtimes.com is blocked in several Arab countries, as is access to the cached copy of the page in Google.

Iran, in addition to blocking Web sites, restricts users' ability to access online content by limiting their Internet speed. In October 2006 the Ministry of Communications and Information Technology ordered ISPs to limit the connection speed they offer to 128 Kb/s in order to hinder users' ability to download foreign cultural products (such as music and films) and to organize political opposition.[9]

Physical restrictions and filtering

An additional mode of control is to regulate the places where users access the Internet. In many countries in the Middle East and North Africa, users primarily go online at Internet cafés. Some governments require these cafés to maintain lists of their patrons and keep an eye on their activities. Yemen and Oman require that computer screens be visible to café managers at all times; indeed, Oman requires that prospective café owners submit a floor plan in their application package. The authorities give specific instructions on how Internet cafés should be designed; these instructions include the height, depth, and width of partitions between computers.

Temporary and event-based blocking

Some countries block Web sites at sensitive political moments. As stated previously, in 2006 Bahrain blocked several Web sites in the run-up to the country's parliamentary elections and Yemen banned access to several media and local politics Web sites ahead of the country's presidential elections.

Another example of temporary blocking mentioned earlier is the banning of access to the VoIP service Skype in Jordan and Google Earth in Bahrain in 2006.

Control without filters

Several countries in the region do not have technical filtering in place, or they selectively filter sensitive content. This, however, does not necessarily mean that there is no media censorship in these countries. Citizens in these countries are able to enjoy unfettered access to the Internet because filtering is either very selective or nonexistent. But sweeping media laws lead to pervasive self-censorship and, in some cases, detention. The intimidating laws discourage users from engaging in political and social conversations online.

Jordan, for example, blocks very few Web sites, but media laws curb the freedom of the press and encourage some measure of self-censorship in cyberspace.[10] Citizens have reportedly been questioned and arrested for Web content they have authored.[11] Similarly, the Egyptian government no longer blocks Web sites, but it has detained people for their online activities. On February 22, 2007, a court in Alexandria sentenced a blogger to four years in prison for "incitement to hate Muslims" and "insulting the president."[12]

The Iraqi government does not block Web sites, but the war there makes it difficult for people to access the Internet and makes it dangerous to express political opinions online.

Internet censorship: The users' response

Users may exploit alternative technologies to circumvent filtering systems when censorship imposed by ISPs restricts access to content. In Saudi Arabia, for example, 93 percent of Internet users regularly try to access blocked Web sites, according to an official at King Abdul Aziz City for Science and Technology (KACST) once responsible for overseeing the country's filtering system.[13]

Many Web sites that discuss sensitive issues and feel that they are likely to be blocked use services such as Yahoo! Groups as part of their contingency plans. Once the Web sites are blocked by ISPs, users continue to exchange content via e-mail. Because it is very difficult for ISPs to filter e-mail discussions, group conversation continues to be virtually uncensored.

Other Web sites and discussion forums post tutorials for their visitors that describe how to use circumvention tools to bypass local filtering systems even before they are actually blocked. One Arabic political Web site's home page once read, "Click here to enter our Web site and click here to learn how to access us once we are blocked."

Another trick used by Internet users is the dissemination of controversial content in a large number of Web sites that are unknown to the ISPs. When the novel *Girls of Riyadh* was banned in Saudi Arabia, for example, the full text was posted in tens of Saudi Arabian forums and blogs that have low visibility. Although this is a violation of copyright, it is also an example of how banned content is being distributed and is evidence that blocking the flow of information is not as easy as was once thought.

Some technologically sophisticated user groups went as far as developing their own circumvention tools. In fact, a special Web browser once emerged on the Internet that enabled users to access blocked Jihadi-oriented Web sites.

Conclusion

Though the Internet is growing rapidly in many countries and high-speed access is spreading, most countries in the Middle East and North Africa maintain control over what citizens can see and say online. Authorities use technology and legal and physical restrictions to limit what users can access online. While filtering is primarily based on religious and cultural concerns, most countries in the region also filter some political content.

Even in countries that filter little or no online content, legal restrictions and extralegal harassment from security agencies can still be used to cow or silence online critics.

In addition, although some countries openly acknowledge practicing Internet filtering of religiously and culturally objectionable content, there is less openness when it comes to blocking of political oppositional content. Other countries deliberately try to obscure the fact that they are filtering content by producing false error messages or time-out messages.

Furthermore, as governments try to prevent people from circumventing filtering, they inflict collateral damage on the Internet, preventing users from using useful and politically neutral services such as privacy tools and online translation services.

In sum, filtering in the MENA region demonstrates an ongoing struggle between the filtering states' desire to integrate into the global economy and their efforts to restrict and prevent access to what they deem to be dissident activities or objectionable materials.

Authors: Helmi Noman, Elijah Zarwan

NOTES

1. AME Info, Arab internet woes, http://www.ameinfo.com/80162.html.
2. International Telecommunication Union, Internet Indicators: Hosts, Users and Number of PCs (2005), http://www.itu.int/ITU-D/icteye/Indicators/Indicators.aspx.
3. AME Info, "DSL Forum acclaims Middle East and Africa broadband growing faster than any region in the world," April 12, 2007, http://www.ameinfo.com/116548.html.
4. Ibid.
5. Reporters Without Borders, "List of the 13 Internet enemies in 2006," November 7, 2006, http://www.rsf.org/article.php3?id_article=19603.
6. Gulf News, "UAE cyber crimes law," February 13, 2006, http://archive.gulfnews.com/uae/uaessentials/more_stories/10018507.html.
7. Saudi Gazettee, Cyber Crime Regulations, April 7, 2007, translated, http://www.saudigazette.com.sa/index.php?option=com_content&task=view&id=28819&Itemid=146.
8. Reporters Without Borders, "Authorities boast of success in Internet filtering," September 15, 2006, http://www.rsf.org/article.php3?id_article=18864.
9. *The Guardian,* "Iran bans fast Internet to cut west's influence," October 18, 2006, http://technology.guardian.co.uk/news/story/0,,1924637,00.html.
10. Reporters Without Borders, Internet Under Surveillance 2004: Jordan, http://www.rsf.org/article.php3?id_article=10737.
11. The Initiative for an Open Arab Internet, "Implacable Adversaries: Arab Government and the Internet (2006): Jordan," http://www.openarab.net/en/reports/net2006/jordan.shtml; Human Rights Watch, "Jordan: Rise in arrests restricting free speech," June 17, 2006, http://www.hrw.org/english/docs/2006/06/17/jordan13574.htm.
12. BBC News, "Egypt blogger jailed for 'insult'," February 22, 2007, http://news.bbc.co.uk/2/hi/middle_east/6385849.stm.
13. Arab News, "Most of kingdom's Internet users aim for the forbidden," October 2, 2005, http://www.arabnews.com/?page=1§ion=0&article=71012&d=2&m=10&y=2005.

Internet Filtering in
Sub-Saharan Africa

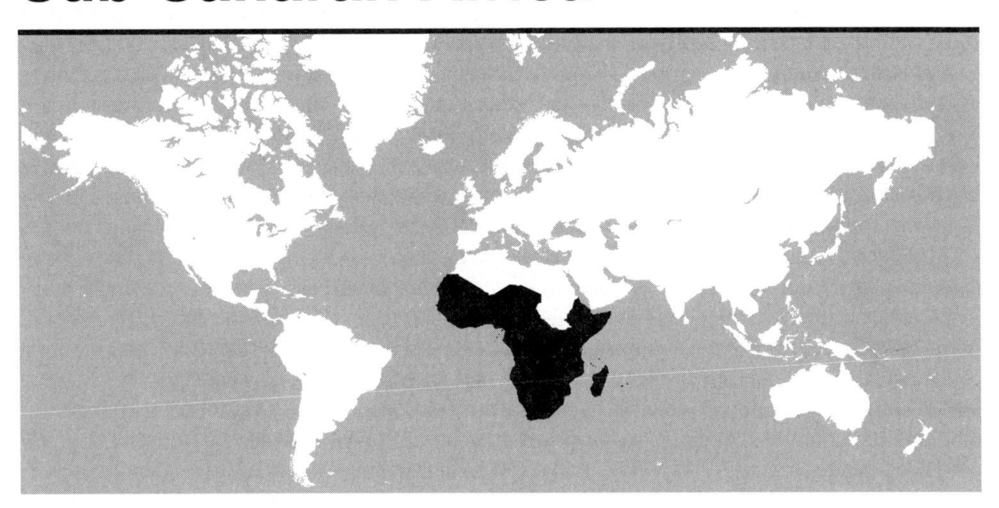

Introduction

Internet penetration in sub-Saharan Africa lags behind much of the rest of the world for a variety of economic, political, and infrastructural reasons. Despite these hurdles, most countries in the region view their future success as inextricably linked to harnessing the Internet's promise for economic development. Internet regulation in Africa, as a result, is primarily focused on infrastructure and access-related issues rather than on content regulation, though countries are making plans to broaden the scope of regulation as the Internet spreads.

Given the current restrictions on the freedoms of expression and the press in sub-Saharan Africa, one would expect similar restrictions on Internet freedom. In one way, this expectation is met: a number of countries in Africa have sought to limit the use of Voice-over Internet Protocol (VoIP) to protect incumbent telecommunications companies. However, ONI unearthed evidence of systematic blocking of Internet content in only one country, Ethiopia.[1] Uganda is also reported, by other sources, to have engaged in one temporary incidence of filtration during the past year. Time will tell whether sub-Saharan countries choose to apply their restrictive laws and practices—targeted originally at traditional media—to the Internet realm as the Internet spreads.

Infrastructure and development

In sub-Saharan Africa Internet penetration rates are exceptionally low. West Africa generally has Internet penetration rates of around 1 percent or lower, with a pocket of countries (Benin, Nigeria, and Togo) maintaining higher Internet usage rates falling between 4 and 6 percent. Eastern and southern African penetration rates are slightly higher on the whole, between 1 and 4 percent. South Africa and Zimbabwe stand out with 11 and 8 percent, respectively.[2]

Various factors contribute to these low Internet penetration rates. Many areas in the region lack the basic infrastructure necessary to support the Internet. Only half of the homes in Botswana[3] and as few as 10 percent in Tanzania[4] have access to electricity.

The poorly developed state of Internet infrastructure is also a formidable problem. In West Africa, a high-capacity cable known as SAT-3 (South African Telecom-3) connects Europe to West Africa to South Africa to India. However, incumbent telecommunication companies, which are usually state owned, often have sole control over the branching unit from this cable to their country. As a result, they often charge exorbitant rates for connection via this cable, between US7,000 and USD15,000 per Mb/s per month depending on the country.[5] There is no equivalent high-bandwidth cable connecting East African countries to the Internet, though there are a number of plans to build one.[6] As a result, many sub-Saharan African Internet service providers (ISPs) are highly reliant on expensive satellite Internet services (not lower than USD1,800 to USD2,000 per Mb/s per month)[7], driving up the price and driving down the availability of Internet services for the people in the region.

Even if sub-Saharan countries were better connected to the rest of the world with more equitable pricing arrangements, it would be difficult and expensive in most cases for them to develop sufficient internal infrastructure to support widespread Internet development. Since populations are highly rural and spread out (in Rwanda, 94 percent of the population lives in rural areas;[8] in Uganda, 85 percent),[9] it is often a better investment for companies to build a cable into another country with more populous cities than to build further into the originating country.[10] Moreover, the endemic poverty and economic degradation of the region makes any significant rural Internet development prohibitively expensive for many countries.[11] In Zimbabwe, for example, with inflation rates reaching nearly 1,600 percent, the government-owned TelOne has had difficulties paying its satellite provider bills to just maintain its *current* level of Internet access.[12] As a result, Internet usage remains low and concentrated in urban areas; in Kenya, 80 percent of Internet users live in Nairobi.[13]

The recent privatization and liberalization of the telecommunications markets was promoted as a means to alleviate the infrastructure problems of Africa. Many countries have made this important step, including Botswana in 1996,[14] Malawi in 1998,[15] South Africa in 2002 (though there was no real competition until 2006)[16], Senegal in 2003,[17] and Kenya in 2004.[18] Despite this liberalization, telecommunications companies that once held monopolies are still dominant,[19] limiting the boost in Internet development expected from deregulation and competition in the Internet services markets.[20]

In Uganda the spread of the Internet has been hampered by tariffs so high as to be prohibitive to the vast majority of the population. For monthly unlimited Internet access, the standard tariff is priced at USD50, along with local phone charges for dial-in users that, for an hour of usage a day, could run from USD31 to USD93 a month (depending on whether the connection was used at peak or off-peak times).[21] Even Internet usage in cybercafés costs around USD6/hour.[22] Compared to Uganda's annual GDP per capita of USD525 in 2005,[23] the cost of Internet use is strikingly disproportionate to the disposal income of most of the population.[24]

The lack of local content may also serve as a disincentive to Internet use. Although English is the official language of Uganda, for instance, very few of the country's inhabitants speak it as a first or even as a second language, although that number is growing.[25] Botswana's own telecommunications regulatory body, the Botswana Telecommunications Authority, has pointed out that "another factor hampering Internet uptake is the lack of indigenous local content."[26]

Nevertheless, many countries actively promote the Internet through ICT policies as an element in their overall development plans. Rwanda, for example, has an ambitious, long-term "ICT-led socio-economic development plan" to transform "Rwanda into a middle-income country by 2020."[27] The plan covers everything from general infrastructure (such as improving electrical power quality), to Internet regulatory issues, and plans for social development (such as the Citizen's Guide to Health Information Services).

Though other countries' plans may not be as ambitious or far-sighted, most countries also have policies to spread Internet use. For example, Nigeria's National Information Technology Development Agency has implemented programs such as the Mobile Internet Unit, the Rural Internet Resource Center,[28] and the Computers for All Nigerians Initiative.[29] In addition, at least thirteen sub-Saharan countries have made huge leaps in providing Internet services by launching Internet exchange points so that Internet traffic can travel easily within their borders.[30] Overall, governments are attempting to shake off these limitations to Internet development and create vibrant ICT systems throughout their countries. In many cases, they have been fairly successful, as Internet usage rates in the region grew by roughly 530 percent between 2000 and 2004.[31]

Legal framework: Freedom of expression and freedom of the press

Before examining the state of Internet content regulation in the region, it is important to understand how countries handle traditional media. Though countries do have protections for free expression and press in sub-Saharan Africa, these protections are often limited and, in a few cases, seem honored most in the breach thereof.

Freedom of expression and press freedoms are subject to constitutional protection in most African countries. "Everyone has the right to free expression, which includes freedom of the press and other media," in South Africa.[32] "Every person shall have the right to freedom of expression" and "the press shall have the right to report and publish freely, within Malawi and abroad, and to be accorded the fullest possible facilities for access to public information" in Malawi.[33] Explicit constitutional exceptions to free speech are common as well, such as an exclusion for hate speech (South Africa),[34] for defamation (Ghana),[35] for anything that may impact the enjoyment of rights by others (Zimbabwe),[36] and broader topics such as public safety and welfare (Rwanda,[37] Kenya,[38] and Ghana[39]).

Legislation designed to limit free speech and the press is found in a number of countries. Some laws protect the government from defamation and insult (Botswana)[40] and restrict the dissemination of obscenity (Liberia),[41] while other laws restrict journalism based on security concerns (Malawi).[42] Fortunately, a number of countries are making efforts to reform their laws to make them less restrictive. Rwanda, for example, has drafted a new, less restrictive press bill to replace its oppressive 2002 Press Law.[43] Other countries in the region, however, are actively pursuing even greater restrictions of free speech. Nigeria, for example, is considering an anti-gay bill entitled the "Same Sex Marriage (Prohibition) Act," which would restrict free speech, association, and assembly relating to homosexuality.[44]

Sub-Saharan governments have also taken significant action against the media under the auspices of antidefamation or national security law. In Nigeria, the leader in media abuses in West Africa,[45] two newspapers were raided in 2004 by the State Security Service for alleged libel of government officials;[46] and in 2006 journalists were charged with an annulled sedition law for questioning the president's new plane purchase.[47] In Côte d'Ivoire, the government has been accused of using the media under its control to promote its own agenda[48] while suppressing oppositional newspapers and arresting journalists, even though the December 2004 press law was supposed to have abolished jail time for journalists.[49] Reporters in Rwanda were threatened by the government after criticizing the administration.[50]

In Uganda there have been reports of journalists being harassed by the military and, in one case, a foreign journalist being excluded from the country altogether.[51] In Kenya there have been numerous reports of journalists harassed by state actors and, in some cases, even jailed.[52] Although Tanzania enjoys widespread press freedom as a general rule, the semiautonomous region of Zanzibar has been marked by threats to the independent press, which has been accused of being "a threat to national unity."[53] The Ethiopian People's Revolutionary Democratic Front arrested seventy-six journalists, politicians, and civil society activists for "'treason,' 'conspiracy' to overthrow the government and 'genocide'" in the ongoing crackdown on opposition to the government following the disastrous May 2005 legislative elections.[54] Self-censorship runs rampant in the Malawi Broadcasting Company because, as a current employee put it, "a mere negative joke about the ruling party can cost someone a job here."[55]

Nor has radio been immune to government intervention. In Zimbabwe the government jammed opposition radio stations in 2006,[56] and in Zambia the government forcibly shut down and revoked the license of a radio station that broadcasted opposition views.[57]

Though some countries, such as Botswana and South Africa, protect free speech and are recognized for their level of freedom, the continent as a whole is characterized by its severe, entrenched restrictions of expression and the press.[58]

Internet content regulation

The regulation of Internet content in sub-Saharan Africa is still in its formative stages. Given the generally low penetration rates across the region, the inchoate nature of Africa's Internet regulatory regimes is not surprising. With a few exceptions, including South Africa, sub-Saharan Africa has just begun to consider options and put together plans for regulating Internet content. In this section, we investigate the current trends and likely futures of Internet content regulation in the region relating to obscene content, defamation, political opposition, security, copyright, and Voice-over IP (VoIP).

Obscene content

Many sub-Saharan African countries have laws that restrict the traditional distribution of obscene materials and empower organizations to enforce those laws. In South Africa all material classified by the Film and Publications Board as XX, including child pornography and violent sexual acts, and X18, including any depictions of explicit sexual conduct, is illegal to distribute.[59] X18 material can be legally distributed, however, if it occurs in a face-to-face manner honoring age restrictions and within a building.[60] The Malawi Censorship Board, established in 1968 under the harsh Banda regime, remains active in restricting pornographic material. For example, in 2002 the board ordered the takedown of a billboard advertisement showing a woman's navel.[61] Zimbabwe similarly restricts pornographic content, with laws making the possession and dissemination of any "indecent or obscene" content (that is, anything "subversive of morality") illegal.[62]

Many countries, however, have not directly applied these laws to the Internet. In many cases, it is still unclear whether they actually could be applied. For example, a nonprofit media organization focusing on gay and lesbian affairs in Africa, known as Mask, argues that Zimbabwe's Censorship and Entertainments Control Act, which regulates obscene content, has not kept pace with technology and, therefore, may not apply to Internet pornography.[63] Similarly, Botswana's Telecommunications Act 1996 makes illegal the transmission "by means of a public telecommunication system, a message or other matter which is offensive or of an indecent, obscene or menacing character,"[64] but the country's ICT policy document questions whether this covers actions such as "exporting" child pornography over the Internet.[65]

Countries have come to the realization that they will need to develop policies to address the social and political ramifications associated with the availability of obscene content online as the Internet spreads through the region. For example, the Malawi ICT for Development (ICT4D) policy calls on the government to "put in place mechanisms that will safeguard girls, boys and women from fraud, misuse of information and immoral behavior brought about by the use of ICTs" and puts the Malawi Censorship Board in charge of "addressing ethical issues of the digital culture in order to ensure the protection of the rights of the vulnerable consumers."[66] Tanzania's ICT policy document comes to a similar conclusion, stating that "the Government will seek to discourage inappropriate use of ICT that is detrimental to our cultural values, ethics, mores, and morality such as viewing pornography."[67]

Countries with greater Internet penetration rates than Malawi and Tanzania, at 0.4 percent and 0.9 percent respectively,[68] are further along in their handling of the issue. Nigeria, for example, with a 4 percent penetration rate,[69] is currently considering the Computer Security and Critical Information Infrastructure Protection Bill 2005, which would explicitly make distribution of child pornography online a crime.[70] In Clause 12 of Ghana's (which has a 2 percent penetration rate)[71] Computer and Computer Related Crimes Act, 2005, there are strict prohibitions for online child pornography.[72] South Africa, with the highest penetration rate in sub-Saharan Africa, at about 11 percent,[73] took the most drastic step of all when, in September 2006, the government notified all pornography sites hosted in South Africa that they must cease posting XX and X18 classified materials by December 31, 2006, or face criminal action under the Film and Publications Act 1996.[74] So far, the vast majority of pornography sites have complied and removed their infringing content, but some remain. The government is currently compiling a list of sites that have refused to remove their content for shut-down and prosecution.[75]

In sum, it appears likely that obscene content on the Internet will be increasingly regulated as Internet development progresses in the region. Obscenity laws that might be applied to Internet content already exist in most sub-Saharan countries. As the South African precedent has shown, decisive action is a possibility.

Defamation

As discussed in the legal framework section, most countries in this region have existing defamation or insult laws restricting what can be broadcast or published. For example, Botswana's Penal Code bans insults directed at its president and flag.[76] As many of these laws in the region are criminal, free expression watchdog groups such as the International Freedom of Expression Exchange (IFEX) have called for their repeal.[77]

Few countries in sub-Saharan Africa, however, have put their defamation laws to use in the Internet sphere. South Africa is an exception, as it often is, with a small amount of case law relevant to civil defamation over the Internet. In the case of *Tsichlas v. Touch Line Media,* the manager (Natasha Tsichlas) of a South African soccer team filed suit against Touch Line Media for anonymous defamatory posts directed at her on Touch Line's Web site, Kick Off. Among the "prayers" of the suit was a requirement for Touch Line to actively monitor posts on Kick Off for defamatory material.[78] The judge found, however, that freedom of speech on the Internet would be significantly curtailed if the hosts of discussion boards were required to self-regulate material posted on their sites.[79] What makes this ruling interesting is that 1) it upheld the principle of limited liability for content hosts under a system of takedown notices similar to the U.S. Digital Millennium Copyright Act (DMCA) for defamatory, copyright infringing, and illegal material, as instituted by the Electronic Communications and

Transactions Act 2002,[80] and 2) it established an Australia-like jurisdictional rule for Internet defamation cases in which publication occurs where the material is experienced.[81] Even though the material was hosted by an ISP in Cape Town, the case was held in the Johannesburg High Court.[82]

One incidence of Internet filtering found in the region was based on defamation in the lead-up to the Ugandan presidential and parliamentary elections on February 23, 2006. On February 16, the government ordered all of the country's ISPs to block the Web site of Radio Katwe, where a user had posted criticism of President Museveni. In addition, on February 18, the government ordered a temporary block on *The Monitor,* an independent daily, though the exact reasoning for the block is unknown. The ISPs complied in both cases, making the sites inaccessible.[83]

In Senegal the French national Christian Costeaux in 2004 was sentenced in absentia to a year in prison and fined 600 million CFA francs (USD1.2 million)[84] for posting an allegedly defamatory article on his tourism Web site www.senegalaisement.com from the Senegalese newspaper *Walfadjiri* that accused aides of the mayor of Ziguinchor of embezzling more than 100 million CFA francs (USD200,000). The site, however, was never blocked.[85]

Finally, in Zimbabwe the government has been cracking down on criminal defamation and insult transmitted by e-mail. In 2005 authorities arrested forty people in a raid on a local Internet café because an e-mail insulting President Robert Mugabe was allegedly sent from the location.[86]

Overall only a small number of countries in the region have begun to apply defamation and insult laws to the Internet. However, as the Internet spreads across the region, there is no reason to doubt that an increasing number of countries will also come to apply their laws to the Internet, as has occurred in the rest of the world.

Political opposition

Though one could imagine other countries in the region doing the same, Ethiopia is so far the only country in sub-Saharan Africa to actively engage in political Internet filtering. ONI research has found that Ethiopia focuses its filtering primarily on political bloggers with oppositional views by blocking two major blog services, blogspot.com and nazret.com. This blanket ban of these blogging domains results in extraordinary overblocking, filtering thousands of Weblogs that have no relevance to politics or Ethiopia. In addition, the government blocks Web sites of opposition parties, sites representing ethnic minorities, sites for independent news organizations, and sites promoting human rights in Ethiopia.[87]

Ethiopia's Internet penetration rate of 0.2 percent[88] is the lowest rate of any country that ONI has identified as implementing an active Internet filtration regime. Other countries such as Zimbabwe, which have higher rates of Internet usage and similarly repressive regimes, filter neither political nor any other content. One possible explanation is location. Ethiopia's proximity to the heavily filtering Middle East and North Africa (MENA) region may influence its decision making on the subject. Whatever the reason, it is striking that a country with such a small population of Internet users would choose to implement an Internet content filtration regime.

As the Internet spreads, there is likely to be a convergence between the regulation of traditional media and the Internet, with countries in the region enacting restrictions on political content online, as they already do offline. This may well take the form of Internet filtration, as it has in many of the other countries which ONI has studied.

Security

Despite limited Internet access in the region, a number of countries have implemented Internet security policies to increase communication interception abilities and curb illegal online activities. In South Africa, parliament passed the Regulation of Interception of Communications and Provisions of Communication-Related Information Act 2002, which requires ISPs to retain data from customers for an as-yet undetermined period of time and makes any Internet system that is unable to be monitored illegal.[89] In addition, the Electronic Communications and Transactions Act 2002 created a legion of cyber inspectors whose job it is to, as Privacy International describes, "inspect and confiscate computers, determine whether individuals have met the relevant registration provisions as well as search the Internet for evidence of 'criminal actions.'"[90]

Zimbabwe's government, on the other hand, has been fighting for years with its High Court for wider powers to monitor and intercept e-mails. As of publication, the court has successfully limited the legal ability of the government to perform these tasks.[91] The raid, mentioned earlier, on a cybercafé in 2005, however, shows that the government appears to be achieving its ends despite these limitations. Furthermore, the government has recently stepped up its surveillance of Internet activity by placing plain clothes agents in cybercafés.[92]

Nigeria has a relatively well developed Internet security regulatory system in place involving agencies such as the Nigerian Cybercrime Working Group.[93] As mentioned earlier, Nigeria is currently considering the Computer Security and Critical Information Infrastructure Protection Bill 2005, which contains provisions to combat cyberterrorism and to allow the government to request that ISPs hold information about users without due process.[94] Additionally, the Nigerian Communications Act 2003 contains vaguely worded "information-gathering powers" in the name of security.[95]

Under the Telecommunications Act 2005 in Ghana, ISPs can be instructed under court order to intercept communications transmitted online and gather all information they can about users.[96] In special cases, the president can grant authorization, avoiding the need to obtain a court order.[97] Clauses 20–24 of the Computer and Computer Related Crimes Act 2005 also delineate specific data retention and Internet communication interception rules for criminal investigations.[98]

In addition to those countries that have already adopted Internet security measures, a handful of countries are formulating strategies to address this issue. Malawi's ICT4D policy, for example, calls for the government to "formulate and enforce laws and regulations that combat cyber crimes; institute mechanisms and laws to curb vandalism and theft of ICT infrastructure; and enact a law to validate digital signatures on documents in relation to the technology on the market today."[99] Botswana, as another example, has acknowledged in its ICT Policy that it lacks "comprehensive legislation in Botswana to deal with data crimes, such as interceptions, modification, data theft, or trafficking in digital signatures or domain names" and has called for further investigation into these topics.[100]

As Internet usage develops in Africa, regulations to ensure increased Internet security will be enacted to address cybercrime and the rights and responsibilities of government investigators, including such topics as investigation powers, surveillance, and data retention laws. Even with limited Internet penetration, a number of countries have already taken or are currently taking significant steps to secure the Internet. The intrusiveness of these measures will likely vary by the repressiveness of the government in question. If South Africa's relatively draconian policies are any indication, however, these security measures are likely to be highly invasive.

Copyright

Copyright protection is generally well established in law in sub-Saharan Africa, covering materials such as written works, music, and videos. Exceptions for "fair use" are also commonplace.

Kenya's Copyright Act 2001, for example, protects "literary works, musical works, artistic works, audio-visual works, sound recordings and broadcasts,"[101] and establishes exceptions to the rights of copyright holders.[102] Even with this legal structure, however, DVD and software piracy is rampant in the region. Sixty percent of all DVDs sold in South Africa and 81 percent of all software in use in Africa is pirated.[103] Few countries have established policies to apply their copyright laws to the Internet.

International pressure is a growing factor in the application and enforcement of copyright laws, particularly in cyberspace; the World Intellectual Property Organization's (WIPO) "Internet treaties" are intended to compel countries to apply copyright protections to the Internet. The WIPO Copyright Treaty (WCT) obligates countries to protect traditional works, and the WIPO Performances and Phonograms Treaty (WPPT) obligates countries to protect producers and performers of sound recorders. WIPO explains that "the treaties thus clarify, first, that the traditional right of reproduction continues to apply in the digital environment, including to the storage of material in digital form in an electronic medium. Second, they clarify that the owners of rights can control whether and how their creations are made available online to individual consumers at a time and a place chosen by the consumer, e.g., at home via the Internet."[104] Only thirteen—including South Africa, Ghana, Nigeria, and Botswana—of the forty-eight sub-Saharan countries included in this overview (but not including Sudan) are parties to these treaties. As a result, there is only moderate international obligation imposed on the region to protect copyright over the Internet.

There has been even less action on the related issue of ISP liability, both in general and for copyright infringement specifically. Only South Africa has done anything on the matter. As mentioned earlier, the Electronic Communications and Transactions Act 2002 establishes blanket liability limitation for South African ISPs through a notice and takedown system similar to that in the DMCA.[105] Botswana, however, has also recognized the need to address this issue in its ICT Policy, in which the writers directly point out the lack of any "appropriate legislative limitation on the liability of Internet Service Providers" and call for the government to "examine the liability of third parties, including Internet Service Providers." [106]

On the whole, online copyright protection in sub-Saharan Africa is still in an embryonic state. As with defamation, however, there does not seem to be any reason that copyright laws will not be applied to the Internet on a widespread basis as the Internet expands. It is unclear, however, what balance will be struck over ISP liability. The one example in the region, South Africa, points to the development of systems with blanket liability, but other models may yet emerge.

VoIP

The introduction of voice over Internet protocol (VoIP) represents a major challenge to sub-Saharan governments with the stiff competition it presents to the incumbent telecommunications companies by offering significantly cheaper calling rates.[107] This challenge has elicited a number of responses. Most countries, such as Botswana, Côte d'Ivoire, Ethiopia,[108] and Malawi, do not allow ISPs to provide VoIP.[109] Only seven sub-Saharan countries—Kenya, Mauritius, Somalia, South Africa, Tanzania, and Uganda[110]—actually allow it. In yet other countries, the policy is less clear. Though Zimbabwe technically allows VoIP, the regulatory agency Potraz has not yet promulgated regulations on the issuance of the particular license that would allow a company to provide VoIP services.[111] In Ghana, while no laws specifically make VoIP illegal, the government chose to protect a duopoly (of Ghana Telecom and Westel) over the international voice gateways by having the National Communications Authority shut down ISPs that offered VOIP services.[112]

Although this duopoly was supposed to end in 2002, in 2003 there were still reports of the national phone company "turning off the lines of those suspected of" using VoIP.[113]

Conclusion

It is striking that a region with such high levels of speech and media restrictions would have only two countries that have engaged in Internet content filtration. The explanation lies partially in the low Internet penetration rates in sub-Saharan Africa; filtration strategies are likely judged to be too expensive, given the limited impact of Internet on access to information. Even if content restrictions were seen to be desirable, there are not enough people online to warrant the expense. Moreover, the governments most likely to institute filtering lack the technical, administrative, and financial resources necessary for implementation (Zimbabwe, for example, fits this mold). However, as the Internet spreads, the balance of costs to benefits may shift, yielding a situation in which a growing number of countries in the region actively filter the Internet. This is by no means inevitable, however. As we see in the experiences of countries around the world, there are no easy answers to the dilemmas that arise with the spread of the Internet—along with its potential as an engine for human and economic development, the Internet is profoundly disruptive, both socially and politically. Filtering is a likely response, but one fraught with many challenges.

Authors: Evan Croen, Jehae Kim, Katie Mapes

NOTES

1. Our finding of limited Internet filtering is based primarily on secondary sources. The ONI tested only in Ethiopia and Zimbabwe during the past year. Sudan, an active Internet filterer as confirmed by ONI testing, is included in the MENA overview.
2. International Telecommunication Union, *World Telecommunication Indicators 2006*.
3. Botswana Telecommunications Authority, "Internet in Botswana," http://www.bta.org.bw/internet.html.
4. Rebecca Ghanadan, "Negotiating reforms at home: Natural resources and the politics of energy access in urban Tanzania," Center for African Studies, 2004, http://repositories.cdlib.org/cgi/viewcontent.cgi?article=1010&context=cas.
5. Fibre for Africa, "Satellite vs. fibre: Different costs for different things," http://fibreforafrica.net/main.shtml?x=4051583&als%5BMYALIAS6%5D=Satellite%2520vs%2520fibre:%2520different%2520costs%2520for%2520different%2520things&als%5Bselect%5D=4051582.
6. Eric Osiakwan and Ethan Zuckerman, Media Berkman, Africa's Internet Infrastructure Video Part I and Part II, April 19, 2007, http://blogs.law.harvard.edu/mediaberkman/.
7. Fibre for Africa, "Satellite vs. fibre: Different costs for different things," http://fibreforafrica.net/main.shtml?x=4051583&als%5BMYALIAS6%5D=Satellite%2520vs%2520fibre:%2520different%2520costs%2520for%2520different%2520things&als%5Bselect%5D=4051582.
8. Government of Rwanda, An Integrated ICT-led Socio-economic Development Plan for Rwanda, 2006-2010: The NICI-2010 Plan, http://www.rita.gov.rw/docs/NICI%202010.pdf.
9. International Telecommunication Union, World Telecommunication Indicators Update, Country Profile: Uganda, April-May-June 2000, http://www.itu.int/ITU-D/ict/cs/uganda/material/CountryProfileUGA.pdf.
10. Eric Osiakwan and Ethan Zuckerman, Media Berkman, Africa's Internet Infrastructure Video Part I and Part II, April 19, 2007, http://blogs.law.harvard.edu/mediaberkman/.
11. World Bank, Africa: Data and Statistics, 2005, $830 GDP per capita for Sub-Saharan Africa, http://devdata.worldbank.org/external/CPProfile.asp?SelectedCountry=SSA&CCODE=SSA&CNAME=Sub-Saharan+Africa&PTYPE=CP.
12. Reuters, "Zimbabwe's Net crawls due to unpaid debt," September 20, 2006, http://news.zdnet.com/2100-9588_22-6117553.html.
13. Wangui Kanina, "Kenya concerned at slow growth of Internet market," Yahoo News, March 28, 2007, http://news.yahoo.com/s/nm/20070328/wr_nm/kenya_telecoms_dc.
14. UNECA, Botswana: NICI Infrastructure, 2002, http://www.uneca.org/aisi/NICI/country_profiles/botswana/botsinfra.htm.
15. UNECA, Malawi: Internet Connectivity, http://www.uneca.org/aisi/nici/country_profiles/malawi/malinter.htm, (last accessed April 20, 2007).
16. Jackie Mackenzie, "The SNO is finally here," August 31, 2006, http://business.iafrica.com/news/984828.htm.

17. Reporters Without Borders, "Internet under surveillance 2004: Senegal," http://www.rsf.org/article.php3?id_article=10727.
18. International Telecommunication Union, Regulatory Newslog, "Privatisation of Telkom Kenya Limited," March 29, 2007, http://www.itu.int/ituweblogs/treg/Privatisation+Of+Telkom+Kenya+Limited.aspx (explaining the slow and limited process of privatizing Telkom Kenya).
19. Some countries have even regressed. In Zimbabwe the government unsuccessfully tried to reinstate TelOne's monopoly status in 2004 after the market was liberalized by the High Court in 1998.
20. See, for example, Privacy International, Internet Censorship Report 2003: South Africa, November 2004, http://africa.rights.apc.org/index.shtml?apc=s21817e_1&x=28050.
21. International Telecommunication Union, The Internet in an Africa LDC: Uganda Case Study, 2001, http://www.itu.int/ITU-D/ict/cs/uganda/material/uganda.pdf.
22. Ibid.
23. Calculated with values from World Bank, *World Development Indicators (WDI) Online.* 2005 figures. Available with subscription at http://go.worldbank.org/DBWYD75Y60.
24. International Telecommunication Union, The Internet in an Africa LDC: Uganda Case Study, 2001, http://www.itu.int/ITU-D/ict/cs/uganda/material/uganda.pdf.
25. Ibid.
26. Botswana Telecommunications Authority, "Internet in Botswana," http://www.bta.org.bw/internet.html.
27. Government of Rwanda: An Integrated ICT-Led Socio-Economic Development Plan for Rwanda 2006-2010: The NICI-2010 Plan, http://www.rita.gov.rw/docs/NICI%202010.pdf.
28. National Information Technology Development Agency of Nigeria, http://www.nitda.gov.ng/project.htm.
29. Nigeria First, Official Web site of the Nigerian Office of Public Communications, Institutional Profiles: NITDA and Nigeria's Quest for IT Advancement, March 15, 2007, http://www.nigeriafirst.org/article_6474.shtml.
30. Network Startup Resource Center, "Internet eXchange Points in Africa," March 2007, http://www.nsrc.org/AFRICA/afr_ix.html.
31. Figure calculated from International Telecommunication Union, *World Telecommunication Indicators 2006* data.
32. Constitution of South Africa, Chapter 2, http://www.info.gov.za/documents/constitution/1996/96cons2.htm#16.
33. Malawi Constitution 1994, http://www.africa.upenn.edu/Govern_Political/mlwi_const.html.
34. Constitution of South Africa, Chapter 2, http://www.info.gov.za/documents/constitution/1996/96cons2.htm#16.
35. Parliament of Ghana, the Constitution, Chapter Twelve, "Freedom and Independence of the Media," http://www.parliament.gh/const_constitution.php#Chapter%2012.
36. Constitution of Zimbabwe, §11, http://www.parlzim.gov.zw/Resources/Constitution/constitution.html.
37. Rwanda Republic, Legal and Constitutional Commission, Constitution of the Republic of Rwanda, http://www.cjcr.gov.rw/.
38. While Kenya's Constitution protects "freedom of expression" and the "freedom to communicate ideas and information," there is a broad exception where the government, acting under color of law, places restrictions "in the interest of defence, public safety, public order, public morality or public health," on defamation, on privileged information, and on government employees. Kenya Constitution, §79, http://www.usig.org/countryinfo/laws/Kenya/Constitution.pdf.
39. Parliament of Ghana, the Constitution, Chapter Twelve, "Freedom and Independence of the Media," http://www.parliament.gh/const_constitution.php#Chapter%2012.
40. IFEX, "IFEX delegation urges Botswana to repeal outdated laws," December 2, 2003, http://www.ifex.org/en/content/view/full/55399/.
41. See quote from §18.1 of Liberian Penal Code in Tukpah, Isaac Vah, "A rebuttal to Theodore Hodge's: Presidential Minister Willis knuckles: Villain or victim?" February 22, 2007, http://www.theperspective.org/articles/2007/0222200703.html.
42. See description of the Official Secrets Act at Press Reference, "Malawi Press, Media, TV, Radio, Newspapers," http://www.pressreference.com/Ky-Ma/Malawi.html.
43. Committee to Protect Journalists, Attacks on the Press in 2006, Africa: Rwanda, http://www.cpj.org/attacks06/africa06/rwa06.html. The press played an incendiary role in the 1994 Rwandan genocide. NPR, On the Media, "We wish to inform you," March 23, 2007, http://www.onthemedia.org/transcripts/2007/03/23/04.
44. Human Rights Watch, "Nigeria: Anti-gay bill threatens democratic reforms," February 28, 2007, http://hrw.org/english/docs/2007/02/28/nigeri15431.htm.
45. International Freedom of Expression Exchange, "Violations against journalists and media rights rise in 2006, reports MFWA," January 18, 2007, http://www.ifex.org/en/content/view/full/80493/.

46. IFEX, "Nigeria: Authorities shut down independent publications," 2004, http://www.ifex.org/en/content/view/full/61264/.
47. Reporters Without Borders, "Sedition case dropped against one journalist, another case adjourned," October 11, 2006, http://www.rsf.org/article.php3?id_article=18141 and CPJ "Nigeria: One journalist still charged in sedition case?" October 12, 2006, http://www.cpj.org/news/2006/africa/nigeria12oct06na.html.
48. BBC News, Country Profile: Ivory Coast, April 16, 2007, http://news.bbc.co.uk/2/hi/africa/country_profiles/1043014.stm.
49. Committee to Protect Journalists, Attacks on the Press in 2006, Africa: Ivory Coast, http://www.cpj.org/attacks06/africa06/ivory06.html.
50. Reporters Without Borders, Rwanda, Annual Report 2007, http://www.rsf.org/country-36.php3?id_mot=203&Valider=OK and Committee to Protect Journalists, Attacks on the Press in 2006, Rwanda, http://www.cpj.org/attacks06/africa06/rwa06.html.
51. Reporters Without Borders, "How to 'kick out' a foreign journalist," March 31, 2006, http://www.rsf.org/article.php3?id_article=16911.
52. See, for example, Reporters Without Borders, "Photographer roughed by president's bodyguards in church," January 9, 2007, http://www.rsf.org/article.php3?id_article=20340; Reporters sans Frontieres, "Tabloid editor jailed for a year because unable to pay libel damages ordered by court," March 7, 2007, http://www.rsf.org/article.php3?id_article=21225.
53. Reporters Without Borders, Tanzania: Annual Report 2006, http://www.rsf.org/article.php3?id_article=17403.
54. Reporters Without Borders, Ethiopia: Annual Report 2007, http://www.rsf.org/article.php3?id_article=20755&Valider=OK.
55. Press Reference, "Malawi press, media, TV, radio, newspapers," http://www.pressreference.com/Ky-Ma/Malawi.html.
56. Reporters Without Borders, Zimbabwe: Annual Report 2007, http://www.rsf.org/article.php3?id_article=20744&Valider=OK.
57. BBC News, "Zambian radio station shut down," August 20, 2001, http://news.bbc.co.uk/1/hi/world/africa/1500031.stm.
58. A graphical representation of press freedom in sub-Saharan Africa can be found at http://www.rsf.org/rubrique.php3?id_rubrique=36. While there are areas of good or satisfactory press freedom, much of the region has "noticeable problems" or is listed as a "difficult situation."
59. Schedule 1, Film and Publications Act 1996, http://www.fpb.gov.za/docs_publications/acts_regulations/acts.asp.
60. See schedule 1, Business Act 1991, http://www.fpb.gov.za/news%5Cmedia_releases%5C2006%5C12_September_2006.asp.
61. Pana, "Malawi: Clampdown on pornography," August 16, 2000, http://www.africafilmtv.com/pages/newsflash/2001/eng2001/nf84.htm; IRIN, "Malawi grapples with sex education," June 3 2002, http://www.irinnews.org/report.aspx?reportid=39867.
62. Censorship and Entertainments Control Act, §§13 and 26, http://www.kubatana.net/docs/legisl/censor_ent_act040501.pdf.
63. Mask, "Zimbabwe," http://www.mask.org.za/index.php?page=zimbabwe.
64. Telecommunications Act 1996, §53, http://www.bta.org.bw/pubs/Botswana%20Telecommunications%20Act%20-%201996.pdf.
65. Botswana's National ICT Policy 2004, Appendix G, Connectivity Laws and Policies, http://www.maitlamo.gov.bw/docs/draft-policies/appx-g-connectivity_laws_and_policies_final_jan_10.pdf.
66. Malawi National ICT for Development (ICT4D) Policy, §§3.3.2 and 4.1.2.1, July 2006, http://www.malawi.gov.mw/Policies/National%20ICT%20Policy%204%20Development%20-%20Draft%20(2).pdf.
67. Ministry of Communications and Transport, National Information and Communications Technology Policy, March 2003, www.tanzania.go.tz/pdf/ictpolicy.pdf.
68. International Telecommunication Union, *World Telecommunication Indicators 2006.*
69. Ibid.
70. Nigerian Cybercrime Working Group, Computer Security and Critical Information Infrastructure Protection Bill 2005, http://www.cybercrime.gov.ng/site/index.php?option=com_content&task=view&id=20&Itemid=56.
71. International Telecommunication Union, ITU ICT Eye: Ghana, ICT Statistics 2005, http://www.itu.int/ITU-D/icteye/DisplayCountry.aspx?countryId=90
72. Ghana Ministry of Communications, The Computer and Computer Related Crimes Act, 2005, http://www.moc.gov.gh/moc/files/Draft%20E%20-%20Legislations/Ghana ComputerCrimesAct010605.pdf.
73. International Telecommunication Union, *World Telecommunication Indicators 2006.*
74. MyADSL, "Local porn website deadline looming," December 5, 2006, http://www.mybroadband.co.za/nephp/?m=show&id=5074.
75. X Biz, "South Africa will now prosecute webmasters of adult dites," February 1,2007, http://www.melonfarmers.co.uk/in07a.htm.

76. Ruth Walden, "Insult laws: An insult to press freedom," Reston, Virginia: World Press Freedom Committee, 2000, www.wpfc.org/Insult%20 Laws-Text.PDF.
77. IFEX, "IFEX delegation urges Botswana to repeal outdated laws," December 2, 2003, http://www.ifex.org/en/content/view/full/55399/.
78. See Prayer 4, *Tsichlas v. Touch Line,* www.up.ac.za/academic/law/docs/RVD110_111June_2004.doc, (accessed April 16, 2007).
79. Reinhardt Buys, "Internet litigation rapidly increasing in SA," August 21, 2005, http://www.mybroadband.co.za/nephp/index.php?m=show&opt=printable&id=590.
80. Electronic Communications and Transactions Act 2002, §75, www.info.gov.za/gazette/acts/2002/a25-02.pdf.
81. *Tsichlas v. Touch Line,* www.up.ac.za/academic/law/docs/RVD110_111June_2004.doc, (accessed April 16, 2007).
82. Estelle Ellis, "Speak no evil, in chatrooms at least," The Star, March 11, 2003, http://www.thestar.co.za/index.php?fSaectionId=327&fArticleId=57154
83. Reporters sans Frontieres, "Net censorship reaches sub-Saharan Africa," February 24, 2006, http://www.rsf.org/article.php3?id_article=16569.
84. Conversion rate of 491.24 CFA francs to USD1 from http://www.fms.treas.gov/intn.html, (last accessed May 1, 2007).
85. Reporters Without Borders, Internet Under Surveillance 2004: Senegal, http://www.rsf.org/article.php3?id_article=10727.
86. APC Blogs, Expression under Repression:WSIS and the Net, 2005, http://blog.apc.org/en/index.shtml?x=2503009.
87. For a more detailed description of the Internet filtration situation in Ethiopia, please refer to Ethiopia's country summary.
88. International Telecommunication Union, *World Telecommunication Indicators 2006.*
89. Privacy International, Internet Censorship Report 2003: South Africa, November 10, 2004, http://africa.rights.apc.org/index.shtml?apc=s21817e_1&x=28050
90. Ibid.
91. For more details of this fight, please refer to the Zimbabwe country summary.
92. http://www.talkzimbabwe.com/default.asp?sourceid=&smenu=84&twindow=&mad=&sdetail=3996&wpage=1&skeyword=&sidate=&ccat=&ccatm=&restate=&restatus=&reoption=&retype=&repmin=&repmax=&rebed=&rebath=&subname=&pform=&sc=1705&hn=talkzimbabwe&he=.com.
93. Nigerian Cybercrime Working Group Web site, http://www.cybercrime.gov.ng/site/index.php.
94. Nigerian Cybercrime Working Group, Computer Security and Critical Information Infrastructure Protection Bill 2005, http://www.cybercrime.gov.ng/site/index.php?option=com_content&task=view&id=20&Itemid=56.
95. Nigerian Communications Act 2003, July 8, 2003, http://ncc.gov.ng/RegulatorFramework/Nigerian%20Communications%20Act,%202003.pdf.
96. Ghana Ministry of Communications, Telecommunications Act 2005, Part 8 "Testing and Inspection" and 9 "Offences," http://www.moc.gov.gh/moc/files/Draft%20E%20-%20Legislations/GhanaTelecomActdraft2.pdf.
97. Ghana Ministry of Communications, Telecommunications Act 2005, Part 11, Article 55, http://www.moc.gov.gh/moc/files/Draft%20E%20-%20Legislations/GhanaTelecomActdraft2.pdf.
98. Ghana Ministry of Communications, The Computer and Computer Related Crimes Act 2005, http://www.moc.gov.gh/moc/files/Draft%20E%20-%20Legislations/GhanaComputerCrimesAct010605.pdf.
99. Malawi National ICT for Development (ICT4D) Policy, §3.3.2, July 2006, http://www.malawi.gov.mw/Policies/National%20ICT%20Policy%204%20Development%20-%20Draft%20(2).pdf.
100. Botswana's National ICT Policy 2004, Appendix G, Legal Policy and e-Readiness, http://www.maitlamo.gov.bw/docs/e-readiness/volume-1/appx-g-legal-e-readiness_july_6_rel_5th_aug_04.pdf.
101. Kenyan Copyright Act 2001, §22, http://portal.unesco.org/culture/admin/file_download.php/ke_copyright_2001_en.pdf?URL_ID=30229&filename=11416612103ke_copyright_2001_en.pdf&filetype=application%2Fpdf&filesize=428054&name=ke_copyright_2001_en.pdf&location=user-S/.
102. Kenyan Copyright Act 2001, §26, http://portal.unesco.org/culture/admin/file_download.php/ke_copyright_2001_en.pdf?URL_ID=30229&filename=11416612103ke_copyright_2001_en.pdf&filetype=application%2Fpdf&filesize=428054&name=ke_copyright_2001_en.pdf&location=user-S/.
103. Southafrica.info, "Fighting fake DVDs—with fakes," May 19,2006, http://www.southafrica.info/ess_info/sa_glance/media/dvd-piracy-190506.htm; iafrica.com, "Software piracy costing Africa billions," October 16, 2006, http://cooltech.iafrica.com/features/286314.htm.
104. World Intellectual Property Organization, The WIPO Internet Treaties, www.wipo.org/freepublications/en/ecommerce/450/wipo_pub_l450in.pdf (last accessed April 24, 2007).
105. Electronic Communications and Transactions Act 2002, §75, www.info.gov.za/gazette/acts/2002/a25-02.pdf.

106. Botswana's National ICT Policy 2004, Appendix G, Legal Policy and e-Readiness," http://www.maitlamo.gov.bw/docs/e-readiness/volume-1/appx-g-legal-e-readiness_july_6_rel_5th_aug_04.pdf.
107. John Yarney, IDG News Service, Info World, "Africa Makes Progress on VOIP," 14 October 2005, at http://www.infoworld.com/article/05/10/14/HNafricavoip_1.html
108. Mark Warren, "Illegal VoIP?" December 8, 2006, http://www.voiploop.com/index.php?option=com_content&task=view&id=1682&Itemid=29.
109. AfrISPA, Voice over IP in Africa, http://www.afrispa.org/voiceip.htm.
110. Russell Southwood, "The future of voice in Africa," January 12, 2007, http://www.itu.int/osg/spu/ni/voice/papers/FoV-Africa-Southwood-draft.pdf.
111. The Sunday Mail, "Potraz yet to come up with VoIP regulations," March 25, 2007, http://www.kubatana.net/html/archive/inftec/070325smail.asp?sector=INFTEC&year=0&range_start=1.
112. Kofi Mangesi, "Ghana ISPs closed down," IT Web, August 2, 2000, http://www.itweb.co.za/sections/telecoms/2000/0008021031.asp
113. G. Pascal Zachary, "Searching for a dial tone in Africa," *The New York Times,* July 5, 2003, http://query.nytimes.com/gst/fullpage.html?res=9E03E5DB1F3AF936A35754C0A9659C8B63&sec=&spon=&pagewanted=1.

Internet Filtering in the
United States and Canada

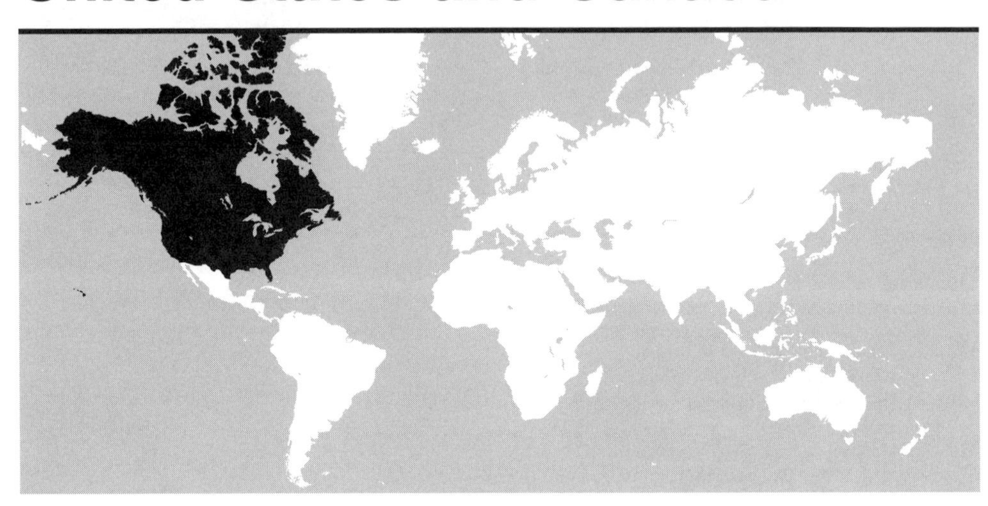

Introduction

Though neither the United States nor Canada practices widespread technical Internet filtering at the state level, the Internet is far from "unregulated" in either state.[1] Internet content restrictions take the form of extensive legal regulation, as well as technical regulation of content in specific contexts, such as libraries and schools in the United States. The pressure to regulate specific content online has been expressed in concerns related to four problems: child-protection and morality, national security, intellectual property, and computer security. In the name of "protecting the children," the United States has moved to step up enforcement of child pornography legislation and to pass new legislation that would restrict children's access to material deemed "harmful." Legislators invoke national security in calls to make Internet connections more traceable and easier to tap. Copyright holders have had the most success in this regard by pressing their claims that Internet intermediaries should bear more responsibility—and more liability—than they have in the past. Those concerned about computer security issues, such as badware and spam, have also prompted certain regulations of the flow of Internet content. In addition, in Canada, although not in the United States, publishing of hate speech is restricted.

Debate on each of these restrictions is heated. Public dialogue, legislative debate, and judicial review have resulted in different filtering strategies in the United States and Canada than those described elsewhere in this volume. In the United States, many government-mandated attempts to regulate content have been barred on First Amendment grounds. In the wake of these restrictions, though, fertile ground has been left for private-sector initiatives. The government has been able to exert pressure indirectly where it cannot directly censor. In Canada, the focus has been on government-facilitated industry self-regulation. With the exception of child pornography, Canadian and U.S. content restrictions tend to rely more on the removal of content than blocking; most often these controls rely upon the involvement of private parties, backed by state encouragement or the threat of legal action.[2] In contrast to those regimes where the state mandates Internet service provider (ISP) action through legal or technical control, most content-regulatory urges in both the United States and Canada are directed through private action.

With only 5.1 percent of the world's population, the United States and Canada are home to 21.1 percent of the world's Internet users. Together their Internet penetration rate is 69.4 percent.[3] Canada and the United States, however, have not kept pace with many other countries in expanding broadband access, slipping in the global ranking of Internet broadband penetration rates to 11th and 16th, respectively, in 2006.[4] These high rates of Internet usage bring with them the ability of citizens to express dissenting points of view, as well as to engage in a large number of other activities (such as accessing pornography) that test a society's dedication to free expression and privacy. Like the states that actively filter the Internet through technical means, Canada and the United States are not immune from the ongoing challenges that these tests pose.

Regulating and filtering obscene and explicit content

It is a truism (i.e., repeated without necessarily being true) that pornographers are the first to embrace every new technology. The first sustained battle over content filtering in the United States broke out over sexually explicit material, particularly because of the perception that it is easily accessible and the fear that it can do harm to minors who access it online.

Canada has tended to act conservatively in response to online obscenity, while legislators in the United States have pursued broader definitions of offenses and mandates on Internet filtering. In its response to online sexually explicit material, Canada has made only *de minimis* amendments to pre-existing law.[5] Legislators have simply revised existing obscenity provisions to encompass online offenses. For example, the passage of the Criminal Law Amendment Act of 2001[6] established online acts of distributing and accessing child pornography and luring a child as crimes.[7] The Criminal Code mandates a system for judicial review of material (including online material) alleged to be child pornography. It does not, however, require ISPs to judge the legality of content posted on their servers or to take corrective action prior to a judicial determination.[8] If a judge determines that the material in question is illegal, ISPs may be required to take it down and to give information to the court to help in the identification and location of the person who posted it.[9]

Many Canadian ISPs, however, have begun to filter content hosted outside of Canada despite regulatory uncertainty in the area. For three days in July 2005, the Canadian ISP Telus blocked access to a Web site run by members of the Telecommunication Workers Union during a labor dispute containing what Telus argued was proprietary information and photographs that threatened the security and privacy of its employees.[10] This unilateral action by Telus broke the "cardinal rule" of Canadian ISPs—that they pass on any and all information without regard for content in exchange for immunity from liability over content. This action also conflicted with Section 36 of the Canadian Telecommunications Act, which states that, without the approval of the Canadian Radio-Television and Telecommunications Commission (CRTC), a "Canadian carrier shall not control the content or influence the meaning or purpose of telecommunications carried by it for the public."[11] Telus, however, argued that content filtering is permitted in the contract it holds with its subscribers, although, to the detriment of their argument, the blocking affected the customers of other ISPs that connect via Telus. The matter was resolved when, though the site was hosted in the United States,[12] Telus was able to obtain court orders from Alberta and British Columbia requiring the Web site operator, who lives and works in Canada, to remove the offending materials.[13]

In August 2006 the Canadian human rights lawyer Richard Warman filed an application with the CRTC to authorize Canadian ISPs to block access to two hate speech sites hosted outside of Canada.[14] The CRTC denied the application, but the decision recognized that, although the CRTC cannot *require* Canadian ISP's to block content it, can *authorize* them to do. However, the CRTC noted that the "scope of this power has yet to be explored."[15]

In November 2006 Canada's largest ISPs launched Project Cleanfeed Canada in partnership with www.cybertip.ca, the nation's child sexual exploitation tipline. The project, modeled after a similar initiative in the United Kingdom, is intended to protect ISP customers "from inadvertently visiting foreign web sites that contain images of children being sexually abused and that are beyond the jurisdiction of Canadian legal authorities."[16] Acting on complaints from Canadians about images found online, www.cybertip.ca analysts assess the reported information and forward potentially illegal material to the appropriate foreign jurisdiction. If a URL is approved for blocking by two analysts, it may be added to the Cleanfeed Canada distribution list. Each of the participating ISPs voluntarily blocks this list without knowledge of the sites it contains, precluding ISP involvement in the evaluation of URLs. Blocked sites fail to load, but attempts to access them are not monitored and users are not tracked.[17]

Since Cleanfeed Canada is a voluntary program, the blocking mechanism is up to the discretion of the ISPs. Sasktel, Bell Canada, and Telus all claim to block only specific URLs, not IP addresses, in an attempt to avoid overblocking.[18] Besides the significant public outcry that would most likely result, overblocking may itself be illegal under the Telecommunications Act mentioned above.

Because accessing child pornography—as well as making it accessible—is unlawful in Canada, the filtering of such content does not infringe on rights of access or speech afforded by the Canadian Charter of Rights and Freedoms. Moreover, because ISP participation in Project Cleanfeed is voluntary, the blocking of sites through the project cannot be said to be state sponsored. However, the project remains controversial for other reasons. First, Cleanfeed Canada has not yet sought or received authorization from the CRTC. Second, the blacklist maintained by www.cybertip.ca remains secret, though necessarily, as publishing a "directory" of child pornography would itself be illegal. This lack of transparency inevitably generates distrust of the list and the process by which it is compiled. Third, the procedure for appealing the blocking of a site may have implications for anonymity.[19] A content owner or ISP customer may complain to the ISP or directly to www.cyber tip.ca, which will reassess the site and, if necessary, obtain an independent and binding judgment from the National Child Exploitation Coordination Centre. It is unclear whether this process might expose the complainant's identity and create the potential for abuse of that individual's rights by the ISP or perhaps even by authorities.

Canada's response to online obscenity and its collaborative filtering initiative look restrained by contrast to the more vigorous regulatory efforts of the United States.

The United States Congress passed the Communications Decency Act (CDA) as part of the Telecommunications Act of 1996. Signed into law by President Clinton in February 1996, the CDA criminalized the transmission of "indecent" material to persons under eighteen and the display to minors of "patently offensive" content and communications.[20] The CDA took aim at both the speakers and service providers of indecent material, although it offered them each safe harbor if they imposed technical barriers to minors' access.

Even before it took effect, the CDA was challenged in federal court by a group of civil liberties and public interest organizations and publishers who argued their speech would be chilled by fear of the CDA's enforcement. The three-judge district court panel concluded that the terms "indecent" and "patently offensive" were so vague that enforcement of either prohibition would violate the First Amendment.[21] "As the most participatory form of mass speech yet developed," Judge Dalzell wrote in a concurring opinion, "the Internet deserves the highest protection from governmental intrusion."[22] The U.S. Supreme Court affirmed this holding in 1997, invalidating the CDA's "indecency" and "patently offensive" content prohibitions.[23] In the landmark case *Reno v. ACLU,* the Court held that CDA was not the "least restrictive alternative" by which to protect children from harm. Rather, parent-imposed filtering could effectively block children's access to indecent material without preventing adults from speaking and receiving this lawful speech.[24]

U.S. lawmakers responded to the Supreme Court's decision in *Reno v. ACLU* by enacting the Child Online Protection Act (COPA)—a second attempt at speaker-based content regulation. In COPA, the Congress directed its regulation at commercial distributors of materials "harmful to minors."[25] The slightly narrower focus of COPA, nicknamed "son of CDA," did not solve the Constitutional problems that doomed the CDA. The district court enjoined COPA on First Amendment grounds.[26] As this volume went to press, the district court had just struck down COPA, finding it void for vagueness and not narrowly tailored to the government's interest in protecting minors. Once again, the court held that criminal liability for speakers and service providers was not the "least restrictive means" to accomplish the government's purpose because the private use of filtering technologies could more effectively keep harmful materials from children.

Plaintiffs successfully argued that CDA and COPA would chill the provision and transmission of lawful Internet content in the United States. Faced with the impossible task of accurately identifying "indecent" material and preemptively blocking its diffusion, ISPs would have been prompted to filter arbitrarily and extensively in order to avoid threatened criminal liability, while writers and publishers felt compelled to self-censor.

Stymied at restricting the publication of explicit material, Congressional leaders changed their focus to the recipient end of the equation. The Children's Internet Protection Act (CIPA) of 2000 forced public schools and libraries to use Internet filtering technology as a condition of receiving federal E-Rate funding. A school or library seeking to receive or retain federal funds for Internet access must certify to the FCC that it has installed or will install technology that filters or blocks material deemed to be obscene, child pornography, or material "harmful to minors."[27] The Supreme Court rejected First Amendment challenges to CIPA, holding that speakers had no right of access to libraries and that patrons could request unblocking.[28] In response, some libraries and schools have rejected E-Rate funding, but most have felt financially compelled to install the filters.

The aftermath of CDA, COPA, and CIPA has left the business of Internet filtering largely to private manufacturers competing for market share. Schools, businesses, parents, and other parties wishing to (or compelled to) block access to certain content have a broad range of competing software packages available to them. Some programs permit access only to whitelists of preapproved sites, but most services generate blacklists of blocked sites through automated screenings of the Web and, in some cases, real-time monitoring. Whatever their configuration, these products and the content they permit and restrict reflect different normative choices about the subjects targeted for filtering. Indeed, it is developers first, and users second, who determine what gets filtered when such software is implemented.

Although CIPA mandates the presence of filtering technology in schools and libraries receiving subsidized Internet access, it effectively delegates blocking discretion to the developers and operators of that technology. The criteria "obscene," "child pornography," and "harmful to minors" are defined by CIPA and other existing legislation, but strict adherence to these (vague) legal definitions is beyond the capacity of filters and inherently subject to the normative and technological choices made during the software design process. Moreover, while CIPA permits the disabling of filters for adults and, in some instances, minors "for bona fide research or other lawful purposes,"[29] it entrusts school and library administrators with deactivating the filters, giving them considerable power over access to online content. Once FCC certification requirements have been met, it is these individuals who shoulder the burden of ensuring access to constitutionally protected material.[30]

In the single known U.S. attempt to install filtering deeper into the network, the Commonwealth of Pennsylvania in 2004 authorized the state attorney general (AG)'s office to force ISPs to block Pennsylvania residents' access to sites the AG's office identified as child pornography. A district court struck this regulation down on First Amendment grounds of overbreadth because the filters' imprecision blocked substantial lawful speech unrelated to child pornography.[31] Since both possession and distribution of child pornography are criminal in the United States, service providers do respond to requests to remove it from their networks and report it to the National Center for Missing and Exploited Children when they encounter it.

Defamation

As in other national contexts, the potential for legal liability for other civil violations, including defamation and copyright, constrains the publishers of Internet content and certain service providers in the United States and Canada. These pressures can have a "chilling effect" on lawful online content and conduct and can threaten the anonymity of users. The content and court adjudication of such laws is "state action," even when the lawsuits and threats are brought by private individuals or entities.

At common law, one crucial factor in determining liability for defamation is the provider's relation to the content—whether the provider functioned as a carrier, distributor, or publisher of the defamatory content. In the United States the common law has been overridden by a federal statute, a holdover portion of the CDA, 47 U.S.C. 230. A key part of the CDA survived judicial scrutiny. Section 230 immunizes ISPs for their users' defamation: "No provider or user of an interactive computer service shall be treated as the publisher or speaker of any information provided by another information content provider."[32] Moreover, the First Amendment shields speakers from liability for much speech about public figures. In Canada ISPs must still find their fit within the traditional categories, where they can escape liability if they are carriers or distributors, transmitting data without discrimination, preference, or regard for content, or may face liability as publishers if they exercise editorial control over material. Thus, while Canadian and U.S. service providers share the right to remove content voluntarily, those in Canada do not have the broad discretion or protection enjoyed by those in the United States, and may be compelled to take down allegedly defamatory content (e.g., postings to message boards) under threat of suit.

Copyright

U.S. copyright law has also evolved more quickly—perhaps even hastily—than Canadian law in addressing the issue of service provider liability and in encouraging removal of infringing material. The "Online Copyright Limitations of Liability Act," a part of the Digital Millennium Copyright Act (DMCA) of 1998,[33] gives service providers a "safe harbor" from liability for their users' copyright infringements provided they implement copyright policies and a notice-and-takedown regime. Where a service provider unknowingly transmits, caches, retains, or furnishes a link to infringing material by means of an automatic technical process, it is protected from monetary liability so long as it promptly removes or blocks access to the material upon notice of a claimed infringement.[34] (The ISPs' CDA 230 immunity discussed above applies primarily in the context of defamation matters and explicitly excludes intellectual property offenses.)

The notice-and-takedown provision has been seen as giving copyright owners—potentially anyone who has fixed an "original work of authorship"—unwarranted leverage over service providers and their subscribers. When a provider is notified of an alleged infringement, risk aversion encourages it to remove or disable access to the specified material, probably without first informing the subscriber. The subscriber may file a counter-notice and have the content restored if the copyright owner does not file a claim in court,[35] but such challenges are rare. Subscribers, like the providers hosting their Web sites, are more likely to concede to takedown pressures, even when an infringement may not actually be occurring. If a subscriber is sued, his or her identity may be subpoenaed, as in cases of defamation, and with similarly little judicial scrutiny.[36] Major search engines such as Google comply with hundreds of removal requests a month, when it is not even clear that provision of a hyperlink would incur copyright liability.[37]

As Canada began to consider amending its copyright laws, it appeared to be following in the footsteps of the United States. In 2004, the House of Commons Standing Committee on Canadian Heritage re-tabled its *Interim Report on Copyright Reform,* which proposed a notice-and-takedown policy similar to that of the DMCA, under which Canadian service providers would be compelled to remove content immediately upon receiving notice of an alleged infringement from a professed copyright holder. The *Report* came under fire from the Canadian Internet Policy and Public Interest Clinic (CIPPIC), Digital Copyright Canada, and the Public Interest Advocacy Centre (PIAC); numerous petitions and critiques have followed, calling for balance between the rights of content creators and fair public use. The government seems to be responding to these inputs as it continues to consider changes to copyright legislation.[38]

In the midst of this period of copyright uncertainty, Canadian ISPs have implemented a notice-and-notice policy for handling copyright infringement. Originally proposed in the now-defunct Bill C-60, which was dropped from the legislative agenda in 2005 with the collapse of the Liberal government,[39] the policy allows copyright owners to send notices to ISPs regarding possible copyright infringement by subscribers. The ISPs then forward the notices to their subscribers requesting them to desist in their illegal activities.[40] Even though the notices do not mean that immediate legal action will follow if infringing activities do not cease, they have been successful in getting significant portions of infringing subscribers to remove their materials.[41]

At present, however, protections against defamation and copyright infringement afforded under U.S. and Canadian law remain in tension with the rights of service providers and Internet users, often giving rise to the censoring and self-censoring of material. Canadian service providers erring on the side of caution may remove content from subscribers' sites, as U.S. providers do when informed of alleged copyright violations. User material is therefore subject to censorship based on unsubstantiated claims. Moreover, because subpoenas offer plaintiffs an avenue for ascertaining subscribers' identities without scrutiny, the potential for misuse of these subpoenas can instill a fear of improper discovery in subscribers that leads to self-censorship. These chilling effects have been well documented,[42] and while they are indirect rather than direct state-mandated filtering, they do constitute real censorship of online speech.[43]

National security, computer security

Security concerns drive many of the state-mandated limitations on the speech and privacy interests of citizens. These security concerns in the United States and Canada take two forms: national security and computer security.

Concerns related to national security have led more to online surveillance by the state than to content filtering. The Bush Administration's warrantless wiretaps are reported to have included taps on major Internet interconnect points and data-mining of Internet communications.[44] Tapping these interconnect points would give the government the ability to intercept all overseas and many domestic communications. At press time, the U.S. government has moved to dismiss lawsuits filed against it and against AT&T by asserting the state secrets privilege; district courts in California and Michigan have refused to dismiss the lawsuits. If the allegations prove to be true, they show that the United States maintains the world's most sophisticated Internet surveillance regime. The Bush Administration is pushing to expand the Communications Assistance to Law Enforcement Act (CALEA) to force providers to give law enforcement wiretap access to electronic communications networks. Attorney General Gonzales has called for data retention laws to force ISPs to keep and potentially produce data that could link Internet subscribers to their otherwise-anonymous communications.[45]

Canadian electronic surveillance, primarily undertaken by the National Defense's secretive Communications Security Establishment (CSE), operates in close cooperation with U.S. and other allied intelligence networks. Although bound by Canadian laws and prohibited from eavesdropping on solely domestic Canadian communications without explicit ministerial approval, the CSE's activities are highly secret and oversight is minimal.

Computer security has led to certain content restrictions in the United States and Canada. Concerns about unwanted messages reaching computers, in various flavors of spam, have prompted content-based restrictions such as the CAN-SPAM Act of 2003 in the United States. In Canada a National Task Force on Spam was convened in 2005 to study the spam problem. While some laws, such as the Personal Information Protection and Electronic Documents Act, were found to at least tangentially apply to spam, the Task Force found a need for legislation directly limiting spam, which has yet to be passed.[46] The U.S. Congress has considered a range of options for limiting the free flow of bits across the Internet to address the problem of bad applications infecting computers, though most of the efforts to filter information based upon content deemed to be a computing security risk are carried out by private firms or individuals on a voluntary basis.[47] Calls are also being made to consider ISP liability in order to contain the worst of "zombie" computers sending spam and distributing badware, in the interest of preserving network safety for other connected PCs. In sum, there is still an active, ongoing discussion about how and why regulation of the flow of obviously malicious code over the Internet might take place.[48]

Conclusion

Although the United States and Canadian Internet are often thought to be relatively free from technical Internet filtering, Internet activity is far from "unregulated." With respect to online surveillance, the United States may be among the most aggressive states in the world in terms of monitoring online conversations. Lawmakers in both countries have imposed Internet-specific regulation that can limit their citizens' access and view of the Internet. In addition, they have empowered private individuals and companies to press Internet intermediaries for content removal or to carry out the filtering in the middle of the network. Although the laws are subject to legislative and judicial debate, these private actions may be less transparent. Governments in both countries, however, have experienced significant resistance to their content restriction policies and, as a result, the extreme measures found in some of the more repressive countries of the world have not gained ground in North America.

Authors: Kevin O'Keefe, John Palfrey, Wendy Seltzer

NOTES

1. See Jack Goldsmith and Tim Wu, "How governments rule the Net," chapter 5, pp. 65–84, in *Who Controls the Internet: Illusions of a Borderless World,* New York: Oxford University Press, 2006.
2. See John Palfrey and Robert Rogoyski, "The Move to the Middle: The Enduring Threat of Harmful Speech to the End-to-End Principle," *Washington University Journal of Law and Policy* 21: 31–65, (2006).
3. Internet World Stats, "Internet usage statistics for the Americas," http://www.internetworldstats.com/stats2.htm.
4. International Telecommunication Union, *World Telecommunication Indicators 2006.*
5. This approach was first recommended in a 1997 study commissioned by Industry Canada. See Internet Content-Related Liability Study, "Conclusion," http://strategis.ic.gc.ca/epic/internet/insmt-gst.nsf/en/sf03316e.html.
6. Passed as Bill C-15a, 1st Session, 37th Parl., 2001.
7. R.S. 1985, c. C-46, §§163.1(3), 163.1(4.1), 172.1.
8. Project Cleanfeed Canada, Frequently Asked Questions, http://www.cybertip.ca/en/cybertip/cf_faq; R.S., 1985, c. C-46, §IV, http://laws.justice.gc.ca/en/showdoc/cs/C-46/bo-ga:l_V//en#anchorbo-ga:l_V.
9. R.S., 1985, c. C-46, §164.1, http://laws.justice.gc.ca/en/showdoc/cs/C-46/bo-ga:l_V//en#anchorbo-ga:l_V.
10. Michael Geist, "Telus breaks ISPs's cardinal rule," Toronto Star, August 1, 2005, http://www.michaelgeist.ca/index.php?option=content&task=view&id=919. OpenNet Initiative, "Telus Blocks Consumer Access to Labour Union Web Site and Filters an Additional 766 Unrelated Sites" http://opennet.net/bulletins/010/.
11. Telecommunications Act, R.S.C., ch. 38, §§27(2), 36, http://www.crtc.gc.ca/eng/LEGAL/TELECOM.HTM; see also Michael Geist, "Telus breaks ISPs's cardinal rule," *Toronto Star,* August 1, 2005, http://www.michaelgeist.ca/index.php?option=content&task=view&id=919.
12. OpenNet Initiative, "Telus Blocks Consumer Access to Labour Union Web Site and Filters an Additional 766 Unrelated Sites" http://opennet.net/bulletins/010/.
13. See "TELUS removes blocking from VFC website," July, 28, 2005, http://www.voices-for-change.ca/news/archive.asp?PagePosition=2.
14. http://www.crtc.gc.ca/PartVII/eng/2006/8646/p49_200610510.htm.
15. http://www.crtc.gc.ca/archive/ENG/Letters/2006/lt060824.htm.
16. See "ISPs and tipline set up battle against Internet child exploitation," November 24, 2006, http://www.cybertip.ca/en/cybertip/cleanfeed_canada.
17. See Project Cleanfeed Canada, Frequently Asked Questions, http://www.cybertip.ca/en/cybertip/cf_faq.
18. "Cleanfeed Canada: What would it accomplish?" http://yro.slashdot.org/article.pl?sid=06/12/15/1624215.
19. See Project Cleanfeed Canada, "Appeal process," http://www.cybertip.ca/en/cybertip/cf_appeal.
20. 47 U.S.C.A. §223(a), §223(d) (Supp. 1997).
21. *ACLU v. Reno,* 929 F. Supp. (E.D. Pa. 1996) at 854-865.
22. *ACLU v. Reno,* 929 F. Supp. (E.D. Pa. 1996) at 883.
23. *Reno v. ACLU,* 521 U.S. 844 (1997).
24. Ibid. "The Government may not 'reduce the adult population . . . to . . . only what is fit for children.' 'Regardless of the strength of the government's interest' in protecting children, 'the level of discourse reaching a mailbox simply cannot be limited to that which would be suitable for a sandbox.'"
25. 47 U.S.C. §231.
26. *ACLU v. Reno,* No. 98-5551 (E.D. Pa. 1999) Memorandum and Order granting preliminary injunction.
27. See Federal Communications Commission, What CIPA Requires, http://www.fcc.gov/cgb/consumerfacts/cipa.html.
28. *United States v. American Library Association,* 539 U.S. 194 (2003).
29. 20 U.S.C. §6777(c); 20 U.S.C. §9134(f)(3); 47 U.S.C. §254(h)(6)(D).
30. For examples of how libraries and schools have responded to CIPA, see Marjorie Heins, Christina Cho, and Ariel Feldman, Internet Filters: A Public Policy Report (2006), pp. 4–7, www.brennancenter.org/dynamic/subpages/download_file_36644.pdf.
31. *CDT v. Pappert,* 337 F.Supp.2d 606 (E.D. Penn. 2004). For an extensive analysis, see Jonathan Zittrain, "Internet points of control," 44 B.C. L. Rev. 653 (2003).

32. 47 U.S.C. §230(c)(1).
33. Pub. L. No. 105-304, 112 Stat. 2860 (1998).
34. 17 U.S.C. §§512(a)-(d).
35. 17 U.S.C. §512(g).
36. 17 U.S.C. §512(h).
37. See Chilling Effects, http://www.chillingeffects.org/dmca512/.
38. See Department of Canadian Heritage: Copyright Policy Branch, Government Statement on Proposals for Copyright Reform, http://pch.gc.ca/progs/ac-ca/progs/pda-cpb/reform/statement_e.cfm.
39. Online Rights Canada, "What are copyright reform and Bill C-60?" December 7, 2005, http://www.online-rights.ca/learn/what_is_c-60/.
40. Michael Geist, "The effectiveness of notice and notice," February 2007, http://www.michaelgeist.ca/content/view/1705/125/.
41. CBC News, "Email warnings deter Canadians from illegal file sharing," February 15, 2007, http://www.cbc.ca/consumer/story/2007/02/14/software-warnings.html.
42. See the work of Chilling Effects Clearinghouse, www.chillingeffects.org.
43. See Wendy Seltzer, "Unsafe harbors: Abusive DMCA subpoenas and takedown demands," http://www.eff.org/IP/P2P/20030926_unsafe_harbors.php#_edn3.
44. James Risen and Eric Lichtblau, "Spy agency mined vast data trove, officials report," *The New York Times,* December 24, 2005.
45. Declan McCullagh, "Gonzales pressures ISPs on data retention," May 2006, http://news.com.com/Gonzales+pressures+ISPs+on+data+retention/2100-1028_3-6077654.html.
46. Michael Geist, "Spam plans," March 15, 2007, http://www.michaelgeist.ca/content/view/1805/125/.
47. Consider, for instance, the interstitial pages that search giant Google places between search results and certain pages on the Internet deemed to host badware that might harm an end-user's computer. See http://stopbadware.org.
48. See, for example, Jonathan Zittrain, *The Future of the Internet and How to Stop It,* chapters 7 and 8 (forthcoming 2007), (discussing various methods for tempering the badware problem through code, law, and social reforms).

Country Summaries

Introduction to the

Country Summaries

The country summaries that follow offer a synopsis of the findings and conclusions of OpenNet Initiative (ONI) research into each of the countries. The summaries also provide a basic framework for considering the factors influencing countries' decision to filter or abstain from filtering the Internet, as well as the impact, relevance, and efficacy of technical filtering in a broader context of Internet censorship.

These summaries cover the countries where ONI conducted both testing and analysis in 2006. As noted in chapter 1, countries selected for in-depth analysis are those in which it was believed that there was the most to learn about the extent and processes of Internet filtering. Many countries known to filter the Internet, including many in Europe and North America, were not subject to testing because their practices are well documented elsewhere. ONI plans to conduct in-depth analysis on an expanded number of countries in future years.

Each country summary includes the summary results of the empirical testing for filtering. The technical filtering data alone, however, do not amount to a complete picture of Internet censorship and content regulation. A wide range of policies relating to media, speech, and expression also act to restrict expression on the Internet and online community formation, as discussed in chapter 2. Legal and regulatory frameworks, including Internet law, the state of Internet access and infrastructure, the level of economic development, and the quality of governance institutions are central to determining which countries resort to filtering and how they choose to implement Internet content controls. Therefore a brief overview of each of these factors is included in the each of the country summaries. Together, these sections are intended to offer a concise, accurate, and unbiased overview of Internet filtering and content regulation.

As described in chapter 1, each country is given a score on a five-point scale presented in the Results-at-a-Glance box. The scores reflect the observed level of filtering in each of four themes:

- **Political:** This category is focused primarily on Web sites that express views in opposition to those of the current government. Content more broadly related to human rights, freedom of expression, minority rights, and religious movements is also considered here.
- **Social:** This group covers material related to sexuality, gambling, and illegal drugs and alcohol, as well as other topics that may be socially sensitive or perceived as offensive.
- **Conflict/security:** Content related to armed conflicts, border disputes, separatist movements, and militant groups is included in this category.
- **Internet tools:** Web sites that provide e-mail, Internet hosting, search, translation, Voice-over Internet Protocol (VoIP) telephone service, and circumvention methods are grouped in this category.

The relative magnitude of filtering for each of the four themes is defined as follows:

- **Pervasive filtering:** Filtering that is characterized by both its *depth*—a blocking regime that blocks a large portion of the targeted content in a given category—and its *breadth*—a blocking regime that includes filtering in several categories in a given theme.

- **Substantial filtering:** Filtering that has either depth or breadth: either a number of categories are subject to a medium level of filtering or a low level of filtering is carried out across many categories.
- **Selective filtering:** Narrowly targeted filtering that blocks a small number of specific sites across a few categories or filtering that targets a single category or issue.
- **Suspected filtering:** Connectivity abnormalities are present that suggest the presence of filtering, although diagnostic work was unable to confirm conclusively that inaccessible Web sites are the result of deliberate tampering.
- **No evidence of filtering:** ONI testing did not uncover any evidence of Web sites being blocked.

The Results-at-a-Glance box also includes a measure (low, medium, or high) of the observed transparency and consistency of blocking patterns. The *transparency* score given to each country is a qualitative measure based on the level to which the country openly engages in filtering. In cases where filtering takes place without open acknowledgment, or where the practice of filtering is actively disguised to appear as network errors, the transparency score is low. In assigning the transparency score, we have also considered the presence of provisions to appeal or report instances of inappropriate blocking. *Consistency* measures the variation in filtering within a country across different ISPs—in some cases the availability of specific Web pages differs significantly depending on the ISP one uses to connect to the Internet.

An aggregate view of the level of development for each country is represented by the results of the first four indexes presented in the Key Indicators box: gross domestic product per capita, life expectancy, literacy rates, and the human development index. The first three measures are drawn from the World Bank development indicators dataset. The GDP measure is expressed in terms of purchasing power parity in constant 2000 international dollars, which captures the ability to purchase a standard basket of consumer goods. Life expectancy can be seen as a proxy for general health, and literacy an imperfect but reasonable indication of the quality of education. The human development index is constructed by the United Nations Development Programme to reflect overall human well-being.

Governance is widely recognized to be a key determinant of economic success and human welfare. We therefore also include two measures of governance: *rule of law* and *voice and accountability*. These indexes are defined and compiled by researchers at the World Bank using an aggregation of the best available data. The authors of the indexes define them in the following way:

- *Rule of Law* includes several indicators which measure the extent to which agents have confidence in and abide by the rules of society. These include perceptions of the incidence of crime, the effectiveness and predictability of the judiciary, and the enforceability of contracts.
- *Voice and Accountability* includes in it a number of indicators measuring various aspects of the political process, civil liberties, political and human rights, measuring the extent to which citizens of a country are able to participate in the selection of governments.

There are a number of similar indicators available that are presented as comparative measures of civil liberties, free expression, and political freedom. We use only those indicators because they have been developed using what we believe to be the most rigorous methodology and without undue bias. Further information is available at the World Bank Governance and Anti-Corruption Web site: www.worldbank.org/wbi/governance.

We also include two measures of Internet accessibility provided by the International Telecommunication Union: the digital opportunity index (DOI) and Internet users as a percentage of the population. The DOI is based on eleven core ICT indicators that are agreed upon by the International Telecommunication Union's Partnership on Measuring ICT for Development. These are grouped in three clusters by type: opportunity, infrastructure, and utilization. The DOI therefore captures the overall potential for and context of Internet availability rather than usage alone. The measure of Internet access, the Internet penetration rate, is simply the percentage of the populace identified as active Internet users.

Internet regulation and filtering practices are often dynamic processes, subject to frequent change, though we expect that the political climate and the aggregate view of the issues reflected in these summaries will change more slowly than the specific instances of filtering. As the context for content regulation and the practice of Internet filtering evolve, updates will be made to the country summaries and new countries may be added. These updates will be available at http://www.opennet.net.

SOURCES FOR KEY INDICATORS

GDP per capita, PPP (constant 2000 international $)

IMF (International Monetary Fund) 2006. *World Economic Outlook Database*. Available at http://www.imf.org/external/pubs/ft/weo/2006/02/data/, converted to 2000 dollars using inflation calculator from http://www.bls.gov/cpi/.

World Bank 2004. *World Development Indicators (WDI) Online*. 2004 figures. Available with subscription at http://web.worldbank.org/WBSITE/EXTERNAL/DATASTATISTICS/0,,contentMDK:20398986~menuPK:64133163~pagePK:64133150~piPK:64133175~theSitePK:239419,00.html.

World Bank 2005. *World Development Indicators (WDI) Online*. 2005 figures. Available with subscription at http://web.worldbank.org/WBSITE/EXTERNAL/DATASTATISTICS/0,,contentMDK:20398986~menuPK:64133163~pagePK:64133150~piPK:64133175~theSitePK:239419,00.html.

Life expectancy at birth (years)

WHO (World Health Organization) 2006. *The World Health Report 2006*, Annex, Table 1. Available at http://www.who.int/whr/2006/en/.

World Bank 2006a. "*Key Development Data & Statistics*," drawn from the *World Development Indicators 2006* (WDI) database. Available at http://www.worldbank.org/data/countrydata/countrydata.html.

Literacy rate (% of people age 15+)

World Bank 2006a. "*Key Development Data & Statistics*," drawn from the *World Development Indicators 2006* (WDI) database. Available at http://www.worldbank.org/data/countrydata/countrydata.html.

World Bank 2006b. *2006 Information & Communications for Development: Global Trends and Policies*. Available at www.worldbank.org/ic4d.

US Department of State 2007a. Bureau of East Asian and Pacific Affairs. *Background Note: North Korea,* January. Available at http://www.state.gov/r/pa/ei/bgn/2792.htm.

US Department of State 2007b. Bureau of East Asian and Pacific Affairs. *Background Note: South Korea,* January. Available at http://www.state.gov/r/pa/ei/bgn/2800.htm.

Human development index (value and ranking)

UNDP (United Nations Development Programme) 2006. *Human Development Report 2006*, Human Development Indicators, Table 1. Available at http://hdr.undp.org/hdr2006/.

Rule of law

World Bank 2006c. *Worldwide Governance Indicators: 1996-2005*. Washington, DC: World Bank, September. Available at http://www.worldbank.org/wbi/governance/govdata.

Voice and accountability

World Bank 2006c. *Worldwide Governance Indicators: 1996–2005*. Washington, DC: World Bank, September. Available at http://www.worldbank.org/wbi/governance/govdata.

Digital opportunity index (value and ranking)

ITU (International Telecommunication Union) 2006. *World Information Society Report 2006*. August. Statistical Annex, Table 1. Available at http://www.itu.int/osg/spu/publications/worldinformationsociety/2006/report.html.

Internet users (% of population)

ITU (International Telecommunication Union) 2005. "Internet Indicators: Hosts, Users, and Number of PCs." 2005 figures. Available at http://www.itu.int/ITU-D/icteye/Indicators/Indicators.aspx#.

Afghanistan

Although the government of Afghanistan, with the help of international donors and private sector partners, continues to build an information communications technology (ICT) infrastructure up from its nominal status of only five years ago, very few Afghans are online. The government does not engage in technical filtering, but it has attempted to regulate media coverage and control published content.

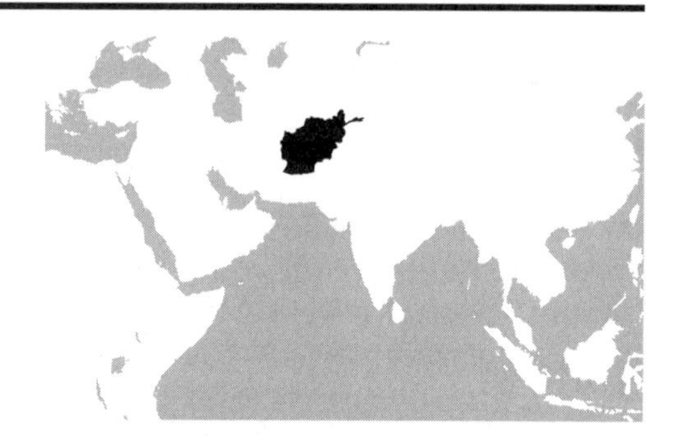

Background

The Islamic Republic of Afghanistan has committed to meeting ambitious goals of achieving genuine security, promoting sustainable economic and social development, strengthening human rights, and promoting the rule of law.[1] However, institutions are being rebuilt in a severely challenged security environment and the gains made since the fall of the Taliban in 2001 continue to be at risk. Other forms of media are gaining some ground (for example, more than thirty independent community radio stations have been established), but journalists continue to be subject to threats, violence, and intimidation.

Internet in Afghanistan

The Internet is one of many sectors in Afghanistan in a process of reconstruction from the dilapidated status it has had since 2001.[2] It was banned by the Taliban in July 2001 because it was thought to broadcast obscene, immoral, and anti-Islamic material, and because the few Internet users at the time could not be easily monitored, as they obtained their telephone lines from Pakistan.[3]

The current government recognizes the Internet as an important source of growth and development for the country, believing that ICT can create opportunities for disadvantaged

RESULTS AT A GLANCE

Filtering	No evidence of filtering	Suspected filtering	Selective filtering	Substantial filtering	Pervasive filtering
Political	●				
Social	●				
Conflict/security	●				
Internet tools	●				

Other factors	Low	Medium	High	Not applicable
Transparency				●
Consistency				●

KEY INDICATORS

		worst → best
GDP per capita, PPP (constant 2000 international $)	1,119	▌
Life expectancy at birth (years)	42	███
Literacy rate (% of people age 15+)	28	█
Human development index (out of 177)	nd	
Rule of law (out of 208)	205	███
Voice and accountability (out of 208)	184	█████
Digital opportunity index (out of 180)	nd	
Internet users (% of population)	0.1	▌

Source (by indicator): IMF 2006; WHO 2006; World Bank 2006a, 2006c, 2006c; ITU 2005
nd = no data available

groups and improve the access of the rural poor to markets.[4]

However, the struggle to make the Internet widely available is an arduous one. With a total of 1,200 Internet subscribers and an average of one Internet user per thousand people for a total of 30,000 estimated users,[5] Afghanistan remains almost completely outside the cybersphere. One major obstacle is the adult literacy rate, which stands around 28 percent (as of 2004).[6] The literacy rate for women is between 9 and 18 percent, the lowest in the world;[7] ongoing violence and intimidation threaten primary school education for girls, and indeed for all children.[8]

A second barrier to Internet use is its cost, which is prohibitive for most Afghans.[9] In part to address these concerns, the Ministry of Communications and Information Technology (MCIT, formerly the Ministry of Communications) contracted two Chinese firms, ZPE and Huawei, to build a digital wireless network in twelve provinces.[10] Internet service providers (ISPs) often choose wireless networks as well, since the number of fixed lines is limited. Still, although the Internet is essentially a luxury of the wealthy, Afghans are interested in the Internet and all it offers.[11]

In 2003 Afghanistan was given legal control of the ".af" domain, and the Afghanistan Network Information Center (AFGNIC) was established to administer domain names. Through a presidential decree, the MCIT was charged with spinning off all telecommunications operations and services to a newly created independent company called Afghan Telecom.[12] Up from five functional ISPs in 2003, in 2006 Afghanistan supported twenty-two Internet hosts and seven main ISPs[13] and a growing number of Internet cafés and telekiosks (public access points located in post offices and at the Kabul airport).[14] It plans to connect the country (and neighboring nations) along major highways with Afghan Telecom's National Optical Fibre Backbone project.[15]

Legal and regulatory frameworks

Freedom of expression is inviolable under the Afghanistan Constitution, and every Afghan has the right to print or publish topics without prior submission to state authorities in accordance with the law.[16] However, the normative limits of the law are clear: under the Constitution no law can be contrary to the beliefs and provisions of the sacred religion of Islam.[17] Mass media law has become increasingly attentive to a more vigorous adherence to this principle. The Media Law decreed by President Hamid Karzai in December 2005, just before the national legislature was formed, included a ban on four broad content categories: the publication of news contrary to Islam and other religions; slanderous or insulting materials concerning individuals; matters contrary to the Afghan Constitution or criminal law; and the exposure of the identities of victims of violence.[18] A draft amendment of the law circulating in 2006 added four additional proscribed categories: content jeopardizing stability, national security, and territorial integrity of Afghanistan; false information that might disrupt public opinion; promotion of any religion other than Islam; and "material which might damage physical well-being, psychological and moral security of people, especially children and the youth."[19]

The independence of the media was also brought into question by the March 2004 Media Law enacted by the transitional government, which handed the Minister of Culture and Information important veto powers (e.g., foreign agencies and international organizations may print news bulletins only after obtaining permission from the Minister)[20] and leadership of a Media Evaluation Commission that reviews appeals of rejections of publishing licenses by the Ministry of Information and Culture.[21] The proposed amendment to the Media Law in late 2006 would dissolve the Media Evaluation Commission and two other regulatory bodies, the National Commission of Radio and Television Broadcast, and an investigation commission that reviewed complaints against journalists and decided which cases should be forwarded to courts for prosecution.[22]

With the approval of the Telecommunications Services Regulation Act in 2005 (Telecom Law), an independent regulatory agency called the Afghanistan Telecom Regulatory Authority (TRA) was created out of the merger of the Telecommunications Regulatory Board and the State Radio Inspection Department (SRID) under the Ministry of Communications.[23] The TRA assumed responsibility for telecommunications licensing as well as promoting sustainable competition for all telecommunications services.

Licensing requirements are straightforward: companies must abide by the law to be licensed by the TRA, and only those with licenses can sell telecommunications services.[24] Of the two types of ISP licenses, transit and national licenses, only transit licenses allow ISPs to establish international connectivity.[25] Part of the TRA mandate is to protect users from the abuse of monopoly market share: companies determined to have "significant market power" must apply to have an amended license[26] and are subject to additional penalties for anti-competitive behavior.[27] A license may be revoked if the licensee has broken the law or has failed to fix repeated breaches in the agreement, has misleading or false information in their application, or does not pay the fee even after a warning.[28]

Under the Telecom Law, ISPs are duty-bound to protect user information and confidentiality.[29] However, the TRA is also authorized to demand the operator or service provider monitor communications between users as well as Internet traffic in order to trace "harassing, offensive, or illegal" telecommunications, although what constitutes these prohibited communications is not specified.[30] Where an issue of national security or a criminal case is involved, operators and service providers must hand over the required information and give the authorities immediate access to their network.[31]

In cases where there is no such immediate need, the TRA still has the right to "relevant information" as long as the TRA has given two weeks' notice.[32] In its Acceptable Use Policy, the AFGNIC prohibits the use of the ".af" domain to make any communications to commit a criminal offense; racially vilify others; violate intellectual property rights; and distribute, publish, or link to pornographic materials that a "reasonable person as a member of the community of Afghanistan would consider to be obscene or indecent."[33] The ban on spam or junk mail also includes unsolicited political or religious tracts along with commercial advertising and other information.[34]

On June 12, 2006, the National Directorate of Security (NDS), Afghanistan's national intelligence agency, issued a list of broadcasting and publishing activities that "must be banned" in light of heightened security problems that could deteriorate public morale.[35] The list of proscribed press activities was quite extensive and attributed negative intention,[36] causality,[37] and morality[38] to reporting on specific issues (primarily terrorism and the Taliban insurgency). President Hamid Karzai denied these were instructions, saying they were merely guidelines and a request for media cooperation.[39] Restricted activities included the publication or broadcasting of exaggerated reports against national unity or peace; decrees, statements and interviews of armed organizations and terrorist groups; and even the proscription against news on terrorism serving as the lead story.[40]

ONI testing results

ONI testing found no evidence of filtering in Afghanistan, although testing was not as extensive there as it was in some other countries.

Conclusion

In a country where 40 percent of the rural population suffers from low-income food deficit and life expectancy is less than forty-five years,[41] the Internet and ICT infrastructure represent only one component of social and economic development needed for the government to meet the goals set forth in the Afghanistan Compact. Only about 0.1 percent of Afghans are online. Testing revealed no evidence of government-led Internet filtering. However, through legal regulation and other acts conducted under the color of authority, Afghanistan may be constraining media coverage in ways that violate the international human rights norms for freedom of expression protected in its own constitution.

NOTES

1. See The Afghanistan Compact, signed at the London Conference on Afghanistan, January 31–February 1, 2006.
2. Sarah Parkes, "Slow road to the digital age: Rebuilding communications in Afghanistan and Iraq," ITU Telecom World 2003 On-Line News Service, October 13, 2003, http://itudaily.com/home.asp?articleid=3101309.
3. *BBC News*, "Taleban outlaw Internet," July 13, 2001, http://news.bbc.co.uk/2/hi/south_asia/1437852.stm.
4. United Nations, *Vision 2020: Islamic Republic of Afghanistan Millennium Development Goals Report 2005* (New York: United Nations), p. 123.
5. International Telecommunication Union, *World Telecommunication Indicators 2006*.
6. World Bank World Development Indicators (2006), http://devdata.worldbank.org/external/CPProfile.asp?PTYPE=CP&CCODE=AFG (accessed April 23, 2007).
7. United Nations, *Vision 2020: Islamic Republic of Afghanistan Millennium Development Goals Report 2005* (New York: United Nations), pp. 21, 123.
8. See Human Rights Watch, Lessons in Terror: Attacks on Education in Afghanistan, July 2006, http://www.hrw.org/reports/2006/afghanistan0706.
9. See *BBC News*, "Afghans plant flag in cyberspace," March 10, 2003, http://news.bbc.co.uk/2/hi/technology/2835799.stm.
10. Internet cafés are expensive, and it would be even more costly for people to purchase a computer and a telephone line to get individual access to the Internet. Amanullah Nasrat, "Internet spreading in Afghanistan," *Kashar World News*, July 17, http://www.kashar.net/technews/complete.asp?id=1725.
11. William Fisher, "In the Arab world, a blog can mean prison," *Daily Star*. (Beirut, Lebanon), March 21, 2005.
12. Ministry of Communications and Information Technology Web site, http://www.moc.gov.af/afghantelecom.asp (accessed March 7, 2007).

13. See Presentation by Gaurab Raj Upadhaya at APOPS Forum (16 APNIC Open Policy Meeting), ICT in Afghanistan, August, 21, 2003, http://www.apnic.net/meetings/16/programme/docs/apops-pres-gaurab-ict-afghan.ppt. See also *CIA World Factbook: Afghanistan*, updated March 15, 2007, https://www.cia.gov/cia/publications/factbook/geos/af.html#Comm; and Ministry of Communications and Information Technology Web site, http://www.moc.gov.af/isp.asp, (accessed April 4, 2007).
14. United Nations Development Programme, Afghanistan: A Country on the Move, http://www.undp.org.af/home/afg_on_the_move.pdf.
15. Ministry of Communications Five-Year Development Plan 1384–1389 (2005–2009), August 13, 2005, http://www.moc.gov.af/Documents/About%20the%20Ministry%20of%20Communications%20-%20FiveYears Plan/Afghanistan%20MoC%205%20yr%20plan.pdf.
16. Article 34, Constitution of the Islamic Republic of Afghanistan, effective January 4, 2004, http://www.oefre.unibe.ch/law/icl/af00000_.html.
17. Article 3, Constitution of the Islamic Republic of Afghanistan, effective January 4, 2004, http://www.oefre.unibe.ch/law/icl/af00000_.html.
18. Amin Tarzi, "Afghanistan: Mass media law comes under scrutiny," Radio Free Europe, February 2, 2007, http://www.rferl.org/featuresarticle/2007/02/ed592517-3307-4092-b556-f9c05aca7b3e.html.
19. Ibid.
20. Article 8(2), Afghan Law on Mass Media, March 2004, http://www.ijnet.org/Director.aspx?P=MediaLaws&ID=233888&LID=1.
21. Article 26, Afghan Law on Mass Media, March 2004, http://www.ijnet.org/Director.aspx?P=MediaLaws&ID=233888&LID=1.
22. Articles 21, 42, Afghan Law on Mass Media, March 2004, http://www.ijnet.org/Director.aspx?P=MediaLaws&ID=233888&LID=1. See U.S. Department of State, Country Reports on Human Rights Practices: Afghanistan, March 6, 2007, http://www.state.gov/g/drl/rls/hrrpt/2006/78868.htm.
23. Telecom Regulatory Board Web site, http://trb.gov.af/telecom%20sector%20profile1.htm (accessed April 4, 2007).
24. Article 13, Telecommunications Services Regulation Act, enacted December 18, 2005, http://trb.gov.af/new%20telecom%20law.htm.
25. Afghan Computer Science Association, IP License, http://acsa.org.af/isplicense.htm.
26. Article 15, Telecommunications Services Regulation Act, http://trb.gov.af/new%20telecom%20law.htm.
27. Articles 21–24, Telecommunications Services Regulation Act, http://trb.gov.af/new%20telecom%20law.htm.
28. Article 18, Telecommunications Services Regulation Act, http://trb.gov.af/new%20telecom%20law.htm.
29. Article 53. Telecommunications Services Regulation Act, http://trb.gov.af/new%20telecom%20law.htm.
30. Article 51, Telecommunications Services Regulation Act, http://trb.gov.af/new%20telecom%20law.htm.
31. Article 52,Telecommunications Services Regulation Act, http://trb.gov.af/new%20telecom%20law.htm.
32. Article 7, Telecommunications Services Regulation Act, http://trb.gov.af/new%20telecom%20law.htm.
33. Article 1.1, Afghanistan Network Information Center, Afghanistan ccTLD (.af) Acceptable Use Policy, October 20, 2004, http://www.nic.af/doc/pdf/afaup201004.pdf.
34. Article 2.1(1), Afghanistan Network Information Center, Afghanistan ccTLD (.af) Acceptable Use Policy, October 20, 2004, at http://www.nic.af/doc/pdf/afaup201004.pdf.
35. See Human Rights Watch Press Release, "Afghanistan: Remove new restrictions on media," June 22, 2006, http://hrw.org/english/docs/2006/06/21/afghan13605.htm.
36. Ibid. For example, those reports that aim to represent that the fighting spirit in Afghanistan's armed forces is weak.
37. Ibid. An example of attributed causality is the prohibition of "Live reports from meetings and ceremonies that disclose confidential governmental and military secrets and cause a deterioration of relations among three branches of state and results in reduction of the prestige of the government and parliament."
38. Ibid. For example, "reports relating to riots and violence which are provocative should not be published and violence should be condemned."
39. Ibid.
40. Ibid.
41. United Nations, *Vision 2020: Islamic Republic of Afghanistan. Millennium Development Goals Report 2005* (New York: United Nations), p. 21.

Algeria

Although Internet access in Algeria is not restricted by filtering, the state controls the Internet infrastructure and regulates content by other means.[1] Internet users and Internet service providers (ISPs) can face criminal penalties for posting or allowing the posting of material deemed contrary to public order or morality, for example, and journalists report being subjected to government surveillance.[2]

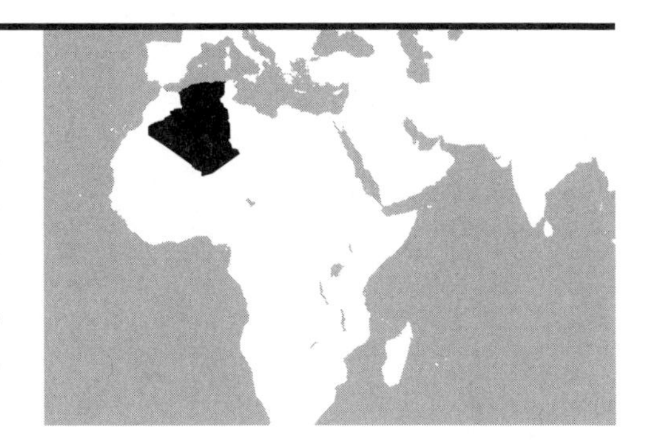

Background

Despite democratic advances made since Algeria held its first contested elections in 2004,[3] the government continues to invite criticism from human rights organizations for repressing dissent. In February 2006, Algeria's cabinet passed the so-called emergency law, which restricts freedom of assembly and threatens imprisonment for those who speak out on atrocities that occurred during the country's civil war.[4] Algerian officials frequently harass journalists and human rights advocates under the guise of security through defamation laws or dubious criminal prosecutions.[5] Religious freedom has declined in recent years as President Bouteflika has pushed through legislation greatly restricting non-Muslim worship.[6] The government has telecommunications regulations in place that require Internet providers to undertake surveillance of Internet content, but watchdog organizations report that there have been no cases of censorship under the regulations thus far.[7]

Internet in Algeria

Algeria first gained Internet connectivity in 1994 under the auspices of the Center for Research on Scientific and Technical Information (CERIST),[8] which by law remained the country's sole ISP until 1998.[9] On August 5, 1998, decree no. 98-257 opened Internet service provision to other

RESULTS AT A GLANCE

Filtering	No evidence of filtering	Suspected filtering	Selective filtering	Substantial filtering	Pervasive filtering
Political	●				
Social	●				
Conflict/security	●				
Internet tools	●				

Other factors	Low	Medium	High	Not applicable
Transparency				●
Consistency				●

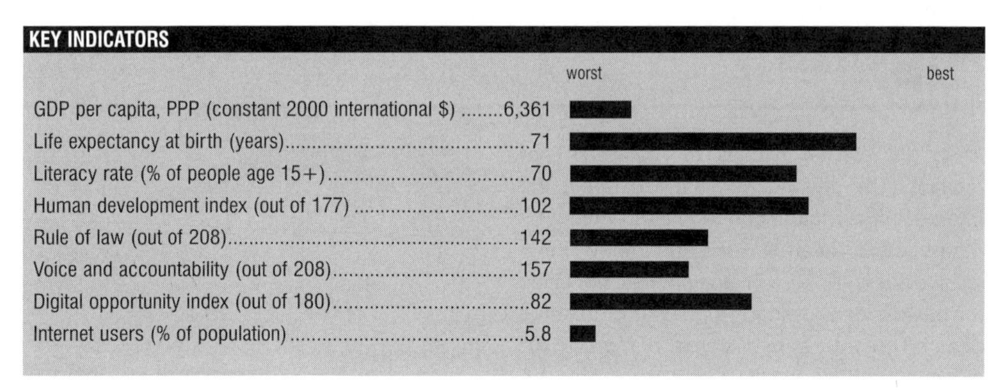

Source (by indicator): World Bank 2005, 2006a, 2006a; UNDP 2006; World Bank 2006c, 2006c; ITU 2006, 2005

providers, but private entry into the market proceeded slowly.[10] Two years later, law no. 2000-03 created the Ministry of Post and Telecommunications (MoPT), which included the Internet regulatory agency Algérie Télécom.[11] Algérie Télécom launched the ISP Djaweb in 2001[12] to extend service beyond universities and research centers.[13] Today, Algérie Télécom lists twenty-six ISP partners operating in the country, including CERIST.[14] CERIST continues to develop the academic, noncommercial Internet[15] under the influence of the state[16] and has created nodes in Algiers, Oran, Constantine, and Ourgla.[17]

The MoPT—the government agency responsible for the Internet in Algeria—has expressed its desire to promote the Internet as a source of investment and job creation.[18] Though Internet penetration has increased dramatically over the past few years, jumping from approximately 1,500 in 1999[19] to nearly 850,000 in 2006,[20] this still represents only 2.6 percent of the population. The government has supported programs that allow users to access the Internet on a "pay-as-you-go" basis, without requiring a monthly subscription.[21] Although most ISPs offer broadband, ADSL, or satellite plans, the prices of these services remain prohibitively high for many Algerians.[22] Consequently, most Algerian Internet users rely on dialup connections and cybercafés for access.

Legal and regulatory frameworks

The establishment of the MoPT in 2000 signaled the government's desire to catch up to some of its neighbors and develop the economic potential of the Internet.[23] At the same time, the government has moved to modernize information-control infrastructure and legislation, including extending criminal penalties for publishing material "contrary to public order" to Internet publications.[24] All connections between the Algerian network and the Internet at large pass through government-controlled content caching servers, an arrangement that reduces bandwidth costs but could also facilitate filtering.[25]

In January 2004 Algeria Telecom announced a deal with Daewoo to introduce high-speed connections. MoPT acts as an independent regulator and is not legally obligated to consult with or inform any other organizations before making decisions.[26] Algeria's network topology is highly centralized,[27] and all Internet connections pass through state-controlled content caching servers before reaching the global Internet.[28] ISPs are privately owned, but must obtain a license from the MoPT.[29] Approximately fifty companies have obtained licenses.[30]

Article 144(b) of the criminal code criminalizes "insulting or defaming" the president, parliament, armed forces, or any other public body, in writing, drawings or speech, through radio, television, electronic, or computer means.[31] Article 14 of a 1998 telecommunications decree makes ISPs responsible for the sites they host, and requires them to take "all necessary steps to ensure constant surveillance" of content to prevent access to "material contrary to public order and morality."[32] Journalists report that it can take up to two days to receive their e-mails, and consequently suspect the government is spying on them.[33] The regulatory framework is under review and MoPT had targeted 2005 as the year for liberalization of various sectors of the telecommunications market.[34]

ONI testing results

Among the most sensitive topics in Algeria are criticism of President Bouteflika and the military,[35] same-sex relationships,[36] and non-Islamic religious worship.[37] Algerians who engage in any of these activities face serious sanctions, including stiff fines and imprisonment. Nonetheless, ONI testing found no evidence that the government filters Internet sites or activity associated with these, or any other, sensitive topics. The government's primary forms of control thus appear to be the access controls and content monitoring regulations noted above.

Conclusion

Although Algeria does not at present filter Internet content, legislation that criminalizes peaceful criticism of the government and requires ISPs to police online content, together with a highly centralized network, could facilitate the filtering of online content in the future.

NOTES

1. Ministre des postes et télécommunications, "Configuration des services de la plate-forme Internet des P&T," http://www.postelecom.dz/service.htm (accessed January 24, 2007).
2. Reporters Sans Frontières, "Internet under surveillance 2004: Obstacles to the free flow of information online," June 22, 2004, http://www.rsf.org/article.php3?id_article=10730, (accessed January 24, 2007).
3. U.S. Department of State, Country Reports on Human Rights Practices 2005: Algeria, http://www.state.gov/g/drl/rls/hrrpt/2005/61685.htm.
4. Human Rights Watch, Algeria's Amnesty Decree, http://hrw.org/english/docs/2006/04/12/algeri13169.htm.
5. U.S. Department of State, Country Reports on Human Rights Practices 2005: Algeria, http://www.state.gov/g/drl/rls/hrrpt/2005/61685.htm; Human Rights Watch, "Algeria: Press freedom at risk despite release of editor," http://hrw.org/english/docs/2006/06/13/algeri13543.htm; Human Rights Watch, "Algeria: Human rights lawyers tried on dubious charges," http://hrw.org/english/docs/2007/02/20/algeri15360.htm.
6. U.S. Department of State, International Religious Freedom Report 2006: Algeria, http://www.state.gov/g/drl/rls/irf/2006/71418.htm.
7. Reporters Sans Frontières "Internet under surveillance 2004: Obstacles to the free flow of information online," June 22, 2004, http://www.rsf.org/article.php3?id_article=10730.
8. Economic Commission for Africa. Algeria: Internet Connectivity, http://www.uneca.org/aisi/nici/country_profiles/Algeria/algerinter.htm.
9. Economic Commission for Africa. Algeria: NICI Policy, http://www.uneca.org/aisi/nici/country_profiles/Algeria/algerpol.htm.
10. Ibid.
11. Ibid.
12. Djaweb, "Presentation," http://www.djaweb.dz/presentation.htm.
13. *Economist Intelligence Unit.* "Algeria: Telecoms and technology background," http://www.ebusinessforum.com/index.asp?layout=newdebi&country_id=DZ.
14. Algérie Télécom, "Les partenaires," http://www.algerietelecom.dz/?p=partenaire.
15. Academic Research Network, "Portail ARN," http://www.arn.dz/index.php?file=infrastructure.
16. GÉANT2, "Academic Research Network & what it gives," http://www.geant2.net/upload/ppt/4_Khelladi_Abdelkader.ppt.
17. Economic Commission for Africa, Algeria: Internet Connectivity, http://www.uneca.org/aisi/nici/country_profiles/Algeria/algerinter.htm.

18. Ministre des postes et télécommunications, "Editorial," http://www.postelecom.dz/secteur1.htm.
19. Economic Commission for Africa, Algeria: Internet Connectivity, http://www.uneca.org/aisi/nici/country_profiles/Algeria/algerinter.htm.
20. Internet World Stats, "Africa Internet usage and population statistics," http://www.internetworldstats.com/africa.htm#dz.
21. Le 1516, "Comment se connecter," http://www.le1516.com/index.html.
22. Djaweb offers 2-megabyte connections for approximately USD1,500 per month. See Djaweb, "Tarifs," http://www.djaweb.dz/tarifs.htm.
23. *Economist Intelligence Unit,* "Algeria: Telecoms and technology background," http://www.ebusinessforum.com/index.asp?layout=newdebi&country_id=DZ.
24. Reporters Sans Frontières, "Internet under surveillance 2004: Obstacles to the free flow of information online," June 22, 2004, http://www.rsf.org/article.php3?id_article=10730.
25. Ministre des postes et télécommunications, "Configuration des services de la plate-forme Internet des P&T," http://www.postelecom.dz/service.htm.
26. *VSAT Case Studies: Nigeria and Algeria*, Research Report prepared by Steve Esselaar and Aki Stavrou on behalf of the LINK Centre for the IDRC, CATIA and GVF, December 2, 2003.
27. Ministre des postes et télécommunications, "Plate forme d'accès à l'Internet," http://www.postelecom.dz/plate.htm.
28. Ministre des postes et télécommunications, "Configuration des services de la plate-forme Internet des P&T," http://www.postelecom.dz/service.htm.
29. Ministre des postes et télécommunications, "Dispositions generales," http://www.postelecom.dz/titre1.htm.
30. Ministre des postes et télécommunications, "Les providers Algeriens," http://www.postelecom.dz/provider.htm.
31. Reporters Sans Frontières, "Internet under surveillance 2004: Obstacles to the free flow of information online," June 22, 2004, http://www.rsf.org/article.php3?id_article=10730.
32. Ibid.
33. Ibid.
34. *VSAT Case Studies: Nigeria and Algeria*, Research report prepared by Steve Esselaar and Aki Stavrou on behalf of the LINK Centre for the IDRC, CATIA and GVF, December 2, 2003.
35. United Nations Commission on Human Rights, 2006, The right to freedom of expression: Report of the Special Rapporteur, Ambehi Ligado, E/CN.4/2006/55/Add.1, March 27.
36. The International Lesbian and Gay Association, World Legal Survey: Algeria, last updated July 31, 2000, http://www.ilga.info/Information/Legal_survey/africa/algeria.htm.
37. *Liberté,* "Les nouvelles sanctions concernant l'exercice illegal du culte: Les évangélistes sous haute surveillance," March 14, 2006, http://www.africatime.com/algerie/nouvelle.asp?no_nouvelle=244719&no_categorie=2; Moharram 1427, Official Journal of the Algerian Republic, "Fixing the conditions and rules for the exercise of religious worship other than Muslim," Ruling no. 06-03 of 29, #12, February 28, 2006, http://www.hrwf.net/religiousfreedom/news/2006PDF/Algeria%202006.doc.

Azerbaijan

The Internet in Azerbaijan remains free from restrictions despite the government's (at times) heavy-handed approach to dealing with political opposition. Azerbaijan has a growing Internet population, helped along by a national strategy to develop the country into an information communications technology (ICT) hub for the Caucasus region. Investment in the ICT sector has been prioritized, with ICT being seen as an essential pillar for diversifying the country's oil-dependent economy—an important policy given that Azerbaijan's rich oil and gas reserves are expected to run out in the next twenty to forty years. Azerbaijan's transition under the charismatic former president, Heidar Aliev, from war and instability in the 1990s left the political opposition weak and fragmented and has led to authoritarian tendencies. The Internet is beginning to surface as an important medium and space for political communication, and there are some indications that restrictions on content may emerge in the future.

Background

After a decade of civil unrest and a disastrous war over the territory of Nagorno-Karabakh, Azerbaijan recovered and stabilized under the strong hand of former President Heidar Aliev (elected in 1993). Since that time, political development in the country has remained dominated by the presidential apparatus. In 2003 Heidar Aliev was succeeded by his son Ilham Aliev in elections whose fairness was questioned by some observers.[1] The first President Aliev strongly promoted information technology (IT) as a pillar for national development, enacting a national ICT strategy in 2003 that set ambitious targets for the development of Internet in government,

RESULTS AT A GLANCE

Filtering	No evidence of filtering	Suspected filtering	Selective filtering	Substantial filtering	Pervasive filtering
Political			●		
Social	●				
Conflict/security	●				
Internet tools	●				

Other factors	Low	Medium	High	Not applicable
Transparency	●			
Consistency	●			

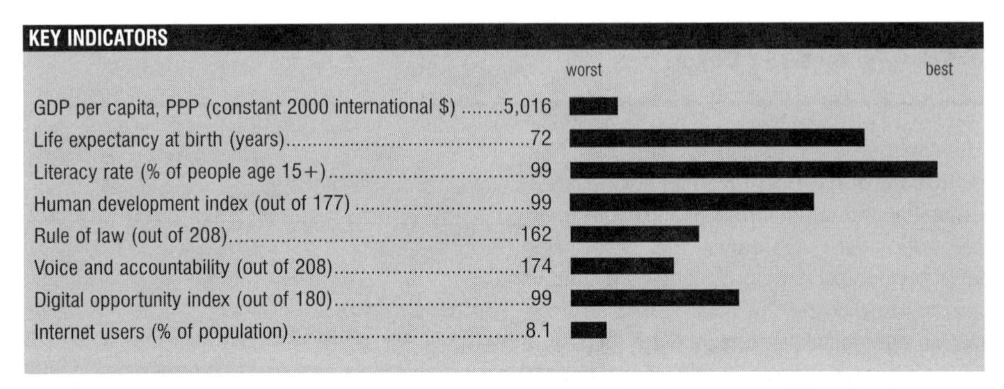

Source (by indicator): World Bank 2005, 2006a, 2006a; UNDP 2006; World Bank 2006c, 2006c; ITU 2006, 2005

education, and the industrial sector.[2] On a popular level, the Internet plays an increasingly important role in daily life, including politics. Opposition groups as well as individuals voicing discontent are now using the Internet as a communication platform, which has prompted a mild crackdown by authorities. Thus far these control measures have been reversed through legal challenges, although the government's concern for maintaining social and political stability suggests that more restrictive measures may be on the way. Azerbaijani hackers are also involved in a longstanding "cyber war" with Armenian hackers over the unresolved Nagorno-Karabakh conflict. No official sanctions have been placed on Azeri hackers, and the attacks do not appear to be a government-organized campaign but rather the work of individuals acting on their own. Web site defacements and Denial of Service (DoS) attacks have led to disruptions in the Azeri Internet, which may prompt the government to act should these attacks begin to affect critical services.

Internet in Azerbaijan

During the Soviet era, Azerbaijan was a major center for IT development, particularly in the area of process control systems. This legacy left the country with a reasonably large and well-developed technical infrastructure, including several research institutes and a political leadership that was savvy about the importance of the ICT sector. Internet development is following the pattern typical of many developing countries, with access centered on the major cities, particularly the capital city Baku. Overall—supported by the government ICT strategy as well as the large Azeri diaspora for whom the Internet is increasingly an important channel for maintaining contact with their homeland—Internet penetration is rising. Between 2004 and 2005 the number of Internet subscribers doubled; it now encompasses around 700,000 users. Official statistics state that penetration is 8 percent, but this figure may be misleading as many Azerbaijanis access the Internet from shared connections, such as their place of work or study, or from cybercafés (with the latter providing access for 21.3 percent of users). PC ownership is low (1.5 per 100), homes account for only 27.5 percent of all Internet users, and broadband penetration stands at 6.5 percent. For connectivity, most individual subscribers rely on mobile telephone (53.7 percent) and dialup (38.4 percent) as their primary means. Official survey results indicate that economic and educational barriers remain the major reason for these low figures, with 34 percent blaming the high cost of computer equipment and 24.3 percent indicating a lack of necessary skills.[3] Economic reasons are particularly important, because the cost of Internet service remains comparatively high for the average citizen: a DSL connection of 64/64Kbps costs around USD40–50 per month and unlimited access costs USD30–35, while the average salary is slightly over USD100 per month.

The Azeri Internet population is young, urban, and mostly male. Over 55 percent of the users are youth in the age range of sixteen to twenty-four, and approximately 70 percent of the users are male. During the 1990s the official language of Azerbaijan switched from Russian to Azeri, and the script from Cyrillic to Latin. As a result the number of Web sites using the Azerbaijani language increased. Azeri is used today on all official government sites and by the major media and general Internet population.

Legal and regulatory frameworks

Azerbaijan has made telecommunications and the Internet national development priorities. As a result, the state plays a leading role in the ICT sector, with the Ministry of Telecommunications Information Technology acting as both regulator and operator. Most services must be licensed,[4] including Voice-over Internet Protocol (VoIP).[5]

Internet provision is highly centralized on two state-owned Internet service providers (ISPs)—BakInterNet and AzTelecomNet—which provide national coverage and re-sell connectivity to the remaining twenty registered ISPs. AzerSat, a joint venture between the Ministry of Communications and Information Technologies and Delta Telecom Ltd, supplies international connectivity to over 85–90 percent of all users. Almost all ISPs use AzerSat's Internet international gateway and only a few possess independent international channels. The exceptions include AzEuroTel, Adanet, AzerOnline, and the nonprofit AZNET/AZRENA project that provides connectivity to the educational and research community and benefits from a satellite channel provided through NATO's "Silk Road" project.[6]

Recently the government has taken steps to liberalize the ISP market. Mandatory state licensing for ISPs is being eliminated, and state influence over domain registration is limited.[7]

From a regulatory perspective, the Internet is treated as mass media[8] and included in the list of telecommunications services regulated by the 2005 Law on Telecommunications. Azeri law does not require mandatory filtering or monitoring of Internet content. However, as Web sites that are critical of unpopular government policies (such as increases in the cost of energy) have emerged, the government has considered introducing a law that will impose restrictions on Web sites with an obscene or antinational character, thereby strengthening existing defamation laws. Content filtering is practiced by AZNET, the education and research ISP, but is regulated by an Accepted Usage Policy and is restricted to filtering out pornographic content. Anecdotal accounts claim that filtering of specific Web sites occurs, which is seemingly the result of informal requests to ISP managers by state officials from the Ministry of National Security, Ministry of Communications and Information Technologies, or the Presidency. These instances have been infrequent, and the resulting public outcry has led to a swift unblocking of the affected sites.

Azerbaijani law does not provide a formal legal foundation requiring Internet surveillance. Nevertheless, surveillance does occur, mainly by sporadic visits of the State Security Services to ISPs. In 2000–01 there was an unsuccessful attempt to adopt the Russian SORM-II model for Internet surveillance, but the project was interrupted because of financial difficulties and opposition from the ISPs and the Internet community.

ONI testing results

ONI tested for content filtering on five ISPs—Adanet, AzerOnline, AzEuroTel, AZnet, and BakInternNet—as well as several end-use locations (such as cybercafés). The results indicate that only AZnet engages in filtering, and that filtering is both limited and backed by an explicit filtering policy. AZnet blocks obscene and erotic content and gay and lesbian sites, as well as certain hacker and dating sites. Some sites with commercial content, such as gambling and drug use, are also inaccessible to users. One religious site was also blocked. Filtering on AZnet is explicit, with the user receiving a blockpage indicating that access is blocked as it violates the network's Accepted Usage Policy. AZnet uses Symantec Gateway Security on a backbone network for virus protection and blocking selected sites. No other instances of persistent filtering were detected on the other ISPs tested, and ONI did not detect the presence of commercial filtering software at any other ISP.

However, ONI did detect the selective blocking of Web sites in early 2007 during protests against the government's unpopular decision to raise consumer prices of basic utilities. The affected sites were temporarily inaccessible from ten Azeri ISPs, and the editor of one of the sites was detained by the police for a few hours, but later released without charge.[9]

At the cybercafé level, many owners impose restrictions that prevent users from downloading large attachments and visiting certain pornographic sites. But these policies are not universal, and they are implemented at the discretion of the cybercafé owner.

At the enterprise level, most employers limit access to the Internet through the use of intelligent firewalls that restrict downloads of files with certain extensions (.mp3, .avi, .mpg, .mov, and so on), as well as access to storage file servers and to the servers of instant communication (ICQ, MSN, Skype, and so on).

The ongoing cyber war between Azeri and Armenian hackers has also caused disruptions to some Web sites and ISPs. In early 2007, five Armenian sites were inaccessible. Users viewed a defaced Web page commenting on the political affiliation of the Nagorno-Karabakh region.[10] At the same time the Web site of the Azerbaijani National TV Channel was taken down.[11] Since most of the allegedly inaccessible sites contained oppositional political content, there are allegations that the Azerbaijani government was involved in the attacks. However, ONI testing could not confirm these suspicions. ONI did not test for political issues related to the proclaimed independence of the Nagorno-Karabakh region.

Conclusion

Azerbaijan's Internet remains for the most part "free and open" as a result of the government's strong interest in developing the country into an "ICT hub" for the region. With the exception of AZnet (which has a declared filtering policy) and the discretion of certain cybercafé owners, ONI did not detect the presence of any systematic policy of Internet filtering. Instances of just-in-time filtering appear to result from "informal" requests by state officials to ISP operators, and these were limited in duration and scope. Moreover, public pressure led to a swift reversal of these policies. That said, the filtering requests appear to have occurred extrajudicially. Given the prospect of increased use of the Internet by Azerbaijani opposition groups and the government's sensitivity to opposition, we may expect to see some attempts to regulate Internet content and further instances of just-in-time filtering affecting opposition Web sites.

NOTES

1. See Human Rights Watch, Azerbaijan: Presidential Elections 2003, http://www.hrw.org/backgrounder/eca/azerbaijan/index.htm (accessed April 4, 2007).

2. Decree no.1146 on the Establishment of National Strategy on Information and Communication Technologies aimed at the Development of the Republic Azerbaijan (2003–2012), signed by the President of the Republic of Azerbaijan on February 17, 2003.
3. Transport, Communication, Information and Communication Technologies, The State Statistical Committee of the Republic of Azerbaijan, http://www.azstat.org/statinfo/transport/en/042.shtml (accessed April 4, 2007).
4. As provided in the Presidential decree no. 861, March 19, 2003.
5. See Day.Az, "Azerbaijani Internet: ISPs can provide VoIP services," January 29, 2007 (in Russian), http://www.day.az/news/hitech/69561.html (accessed April 4, 2007).
6. AzEuroTel, AzerOnline started commercial activity as telecommunications companies and thus managed to establish a relatively wide network infrastructure. AzEuroTel and Adanet have additional satellite channels to Russia. AzerOnline has an additional channel to Turkey.
7. The assignment of domain names is controlled by AzNic Ltd, a joint venture between three Azeri firms. Since 2002, the number of registered domain names has rapidly increased, with approximately 3,000 first-level and 5,000 second-level domains registered under the ".az" domain.
8. Under the provisions of the Law on Mass Media adopted on December 7, 1999.
9. Day.Az, "In Azerbaijan: The author of a website protesting price increases is arrested," January 15, 2007 (in Russian), http://www.day.az/news/politics/68040.html (accessed April 4, 2007).
10. Day.Az, "Azerbaijan hacker breaks five Armenian websites," January 29, 2007 (in Russian), http://www.day.az/news/hitech/68996.html (accessed April 4, 2007).
11. Day.Az, "Azerbaijani public television website defaced by Armenian secret service," January 22, 2007 (in Russian) http://www.day.az/news/hitech/68493.html (accessed April 4, 2007).

Bahrain

Bahrain is one of the most connected countries in terms of the Internet in the Middle East and maintains a liberal Internet filtering regime relative to the region. The government prevents its citizens from accessing a small number of Internet sites, which are mostly related to pornography or gays and lesbians. The state also blocks access to a number of Bahraini political Web sites that criticize the government or the ruling family.

Background

The royally decreed political reforms of 2001–2002, which reinstituted the legislature and declared protection of personal freedoms, have improved the state of human rights in Bahrain. A small, rich, majority Shiite (estimated to be 70 percent of the population)[1] but Sunni-led state in a dangerous neighborhood, Bahrain attempts to delicately balance its policies to preserve the government's power. It has partnered with the United States in the war on terror and receives assistance accordingly (USD17.3 million requested for military and counterterrorism assistance in FY2007).[2] In addition, the United States keeps "important air assets" and its headquarters for the U.S. Navy's Fifth Fleet in Bahrain, further exhibiting Bahrain's importance to U.S. interests in the region.[3] Despite its reforms and U.S. partnership, Bahrain continues to act and pass laws contrary to its supposed democratization. Decree no. 56 of 2002 grants blanket immunity to government officials suspected of human rights abuses committed before 2001.[4] Law no. 32 of 2006 requires meeting organizers to send three days' notice of meetings to Public Safety to receive authorization; this law was invoked as justification for the use of rubber bullets and teargas by police to break up a meeting of the Movement of Liberties and Democracy on

RESULTS AT A GLANCE

Filtering	No evidence of filtering	Suspected filtering	Selective filtering	Substantial filtering	Pervasive filtering
Political				●	
Social			●		
Conflict/security	●				
Internet tools			●		

Other factors	Low	Medium	High	Not applicable
Transparency	●			
Consistency			●	

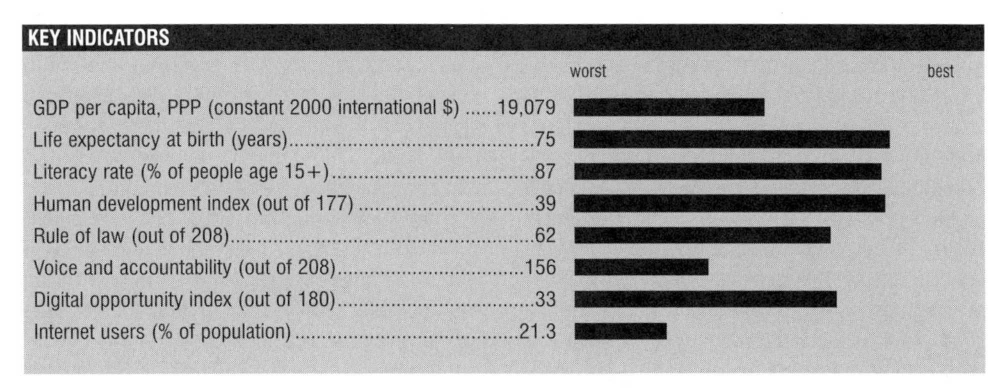

Source (by indicator): World Bank 2004, 2006a, 2006a; UNDP 2006; World Bank 2006c, 2006c; ITU 2006, 2005

September 22, 2006.[5] Press law no. 47 of 2002 allows for prosecution of journalists who are found to report against the king or Islam, advocate change in the government, or generally threaten national security.[6] Not surprisingly, this law has created a culture of self-censorship in the media.[7] And finally, the Supreme Court in October 2006 banned any mention in the media of a scandal known as "Bandargate," which involved the royal family and other politicians. Two journalists were threatened anonymously over the phone for writing on the subject.[8] It appears, as Reporters Without Borders argues, that the democratization trend has "quickly faded before the demands of the country's Shiite majority for a voice."[9]

Internet in Bahrain

Bahrain has one of the highest Internet penetration rates in the region. As of 2005, the United Arab Emirates led the Arab world in Internet penetration, followed by Bahrain and Qatar.[10] According to the ITU, there are an estimated 152,700 Internet users out of a population of 738,874,[11] and approximately 121,000 Bahrainis own computers.[12] Bahrain is also unique in that Internet telephony is legal, unlike in some other Gulf Arab states.[13] Batelco, a state-owned company, functioned as a monopoly (and in practical terms it still does) over Internet access in Bahrain until the Telecommunications Law of 2002, which attempted to inject competition in the Internet service provider (ISP) market.[14] As a result, a number of additional ISPs have sprung up in Bahrain,[15] though none has yet seriously challenged Batelco.[16] Resistance to Batelco's continued dominance has been increasing, however, as a number of disgruntled consumers have begun to voice their displeasure on the site www.boycottbatelco.com in response to the company's recent decision to place a monthly usage limit on their ADSL packages and charge more for higher limits.[17]

Legal and regulatory frameworks

Bahrain started democratization efforts in 2002 by adopting a new constitution, which reinstated a legislative body with one elected chamber. The constitution mentions the right to free speech and free press,[18] but press law no. 47 of 2002 superseded these protections and was used to prosecute, detain, and expel journalists. Three www.bahrainonline.org moderators were detained under this law for two weeks in March 2005 after they were charged with defaming the king.[19]

Bahrain's Internet is regulated by legal infrastructure governing both access and available content. The Telecommunications Regulatory Authority (TRA), created by the 2002 Telecommunications Law, is tasked with liberalizing Bahrain's telecommunications market. More specifically, the TRA seeks "to protect the interests of subscribers and users of telecommunications services and maintain effective and fair competition between established and new entrants to the telecommunications market of the Kingdom of Bahrain."[20] Two of the major initiatives set forth in the National Telecommunications Plan of 2003, produced by the Minister of Transportation in accordance with the Telecommunications Law of 2002, are the continued introduction of competition into the market through a liberal licensing regime and the eventual divestment of the state's shareholdings in Batelco.[21] The updated list of licensed entities is available at www.tra.org.bh/en/LicensingCurrent.asp.

However, the Telecommunications Law of 2002 also contains penalties for illicit use of the network, including the transmission of messages that are offensive to public policy or morals.[22] A stipulation in the law allows "security organs to have access to the network for fulfilling the requirements of national security."[23] Further, in 2005, the Ministry of Information decreed that all Web sites within Bahrain and all sites external to Bahrain containing content involving Bahraini affairs must register with the government.[24] This rule met with widespread resistance and has not been put into practice.[25]

ONI testing results

ONI ran in-country tests in 2006 on Bahrain's ISP, Batelco, using dialup as well as broadband access points. Batelco was found to institute limited Internet filtering compared with the other Gulf States. The testing found a broad range of topics to be subject to filtering, including pornography; gay and lesbian discussion; proxy and anonymizing servers; Web sites that attempt to convert Muslims to Christianity; Web sites that are critical of the Bahraini government, parliament, and the ruling family, such as www.bahraintimes.org and www.vob.org; and the Web site www.rezgar.com, which has secular leftist Arabic content. However, the vast majority of sites with content similar to these blocked sites were not blocked, indicating that the filtration regime is not comprehensive.

Unlike most of the Gulf States' ISPs, Batelco is not transparent about its blocking policy; users do not always get a blockpage when they try to access banned Web sites. For some blocked Web sites, users receive error messages such as "The page cannot be displayed."

According to Reporters Without Borders, Bahrain blocked access in October 2006 to several Web sites that were critical of the government. Among these Web sites is the Bahrain Center for Human Rights (www.bahrainrights.org) and the popular blog www.mahmood.tv, which openly criticizes the government and parliament members.[26] Both sites were found to be accessible during ONI's testing in November.

In August 2006, Bahrain banned access to Google Earth for three days. Soon after the blocking of Google Earth, cyberactivists circulated via e-mail a PDF file that annotated Google Earth screenshots of Bahrain to highlight what they claimed as the inequity of land distribution in Bahrain.[27]

Conclusion

Despite the broad range of topics that are filtered, Bahrain allows for relatively unfettered access to the Internet, especially compared with its neighbors. ONI found only very limited filtering of pornography; gay and lesbian material; content related to the conversion of Arab Muslims to Christianity; criticism of the Bahraini government, parliament, and royal family; and secular leftist Arabic content. This extremely light blocking indicates that this filtration effort is likely symbolic rather than an attempt to completely impair the ability of Bahrainis to access certain types of Internet content. For each blocked site, there are numerous similar sites that are not blocked by the government.

In 2006, however, Bahrain temporarily blocked a Bahraini human rights Web site, a popular blog run by a Bahraini citizen, and Google Earth. Even though the ban on these Web sites and services did not last long, this might indicate an intensification of the state's comparatively liberal, yet not transparent, filtering policy. In addition, given the state's close relationship with Batelco and its comprehensive regulatory structure, the government could quickly introduce new filtration if it wished.

NOTES

1. Chaieb, Mounira, "Young in the Arab World: Bahrain," BBC News, February 2005, http://news.bbc.co.uk/2/hi/middle_east/4229337.stm.
2. Human Rights Watch, World Report 2007: Bahrain, 2007, http://hrw.org/englishwr2k7/docs/2007/01/11/bahrai14699.htm.
3. Ibid.
4. Ibid.
5. Ibid.
6. Initiative for an Open Arab Internet, "Bahrain," http://www.openarab.net/en/reports/net2006/bahrain.shtml.
7. Reporters Without Borders, Bahrain: Annual Report 2007, 2007, http://www.rsf.org/article.php3?id_article=20752&Valider=OK.
8. Ibid.
9. Ibid.
10. "Online population in Arab world exceeds 26 million in 2005—Madar Research," Al-Bawaba News, September 18, 2006, http://www.ameinfo.com/96622.html.
11. Internet World Stats, Middle East Internet Usage Stats and Population Statistics, http://www.internetworldstats.com/middle.htm#bh.
12. International Telecommunication Union, *World Telecommunication Indicators 2006*.
13. Maricelle Ruiz, "Internet telephony is illegal throughout the Middle East," July 2006, http://www.ibls.com/internet_law_news_portal_view.aspx?s=latestnews&id=1538.
14. Legislative decree no. 48 of 2002 Promulgating the Telecommunications Law, October 23, 2002, http://www.tra.org.bh/en/pdf/Law_48_of_2002.pdf.
15. Telecommunications Regulatory Authority, "Licensing," February 2007, http://www.tra.org.bh/ar/licensingCurrent.asp.
16. Initiative for an Open Arab Internet, "Bahrain," http://www.openarab.net/en/reports/net2006/bahrain.shtml.
17. See www.boycottbatelco.com.
18. See Article 24 of the Bahrain Constitution, http://www.oefre.unibe.ch/law/icl/ba00000_.html.
19. Human Rights Watch, "False freedom: Online censorship in the Middle East and North Africa: Summary," November 2005, http://hrw.org/reports/2005/mena1105/2.htm.
20. TRA, "Introduction to the TRA," http://www.tra.org.bh/en/home.asp?dfltlng=1.
21. See http://www.opennetinitiative.net/studies/bahrain/#toc1.
22. Legislative decree no. 48 of 2002 Promulgating the Telecommunications Law, § 75, http://www.tra.org.bh/en/pdf/Law_48_of_2002.pdf.
23. Legislative decree no. 48 of 2002 Promulgating the Telecommunications Law, § 78, http://www.tra.org.bh/en/pdf/Law_48_of_2002.pdf.
24. See Initiative for an Open Arab Internet, "Bahrain," http://www.openarab.net/en/reports/net2006/bahrain.shtml.
25. Reporters Without Borders, Bahrain: Annual Report 2007, 2007, http://www.rsf.org/article.php3?id_article=20752&Valider=OK.
26. Reporters Without Borders, "Authorities block access to influential blog covering Bandargate scandal" and "Website blocked one month ahead of parliamentary elections," October 2006, http://www.rsf.org/article.php3?id_article=19487.
27. William Wallis, "Overhead view stirs up Bahrain," *Los Angeles Times*, December 4, 2006, http://www.latimes.com/business/la-ft-bahrain4dec04,1,1102601.story?coll=la-headlines-business&ctrack=1&cset=true. The file can be found at http://www.ogleearth.com/BahrainandGoogleEarth.pdf.

Belarus

Internet content in Belarus remains largely accessible to users despite the declared policy of selective filtering. Access remains largely centralized, as the government aims to retain firm control over the Internet. Self-censorship by online media is encouraged by the political climate, as opposition leaders and independent journalists are frequently detained and prosecuted.

Background

Under President Lukashenka's authoritarian rule, Belarus has been criticized for its repressive and increasingly authoritarian tendencies. The economy and political system remain highly centralized, with executive authority vested in the office of the president. Charges of election fraud have been widespread. Human rights organizations are heavily critical of the regime, including the steady increase in the control over information that has occurred over the past few years. Nevertheless, Lukashenka remains genuinely popular with many citizens, particularly the middle-aged and rural populations who have benefited most from his protectionist economic policies and the overall stability that Belarus has enjoyed (which contrasts with that of Ukraine and other Commonwealth of Independent States (CIS) countries).

Steady economic growth in Belarus has stimulated the development of telecommunications in recent years. However, because of excessive regulation and state control of major participants in the telecommunications industry, the development of telecommunications remains low compared with the rest of the region. The state retains a dominant position over the telecommunications sector, with all Internet connections passing through the state-owned operator Beltelecom. The top-level domain is managed

RESULTS AT A GLANCE

Filtering	No evidence of filtering	Suspected filtering	Selective filtering	Substantial filtering	Pervasive filtering
Political		●			
Social		●			
Conflict/security	●				
Internet tools	●				

Other factors	Low	Medium	High	Not applicable
Transparency				●
Consistency				●

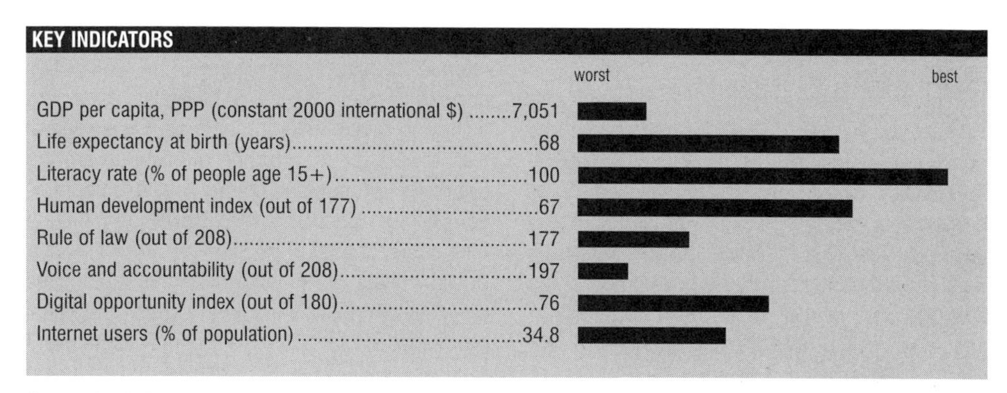

Source (by indicator): World Bank 2005, 2006a, 2006a; UNDP 2006; World Bank 2006c, 2006c; ITU 2006, 2005

and administered by the Belarus equivalent of the U.S. National Security Agency. Taking into account both the increase in Internet users and the potential of the Internet to spread political ideas, the government is adopting restrictive policies, monitoring content, and placing temporary limitations on access to politically sensitive Web sites. During the February 2006 presidential elections, ONI documented the use of just-in-time methods against opposition Web sites, which included domain name system tampering, network disconnection, and allegations of DoS attacks.[1]

Internet in Belarus

The cost of access in Belarus has recently decreased dramatically. This has seen a stark rise in usage and a growing Internet service provider (ISP) market. As of 2005, Belarus had an Internet penetration rate of approximately 30 percent. Nevertheless, prices remain higher than those of neighboring countries, and home access is not affordable for most of the population. In 2005 the cost of Internet access through the state-owned provider Beltelecom was USD0.68 per hour, while an ADSL connection cost USD340 per month—placing the latter beyond the reach of most citizens, given that the average salary was around USD230 in 2005.[2]

The most active Internet users in Belarus are in the twelve- to thirty-four-year-old age range, although some 30 percent of Belarusians in this age group have never used the Internet.[3] The level of computer penetration in the country remains low. In 2005, 58 percent of schools in Belarus had computers and only 25 percent of the schools had Internet access. The popularity of cybercafés has fallen in recent years, as most users prefer to access the Internet from home or work. Russian is the most widely used language by Belarussians on the Internet, followed by Belarusian and then English.

As Internet usage has risen, related services have developed into fast-growing and profitable businesses in Belarus. In 2005 there were thirty-two ISPs active in the country.[4] The state-controlled ISP Beltelecom holds the biggest market share, with 187 public Internet access points in the country. All ISPs are required to connect through Belpak, Beltelecom's Internet subsidiary. Beltelcom has a legal monopoly on the external channels of communications. As a consequence, all other ISPs run their traffic through

Beltelecom's infrastructure, often at very high prices.[5]

In recent years, broadband Internet services have begun to develop rapidly. Beltelecom has announced plans to establish 250,000 ADSL connections by the year 2010. Beltelecom holds a monopoly over the fixed-line infrastructure and services. Despite the formal liberalization of the Belarusian mobile market, the government owns a significant stake in all four operators.[6] Beltelecom is the only operator licensed to provide Voice-over Internet Protocol (VoIP) services in Belarus. The high prices maintained by the monopolist operator encourage the emergence of illegal VoIP providers, which are criminally prosecuted. Under decree of the Ministry of Communication and Information, IP telephony can be used only for noncommercial purposes. [7]

Legal and regulatory frameworks

The Ministry of Communications and Informatization is the main regulatory authority of the telecommunications sector. The ministry is frequently accused of placing unjustified limitations on commercial operators to reinforce Beltelecom's monopoly. Actual information communications technology (ICT)-related policy appears to be mostly created on an ad-hoc basis by President Lukashenka and his administration. The President frequently holds special meetings to issue directives regarding ICT regulation and the implementation of particular policies. The Security Council, chaired by the President, decides on a wide range of questions related to the security of the regime, including information security. Additionally, a number of state entities have significant power to influence and control the Internet. The State Center for Information Security, under the supervision of the President and initially a subdivision unit of the special security services (KGB), is a specialized body responsible for protecting state secrets. The Center also manages the administration of the country's top-level domain (".by").

Although Belarus lacks a well-developed Internet regulatory framework, the authorities appear to be pursuing a legislative basis for achieving control over the Internet. Conscious of the popularity of Internet publishing among opposition groups and private media, authorities compel self-censorship through frequent threats and prosecutions. In addition, in order to avoid public debate of pending measures, authorities often delay publishing laws before their final promulgation. In 2005 the Security Council of Belarus drafted a document entitled "The Conception on Information Security," which was then revised at the President's directive to take into consideration the new challenges to national security posed by ICT. This text is not available to the public.

Officially, Internet filtration and monitoring telecommunications networks is illegal in Belarus. However, authorities conduct surveillance of Internet activities under the pretext of protecting national security. In 2001 a Presidential Decree extended the concept of "national security" to include the Internet as a potential threat.

The special bodies of the Ministry of Internal Affairs and the KGB have the right to seize data distributed through any channel of communication[8] in order to fight criminal activity and guarantee state security. This law establishes the right of the KGB to obtain any data considered to be "relevant" from state entities and from private or public organizations, and also gives the KGB unlimited access to the information systems (including log files and so on) of communication providers.

Belarus does not have systems monitoring Internet traffic analogous to the Russian SORM-II. However, it is likely that the Belarusian and Russian special services cooperate in this sphere. Over 70 percent of Belarusian Internet traffic goes through Russia and part of it is processed through the Russian system SORM-II. Some providers confirm that the authorities have unofficially requested that all user logins be kept for up to one month and be turned over to the security services on demand.

Extensive governmental regulation, a strict licensing regime, and the state-owned Beltelecom monopoly are major impediments to the development of Internet services in Belarus. Beltelecom is under the direct supervision of the Ministry of Communications and Informatization. This may change in 2007 because of the World Trade Organization (WTO)'s accession requirements, which demand that Beltelecom be privatized and end its monopoly on external communication channels. The ministry has agreed to privatize Beltelecom, but it is likely that the government will become the controlling shareholder. The ministry has also declared that Beltelecom's control over external communication channels will remain after privatization, with licenses given only to those operators that have built their own external communication infrastructure.

E-commerce is regulated by the state. All Internet retailers are legally obliged to register domain names with the State Center of Information Security, as well as to obtain a license for retail trade by e-commerce activities. International electronic payment systems are seriously limited in Belarus. All international monetary transfers must occur through banks that notify the tax authorities of all fund transfers from abroad.

ONI testing results

ONI tested five leading ISPs: Aichyna, Belinfonet, Beltelecom, BN, and Solo. The testing could not ascertain blocking, although filtering of content is suspected, given the government's declared policy of blocking selected Web sites. The filtering of gay and lesbian Web sites has been an official policy since the beginning of 2005, on the basis that they contain pornographic material. Interestingly, these Web sites were inaccessible on all ISPs except for the state-owned Beltelecom.

ONI suspects that Internet filtering in Belarus has a deliberate but episodic character. Beltelecom's control over external connection channels allows for the creation of an effective system for regulating Internet traffic. During presidential elections, access to opposition and independent media sites appear to have been temporarily blocked. In 2006 ONI documented just-in-time tampering (indirect filtering) of opposition sites. Some specialists have suggested that, during presidential elections, Beltelecom established so-called shaping practices—that is, deliberately slowing down access to specific IP addresses. Beltelecom allegedly received special "requests" by authorities to block certain Web sites for a limited period.

Self-censorship by Internet users has become a pervasive phenomenon. In 2005 the popular Belarusian portal www.tut.by refused to put up banners advertising opposition Web sites. It is unknown whether this activity was a result of pressure by the authorities or merely an attempt to protect its own business.

ONI researchers confirmed that most cybercafés restrict access to sites containing pornography, terrorist material, and proxy-related material. Cybercafés install software that either blocks URLs within the list of forbidden sites or alerts the administrator if such a URL is visited. The restricted URL list includes Web sites forbidden for distribution by the Republic Committee on Prevention of Pornography, Violence, and Cruelty Propaganda. Administrators often require passport identification of customers. Some cybercafés also limit the volume of Internet traffic and decrease the download speed when exceeded. On the request of state security services, administrators keep the logs of users' network activity.

Conclusion

As Internet use in Belarus has risen significantly in recent years, the government seems intent to extend its firm control over all forms of information flows within the country. All ISPs in Belarus must connect to the Internet through channels of the state-owned ISP Beltelecom, thus facilitating government's control over all traffic. The president has established a strong and elaborate information security policy, and has declared his intention of exercising strict control over the Internet under the pretext of national security. Based on periodic testing, ONI suspects sporadic but sophisticated blocking of Internet content related to political events in the country.

NOTES

1. See OpenNet Initiative, ONI Internet Watch 001, The Internet and Elections: The 2006 Presidential Elections in Belarus (and Its Implications), April 2006, http://www.opennet.net/belarus/.
2. See Weekly Digest of Belarusian News, "Embassy of the Republic of Belarus in the USA," http://www.belarusembassy.org/news/digests/092505.htm.
3. Paul Budde Communications Pty Ltd., 2006: Belarus – Telecoms Market Overview & Statistics, April 2, 2006.
4. According to the statement of Ivan Rak, Deputy Minister of Communications and Informatization, more than fifty providers in 2005 had licenses for providing Internet access services. See "Belarus coordinated with WTO questions regarding liberalization of the telecommunications market," April 7, 2005, ByBanner.com, http://www.bybanner.com/article/779.html (accessed April 29, 2007) (in Russian).
5. The only exception is the network of the National Academy of Sciences of Belarus (BasNet), which has its own satellite channel.
6. See Paul Budde Communications Pty, Ltd., 2006: Belarus – Telecoms Market Overview & Statistics, April 2, 2006 at 10.
7. ByBanner.com, "The Ministry of Communications and Informatization allows Skype," March 3, 2006, http://www.bybanner.com/article/1747.html (accessed April 29, 2007) (in Russian).
8. Two legislative acts are significant in this respect: The Law on Operational and Investigative Activity of July 9, 1999, http://pravo.by/webnpa/text_txt.asp?RN=H19900289 (accessed May 2, 2007) and The Law on the Authorities of State Security of the Republic of Belarus of December 3, 1997, http://www.kgb.by/legaltexts/act01/ (accessed May 2, 2007) (both in Russian).

China (including Hong Kong)

China continues to expand the largest and most sophisticated filtering system in the world, despite the government's occasional denial that it restricts any Internet content.[1] As the Internet records extraordinary growth in services as well as users, the Chinese government has undertaken to limit access to any content that might potentially undermine the state's control or social stability, a goal also underlying President Hu Jintao's call, in January 2007, for officials to promote "healthy" online culture.[2]

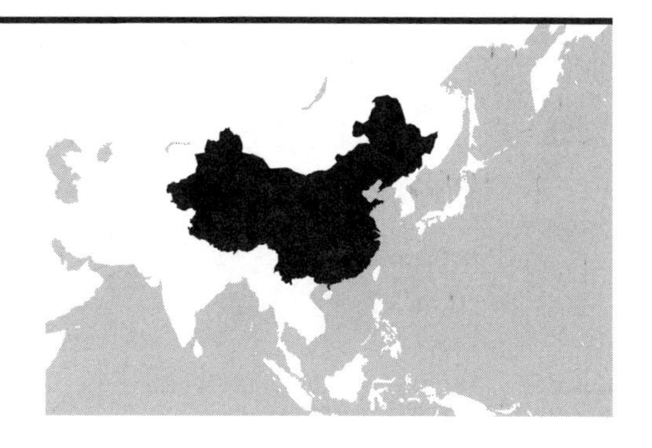

Background

The government's strenuous commitment to achieving strict supervision of the Chinese Internet showed no signs of abating in 2006, a year beginning with the introduction of Internet police cartoon mascots (Shenzhen's *Jingjing* and *Chacha*) and closing with regulations, cautiously welcomed, that allow foreign reporters to travel throughout the country and conduct interviews without prior official consent through the 2008 Olympic Games. At least eight cyber-dissidents were sentenced to prison terms in 2006.[3]

Expectations that political participation and greater government transparency and accountability would be inevitable windfalls of nearly thirty years of economic reform have been largely deflated. The government under the leadership of Hu Jintao has responded in part to sharp increases in "mass incidents" of public disorder, rampant social and economic inequalities, breakdowns in social services and public infrastructure, and growing social unrest with increased restrictions and harsh treatment of lawyers, journalists, and civil society activists. At the same time, its Herculean effort to tame the Internet activities and expression of over 100 million citizens to levels considered appropriate is

RESULTS AT A GLANCE

Filtering	No evidence of filtering	Suspected filtering	Selective filtering	Substantial filtering	Pervasive filtering
Political					●
Social				●	
Conflict/security					●
Internet tools				●	

Other factors	Low	Medium	High	Not applicable
Transparency	●			
Consistency		●		

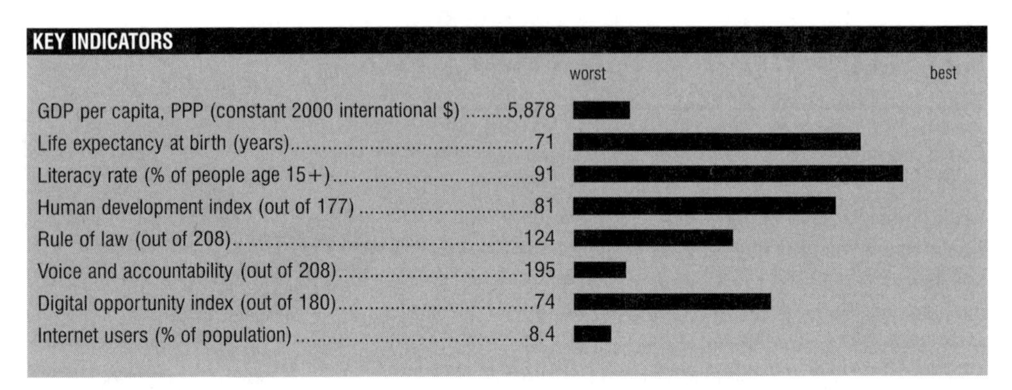

Source (by indicator): World Bank 2005, 2006a, 2006a; UNDP 2006; World Bank 2006c, 2006c; ITU 2006, 2005
Key indicator data refers to China only.

achieving greater success and efficacy, largely as a result of self-censorship and monitoring controls placed at every point of access. As one commentator noted, "while China is the world's biggest jailer of journalists, China is also writing the manual on how to control your press and citizen media—and hence your national discourse—while jailing a minimum number of people."[4]

Internet in China

From 2005 to the end of 2006, the number of Internet users grew from 94 million to 137 million.[5] The countrywide Internet penetration rate is now 10.5 percent, but this rate varies regionally—while a quarter or more of residents in major cities such as Tianjin are online, in poorer and western provinces the rate is usually less than 10 percent.[6] Gender and age are also important demographic factors, with male users significantly outnumbering women (58.3 percent to 41.7 percent) and eighteen- to twenty-four-year-olds comprising over 35 percent of all Internet users.[7] While 76 percent of users in China connect from home, 30 percent of users also use Internet cafés as a main access location.[8] Not only do Chinese users cite the Internet as the most important source for information, more important than television and newspapers, but they also have access to a wide variety of well-developed Internet services such as search engines, Bulletin Board Services (BBS), online video, blogging, and booming business-to-customer e-commerce.[9] China has the largest number of Voice-over Internet Protocol (VoIP) users in the world.[10] In March 2006, Tom Online (which formed a joint venture with Skype), announced that the government would issue no licenses for paid computer-to-telephone service (known as SkypeOut) for two years,[11] reportedly because of concerns about the financial losses to the core businesses of the major telecom carriers.[12] Only China Netcom and China Telecom were permitted to offer pilot commercial VoIP services in selected cities.[13]

Physical access to the Internet is controlled by the Ministry of Information Industry (MII), the main regulatory organ of the telecommunications sector, and is provided by seven state-licensed Internet access providers (IAPs) (with three IAPs under construction), each of which has at least one connection to a foreign Internet backbone.[14] IAPs peer at three Internet exchange points (IXPs) run by the state. IAPs grant regional

Internet service providers (ISPs) access to backbone connections. In November 2006 the Ministry of Public Security announced the completion of the essential tasks of constructing the first stage of its "Golden Shield" project, which is a digital national surveillance network with almost complete coverage across public security units nationwide.[15]

By sheer scope and range of topics—from online novels to video satires—discussion and expression over the Internet is flourishing. A major development has been the explosion of the Chinese blogosphere, which reached 20.8 million blogs at the end of 2006.[16] The growth of the Internet, in tandem with other technologies such as short messaging services, has also engendered a phenomenon of increasingly relevant "public opinion" in China, where incidents not necessarily prioritized by traditional media receive national attention and frequently lead to calls for government action and response. At times, online activity has tested this relationship between citizens and government on a range of sensitive issues.

Legal and regulatory frameworks

Although China's constitution formally guarantees freedom of expression and publication,[17] as well as the protection of human rights, legal and administrative regulations ensure that the Chinese Communist Party will be supported in its attempt at strict supervision of all forms of media. Government ministries and Party organs also use both formal and informal controls, including policies and instructions, editor responsibility for content, economic incentives, defamation liability, intimidation, and other forms of pressure to discipline media.[18]

Many of these formal and informal controls have been extended to Chinese cyberspace, though the greater range of nonstate actors makes legal regulation over the Internet a more complex effort. China's legal control over Internet access and usage is multilayered and achieved by distributing criminal and financial liability, licensing and registration requirements, and self-monitoring instructions to nonstate actors at every stage of access, from the ISP to the content provider and the end user. The Internet has been targeted for monitoring since before it was even commercially available,[19] and the government seems intent on keeping regulatory pace with its growth and development. For example, over half of the 137 million Internet users in China were found to have visited video sharing sites,[20] and in August 2006 the State Administration for Radio, Film, and Television (SARFT) announced it would be issuing regulations subjecting all online video content to its inspection.[21]

ISPs are required to record important data (such as identification, length of visit, and activities) about all of their users for at least sixty days and to ensure that no illegal content is being hosted on their servers.[22] Internet content providers, such as BBS and other user-generated content sites, are directly responsible for what is published on their service.[23] Internet access through cybercafés is also heavily regulated: all cafés are required to install filtering software, ban minors from entering, monitor the activities of the users, and record every user's identity and complete session logs for up to sixty days.[24] Getting a permit for a café is a complex process, and at any time one of at least three state departments have jurisdiction to deem a cybercafé to be inadequately self-policing and shut it down.[25] All services providing Internet users with information via the Internet that fail sufficiently to monitor their sites and report violations to the proper authorities also face serious consequences, including shutdown, criminal liability, and license revocation.[26]

New subscribers to ISPs themselves have been expected to register with their local police bureaus since 1996.[27] In October 2006, the Internet Society of China recommended the drafting of regulations that would require all individuals to register actual personal identifying data with Web site operators in order to open a blog or make comments on bulletin boards, a change from current requirements where individuals must register real names with Web sites but not blog-hosting services.[28] State media reported that 83.5 percent of respondents in a survey conducted by *China Youth Daily* opposed the proposed real-name registration system.[29]

Underlying all regulation of the Internet is a pantheon of proscribed content. Citizens are prohibited from disseminating between nine and eleven categories of content that appear consistently in most regulations;[30] all can be considered subversive and trigger fines, content removal, and criminal liability.[31] Illegal content, although broadly and vaguely defined, provides a blueprint of topics the government considers sensitive, from endangering national security to contradicting officially accepted political theory; more recently illegal content includes conducting activities in the name of an illegal civil organization inciting illegal assemblies or gatherings that disturb social order. One prominent application of these rules was the July 2006 shutdown of the online forum Century China (*Shiji Zhongguo*), a site with over 30,000 registered members and hundreds of thousands of readers co-sponsored by the Chinese University of Hong Kong's Institute of Chinese Studies.[32]

Technical filtering associated with the "Great Firewall of China" is only one tool of information control among more blunt and frequently applied methods such as job dismissals; Web site and blog closures and deletions; and the detention of journalists, writers, and activists. In 2006, fifty-two individuals were known to be imprisoned for online activities, among them several writers and journalists who were convicted in part because of the disclosure of their personal e-mail accounts by Yahoo's Chinese partner.[33] Web sites can be closed not only for a broad array of taboo topics, but also from asking the wrong questions in opinion polls.[34] In June 2006, the Information Office under the State Council and the MII embarked on a period of "strict supervision" of search engines, chat rooms, and blog service providers to curb the circulation of "harmful" information online.[35] According to the *South China Morning Post*, official statistics show that in 2006 authorities had shut down hundreds of liberal Web sites and forums and ordered eight search engines to filter "subversive and sensitive" content based on about 1,000 keywords.[36]

Because many of the laws defining illegal content are vaguely worded and have been inconsistently enforced, they provide the government with almost endless authority to control and censor content while discouraging citizens from testing the boundaries of these areas. Further, for a wide range of reasons—from economic incentives and demographic factors of the online community to the dragnet of legal liability—the impact of self-censorship is likely enormous and increasingly public, if difficult to measure. On April 9, 2006, fourteen major Web portals including www.sina.com, www.sohu.com, www.baidu.com, www.tom.com, and Yahoo's Chinese Web site issued a joint declaration calling for the Internet industry to censor "unhealthy" and "indecent" information that is "severely harmful to society," voluntarily accept supervision, and strengthen "ethical" self-regulation.[37] Their proposal sparked a flurry of similar pledges across China, from legal Web sites to blog hosting services, and with targeted content extended to include Party secrets and information affecting national security.[38]

ONI testing results

China employs targeted yet extensive filtering of information that could have a potential impact on social stability and the Party's control over society, and is therefore predominantly focused on Chinese-language content relating to domestic issues. For the government, information constituting a threat to public order extends well beyond well-publicized sensitive topics, such as the June 1989 crackdown and the Falun Gong spiritual movement (both of which are methodically blocked), and includes independent media and dissenting voices, human rights, political reform, and circumvention tools.

Testing was conducted on two backbone providers, the state-owned telecoms China Netcom (CNC) and China Telecom (CT), which between them provide coverage nationwide. Because both control access to an international gateway, URL filtering and domain name system (DNS) tampering implemented by CNC and CT affect all users of the network regardless of ISP. China also uses IP blocking at these international gateway to block access to at least 300 IP addresses, which are remarkably similar across both backbone ISPs. Though China does not employ keyword blocking on the body content of any given page, it filters by keywords that appear in the host header (domain name) or URL path.

Although there is almost complete correlation in blocking between CNC and CT, there are some gaps within certain families of Web sites. The English and Chinese versions of Wikipedia continue to be closely monitored by media and rights groups, and at time of testing the site www.wikipedia.org was accessible on both ISPs, while Chinese-language Wikipedia (zh.wikipedia.org) was inaccessible only on China Telecom. Certain bloggers, including Zeng Jinyan, the wife of activist Hu Jia (zengjinyan.spaces.live.com) were also blocked solely on CT.

As an example of targeted filtering, of the major international news organizations, only the BBC (news.bbc.co.uk) is blocked by both ISPs. The main Web site of the U.S. government–sponsored Voice of America news service, along with the *Epoch Times* (the newspaper published by the Falun Gong), are the other media outlets on the global list filtered by CNC and CT. The situation changed entirely, however, with Chinese-language media outside mainland China. From Hong Kong's *Apple Daily*, *Ming Pao* and *Sing Tao Daily* newspapers to the U.S.-based *World Journal* and *Chinesenewsnet,* a significant number of independent media representing different points on the political spectrum were filtered. The Taiwan newspaper *China Times* (www.chinatimes.com.tw), although blocked at time of testing, was reportedly accessible in early 2007.[39] Further, news in languages spoken by ethnic minorities in contested regions was also blocked, but with less uniformity. While Radio Free Asia (RFA)'s Uyghur service (www.rfa.org/uyghur) was blocked by both ISPs, RFA's main site and its Tibetan service were inaccessible only on China Telecom.

China filters a significant portion of content specific to its own human rights record and practices. As such, only a few global human rights sites, including Amnesty International, Article 19, and Human Rights First were blocked or suspected to be blocked. Thus, although China is a member of the International Labor Organization, which along with other U.N. bodies are accessible to mainland users, the Web site of the *China Labour Bulletin* (www.clb.org.hk/public/main) and other Chinese labor rights watchdogs are blocked. Similarly, the Web site of the Congressional-Executive Commission on China (www.cecc.gov) is filtered, but the U.S. Commission on International Religious Freedom (www.uscirf.gov), which has a broader mandate but has published critical reports on China, remains accessible. While blocked content mostly originates from overseas organizations and individuals (including those from Hong Kong), some organizations within China are also filtered (such as the rights defender network www.gmwq.org/web/index.asp).

Certain targets for blocking cut across political and social lines of conflict. The consistent filtering of Web sites supporting greater autonomy and rights protection for the Uyghur (www.uyghurcongress.org), Tibetan (www.savetibet.org), and Mongolian (www.innermongolia.org) ethnic minorities is not surprising, as these issues have already been excluded from official discourse inside China.

The government has long characterized the Muslim Uyghur community as presenting a separatist threat, and has blocked not only the site of the Uyghur American Association (whose president, Rebiya Kadeer, is an exiled former political prisoner and human rights activist) but has also blocked a substantial number of sites on Islam in Arabic, including those presenting extremist viewpoints (www.alumah.com).

China filters a significant number of sites presenting alternative or additional perspectives on its policies toward Taiwan and North Korea. For example, the main portal of the Taiwanese government (www.gov.tw) as well as its Mainland Affairs Council were among the many official sites blocked, along with the Democratic Progressive Party (DPP) of Taiwan (www.dpp.org.tw).

Other topics bridging the political-social divide, such as corruption, were not treated uniformly. Among the limited anticorruption Web sites filtered was the New Threads site (www.xys.org), run by the scientist Fang Shimin and focusing on academic fraud. The only HIV/AIDS-related site to be filtered was the English-language China AIDS Survey (www.casy.org), a site not updated since 2005. All other content relating to public health, women's rights, reproductive health, the environment, and development that ONI tested was accessible.

Of blocked Web sites, the major exceptions to the focus on politically sensitive topics specific to China are circumvention tools and pornography. A portion, though not a majority, of proxy tools and anonymizers in both the Chinese (www.gardennetworks.com) and English language (www.peacefire.org) was blocked. The circumvention tool Psiphon (psiphon.civisec.org) is also blocked. Both ISPs also blocked a substantial amount of pornographic content.

While the IP address of the blog search engine Technorati was blocked by both ISPs, at time of testing no blog hosting service was blocked by either ISP. However, though Google's Blogspot domain (www.blogger.com) was accessible, all individual Blogspot blogs tested were accessible on China Netcom and blocked or inaccessible on China Telecom. Ongoing ONI testing has confirmed that Blogspot has been blocked for several years in China, with periods of intermittent accessibility.

Hong Kong

ONI also conducted testing on two ISPs in Hong Kong, City Telecom (HK) Limited and PCCW, and found no evidence of filtering. However, the mainland government blocks a significant amount of content originating from its own special administrative region. In addition to many independent newspapers, sites operating out of Hong Kong that focus on political reform and governance—even those not focusing on mainland affairs but instead on exclusively local issues (such as the Hong Kong Human Rights Monitor)—are blocked across most of the categories where filtering occurs. Thus, Hong Kong–based alternative media, grassroots NGOs and coalitions (www.alliance.org.hk), religious organizations, and legitimate political parties (www.dphk.org) are all affected.

Conclusion

As China's Internet community continues to grow exponentially, the government continues to refine its technical filtering system while deputizing a range of actors, including users, ISPs and content providers, to limit the ability of its citizens to access and post content the state considers sensitive. A complex, overlapping system of legal regulation, institutionalized practices, and informal methods has been extended from print and broadcast media to the Internet. A consistent feature of regulation of the Chinese Internet has been the lack of transparency, which has long been a hallmark of the government's management and suppression of information.

NOTES

1. Gillian Wong, "China defends Web controls following online protest," Associated Press, November 8, 2006, http://news.com.com/China+We+dont+censor+the+Internet.+Really/2100-1028_3-6130970.html.
2. Xinhua News Agency, "President Hu Jintao asks officials to better cope with Internet," January 24, 2007, http://news.xinhuanet.com/english/2007-01/24/content_5648674.htm.
3. Reporters Without Borders, "Cyber-Dissident gets three-year sentence on day French President arrives," October 25, 2006, at http://www.rsf.org/article.php3?id_article=19435.
4. RConversation, "Online journalists in jail... and Chinese innovation..." December 8, 2006, http://rconversation.blogs.com/rconversation/2006/12/online_journali.html.
5. China Internet Network Information Center, Nineteenth Statistical Report on the Development of the Internet in China (*di shijiu zhongguo hulian wangluo fazhan zhuangkuang tongji baogao*), issued January 23, 2007 (in Chinese).
6. Ibid.
7. Ibid.
8. Ibid.
9. Analysis International, "China B2C E-Commerce Market Size Reached RMB 854 million in Q1 2006," Report released May 26, 2006, Beijing, http://english.analysys.com.cn/3class/detail.php?id=210&name=report&daohang=%E4%BA%A7%E4%B8%9A%E5%88%86%E6%9E%90&title=China%20B2C%20E-Commerce%20Market%20Size%20Reached%20RMB%20854%20Million%20in%20Q1%202006.
10. SinoCast China IT Watch, "Skype has over 25mn users in China," November 28, 2006.
11. Alison Maitlin, "Skype says texts are censored by partner in China," *The Financial Times,* April 18, 2006.
12. Paul Budde Communications Pty Ltd., China: Infrastructure: IP Networks, July 30, 2006.
13. SinoCast China IT Watch, "China to issue its first VoIP license," March 13, 2006.
14. China Internet Network Information Center, Nineteenth Statistical Report on the Development of the Internet in China, (*di shijiu zhongguo hulian wangluo fazhan zhuangkuang tongji baogao*), issued January 23, 2007 (in Chinese).
15. Ministry of Public Security, "National Development and Reform Commission issues national approval for the 'Golden Shield' construction project at management conference," (*guojia fazgaiwei zhuchi zhaokai dahui tongguo "jinzhi gongcheng" jianshe xiangmu guojia yanshou*), November 17, 2006, http://www.mps.gov.cn/cenweb/brjlCenweb/jsp/common/article.jsp?infoid=ABC00000000000035645. See Greg Walton, China's Golden Shield: Corporations and the Development of Surveillance Technology in the People's Republic of China, a Rights and Democracy Report, October 2001, http://www.ichrdd.ca/english/commdoc/publications/globalization/goldenShieldEng.html.
16. People's Daily Online, "China has 20.8 million bloggers," January 11, 2007, http://english.people.com.cn/200701/11/eng20070111_339952.
17. http://www.cecc.gov/pages/virtualAcad/exp/explaws.php#protectivelaws.
18. See Benjamin Liebman, "Watchdog or demagogue? The media in the Chinese legal system," *The Columbia Law Review,* January 2005, p. 41.
19. See the Regulations of the People's Republic of China for the Safety Protection of Computer Information Systems (*Zhonghua renmin gongheguo jisuanji xitong anquan baohu tiaoli*), issued by the State Council on February 18, 1994.
20. Xinhua News Agency, "Webcasting casts spell on Chinese Internet users," February 3, 2007, http://www.chinadaily.com.cn/china/2007-02/03/content_800573.htm.
21. Xinhua News Agency, "China to issue new regulations to censor online video programs," August 16, 2006.
22. Ibid.
23. Article 13, Rules on the Management of Internet Electronic Bulletin Services (*Hulianwang dianzi gonggao fuwu guanli guiding*), issued by the Ministry of Information Industry on October 7, 2000.
24. Articles 19, 21, and 23, Regulations on the Administration of Business Sites Providing Internet Services (*Hulianwang shangwang fuwu guanye changsuo guanli tiaolie*), issued by the State Council on September 29, 2002, effective November 15, 2002.
25. Ibid.
26. Article 20, Measures for Managing Internet Information Services (*Hulianwang xinxi fuwu guanli banfa*), issued by the State Council on September 25, 2000, effective October 1, 2000.
27. Human Rights Watch Backgrounder, Freedom of Expression and the Internet in China, http://www.hrw.org/backgrounder/asia/china-bck-0701.htm; Alfred Hermida, "Behind China's Internet red firewall," BBC News Online, September 3, 2002, http://news.bbc.co.uk/1/low/technology/2234154.stm.

28. Xinhua News Agency, "China strives to regulate cyberspace," December 12, 2006.
29. Xinhua News Agency, "Real-name online registration system meets opposition in China," January 8, 2007.
30. The nine types of content that have been illegal to produce or disseminate since the earliest Internet regulations are: 1) violating the basic principles as they are confirmed in the Constitution; 2) endangering state security, divulging state secrets, subverting the national regime, or jeopardizing the integrity of national unity; 3) harming national honor or interests; 4) inciting hatred against peoples, racism against peoples, or disrupting the solidarity of peoples; 5) disrupting national policies on religion, propagating evil cults and feudal superstitions; 6) spreading rumors, disturbing social order, or disrupting social stability; 7) spreading obscenity, pornography, gambling, violence, terror, or abetting the commission of a crime; 8) insulting or defaming third parties, infringing on legal rights and interests of third parties; and 9) other content prohibited by law and administrative regulations. Two categories of prohibited content were added in Article 19 of the Provisions on the Administration of Internet News Information Services (Internet News Information Services Regulations) (*hulianwang xinwen xinxi fuwu guanli guiding*), promulgated by the State Council Information Office and the Ministry of Information Industry on September 25, 2005. These two additional categories are 1) inciting illegal assemblies, associations, marches, demonstrations, or gatherings that disturb social order; and 2) conducting activities in the name of an illegal civil organization. Translation is available at http://www.cecc.gov/pages/virtualAcad/index.phpd?showsingle=24396.
31. See, for example, Rules of the NPC Standing Committee on Safeguarding Internet Security (*Quanguo renda changweihui guanyu weihu hulianwang anquan de guiding*), issued by the NPC Standing Committee on December 28, 2000.
32. Raymond Li and Kristine Kwok, "Popular forum rushes to go offline after closure order: Mainland authorities shut-down Century China, a multi-forum site with over 30,000 registered users," *South China Morning Post*, July 25, 2006, reprinted at http://www.asiamedia.ucla.edu/article.asp?parentid=49836; *South China Morning Post*, "Chinese Communist Party launches website to promote social stability," October 20, 2006. The official order cited Article 19 of the 2005 Internet News Information Services Regulations.
33. Reporters Without Borders, Press Release, "Bad start to year for online free expression," January 12, 2007, http://www.rsf.org/article.php3?id_article=20383; Reporters Without Borders, Press Release, "Verdict in cyberdissident Li Zhi case confirms implication of Yahoo!" February 27, 2006, http://www.rsf.org/article.php3?id_article=16579; Reporters Without Borders, Press Release, "Cyber-dissident convicted on Yahoo! information is freed after four years," November 9, 2006, http://www.rsf.org/article.php3?id_article=8453. See also Human Rights in China Case Highlight, Shi Tao and Yahoo, at http://hrichina.org/public/highlight/index.html.
34. For example, the Web site China Consultation Net was reportedly shut down on August 3, 2006, after 80 percent of respondents answered "yes" to a poll asking whether competitive elections should be held to choose the general secretary of the Chinese Communist Party. Kristine Kwok, "Mainland censors shut down pair of websites," *South China Morning Post,* August 5, 2006. Also, a Netease editor was reportedly fired for an online survey where 64 percent of respondents replied they would not like to be born Chinese in the next life. Howard French, "Beijing's growing urge to dominate the media," *The International Herald Tribune,* September 22, 2006. In another case, authorities withdrew the operating license of the Web site polls (*Zhongguo guoqing zixun*) on August 3, 2006. The Web site (http://www.s007s.com/) had recently asked visitors to "cast votes" on the question: "Do you think the General Secretary of the Communist Party of China should be chosen from among several candidates in differential voting?" Nearly 75 percent of those polled had answered "yes." (See IFEX, "Websites shut down amid Internet crackdown," http://ifex.org/alerts/content/view/full/76210/).
35. Xinhua News Agency, "China to tighten supervision over blogs, online search engines," June 29, 2006. See also Howard French, "Chinese discuss plan to tighten restrictions on cyberspace," *The New York Times,* July 4, 2006.
36. Raymond Li and Kristine Kwok, "Popular forum rushes to go offline after closure order: Mainland authorities shut-down Century China, a multiforum site with over 30,000 registered users," *South China Morning Post,* July 25, 2006, reprinted at http://www.asiamedia.ucla.edu/article.asp?parentid=49836.
37. People's Daily, "Fourteen Beijing websites issue joint proposal for civilized management of the Internet," (*Beijing 14 jia wangzhan lianhe xiang hulianwangjie fachu wenming banwang changyishu*), April 10, 2006, http://politics.people.com.cn/GB/1026/4283453.html.

38. See Congressional-Executive Commission on China, "Internet Operators in China Agree to Support Hu Jintao, Marxism, and the Party," China Human Rights and Rule of Law Update, May 2006, http://www.cecc.gov/pages/virtualAcad/index.phpd?showsingle=48987.
39. Reuters, "China unblocks Taiwan newspaper Web sites," February 5, 2007, http://news.zdnet.com/2100-9588_22-6156197.html.

Cuba

Internet use is severely restricted in Cuba. A combination of Cuban government policy, the U.S. trade embargo, and personal economic limitations prevents the vast majority of Cuban citizens from ever accessing the Internet. The few who gain access are limited by extensive monitoring and excessive penalties for political dissent expressed on the Internet, leading to a climate of self-censorship. Access probably is restricted even further by the U.S. government's sponsorship of reverse filtering, which encourages Web sites to prevent access from Cuba and other countries.

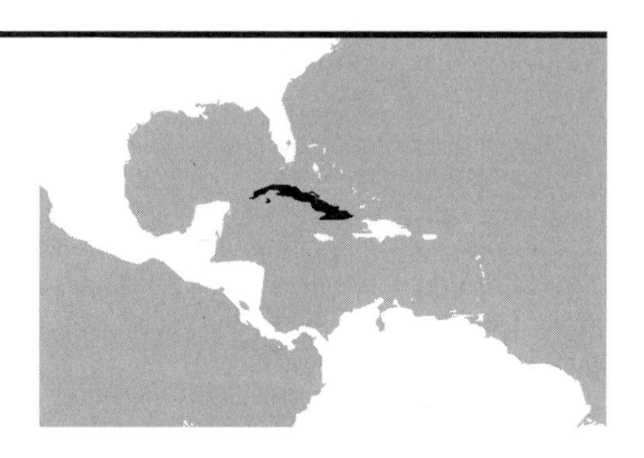

Internet in Cuba

In October 1996 Cuba first connected full time to the Internet, and in 1998 Cuba had only a single 64-Kbps satellite connection run by Sprint in Florida and allowed by an exception for communications to the U.S. trade embargo.[1] More recent legislation forbids U.S. investment in Cuban telecommunications and hampers acquisition of Cuban IP addresses; these policies, as well as Cuba's own economic policies, have hindered connectivity.[2] Currently Cuba still uses its satellite connection with a 65 Mb/s upload bandwidth and a 124 Mb/s download bandwidth for the entire country.[3]

In 1998, out of a population of eleven million, approximately 200 government-approved scientists, medical researchers, and government officials had Internet access from their desktops and 5,000 had e-mail addresses, used on Cuban intranets that remained entirely within the country.[4] By 2000 there were 6,000 computers linked to the Internet and approximately 80,000 Cubans possessed e-mail accounts, but only half of those accounts had full Internet access—accounts were selectively granted by the government, and development focused on government and tourism efforts. The country had only a single Internet café and banned personal computing purchases.[5] Currently Cuba has approximately 480,000 email accounts[6] and 190,000 regular Internet users (less than 2 percent of the population).[7] The cost of public Internet access (approximately USD4.50 per hour, or half the average monthly wage) and the very slow connections prohibit most Cubans from using the international Internet connections; most Cubans choose the national intranet instead (approximately USD1.50 per hour).[8] In 2005 Cuban computer ownership was 3.3 per 100 inhabitants.[9] An unknown number of Cubans illegally access the Internet through black market purchases of access or illegally shared authorized connections.[10]

Although the Cuban people primarily use connections to send e-mail, the Cuban government hopes to use the Internet to spread its political messages, promote tourism, and improve the efficiency of medical services.[11]

ONI did not carry out empirical testing for Internet filtering in Cuba for this report.

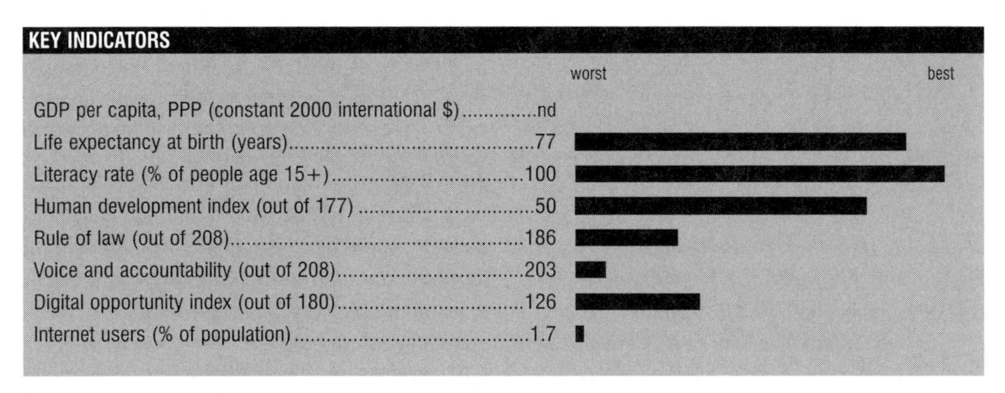

Source (by indicator): World Bank 2006a, 2006a; UNDP 2006; World Bank 2006c, 2006c; ITU 2006, 2005

Legal and regulatory frameworks

The Cuban executive branch controls governmental power, the law criminalizes dissent and permits imprisonment and surveillance without cause, and the court system lacks independence, preventing fair trials with adequate defense.[12]

Upon the arrival of a Cuban Internet connection, the government declared Internet access a "fundamental right" of the Cuban people.[13] However, Cuban Internet use also has been restricted since its beginning, with the 1996 Decree-Law 209 requiring accreditation for Internet use and outlawing Internet use "in violation of Cuban society's moral principles or the country's laws" as well as e-mail messages that "jeopardize national security."[14] All Internet access requires government authorization, and the Cuban Ministry of Computer Technology and Communications has overseen Internet and computer use since January 2000.[15]

In 1998 the Centro Nacional de Intercambio Automatizado de Información (CENIAI) was the only Cuban Internet service provider (ISP).[16] By 2000 the International Telecommunication Union reported full competition in the Cuban ISP market.[17] This level of competition is a contrast to the monopolies in the various telephone, data, and television markets;[18] however, all ISPs were under government control and oversight, and of the ISPs, only CENIAI provided personal Internet access to Cuban citizens.[19] All services, including ISPs, are subject to licensing.[20]

In terms of hardware restrictions, purchases of computers were limited to foreign nationals and government officials in 1998.[21] Since 2002, purchases by private individuals of computers, printers, and other hardware have been banned by a ministry of domestic commerce decree, and modem sales were banned earlier.[22]

Reporters Without Borders considers Cuba "one of the world's 10 most repressive countries [in regard to] online free expression" because of the highly limited access and the severe punishment of illegal Internet use, including "counter-revolutionary" usage.[23] The restrictions stem from the strong desire of the Cuban government to prevent attacks upon its political ideology from broad access to contrary views.[24]

The restriction of access to the Internet as a whole is the most significant governmental control. In addition to government prohibition of private computer sales, the Cuban police have seized numerous already-owned private computers and modems, claiming that the machines were illegal or were used against the government.[25] The lack of private materials forces most Cubans to use public access points. These sites generally require ID and registration, and many only access national e-mail and Cuban intranets; the government limits use of most hotel and cybercafé Internet connections to foreign tourists.[26] Additionally, the Cuban government openly prohibits the use of IP telephony.[27]

The government further restricts Internet use by having all legal Cuban Internet traffic pass through state-run ISPs, which use software to detect politically dissident information.[28] This filtering includes the monitoring of e-mail messages prior to their being sent or received.[29] Tests and investigation by Reporters Without Borders found that very few Web sites are actually blocked from access, but e-mail and word processing programs automatically close for "state security reasons" upon detecting mention of dissidents or other politically sensitive issues.[30]

For those who gain Internet access and use it illegally, the penalties are severe. In 2002 thirty-one people were sanctioned for improper Internet use or use of e-mail addresses that did not belong to them.[31] Penalties for Internet violations include twenty years in prison for "counter-revolutionary" article writing and five years for connecting illegally.[32] Twenty-four independent journalists currently are serving prison sentences in Cuba of up to twenty-seven years for Internet activity.[33]

The harsh penalties and pervasive monitoring, particularly when combined with requirement of name and ID for access, makes free Internet usage difficult and dangerous. E-mail users restrict the contents of their messages because of fear of state monitoring.[34] Cuban Internet policies lead to self-censorship.

Reported reverse filtering by the United States

Historically the U.S. government has placed considerable emphasis on influencing Cuban communications, creating specific policies for these technologies and spending considerable time and resources on anti-Castro radio and television programming, such as TV Martí.[35] The United States exerts some open control over the Cuban Internet, preventing U.S. investors from spending on the Cuban telecommunications market, requiring special U.S. Department of Treasury licensing for Cuban satellite connections, and prohibiting the direct sale of U.S. hardware and software.[36]

However, the United States is also suspected of engaging in less-public controls by reverse filtering and the promotion of reverse filtering. In a memo of April 15, 1994, the National Science Foundation (NSF) included Cuba on a list of countries to block from using NSF servers, a policy reversed several months later under pressure from anti-Castro politicians who wanted to use information technology to sway the population against the Cuban government.[37] Although this particular block is no longer in effect, it does set a precedent for U.S. governmental interest in using route-filtering to prevent Cuban access. More recently, in 2004, a report was made of a private Web site being requested by the U.S. government to refrain from conducting business with Cuba, among other countries.[38]

Conclusion

Cuba does not have the resources to provide Internet access for all of its citizens, particularly considering the higher prices caused by the U.S. trade embargo. However, the resources the government does devote to Internet development do not promote broad and open access. Government monitors, harsh penalties, and self-censorship discourage the transfer of politically sensitive information, and access is limited to government-approved individuals. The approved Cuban users may also be limited by reverse filtering. The Cuban Internet environment obstructs freedom of information and freedom of expression.

NOTES

1. Patrick Symmes, "Che is dead," *Wired,* http://www.wired.com/wired/archive/6.02/cuba.html, (accessed April 8, 2007).
2. Geoffry L. Taubman, Keeping Out the Internet? Non-Democratic Legitimacy and Access to the Web, www.firstmonday.org/issues/issue7_9/taubman/index.html, (accessed April 10, 2007).
3. Amaury E. Del Valle, Estados Unidos Bloquea Internet en Cuba (I), http://www.juventudrebelde.cu/cuba/2006-11-02/estados-unidos-bloquea-internet-en-cuba-l/, (accessed April 10, 2007).
4. Patrick Symmes, "Che is dead," Wired, http://www.wired.com/wired/archive/6.02/cuba.html.
5. Geoffry L. Taubman, Keeping Out the Internet? Non-Democratic Legitimacy and Access to the Web, www.firstmonday.org/issues/issue7_9/taubman/index.html.
6. Reporters Without Borders, Cuba, http://www.rsf.org/article.php3?id_article=10611, (accessed April 2, 2007).
7. *Los Angeles Times,* "Cuba inches into the Internet age," http://www.latimes.com/technology/la-fg-cubanet19nov19,1,2828501.story?ctrack=1&cset=true, (accessed April 5, 2007).
8. Reporters Without Borders, "Going online in Cuba: Internet under surveillance," http://www.rsf.org/IMG/pdf/rapport_gb_md_1.pdf, (accessed April 8, 2007).
9. Ibid.
10. Reporters Without Borders, Cuba, http://www.rsf.org/article.php3?id_article=10611.
11. Geoffry L. Taubman, Keeping Out the Internet? Non-Democratic Legitimacy and Access to the Web, www.firstmonday.org/issues/issue7_9/taubman/index.html.
12. Human Rights Watch, World Report 2006, p. 187, http://hrw.org/wr2k6/wr2006.pdf.
13. Patrick Symmes, "Che is dead," *Wired,* http://www.wired.com/wired/archive/6.02/cuba.html.
14. Reporters Without Borders, "Going online in Cuba: Internet under surveillance," http://www.rsf.org/IMG/pdf/rapport_gb_md_1.pdf.
15. Reporters Without Borders, Cuba, http://www.rsf.org/article.php3?id_article=10611.
16. Patrick Symmes, "Che is dead," *Wired,* http://www.wired.com/wired/archive/6.02/cuba.html.
17. International Telecommunication Union, Trends in Telecommunication Reform 2000-2001, p. 193, http://www.ituarabic.org/arabbook/2004/GTTR-2000.pdf; see also http://www.cuba.cu/sitios.php?idr categoria=8&base=0 (listing Cuban Internet providers).
18. Ibid.
19. Dana Bomkamp and Maria Soler, Information Technology in Cuba, http://www.american.edu/carmel/ms4917a/Internet%20Diffusion.htm.
20. International Telecommunication Union, Trends in Telecommunication Reform 2000–2001, pp. 165, 193, http://www.ituarabic.org/arabbook/2004/GTTR-2000.pdf.
21. Patrick Symmes, "Che is dead," *Wired,* http://www.wired.com/wired/archive/6.02/cuba.html.
22. Reporters Without Borders, Cuba, http://www.rsf.org/article.php3?id_article=10611.
23. Ibid.
24. Geoffry L. Taubman, Keeping Out the Internet? Non-Democratic Legitimacy and Access to the Web, www.firstmonday.org/issues/issue7_9/taubman/index.html.
25. Patrick Symmes, "Che is dead," *Wired,* http://www.wired.com/wired/archive/6.02/cuba.html.
26. Reporters Without Borders, Cuba, http://www.rsf.org/article.php3?id_article=10611.
27. International Telecommunication Union, IP Telephony Workshop Background Issues Paper, p. 22, www.itu.int/osg/spu/ni/iptel/workshop/iptel.pdf.
28. Geoffry L. Taubman, Keeping Out the Internet? Non-Democratic Legitimacy and Access to the Web, www.firstmonday.org/issues/issue7_9/taubman/index.html.
29. Reporters Without Borders, Cuba, http://www.rsf.org/article.php3?id_article=10611.
30. Reporters Without Borders, "Going online in Cuba: Internet under surveillance," http://www.rsf.org/IMG/pdf/rapport_gb_md_1.pdf.
31. Reporters Without Borders, Cuba, http://www.rsf.org/article.php3?id_article=10611.
32. Reporters Without Borders, "Going online in Cuba: Internet under surveillance," http://www.rsf.org/IMG/pdf/rapport_gb_md_1.pdf.
33. Ibid.
34. *Los Angeles Times,* "Cuba inches into the Internet age," http://www.latimes.com/technology/la-fg-cubanet19nov19,1,2828501.story?ctrack=1&cset=true.
35. Patrick Symmes, "Che is dead," *Wired,* http://www.wired.com/wired/archive/6.02/cuba.html.
36. Amaury E. Del Valle, Estados Unidos Bloquea Internet en Cuba (I), http://www.juventudrebelde.cu/cuba/2006-11-02/estados-unidos-bloquea-internet-en-cuba-l/.
37. Patrick Symmes, "Che is dead," *Wired,* http://www.wired.com/wired/archive/6.02/cuba.html.
38. "Reverse Filtering" (post) http://ice.citizenlab.org/?p=7.

Egypt

Currently there is no evidence of Internet filtering in Egypt, although a small group of politically sensitive Web sites have been blocked in the past. Online writers and bloggers have been harassed and detained for their activities online and offline. Current laws allow jail terms for journalists, editors, and online writers.

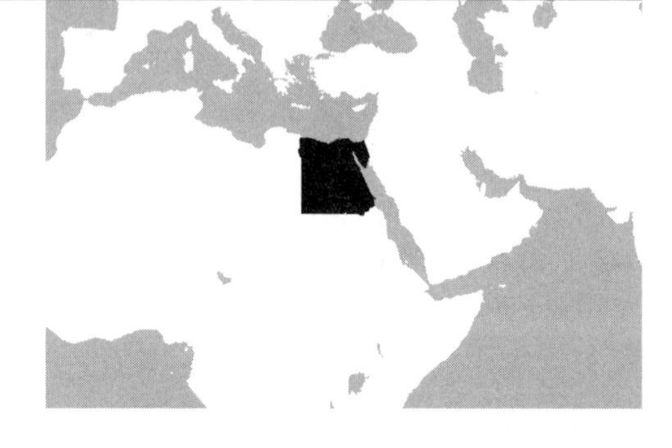

Background

Freedom of the press and freedom of expression have traditionally faced severe limits in Egypt, particularly in the spheres of religion and politics. July 2006 amendments to Egypt's Press Law left intact provisions that criminalize criticizing the president or the leaders of foreign countries and "spreading false news." Although local bloggers and human rights organizations now routinely use the Internet to cross the "red lines" that formerly circumscribed public speech, the Egyptian government monitors online communications and, in some cases, has harassed and detained people for their online activities. Though no laws specifically empower the Egyptian government to filter Web sites, provisions of the Penal Code and the Emergency Law (effective since 1981)[1] provide the government with broad authority to restrict and monitor communications. Although many journalists do criticize the government without repercussion, the government detained and beat several journalists in 2006.

Internet in Egypt

Since introducing Internet service in 1993, the Egyptian government has embarked on an ambitious program to expand Web access. The country has about five million Internet users, making up approximately 6.75 percent of the total population.[2] The government's "Free Internet Program," which allows any Egyptian with a computer, a modem, and a phone line to

RESULTS AT A GLANCE

Filtering	No evidence of filtering	Suspected filtering	Selective filtering	Substantial filtering	Pervasive filtering
Political	●				
Social	●				
Conflict/security	●				
Internet tools	●				

Other factors	Low	Medium	High	Not applicable
Transparency				●
Consistency				●

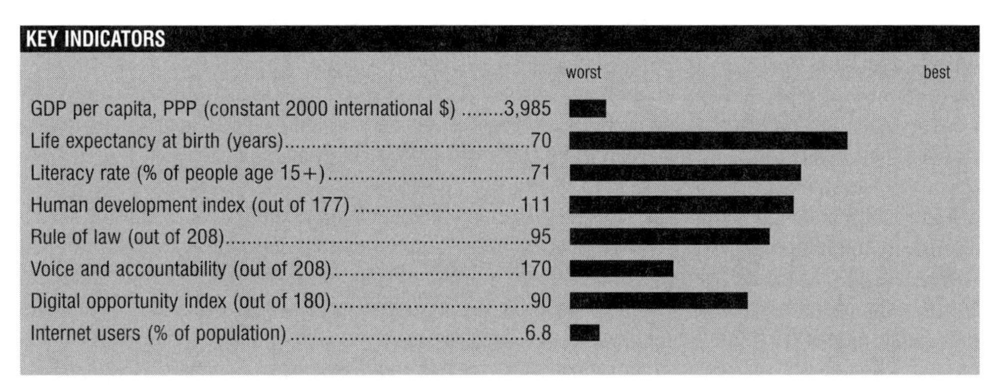

Source (by indicator): World Bank 2005, 2006a, 2006a; UNDP 2006; World Bank 2006c, 2006c; ITU 2006, 2005

access the Internet for the price of a local phone call, has led to a sharp rise in Internet use and has served as a model for other developing countries.[3] As only 3.78 percent of people own personal computers,[4] most users gain access through one of Egypt's four hundred[5] Internet cafés, a Mobile Internet Unit,[6] or nearly 1,300 public information technology clubs.[7] These clubs allow users to access the Internet for a small fee; they are affiliated with the Ministry of Communication and are located in public buildings such as schools.[8] In 2004 the "PC for Every Home" initiative helped 120,000 people obtain a personal computer through a combination of low-cost hardware and government financing.[9] The government is experimenting with WiMax technology that could provide vast areas of the countryside with high-speed, wireless access.[10] Despite all of these efforts, Internet access remains most prevalent in the cities.[11]

Egypt boasts the largest fixed-line communications network in the Arab world. Where many nations in the region are serviced by state-owned companies or monopolies, Egypt has licensed four Internet carriers and eight data service providers, along with hundreds of Internet service providers (ISPs).[12] Service is currently provided by 211 ISPs, the largest of which are LINKdotNET, a private company founded in 1992, and TEData, the Internet arm of the giant Telecom Egypt, slated for privatization in 2007 whose shares are 80 percent owned by the Egyptian government and 20 percent free float. In 2004, the government, along with nine companies, introduced ADSL service to Egypt. As of 2006, the service had approximately 130,000 subscribers at an average monthly cost of 95 Egyptian pounds (USD17).[13] In 2007, the National Telecommunications Regulatory Authority (NTRA) is expected to issue two new licenses for international telecommunications services, and the country has recently liberalized the Voice-over Internet Protocol (VoIP) market.[14]

Legal and regulatory frameworks

Despite Egypt's progressive attitude toward industry regulation, the Egyptian government continues to rely on legal and extralegal measures to restrict the flow of information. Egypt's Emergency Law allows authorities to detain individuals without charge or trial for prolonged periods of time and to censor, confiscate, and close down any publication that the Ministry of Interior sees fit.[15] This law has been renewed for successive three-year periods since President Anwar Sadat's assassination in 1981. Although in his re-election campaign Hosni Mubarak said he would replace the Emergency Law with an anti-terrorism law, in May of 2006 parliament extended the Emergency Law for another two years while the government drafts the new law.[16]

Much-anticipated amendments to Egypt's Press Law, which Mubarak signed in July 2006, struck many of the old law's most controversial provisions. However, it left intact prison sentences for journalists who criticize the president or foreign leaders, or who "spread false news,"[17] as mentioned previously. The laws cover print and "other" publications, which courts have interpreted as including online writings.[18]

Although there is no law that explicitly empowers the government to block Web sites, a 2006 court decision maintained that the Ministry of Communications & Information Technology is permitted to "block, suspend or shut down any website liable to pose a threat to national security."[19] This ruling gives the Department for Confronting Computer and Internet Crime, a special unit within the Ministry of Interior, additional tools to pursue Web sites deemed "threatening," and some worry that such pursuit is escalating.[20]

In January 2007, the Ministry of Interior announced plans to propose an international initiative to combat terrorism online. No Web sites are currently blocked outright, but security officials monitor data traffic, including e-mail, blogs, bulletin boards, and other Web sites. Internet café owners have reported that security officials have instructed them to keep lists of their customers and the customers' identification numbers.[21] Furthermore, Internet café owners must seek a license from the Ministry of Telecommunications;[22] those without licenses can be shut down.[23] Owners are sometimes given lists of people who are to be banned from using their cafés, and they are always supposed to check IDs; some places have signs that "announce 'No entry to political or sexual sites by order of the State Security.'"[24]

The government has arrested writers for their online activities. In 2003, for example, State Security officers detained activist Ashraf Ibrahim on charges of "spreading false news" for e-mailing accounts and photographs of police violence at anti-war demonstrations to international human rights organizations. On February 22, 2007, a criminal court in Alexandria sentenced 22-year-old blogger Abd al-Karim Nabil Sulaiman to four years in prison on charges of "vilifying Islam" and "insulting the president."[25] The Egyptian government has also used the Internet to entrap men engaged in consensual homosexual conduct. Though it is not officially against the law to engage in homosexual acts, dozens of men have been charged with "debauchery" or "distributing obscene material" after chatting with police who were posing as gay men online.[26]

ONI testing results

ONI conducted in-country tests in fall 2006 and found no evidence of Internet filtering in Egypt. In 2005, most ISPs blocked www.ikhwanonline.com, the official site of the Muslim Brotherhood, Egypt's largest opposition movement. At one time, the popular ISP LINKdotNET blocked www.alshaab.com, the Web site of the Labor Party's biweekly newsletter, but no longer does.[27]

Though there have been reports that Web sites for the Muslim Brotherhood are regularly blocked, neither the official Web site for the Muslim Brotherhood, www.ikhwanonline.com, nor the unofficial www.ikhwanweb.com, were blocked when the testing was conducted.

A number of ISPs also offer optional filters that block pornography; TEData offers a "Family Internet" plan that filters pornography and dating sites. Some of these packages restrict blogs and other Web sites as well.[28]

Conclusion

Internet users in Egypt have unfettered access to the Internet but the government monitors online activities and has prosecuted online writers. Bloggers have reported instances of harassment and intimidation on the part of security forces.

NOTES

1. BBC World News. Country Profile: Egypt. Last updated January 22, 2007, http://news.bbc.co.uk/2/hi/world/middle_east/country_profiles/737642.stm.
2. International Telecommunication Union, *World Telecommunication Indicators 2006*.
3. Deborah L. Wheeler, "The Internet in the Arab world: Digital divides and cultural connections," Jordan's Royal Institute for Inter-Faith Studies, June 16, 2004, http://www.riifs.org/guest/lecture_text/Internet_n_arab world_all_txt.htm; Ministry of Communications and Information Technology, Arab Republic of Egypt, "Telecom Reform Milestones," http://www.mcit.gov.eg/tele_Mileston.aspx.
4. International Telecommunication Union, *World Telecommunication Indicators 2006*.
5. Deborah L. Wheeler, "The Internet in the Arab world: Digital divides and cultural connections," Jordan's Royal Institute for Inter-Faith Studies, June 16, 2004, http://www.riifs.org/guest/lecture_text/Internet_n_arabworld_all_txt.htm.
6. Set up with the help of the United Nations Development Programme. Arab Republic of Egypt Ministry of Communications and Information Technology, "Access for All," http://www.mcit.gov.eg/ict_access.aspx.
7. Information & Decision Support Center, "Statistical Indicators about Egypt," January 16, 2007, http://www.idsc.gov.eg/Indicators/IndicatorsResult.asp?rIssueCategory=1&MainIssues=107&IndicatorSector=62&IndicatorClass=&Cond=OR.
8. Arab Republic of Egypt Ministry of Communications and Information Technology, "Access for All," http://www.mcit.gov.eg/ict_access.aspx.
9. Egypt at WSIS, "Partnering for Success," http://www.egyptatwsis.com.eg/pppict.asp; Colleen Taylor, "Intel unveils WiMax network in rural Egypt," Electronic News Network, December 18, 2006, http://www.edn.com/article/CA6400848.html?partner=enews.
10. CIT Egypt, News Room, "Egypt to be the first in the WIMAX technology," March 1, 2007, http://www.citegypt.com/press.asp?delta=88.
11. The Mobile Internet Unit is a bus equipped with computers that can travel to remote areas to provide temporary Internet access. Arab Republic of Egypt Ministry of Communications and Information Technology, "Access for All," http://www.mcit.gov.eg/ict_access.aspx; Deborah L. Wheeler, "The Internet in the Arab world: Digital divides and cultural connections," Jordan's Royal Institute for Inter-Faith Studies, June 16, 2004, http://www.riifs.org/guest/lecture_text/Internet_n_arab world_all_txt.htm.
12. TradeArabia, "Egypt telecom set for massive growth," January 7, 2007, http://www.tradearabia.com/tanews/newsdetails_snIT_article117205_cnt.html.
13. Human Rights Watch, "False freedom: Online censorship in the Middle East and North Africa: Egypt," November, 2005, http://hrw.org/reports/2005/mena1105/4.htm.
14. Paul Budde Communication Pty Ltd., "Egypt: Convergence, broadband and Internet markets," December 6, 2006, http://www.budde.com.au/publications/annual/contents/2006-African-Broadband-and-Internet-Markets-3916.html.
15. Egyptian Organization for Human Rights, Journalism in Egypt: Caught Between Laws and the Government, July 12, 2006, http://www.eohr.org/report/2006/re0821.shtml.
16. Al Jazeera, "Egypt extends emergency law," May 4, 2006, http://english.aljazeera.net/News/archive/archive?ArchiveId=22418.
17. Emergency Law 147/2006.
18. On January 22, 2007, for example, Alexandrian blogger Abd al-Karim Sulaiman Nabil was sentenced to four years in prison for "insulting Islam and the President." See Associated Press, "Egyptian blogger's lawyers appeal his four years prison sentence," February 26, 2007, http://www.iht.com/articles/ap/2007/02/26/africa/ME-GEN-Egypt-Blogger.php.
19. Sarah El Sirgany, "Al-ahram reverses Internet block on blogs," *The Daily Star,* August 15, 2006, http://www.dailystaregypt.com/article.aspx?ArticleID=2615; and Reporters Without Borders, "Court upholds government's claim to be able to block opposition Websites," June 27, 2006, http://www.rsf.org/article.php3?id_article=18136.
20. Negar Azimi, "Bloggers against torture," *The Nation,* February 6, 2007, http://www.thenation.com/doc/20070219/azimi.
21. The Initiative for an Open Arab Internet, "Egypt," November 2005, http://www.openarab.net/en/reports/net2006/egypt.shtml and Human Rights Watch, "False freedom: Online censorship in the Middle East and North Africa: Egypt," November 2005, http://hrw.org/reports/2005/mena1105/4.htm.

22. The Initiative for an Open Arab Internet, "Egypt: Disputes between Ministries of Telecommunications and Culture on Internet cafes are settled; Ministry of Telecommunications is entitled to license Internet cafes," http://openarab.net/en/net/net1912.shtml.
23. Human Rights Watch, "False freedom: Online censorship in the Middle East and North Africa: Egypt," November 2005, http://hrw.org/reports/2005/mena1105/4.htm.
24. Negar Azimi, "Bloggers against torture," The Nation, February 6, 2007, http://www.thenation.com/doc/20070219/azimi.
25. Carolynne Wheeler, "Young blogger jailed in Egypt; Chill envelops online dissent," Globe and Mail, February 3, 2007, http://www.theglobeandmail.com/servlet/story/LAC.20070302.EGYPT02/TPStory/TPInternational/Africa/.
26. BBC News, "Egypt crackdown on homosexuals," March 6, 2002, http://news.bbc.co.uk/2/hi/programmes/crossing_continents/1858469.stm; Human Rights Watch, "False freedom: Online censorship in the Middle East and North Africa: Egypt," November 2005, http://hrw.org/reports/2005/mena1105/4.htm; and U.S. State Department, Human Rights Country Report: Egypt 2005, http://www.state.gov/g/drl/rls/hrrpt/2005/61687.htm.
27. Human Rights Watch, "False freedom: Online censorship in the Middle East and North Africa: Egypt," November 2005, http://hrw.org/reports/2005/mena1105/4.htm.
28. The Initiative for an Open Arab Internet, "Egypt," November 2005, http://www.openarab.net/en/reports/net2006/egypt.shtml.

Ethiopia

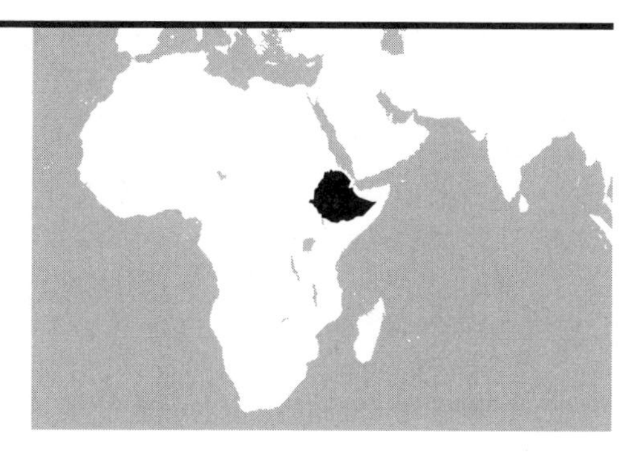

Opponents of the current political regime have increasingly used online media to criticize the government, and Ethiopia has responded by implementing a filtering regime that blocks access to popular blogs and the Web sites of many news organizations, dissident political parties, and human rights groups. However, the filtering is not comprehensive, and much of the media content that the government is attempting to censor can be found on sites that are not banned.

Background

Ethiopia's record on human rights and political openness took a turn for the worse after the legislative elections of May 2005. Though originally hailed by the U.S. State Department as "a milestone in creating a new, more competitive multiparty political system in one of Africa's largest and most important countries,"[1] the elections were quickly followed by protests and riots by opposition parties alleging voter intimidation and rigging by the Ethiopian People's Revolutionary Democratic Front (EPRDF).[2] Ethiopian police in turn arrested more than 10,000 people in Addis Ababa during the protests. According to Human Rights Watch, most were released within a month, but hundreds remained locked up. The government recently released 400 prisoners in March of last year, but it is unknown how many are still imprisoned.[3] The EPRDF continued its crackdown on opposition by arresting seventy-six "politicians, journalists, and civil society activists"[4] and charging them with "'treason', 'conspiracy' to overthrow the government and 'genocide.'"[5] In foreign affairs, Ethiopia is involved in a border dispute with Eritrea, the subject of a war between the two states from 1998 to 2000.[6] Ethiopia, on behalf of the U.N.-recognized transitional government of Somalia, has also

RESULTS AT A GLANCE

Filtering	No evidence of filtering	Suspected filtering	Selective filtering	Substantial filtering	Pervasive filtering
Political				●	
Social			●		
Conflict/security			●		
Internet tools			●		

Other factors	Low	Medium	High	Not applicable
Transparency	●			
Consistency			●	

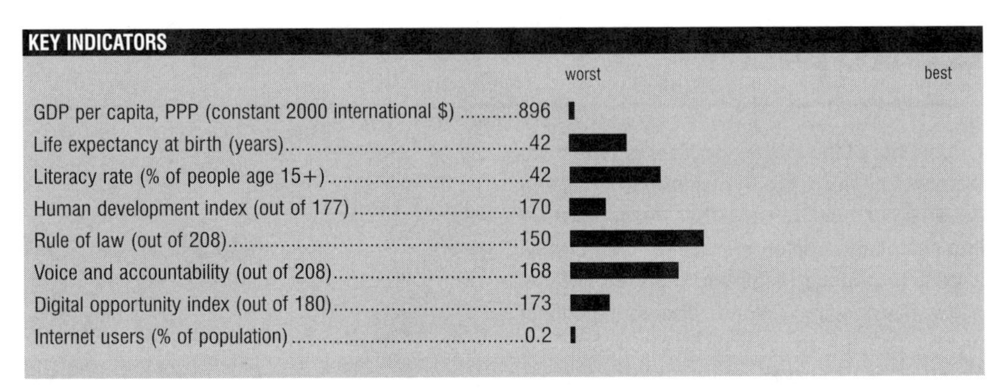

Source (by indicator): World Bank 2005, 2006a, 2006b; UNDP 2006; World Bank 2006c, 2006c; ITU 2006, 2005

entered into conflict with the Union of Islamic Courts (UIC). The UIC is an Islamist group vying for control of Somalia.[7] Because of the rise of Islamist extremism in Somalia, the United States views Ethiopia as an important ally in the global war on terror. As a result, Ethiopia still receives the largest amount of U.S. aid in sub-Saharan Africa despite U.S. disapproval of the repression following the elections of May 2005.[8] Self-censorship in the media is driven by the fear of government reprisal. Foreign journalists have difficulties acquiring authorization to work in Ethiopia, and an Associated Press reporter was sent out of the country in early 2007 after "tarnish[ing] the image of the country."[9] The government has reportedly entered into censorship of blogs and opposition Web sites, though it officially denies doing so.[10]

Internet in Ethiopia

Ethiopia lags behind much of Africa in Internet availability and is currently attempting a broad expansion of access throughout the country, though efforts have been hampered by the largely rural makeup of the Ethiopian population and the government's refusal to permit any privatization of the telecommunications market. Only 113,000 people had Internet access in 2005, for a penetration rate of 0.2 percent, one of the lowest in Africa.[11] The state-owned Ethiopian Telecommunications Corporation is the sole Internet service provider (ISP) in the country. Internet cafés are a major source of access in urban areas, and an active community of bloggers and online journalists now plays an important role in offering alternative news sources and venues for political dialogue. However, three-quarters of the country's Internet cafés are in the capital city, and even there access is often slow and unreliable.[12]

In 2005, Ethiopia announced plans to spend hundreds of millions of dollars over the next three years to connect all of the country's schools, hospitals, and government offices—and most of its rural population—to broadband Internet via satellite or fiber-optic cable.[13] Currently satellite Internet is available to some large corporations, but individuals are not permitted to have private satellite connections. The Ethiopian Telecommunications Commission (ETC) also bans the use of Voice-over Internet Protocol (VoIP) in Internet cafés and by the general population.[14]

Legal and regulatory frameworks

The Ethiopian government maintains strict controls over access to the Internet and online media, despite constitutional guarantees of freedom of the press and free access to information.

The state-owned ETC and the Ethiopian Telecommunication Agency (ETA) have exclusive control of Internet access throughout the country. The ETA is not an independent regulatory body and its staff and telecommunications policies are controlled by the national government.[15] It grants the ETC a monopoly license as Ethiopia's sole ISP and seller of domain names under the country code top-level domain, ".et". Internet cafés and other resellers of Internet services must be licensed by the ETA and purchase their access through the ETC.[16] Individual purchasers must also apply for Internet connections through the ETC. Though Ethiopia has considered some limited privatization of the telecommunications market, these plans are on hold indefinitely despite acknowledgments that the ETC has not been an effective service provider.[17]

In the face of political turmoil over the last two years, the ruling party in Ethiopia has become an increasingly active censor. In mid-2006 the government cut off access within the country to online publications run by political dissidents and to all blogs hosted by www.blogspot.com (the ETC claimed that the blockage was a technical glitch but offered no further explanation).[18] The government has also banned reporters for the state-run news agency from using the Internet at all and now frequently jails journalists, including online journalists, for charges including treason; most private news outlets have now been shut down.[19] The Committee to Protect Journalists named Ethiopia one of the top four jailers of reporters in the world in 2006.[20]

In late December 2006, the ETA began requiring Internet cafés to log the names and addresses of individual customers, apparently as part of an effort to track users who engaged in illegal activities online. The lists are to be turned over to the police, and Internet café owners who fail to register users face prison.[21] Bloggers believe that their communications are being monitored.[22] The state maintains the right to cut off Internet access to resellers or customers who do not comply with security guidelines. In practice, it has shut down Internet cafés in the past for offering VoIP services and other policy violations.[23]

ONI testing results

ONI conducted testing on Ethiopia's sole ISP, the ETC. The ETC blocking effort appears to focus on independent media, blogs, and political reform and human rights sites, though the filtering is not very thorough and many prominent sites that are critical of the Ethiopian government remain available within the country.

The prime target of Ethiopia's filtering is political bloggers, many of whom oppose the current regime. Ethiopia blocks all the blogs hosted at www.blogspot.com and at www.nazret.com, a site that aggregates Ethiopian news and has space for blogs and forums. Though many of the filtered nazret blogs are critical of the government, the scope of the filtering is wide: one blocked blogger wrote solely about the 2006 World Cup. The blogspot-hosted sites that are blocked include Ethiopian and international commentators on politics and culture, including popular blogs EthioPundit and Enset.

The Web sites of opposition political parties appeared to be a priority for blocking (www.kinijit.org, www.hebret.com, and others), as did pages for groups that represent ethnic minorities within Ethiopia (www.anaukjustice.org, www.oromia.org). Although women's rights groups in general were not filtered, the ETC did block one Web site aimed at connecting women involved in politics in Asia (www.onlinewomeninpolitics.org).

Many independent news sites covering Ethiopian politics or compiling international and local coverage were blocked, including Cyber-

Ethiopia, the Tensai-Ethiopia radio site, Ethio-Media, EthioX, and EthioIndex. But some media sites carrying news and editorials that are unfavorable to the Ethiopian government remained available, including Addis Voice and Ethiopian Review, which had been blocked as part of the ETC's initial filtering of blogs and media sites in 2006.[24] International news sites such as CNN and Voice of America radio were not blocked.

Some human rights sites focusing specifically on Ethiopia were filtered. The Ethiopian Democratic Action League, which advocates for political prisoners, was blocked, as was a page calling for the freedom of jailed opposition leader Yacob Haile-Mariam (www.freeyacob.com) and a site about the imprisonment of human rights activist Mesfin Woldemariam (www.mesfinwolde mariam.org). However, information about these and other imprisoned dissidents is available via a number of human rights pages that are not blocked, including Human Rights Watch, Amnesty International, and various Ethiopian-focused rights groups. Reporters Without Borders, which has chronicled Ethiopian Internet filtering on its Web site (www.rsf.org), is not banned.

ONI testing found that search engines, including Google, Yahoo, MSN, and others, were available in Ethiopia, and no e-mail sites have been blocked. Though VoIP has been banned within the country, sites offering that service, such as Skype, were not filtered. The ETC did not block censorship circumvention tools such as www.anonymizer.com, and Internet users within Ethiopia appear to have found alternative means of accessing banned sites.[25]

Conclusion

Ethiopia's current approach to filtering can be somewhat spotty, with the exception of the blanket block on two major blog hosts. Much of the banned political and human rights–related content is available at sites that are not blocked. The authors of the blocked blogs have in many cases continued to write to an international audience, apparently without sanction. But Ethiopia is increasingly jailing journalists and the government has shown an increasing propensity toward repressive behavior online; it seems likely that the trend will be more extensive censorship as Internet access expands across the country. When the ETC becomes more sophisticated as an ISP, its filtering regime may become broader and more comprehensive, particularly if the Ethiopian political situation remains unstable.

NOTES

1. Sean McCormack, "Ethiopian Elections," U.S. State Department, http://www.state.gov/r/pa/prs/ps/2005/53355.htm.
2. ABC News, "One student killed in Ethiopian election protest," June 2005, http://www.abc.net.au/news/newsitems/200506/s1385945.htm.
3. Human Rights Watch, World Report 2007: Ethiopia, 2007, http://hrw.org/englishwr2k7/docs/2007/01/11/ethiop14704.htm.
4. Ibid.
5. Reporters Without Borders, Ethiopia: Annual Report 2007, 2007, http://www.rsf.org/article.php3?id_article=20755&Valider=OK.
6. BBC News, "Country Profile: Eritrea," December 2006, http://news.bbc.co.uk/2/hi/africa/country_profiles/1070813.stm.
7. BBC News, "Ethiopian army faces Somali test," December 2006, http://news.bbc.co.uk/2/hi/africa/6208759.stm.
8. Human Rights Watch, World Report 2007: Ethiopia, 2007, http://hrw.org/englishwr2k7/docs/2007/01/11/ethiop14704.htm.
9. Reporters Without Borders, Ethiopia: Annual Report 2007, 2007, http://www.rsf.org/article.php3?id_article=20755&Valider=OK.
10. Ibid.
11. Internet World Stats Africa, Internet Usage Statistics, http://www.internetworldstats.com/stats1.htm.
12. Lynn Hartley and Michael Murphree, "Influences on the partial liberalization of Internet service provision in Ethiopia," *Critique,* Fall 2006, http://lilt.ilstu.edu/critique/fall2006docs/Influences on the Partial Liberalization of Internet Service.pdf.

13. Ian Limbach, "Waking up to a laptop revolution: From grand infrastructure projects to small grass-roots initiatives in education, technology is bringing about change in the developing world," *Financial Times,* March 29, 2006; Xinhua News Agency, "Ethiopian firm launches standard virtual internet service," February 6, 2006.
14. Economic Commission for Africa, Consultative workshop held on Ethiopia's new broadband initiative, March 29, 2005, http://www.uneca.org/eca_resources/news/032905disd_dna.htm; Ethiopian Telecommunication Agency, Telecommunication (Amendment) Proclamation No. 281/2002, July 2, 2002, http://www.eta.gov.et/Button/Scan/Telecom Proc 281_2002 (amendment) NG.pdf.
15. Lynn Hartley and Michael Murphree, "Influences on the partial liberalization of Internet service provision in Ethiopia," *Critique,* Fall 2006, http://lilt.ilstu.edu/critique/fall2006docs/Influences on the Partial Liberalization of Internet Service.pdf.
16. Ethiopian Telecommunication Agency, About ETA, http://www.eta.gov.et/Button/About%20ETA.html; Ethiopian Telecommunication Agency, Licenses, http://www.eta.gov.et/Button/Licenses.htm.
17. Balancing Act News Update, "Ethiopia puts off privatising ETC but rolls out nationwide fibre network," Issue 272, September 19, 2005, http://www.balancingact-africa.com/news/back/balancing-act_272.html.
18. Reporters Without Borders, "Three more sites unaccessible, government denies being involved," May 29, 2006, http://www.rsf.org/article.php3?id_article=17783.
19. Julia Crawford, "'Poison,' politics, and the press: In Ethiopia's toxic political climate, Zenawi's government sweeps up journalists and shuts down newspapers," Committee to Protect Journalists, April 28, 2006, http://www.cpj.org/Briefings/2006/DA_spring_06/ethiopia/ethiopia_DA_spring_06.html.
20. Committee to Protect Journalists, "Internet fuels rise in number of jailed journalists," December 7, 2006, http://www.cpj.org/Briefings/2006/imprisoned_06/imprisoned_06.html.
21. Groum Abate, "Ethiopia Internet cafes start registering users," The Capital, December 27, 2006, http://nazret.com/blog/index.php?title=ethiopia_internet_cafes_start_registerin&more=1&c=1&tb=1&pb=1.
22. Weichegud! ET Politics, "Clarification," December 13, 2006, http://weichegud.blogspot.com/2006_12_01_weichegud_archive.html.
23. Groum Abate, "Ethiopia Internet cafes start registering users," The Capital, December 27, 2006, http://nazret.com/blog/index.php?title=ethiopia_internet_cafes_start_registerin&more=1&c=1&tb=1&pb=1.
24. See, for example, CyberEthiopia, "Internet repression in Ethiopia," September 2006, http://www.cyber-ethiopia.com/net/docs/internet_repression_in ethiopia.html.
25. Carpe Diem Ethiopia, "Blogging in the valley of the shadow of death," June 2, 2006, http://carpediemethiopia.blogspot.com/2006/06/blogging-in-valley-of-shadow-of-death.html; Meskel Square, "The magnificent eight," June 1, 2006, http://www.meskelsquare.com/archives/2006/06/the_magnificent_1.html.

India

As a stable democracy with strong protections for press freedom, India's experiments with Internet filtering have been brought into the fold of public discourse. The selective censorship of Web sites and blogs since 2003, made even more disjointed by the non-uniform responses of Internet service providers (ISPs), has inspired a clamor of opposition. Clearly government regulation and implementation of filtering are still evolving.

Background

India is the world's second most populous nation, with a population of over one billion. India generally respects the right to free speech and the right to publish sensitive materials. A wide array of political, social, and economic beliefs is represented by the Indian media, generally without repercussion.[1] However, targeted censorship around issues of political and social conflict is a reality, particularly in areas of unrest. With the political turmoil present in the continuing dispute with Pakistan over Kashmir as well as fighting between religious groups, and issues between castes, the state takes an interest in censoring offensive material that could induce violence. Rarely are journalists detained on censorship issues, and they are often quickly released if held. Most violent attacks on journalists are carried out by religious or ethnic groups, with occasional harassment by state authorities.[2]

Internet in India

With an estimated forty-eight million users, the Internet community in India is the fifth largest in the world, although Internet users formed only about 4.3 percent of the country's population in 2005.[3] Access is gradually expanding from the most heavily populated urban centers, currently 41 percent of users, to small cities and towns.[4]

RESULTS AT A GLANCE

Filtering	No evidence of filtering	Suspected filtering	Selective filtering	Substantial filtering	Pervasive filtering
Political	●				
Social	●				
Conflict/security			●		
Internet tools			●		

Other factors	Low	Medium	High	Not applicable
Transparency			●	
Consistency		●		

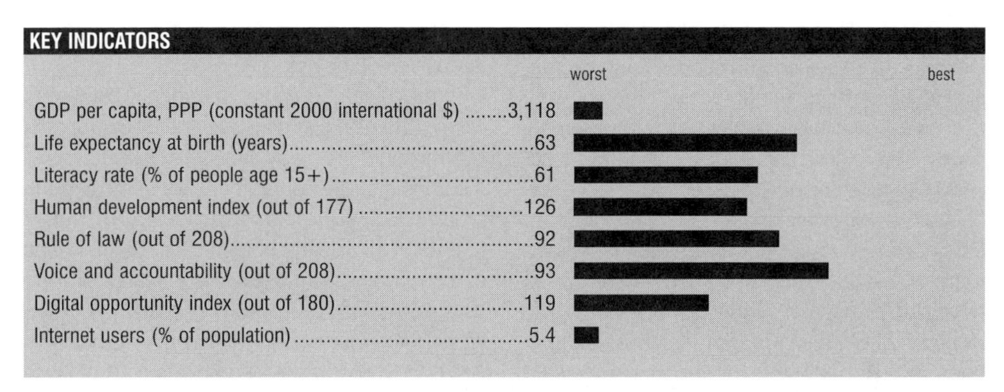

Source (by indicator): World Bank 2005, 2006a, 2006a; UNDP 2006; World Bank 2006c, 2006c; ITU 2006, 2005

Because 71 percent of the population lives in rural areas, and because the gap between rural and urban teledensity is increasing, the majority of Indians are shut out of the Internet.[5] In decreasing order of popularity, points of access are cybercafés, home, work or business, and schools, with cybercafés remaining the most popular option.[6] An estimated 38 percent of all Internet users in India are "heavy users" and spend an average of 8.2 hours per week on the Internet.[7] A Windows Live Spaces report on a thriving blogging community in India, estimated at 14 percent of Internet users, found that a vast majority of bloggers are men under the age of thirty-five; this conforms to the demographic snapshot of Internet users as predominantly male, middle class, and young.[8]

There are 153 ISPs in operation today, although the majority market share (62 percent) remains with the public-sector corporations Bharat Sanchar Nigam Limited (BSNL) (43 percent) and Mahanagar Telephone Nigam Limited (MTNL) (19 percent).[9] In the mid-1980s two state-owned corporations were formed to provide limited telecom services—Videsh Sanchar Nigam Limited (VSNL) for international long distance, and Mahanagar Telephone Nigam Limited (MTNL) for Mumbai and Delhi. In 1995 VSNL was the first to provide Internet services in India, and it was privatized in 2002. The first Action Plan of the National Task Force on Information Technology and Software Development, created in May 1998, sought to create Internet access nodes in all district headquarters by January 2000. The government began allowing ISPs to legally handle Voice-over Internet Protocol (VoIP) in April 2002. As of March 2006, 134 ISPs were authorized to offer Internet-based telephony services, but only 32 were actually providing the service.[10]

In January 2007 the Department of Telecommunications (DOT) announced that it would be installing filtering mechanisms at India's international gateways. The head of the Internet Service Providers Association of India (ISPAI) stated that these new "landing stations" would be able to both engage in centralized filtering of Web sites and blocking of VoIP telephony services such as Yahoo, MSN, and Skype (and many more) that have not technically been approved to provide these services in India.[11]

Legal and regulatory frameworks

India guarantees freedom of speech and expression in its constitution, but reserves the authority to impose reasonable restrictions in the interests of the sovereignty and integrity of India, state security, foreign relations, public order, decency, or morality; or in relation to contempt of court, defamation, or incitement to an offense.[12] Each form of media—print, film, and television—is governed by its own regulatory apparatus. For example, the Press Council of India (PCI), a quasi-judicial body with two-thirds membership of representatives from print media, has a mandate to protect the independence of the press. The PCI adjudicates complaints against the media, issues normative guidelines, and performs a public education function.[13] In contrast, films cannot be exhibited without certification of a board appointed by the central government.[14] Private FM radio station ownership was legalized in 2000, but ownership licenses were granted only for stations airing entertainment or educational content; commercial and community FM radio stations are not allowed to broadcast news and current affairs.[15] The state still controls all AM radio stations.

Until the late 1990s, the Indian government had control over all aspects of the telecommunications sector—policy, regulation, and operations.[16] The New Internet Policy introduced in November 1998 allowed private companies to apply for licenses to become ISPs and either lease transmission network capacity or build their own, thereby ending the monopoly over domestic long distance networks of the Department of Telecoms. Most, however, opted to use the lines already established by the government.[17]

In June 2000 the Indian Parliament created the IT Act to provide a legal framework to regulate Internet use and commerce, including digital signatures, security, and hacking. The act criminalizes the publishing of obscene information electronically, and grants police powers to search any premises without a warrant and arrest individuals in violation of the act. [18]

The Indian Computer Emergency Response Team (CERT-IN) was set up by the Department of Information Technology under the IT Act to implement India's filtering regime.[19] By stretching the prohibition against publishing obscene content to include the filtering of Web sites, CERT-IN was empowered in 2003 to review complaints and act as the sole authority for issuing blocking instructions to the Department of Telecommunications (DOT).[20] Only specified individuals or institutions can make official complaints and recommendation for investigation to CERT-IN, a list that is limited to high-ranking government officials, the police, government agencies, and "any others as may be specified by the Government."[21] Many have argued that giving CERT-IN this power through executive order violates constitutional jurisprudence holding that specific legislation must be passed before the government can encroach on individual rights. The blocking mechanism created under the Act provides for no review or appeal procedures, except in court, and is permanent in nature. When CERT-IN has issued orders to block specific Web sites, no communication has been made to the public beforehand.[22]

Another basis for filtering was demonstrated with the blocking of the site www.hinduunity.org on April 28, 2004, reportedly ordered by the Mumbai police on the grounds that it contained inflammatory anti-Islamic material.[23] Police commissioners, who can exercise the powers of executive magistrates in times of emergency, can block Web sites containing material constituting a nuisance or threat to public safety under Section 144 of the Code of Criminal Procedure.[24] While major and small ISPs immediately complied with the blocking request, one of the nation's largests ISPs, Sify, refrained from blocking the Web site, arguing that only CERT-IN had the authority to issue blocking orders.[25]

Filtering can also be mandated through licensing requirements. For example, ISPs seeking licenses to provide Internet services with the DOT "shall block Internet sites and/or individual subscribers, as identified and directed by the Telecom Authority from time to time" in the interests of "national security."[26] License agreements also require ISPs to prevent the transmission of obscene or otherwise "objectionable material."[27]

The proposed amendment to the IT Act brought before Parliament on December 15, 2006, aims to address growing concerns about information security and data theft that threaten the vitality of India as an outsourcing hub.[28] Under specific conditions, the bill absolves intermediaries (including cybercafés) of responsibility for making available information or links created by third parties.[29] The government has also created "guidelines" for ISPs to follow, such as the monitoring of subscriber traffic by keyword and the disclosure of dynamic IP addresses of clients by ISPs.[30]

According to the Right to Information Act passed in 2005, designated government officers are required to respond to requests for information within thirty days.[31] Although it is not clear whether information about the blocking of Web sites falls within the exceptions listed in the Act,[32] which include information relating to national security and state sovereignty, individuals have filed RTI requests seeking greater transparency in the filtering process.[33]

ONI testing results

Results from ONI testing reveal that Indian ISPs selectively filter sites identified by government authorities as relating to national unity and state security. ONI conducted testing on Bharti, Direct, Reliance, YOU Telecom (formerly known as Iqara), Pacenet, and VSNL. Variations in blocking among ISPs of the same limited range of sites suggest that CERT-IN and the DOT continue to rely on ISPs to implement filtering instructions. Although obscene information is the only type of content to be made illegal under the IT Act, ONI found no evidence that pornography is filtered in India. Rather, nearly all the sites filtered had already been reported publicly as blocked at some time.

The only site made inaccessible by all ISPs tested was the Hindu Unity Web site (www.hinduunity.org). (A number of different URLs direct to this site; these URLs were blocked with varying consistency between ISPs.) Further evidence that filtering has yet to be implemented through a uniform process can be found in the inconsistencies in filtering of the Web sites named in the CERT-IN blocking order following the bombings of suburban trains in Mumbai on July 11, 2006. On July 13, 2006, CERT-IN ordered access to seventeen Web sites blocked, reportedly because the attackers were believed to have communicated via the blogosphere. The Web sites that were ordered to be blocked included "American right-wing" sites (www.mypetjawa.mu.nu; www.mackers-world.com), Hindu extremist or "Hindutva" sites, and a defunct Web site supporting the formation of a "Dalit" homeland within India (www.dalitstan.org).[34]

Among the ISPs, Bharti, YOU Telecom, Reliance, and VSNL blocked the majority of sites included on the July 13 CERT-IN order. In this context, the personal Web site of a member of the Hindutva party VHP (and a university student in Indiana), www.rahulyadav.com, was filtered almost certainly because it was included in the July CERT-IN order, but the actual Web site of the VHP party (www.vhp.org) was available on all ISPs tested.

In 2006, filtering requests were also generated by individuals protesting content they considered offensive or obscene. In response to a Public Interest Litigation (PIL) petition calling for the ban of the social networking site Orkut for hosting a "We Hate India" community, the Bombay High Court had directed the Maharashtra government to issue notice to Google for "alleged spread of hatred about India" on Orkut.[35] A month later, in response to protests over an "anti-Shivaji" community on Orkut, Pune police banned Orkut, temporarily shut down cybercafés where users were found to be using the site, and began an investigation under the IT Act and penal code provisions for obscene publications and religious insult.[36] In December 2006, a government official made a similar blocking request after reportedly "obscene" material about "Hindu girls" was posted on Orkut.[37] However, none of these efforts resulted in a comprehensive ban on Orkut, for though it was intermittently available in Pune it was nevertheless accessible on all ISPs tested.

ONI testing determined that filtering occurred at the ISP level, with considerable variation between ISPs. Direct, Pacenet, and VSNL blocked more of the tested URLs than did other ISPs. Filtering focused primarily on Web sites seen as a threat to national security, as well as sites offering untraceable communication such as the VoIP site www.hotfoon.com and the SMS gateway www.clickatell.com. Other sites, such as www.kahane.org, appear to have been blocked only because they shared an ISP address with a targeted site.

In contrast to the collateral blocking of Web sites in August 2003[38] and July 2006, where ISPs in both incidents responded to CERT-IN orders by cutting off access to parent Web sites including Google's www.blogspot.com, www.typepad.com, and Yahoo!'s www.geocities.com, banned Web site owners continue to migrate their content successfully to other domains. For example, while ISPs are clearly blocking on the subdomain level (for example, the site www.princesskimberley.blogspot.com is filtered on four ISPs tested), the reportedly banned Maoist Web site www.peoplesmarch.com was accessible in other forms (www.peoplesmarch.wordpress.com, www.naxalrevolution.blogspot.com) on all ISPs at time of testing.

Conclusion

Amidst widespread speculation in the media and blogosphere about the state of filtering in India, the sites actually blocked indicate that while the filtering system in place yields inconsistent results, it nevertheless continues to be aligned with and driven by government efforts. For example, efforts to block certain communities on Orkut, and in some instances the entire site altogether, have been initiated largely by individuals, but the government response has not resulted in the systematic blocking of Orkut by the ISPs that ONI tested. Government attempts at filtering have not been entirely effective, as blocked content has quickly migrated to other Web sites and users have found ways to circumvent filtering. The government has also been criticized for a poor understanding of the technical feasibility of censorship and for haphazardly choosing which Web sites to block. The amended IT Act, absolving intermediaries from being responsible for third-party created content, could signal stronger government monitoring in the future.

NOTES

1. U.S. Department of State, Country Reports on Human Rights Practices, 2006: India, http://www.state.gov/g/drl/rls/hrrpt/2006/78871.htm.
2. Ibid.
3. Paul Budde Communication Pty Ltd., India-Key Statistics and Telecommunications Market Overview, 2006, p. 2.
4. Internet and Mobile Association of India, Internet in India: 2006, http://www.iamai.in/research_index.php3.
5. Telecom Regulatory Authority of India, Indian Telecom Services Performance Indicators April-June 2006, October 2006, p. 40, http://www.trai.gov.in/Reports_content.asp?id=29.
6. Internet and Mobile Association of India, Internet in India: 2006, p. 16, http://www.iamai.in/research_index.php3.
7. Internet and Mobile Association of India, Internet in India-2006, p. 37, http://www.iamai.in/research_index.php3.
8. The Press Trust of India, "Indians prefer good-old diary to blogs," November 27, 2006.

9. Telecom Regulatory Authority of India (TRAI), Indian Telecom Services Performance Indicators April-June 2006, http://www.trai.gov.in/Reports_content.asp?id=29; http://www.indiabroadband.net/bsnl-broadband/2676-bsnl-top-isp-india.html.
10. TRAI Annual Report 2005–2006, http://www.trai.gov.in/traiannualreport.asp.
11. Indrajit Basu, "Security and censorship: India to clip the wings of Internet," January 16, 2007, Digital Communities, http://www.govtech.net/digitalcommunities/story.php?id=103332.
12. Article 19, The Constitution (Ninety-Third Amendment) Act, January 20, 2006, http://lawmin.nic.in/coi.htm.
13. Articles 13–15, Press Council Act 1978, http://presscouncil.nic.in/act.htm.
14. Parishi Sanjanwala, Internet Filtering in India, unpublished paper.
15. Ministry of Information and Broadcasting, Policy Guidelines for Setting up Community Radio Stations in India, http://mib.nic.in/welcome.html; Ministry of Information and Broadcasting, Grant of Permission Agreement, http://mib.nic.in/fm/fmmainpg.htm. See also Subramanian Vincent, "Community radio gets its day," India Together, November 18, 2006, http://www.indiatogether.org/2006/nov/sbv-cradio.htm.
16. See Thankom G. Arun, Regulation and competition: Emerging issues in an Indian perspective, Centre on Regulation and Competition, Working Paper Series No. 39, October 2003, www.competition-regulation.org.uk/publications/working_papers/wp39.pdf.
17. Peter Wolcott, The Provision of Internet Services in India, http://mosaic.unomaha.edu/India_2005.pdf.
18. The Information Technology Act, Article 67, 2000. Under the Act, anyone who publishes "any material which is lascivious or appeals to the prurient interest or if its effect is such as to tend to deprave and corrupt . . ." is subject to a fine and up to five years in prison. See http://www.sarai.net/journal/pdf/133-135%20(bill).pdf.
19. Notification no. GSR. 181(E), dated February 27, 2003; Notification GSR 529 (E) of July, 2003.
20. Ibid.
21. Notification no. GSR. 181(E) dated February 27, 2003, cited in Parishi Sanjanwala, Internet Filtering in India, p. 18.
22. Shivam Vij, "The discreet charms of the nanny state," National Highway, October 2006, at http://www.shivamvij.com/2006/10/the-discreet-charms-of-the-nanny-state.html.
23. Priya Ganapati, Mumbai Police Gag hinduunity.org, http://us.rediff.com/news/2004/may/26hindu.htm. See also Parishi Sanjanwala, Internet Filtering in India, p. 58.
24. Article 144, Code of Criminal Procedure, 1973, http://www.delhidistrictcourts.nic.in/CrPC.htm.
25. Priya Ganapati, Mumbai Police Gag hinduunity.org, http://us.rediff.com/news/2004/may/26hindu.htm (accessed May 23, 2006).
26. Schedule C, Section 1.10.2, Government of India, Ministry of Communications and Information Technology, Department of Telecommunications Telecom Commission, License Agreement for Provision of Internet Service (including Internet Telephony), http://www.dot.gov.in/ispt/isptindex.htm.
27. Schedule C, Section 1.12.9, Government of India, Ministry of Communications and Information Technology, Department of Telecommunications Telecom Commission, License Agreement for Provision of Internet Service (including Internet Telephony), http://www.dot.gov.in/ispt/isptindex.htm.
28. The Press Trust of India, "India to amend IT Act for greater data protection, privacy," October 16, 2006.
29. Article 79, The Information Technology (Amendment) Bill (2006), Bill 96 of 2006, http://www.mit.gov.in/.
30. See Ministry of Communications, Department of Telecom, Guidelines and General Information for Setting up of Submarine Cable Landing Stations for International Gateways for Internet, 2000, www.dot.gov.in/isp/landing_station.doc.
31. Article 7(1), Right to Information Act (2005), at http://www.righttoinformation.info/index.htm.
32. Article 8, Right to Information Act (2005), at http://www.righttoinformation.info/index.htm.
33. National Highway, "Internet censorship in India," http://www.shivamvij.com/2006/09/internet-censorship-in-india-an-rti-application.html.
34. A scanned copy of the July 13, 2006, CERT-IN order is available at http://photos1.blogger.com/blogger/507/157/1600/Indian_censored_list.jpg.
35. http://www.business-standard.com/general/printpage.php?autono=261383.
36. India Daily, "Orkut blocked in Pune, PIL filed against it for running anti Shivaji community," November 24, 2006, http://www.indiadaily.org/entry/orkut-blocked-in-pune-pil-filed-against-it-for-running-anti-shivaji-community/; Press Trust of India, "Orkut forum on Shivaji Maharaj blocked," November 18, 2006, http://www.expressindia.com/fullstory.php?newsid=77287.
37. Ganesh Kanate, "Patil wants bar on Orkut," DNA, December 9, 2006, http://www.dnaindia.com/report.asp?NewsID=1068353.
38. In August 2003, CERT-IN issued an order to ISPs to block the mailing list "kynhun" on Yahoo! Groups of the militant outfit Hynniewtrep National Liberation Council. See http://pib.nic.in/archieve/lreleng/lyr2003/rsep2003/22092003/r2209200314.html.

Iran

Since 2000—in the midst of a media crackdown that has seen the judiciary close more than 100 publications, inspiring widespread self-censorship—the Islamic Republic of Iran has installed one of the most extensive technical filtering systems in the world. Iranian authorities have detained dozens of people for publishing material online.[1] In addition, Iran has moved to contain the Internet within heightened and more explicit regulation, accommodating aggressive online censorship policies through a complex system of political networks and their affiliated government institutions.

Background

Regulation of freedom of expression in Iran is extensive and the parameters of prohibited conduct are vague and ambiguous, or simply undefined. It is prohibited to publish sensitive information and matters relating to atheism without prior approval, and media cannot promote social discord or divisions, dissent against state interests, insult Islam or public officials, or quote from deviant parties or parties opposed to Islam.[2] Compared with the constitutionally mandated state control of radio and television,[3] and the repression against independent papers and reformist voices in print media, the space initially afforded to free expression online was a unique phenomenon for Iran. However, after several years of relative openness in Iranian cyberspace, bloggers, journalists, and others began to be targeted, detained, and even tortured for their online activities. And zealous new legislation places sweeping controls over what people may post to the Internet.

Internet in Iran

The Internet in Iran has experienced the most explosive growth of the countries in the Middle

RESULTS AT A GLANCE

Filtering	No evidence of filtering	Suspected filtering	Selective filtering	Substantial filtering	Pervasive filtering
Political					●
Social					●
Conflict/security				●	
Internet tools					●

Other factors	Low	Medium	High	Not applicable
Transparency		●		
Consistency		●		

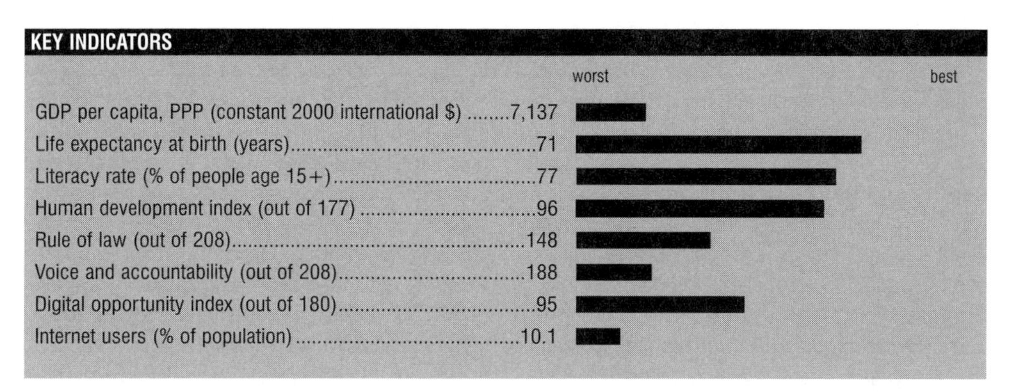

Source (by indicator): World Bank 2005, 2006a, 2006a; UNDP 2006; World Bank 2006c, 2006c; ITU 2006, 2005

East, with an increase of 2,900 percent between 2000 and 2005.[4] Today an estimated 7.2 million people are online in Iran, and there are approximately 400,000 blogs in Farsi.[5] Yet even as the government continues to promote the Internet as an engine of economic growth, one Iranian official recently boasted that Iran has censored ten million Web sites, and that the judiciary requests an additional 1,000 sites to be blocked every month.[6]

On October 11, 2006, an order reportedly issued to Internet service providers (ISPs) by the Ministry of Communications and Information Technology (MCIT) made providing Internet services—for use in private or public places—at a speed higher than 128 kilobytes per second illegal, reportedly with the aim of hindering users' ability to download foreign cultural products (such as music and films) and organize political opposition.[7] Such an about-face contradicts Iran's fourth Five-Year Development Plan, which calls for 1.5 million high-speed Internet ports throughout the country.[8]

At that time, some 250,000 users were using broadband services, with demand growing sharply.[9] Though the order applies to both public and home use, high-speed Internet services are most commonly available for commercial and office use. Over the previous two years eleven companies had been licensed to provide such high-speed services free from government competition and had invested significant capital in importing the required machinery and setting up the required infrastructure.[10] The ban on high-speed Internet services has pushed these private sector companies to the verge of bankruptcy. Furthermore, projects such as the USD6 million Internet television project of the ITC (Information Technology Company), as well as virtual surgery lab projects and e-universities and many more scientific and commercial projects being implemented in the country may be doomed to fail.[11]

Members of parliament are campaigning against the broadband ban and have started a Web site, www.more-speed-more-progress.ir, which is hosted on Iranian government servers.[12] Although the head of the Public Relations Department of the Regulation Organization said that the ban would be lifted in four months' time, after the government had had a chance to put in place measures to more effectively monitor Internet use, the Minister of the MCIT stated that "positive" results could make the measure permanent. Individuals can file a written complaint with the Regulation Organization and those who can demonstrate that they will use broadband access for "legitimate" purposes may be allowed to circumvent the ban.[13]

Legal and regulatory frameworks

As with all print media under the Press Law, Internet content providers are subject to two complementary sets of requirements: they must produce content within state-defined objectives and they must refrain from producing state-defined types of illegal material.[14] Thus, through the judiciary, parliament, and the executive bodies who all exercise the authority to make law, content providers are encouraged to promote genuine Islamic culture while being warned against fomenting social discord or encouraging dissent against state interests.

The legal status of blogs and Web sites in Iran has been contested, but starting in 2006 the government took additional steps to bring them firmly in hand. First, framing regulations to systematize control and management of Internet activity were issued by the government and signed by the vice president on November 26, 2006. Second, the Bill of Cyber Crimes' Sanctions (Cyber Crimes Bill) prepared by the Judiciary's Committee for Combating Cyber Crimes on October 12, 2006, was slated to be signed in to law by parliament;[15] this bill applies to all forms of electronic writings and graphics and generally any activity within the realm of cyberspace.

The November 2006 regulations were a response to a directive of the Supreme Cultural Revolution Council (SCRC) to manage Internet activity "while considering individual rights and safeguarding Islamic, national and cultural values."[16] The Ministries of Islamic Culture and Guidance (MICG), Justice, and Information are the main governmental bodies responsible for leading this effort, and the MICG was given the duty to create an infrastructure to systematize management and stamp out illicit and immoral content.[17] All activities of Web sites and blogs that do not obtain a license from the MICG are considered illegal. On January 1, 2007, the MICG issued a notice requiring all owners of blogs and Web sites to register by March 1, provide detailed personal information, and abstain from posting certain types of content.[18] An official from the Telecommunications Ministry claimed that enforcement would be impracticable.[19]

The Cyber Crimes Bill makes ISPs criminally liable for the content they carry, effectively shifting the burden of censoring Web sites and potentially e-mail correspondence on to their shoulders. Under the Cyber Crimes Law, ISPs that do not abide by government regulations (including filtering regulations) may be temporarily or permanently suspended, depending on the graveness of the offense, and their owners could face prison terms.[20] Article 18 of the bill requires ISPs to ensure that "forbidden" content is not displayed on their servers, that they immediately inform law enforcement agencies of violations, that they retain the content as evidence, and that they restrict access to the prohibited content. The bill also includes provisions for the protection and disclosure of confidential data and information as well as the publishing of obscene content.

Until the introduction of the Cyber Crimes Bill, the most relevant statute governing the activities of blogs and Web sites was the 2000 Press Law. Although experts argued to the contrary, through the Press Law electronic publications were subsumed into the definition of press publications.[21] As such, Iranians were theoretically required to first obtain a license to publish a Web site or a blog and were subject to the Press Law. Among the Press Law's broad prohibitions on speech are articles that prohibit "promoting subjects that might damage the foundation of the Islamic Republic ... offending the Leader of the Revolution ... or quoting articles from the deviant press, parties or groups that oppose Islam (inside and outside the country) in such a manner as to propagate such ideas."[22] Other provisions prohibit insulting Islam or senior religious authorities.[23] The Press Supervisory Board under the Ministry of Islamic Culture and Guidance had absolute power to revoke licenses, ban publications, and refer complaints to a special Press Court.[24]

As "publications" under the Press Law, blogs and Web sites that did not obtain licenses became subject to stricter "General Laws." As a part of the "General Laws," the Penal Code places further restrictions on speech. The Penal Code incorporates content-based crimes such as propaganda against the state (while leaving "propaganda" undefined).[25] Similarly, Article 513 allows for the death penalty or imprisonment of up to five years for speech deemed to be an "insult to religion," but leaves "insult" undefined.[26] Article 698 provides maximum sentences of two years imprisonment or seventy-four lashes for those convicted of intentionally creating "anxiety and unease in the public's mind," spreading "false rumors," or writing about "acts which are not true."[27] Article 609 criminalizes criticism of state officials in connection with carrying out their work, and calls for a fine, seventy-four lashes, or between three and six months in prison as punishment for such "insults."

ISPs and subscribers are also subject to prohibitions on twenty types of activities, where insulting Islam and religious leaders and institutions, as well as fomenting national discord and promoting drug use or obscenity and immoral behaviors, are prominent.[28]

The Committee in Charge of Determining Unauthorized Sites is legally empowered to identify sites that carry prohibited content.[29] Established in December 2002 (some reports state June 2003), this Committee notifies the MICT of criteria for identifying unauthorized Web sites and what sites shall be blocked. The SCRC oversees committee members from the Ministry of Culture and Islamic Guidance, the Intelligence and Security Ministry, and the Sound and Vision Organization (Islamic Republic of Iran Broadcasting).[30]

In February 2007 the online conservative journal Baztab (www.baztab.com) became the first site reported to have been blocked by the November 2006 regulations. According to a government official, Baztab not only failed to apply for a license, but it also violated the regulations by disclosing state secrets and other confidential military information, insulting government officials, and publishing false news.[31] However, the Supreme Court of Iran ruled against the filtering of Baztab and it was made accessible inside Iran again.[32] This incident sparked a debate within Iranian legal and media circles over the authority of the Committee in Charge of Determining Unauthorized Sites, and whether as an executive body (government) it was improperly involved in making legislative or judicial decisions according to the constitution.[33]

However, not all filtering occurs through this body. The Internet Bureau of the Judiciary also orders ISPs to block sites through court orders, which are considered a form of lawful punishment imposed on legal entities.[34] Tehran Prosecutor General Saeed Mortazavi, who has led harsh crackdowns on media and has also been implicated in cases of the torture of detainees, including twenty-one bloggers arrested in 2004, has also ordered that certain sites be censored.[35] In May 2006 the MICT announced the formation of a central filtering office, reportedly to filter illegal content, identify Internet users, and keep a record of the sites they visit.[36] The MICT subsequently denied having such tracking capabilities, saying its primary objective was to block pornography.[37]

In 2001 the SCRC declared that the government was taking control of all access service providers (ASPs).[38] ISPs were required to obtain bandwidth from these ASPs and also to employ filtering systems to block access to immoral, political, and other "undesirable" content while storing user data and reporting to the ICT

Ministry.[39] ISPs in which the government owns a share, such as the popular Pars Online, reportedly filter some sites at their own discretion over and above what is required by the regulations.

ONI testing results

ONI conducted testing on seven ISPs: APN, Dana Fajr, Datak, Jahan Nama Co., Pars Online, Shatel, and Tarashe. ONI testing confirmed that Iran employs the greatest degree of filtering of all the countries tested, in both scope and depth of content. Iran uses a filtering proxy that displays a blockpage when accessing blocked content. Heavily filtered types of content include pornography, provocative attire, and circumvention tools, which is characteristic of states that use commercial software such as SmartFilter. ONI testing also found significant blocking of content related to homosexuality, particularly if it had any connection to Iran; Farsi-language news sites; and opposition political sites.

A majority of circumvention tools were blocked by all ISPs, including www.peoplesproxy.com and www.guardster.com. Compared with anonymizers and proxies, filtering of other Internet tools was more selective but nevertheless occurred in all categories tested. Certain multimedia sharing sites, such as www.metacafe.com and www.photobucket.com, were completely blocked, while others were less consistently filtered: the popular photo-sharing Web site Flickr was blocked on four ISPs at time of testing, while the video-sharing site YouTube was blocked on only two. Also filtered in limited numbers were social networking sites, but at the time of testing popular social networking sites such as Myspace and Orkut were universally available. Some Farsi-language forums for discussing movies (www.aghaghi.com) and music (www.roozi.com) were also filtered.

Only a limited number of search engines were filtered, and then, only on some ISPs. Among them were www.163.com and the Chinese site www.sina.com. However, on certain ISPs—including Shatel, Datak, and Pars Online, keywords in URL paths were blocked, most often affecting queries in search engines (e.g., http://128.100.171.12/key.php?word=torture).

Of blog-hosting sites tested, only one, www.livejournal.com, was blocked by all ISPs. A limited number of other sites, including www.xanga.com and the blog search engine www.technorati.com, were blocked by multiple ISPs. Instead, filtering targeted individual blogs. A substantial number, though not a majority, of individual blogs hosted by Blogspot and others were filtered; these blogs spanned subjects such as religion, women's rights, political reform, and reproductive health. All seven ISPs chose to filter the same individual blogs, which all happened to be hosted on Blogspot. Very few of the individual blogs hosted on Persian-language services, such as Blogfa and www.persianblog.com, were filtered by any ISP.

Iran is among the most successful blockers of pornographic Web sites in countries where ONI conducted testing. Esmail Radkani, of Iran's quasi-official Information Technology Company, claimed in a recent interview that 90 percent of the ten million filtered sites were deemed to contain "immoral" content.[40] This assertion was supported by ONI's tests. With very few exceptions, all of the pornography and provocative attire sites tested were blocked by all ISPs. Further, no pornography site tested was blocked by fewer than five ISPs. The government does not filter content regarding drugs, alcohol, gambling, or dating as universally, though a substantial number of sites in these categories are blocked as well.

Outside of "immoral" content, independent and dissenting voices are filtered across a range of issues pertaining to Iran, including political reform, criticism of the government, reporting on human rights issues, and minority and women's rights. Filtering in these areas, across nongovernmental organizations (NGOs), blogs, and thematic Web sites, is inconsistent and limited when content is provided solely in English, and much more substantial and complete across ISPs for sites relating to Iran or in Farsi. For example, while no independent media sites or newspapers available only in English were filtered across all ISPs tested, a large majority of similar sites relating to Iran or composed in Farsi were consistently blocked, such as www.iranvajahan.net and the publisher www.kayhanpublishing.uk.com.

All seven ISPs tested blocked access to almost the same list of human rights, political reform, and opposition sites. All ISPs kept access to international watchdogs such as Amnesty International and FIDH open, but unilaterally blocked Iran-focused groups such as the Society for the Defense of Human Rights in Iran (www.polpiran.com) and the online magazine Siah Sepid (www.siahsepid.com).

For remaining content categories, the considerable variation among the sites blocked by the ISPs suggests that they are exercising some control over the implementation of filtering. There is no discernible pattern in the content of sites blocked only by one ISP. For example, Pars Online, the largest private provider of Internet services in Iran, is the only ISP to block such disparate sites as www.boingboing.net, the *International Herald Tribune,* and the teen sexual health site www.teenhealthfx.com. Tarashe is the only ISP that blocked the e-mail service Hushmail and the *Times of India* newspaper.

Overall, the greatest overlap in filtering occurred with Jahan Nama, Pars Online, Datak, and Shatel. Together, these ISPs filtered a range of Web sites in common, including a substantial number of lesbian, gay, bisexual, and transgender (LGBT) rights organizations (including www.gmhc.org and www.iglhrc.org), NGOs focusing on free expression and access to information, dating services, and alcohol and drug sites.

Conclusion

Iran continues to maintain the most extensive filtering regime of any country ONI has studied. As filtering and censorship policies evolve, government officials and citizens have pushed back against many of the more extreme measures, including the ban on high-speed Internet in 2006. New developments may provide opportunities to contest these policies further. The draft Cyber Crimes Bill prohibits any blocking or investigation of data without a warrant issued by a court after evidence of suspicious activity. When this provision becomes law, it could potentially be used to impede the arbitrary closures and blocking of Web sites.

NOTES

1. Human Rights Watch Report, False Freedom: Online Censorship in the Middle East and North Africa: Iran, November 2005, http://hrw.org/reports/2005/mena1105/5.htm#_Toc119125727.
2. Ibid.
3. Article 44(2), Constitution of the Islamic Republic of Iran, amended July 28, 1989, http://www.oefre.unibe.ch/law/icl/ir00000_.html.
4. See U.N. Office on Drugs and Crime, http://www.unodc.org/iran/en/crime_prevention.html (accessed April 30, 2007); Telecommunications Company of Iran, Annual Report 2005, http://irantelecom.ir/eng.asp?sm=3&page=17&code=5.
5. BBC Monitoring International Reports, "Iran press Iranian activists oppose regulation of websites, weblogs," January 2, 2007 (citing text of report by E'temad-e Melli).
6. Interview with Esma'il Radkani, Iranian Communication and Information Technology News Agency, September 11, 2006, http://citna.ir/435.html.
7. *The Guardian,* "Iran bans fast internet to cut west's influence: Service providers told to restrict online speeds Opponents say move will hamper country's progress," October 18, 2006.
8. See the Information Technology Company Web site, http://www.dci.ir/english/moarefi.htm.
9. BBC Persian, "Speed reduced for high speed Internet in Iran," October 20, 2006, http://www.bbc.net.uk/persian/science/story/2006/10/061020 fb_rsh_adsl.shtml.
10. Ibid.
11. Ibid.

12. *The Guardian,* "Iran bans fast internet to cut west's influence," October 18, 2006.
13. Information Technology News Agency, "Speed limitation becomes permanent," October 20, 2006, http://www.itna.ir/archives/news/005618.php (in Persian).
14. Article 19, Regulation of the Media in the Islamic Republic of Iran, March 2006.
15. Bill on Punishment for Cyber Crimes (Cyber Crimes Bill), October 12, 2006, http://www.hoqouq.com/law/article522.html.
16. BBC Monitoring Middle East, translation of report of Kayhan newspaper, "There were sound grounds for banning Baztab website: MP," February 25, 2007.
17. Ibid.
18. BBC Monitoring International Reports, citing text of report by E'temad-e Melli, "Iran press Iranian activists oppose regulation of websites, weblogs," January 2, 2007. See also Omid Memarian, "Bloggers rebel at new censorship," Inter Press Service News Agency, January 10, 2007, reporting that prohibited content includes criticism of religious figures, sexual matters, content considered offensive to the Ayatollah Khomeini, or content slanderous of Islamic law.
19. BBC Monitoring International Reports, citing text of report by E'temad-e Melli, "Iran press Iranian activists oppose regulation of websites, weblogs," January 2, 2007.
20. Article 21, Bill on Punishment for Cyber Crimes (Cyber Crimes Bill), October 12, 2006, http://www.hoqouq.com/law/article522.html.
21. Note 2 of Article 1 of Iran's Press Law (as amended in 2000) defines electronic publications as "publications regularly published under a permanent name, specific date and serial number ... on different subjects such as news, commentary, as well as social, political, economic, agricultural, cultural, religious, scientific, technical, military, sports, artistic matters, etc via electronic vehicles." Publications must also have obtained "publication licenses from the Press Supervisory Board in the Ministry of Cultural and Islamic Guidance," otherwise they "fall out of the scope of the Press law and become subject to General Laws."
22. Press Law, Article 6.
23. Press Law, Articles 26–27.
24. Article 19, Memorandum on Regulation of the Media in the Islamic Republic of Iran, March 2006.
25. Islamic Penal Code of Iran, May 22, 1996, Article 500 states that "anyone who undertakes any form of propaganda against the state ... will be sentenced to between three months and one year in prison"; unofficial translation at http://mehr.org/index_islam.htm.
26. Islamic Penal Code of Iran, May 22, 1996, Articles 513–15; unofficial translation at http://mehr.org/index_islam.htm.
27. Islamic Penal Code of Iran, May 22, 1996, Articles 697, 698, 700; unofficial translation at http://mehr.org/index_islam.htm. Chapter 27 of the Penal Code on libels and revilements, and more specifically Article 698, only stipulate punishment for statements made in the press and does not cover writings in electronic format in computer and other communication systems. As seen earlier, "press" is defined in Article 1 of the Press Law, but the general articulation of Article 698 subsumes both publications with and without licenses.
28. Human Rights Watch Report, False Freedom: Online Censorship in the Middle East and North Africa: Iran, November 2005, http://hrw.org/reports/2005/mena1105/5.htm.
29. Decree on the Constitution of the Committee in Charge of Determining Unauthorized Websites, Official Gazette No. 16877.
30. Bill Samii, "Iran: Government strengthens its control of the Internet," Radio Free Europe, September 29, 2006, http://www.rferl.org/featuresarticle/2006/09/e6ed377e-7618-479d-8e0e-b2917d6f9f92.html?napage=2.
31. BBC Monitoring Middle East, translation of report of Kayhan newspaper, "There were sound grounds for banning Baztab website: MP," February 25, 2007.
32. http://www.theage.com.au/news/Technology/Iran-lifts-ban-on-conservative-website/2007/03/20/1174153008276.html.
33. Iranian Student News Agency (ISNA), Interview with Head of Iran Law Society, February 16, 2007, http://www.isna.ir/Main/NewsView.aspx?ID=News-877388.
34. Information Technology News Agency "Report of an ISP closure by judicial system agents," http://www.itna.ir/archives/article/000665.php (accessed April 30, 2007).
35. Human Rights Watch Press Release, Iran: Prosecute Torturers, Not Bloggers, December 12, 2006, http://hrw.org/english/docs/2006/12/12/iran14824.htm . See also Human Rights Watch Report, Like the Dead in Their Coffins: Torture, Detention, and the Crushing of Dissent in Iran, June 7, 2004, http://www.hrw.org/campaigns/torture/iran/.
36. Bill Samii, "Iran: Government strengthens its control of the Internet," Radio Free Europe, September 29, 2006, http://www.rferl.org/featuresarticle/2006/09/e6ed377e-7618-479d-8e0e-b2917d6f9f92.html?napage=2.
37. Ibid.

38. Iran CSOs Training & Research Center, A Report on the Status of the Internet in Iran, November 2005, 8, www.genderit.org/upload/ad6d215b74e2a8613f0cf5416c9f3865/A_Report_on_Internet_Access_in_Iran_2_.pdf.
39. Ibid. See also Human Rights Watch Report, False Freedom: Online Censorship in the Middle East and North Africa: Iran, November 2005, http://hrw.org/reports/2005/mena1105/5.htm.
40. Ibid.

Iraq

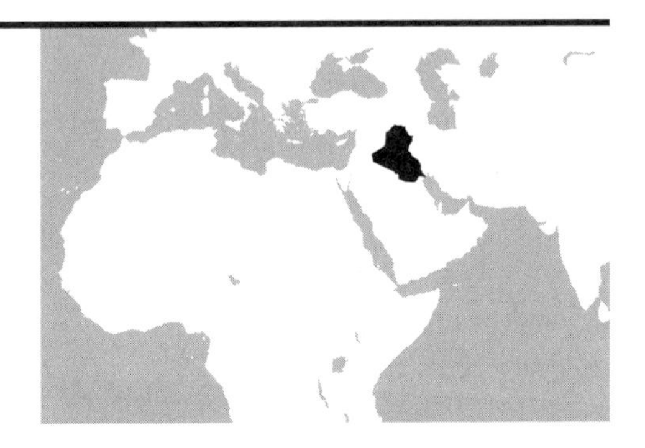

Iraq does not have a declared Internet filtering policy, nor is there evidence of Internet filtering practiced by the state Internet service provider (ISP). However, ongoing conflict and deteriorating conditions prevent many Iraqis from accessing the Internet.

Background

Although the Iraqi constitution, ratified in October 2005, guarantees the freedoms of "expression, press, printing, advertisement, media, publication, assembly, and peaceful demonstration,"[1] on February 13, 2007, Iraqi Prime Minister Nuri al-Maliki gave far-reaching martial law powers to military commanders. These include the power to conduct searches and seizures without warrants, to arrest, detain, and interrogate people, and to monitor, search, and confiscate "all mail parcels, letters, cables, and wire and wireless communications devices," and to restrict all public gatherings, including "centers, clubs, organizations, unions, companies, institutions, and offices."[2] The Iraq war has had immense impact on the country. A joint Johns Hopkins-MIT study in October 2006 estimates that Iraq has suffered 650,000 "excess deaths" (600,000 of them violent deaths) since the onset of the Iraq war in 2003 and that Iraqi death rates are continually rising.[3] Reporters Without Borders reports that 150 journalists and media assistants have died since the start of the war.[4] As of October 2007, the U.S. military has suffered 3,800 deaths in the initial invasion and subsequent efforts to stabilize the country.[5]

Internet in Iraq

Saddam Hussein placed severe restrictions on Iraqis' ability to receive and impart information. The press and broadcast media were tightly controlled, as was access to the Internet. In 2002,

RESULTS AT A GLANCE

Filtering	No evidence of filtering	Suspected filtering	Selective filtering	Substantial filtering	Pervasive filtering
Political	●				
Social	●				
Conflict/security	●				
Internet tools	●				

Other factors	Low	Medium	High	Not applicable
Transparency				●
Consistency				●

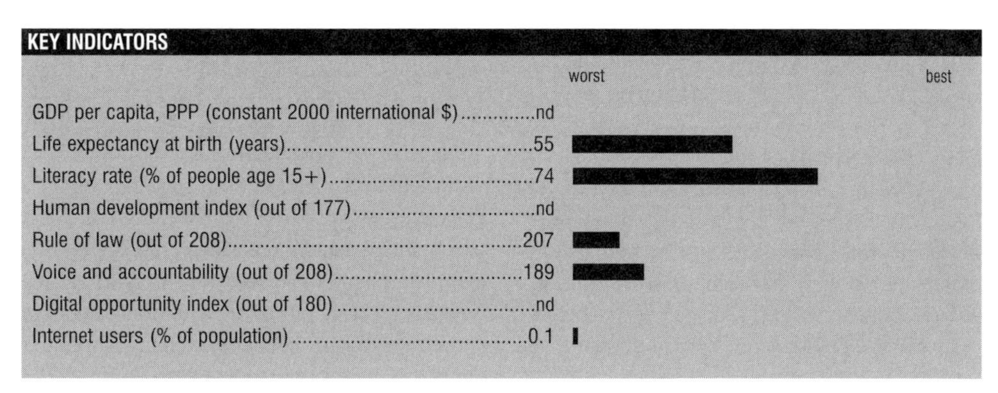

Source (by indicator): WHO 2006; World Bank 2006a, 2006c, 2006c; ITU 2004
nd = no data available

there were only 45,000 Internet users, many of them state officials, in Iraq.[6] But after the U.S. invasion Internet cafés spread throughout the country. There are no recent studies that enumerate Internet users in Iraq, but the ITU estimated the number to have reached 50,000 users by the end of 2004.[7]

The U.S. Defense Department spent more than USD165 million to set up cybercafés in Iraq. In 2004, Iraq contained 36 cafés, and by July 2006 it had more than 170.[8] According to a report from the London-based pan-Arab daily *Al-Sharq al-Awsat*, approximately 45 percent of those who frequent Internet cafés are women, and both male and female users use matchmaking Web sites to find future spouses from outside Iraq so that they can get out of the country.[9]

Iraq's primary ISP, uruklink, provides dialup, DSL, and wireless connection services,[10] and it also provides wireless high-capacity data and voice communications to government sites through the Baghdad Wireless Broadband Network in Baghdad.[11]

There is a very active Iraqi blogging community that blogs mostly about the experience of living in a war zone and the effects of the conflict on citizens' lives. Several Iraqi bloggers have caught the attention of international media. The blog Baghdad Burning, for example, was adapted for stage[12] and a book.[13] The blog was started in 2003 by a twenty-four-year-old Iraqi girl using the pseudonym Riverbend.

In November 2005, Iraq secured the .iq ccTLD, and entrusted its administration to Iraq's National Communications and Media Commission (NCMC). [14]

Legal and regulatory frameworks

The new Iraqi government has placed few restrictions or regulations on the Internet in Iraq. Article 36 of the Iraqi constitution guarantees freedom of expression. The sweeping powers Prime Minister al-Maliki granted to the military in the name of security did not place unusual restrictions on the right to free expression, but they did circumscribe the corollary rights to privacy and assembly.[15] Provisions of the penal code restricting freedom of the press remain on the books; these provisions have been used to sentence journalists to long prison sentences in recent years.[16]

Iraqi security forces detained at least thirty reporters over the course of 2006, with four still held without charge at year's end.[17] The Committee to Protect Journalists reports that

U.S. forces have detained dozens of journalists since the war began in 2003. Though they quickly released most, at least eight were detained for weeks or months. On September 7, 2006, the Iraqi government closed the Baghdad bureau of the pan-Arab satellite news station Al-Arabiyya for one month on the grounds that its reporting amounted to "incitement." Iraqi networks were ordered not to show scenes of violence. The government continued to prevent the pan-Arab satellite station Al-Jazeera from reopening its Baghdad bureau. In 2004, Reporters Without Borders stated that "[t]he United States has total control of the country's telecommunications system."[18]

ONI testing results

ONI conducted in-country tests in 2006 on Iraq's ISP uruklink. The tests revealed no evidence of filtering for any of the categories tested.

On the other hand, the BBC reported that the "Pentagon is keeping a close eye on what its troops post online, with special attention being paid to videos that show the aftermath of combat." The BBC added that, "[o]ne soldier who served in Iraq in 2005 told the BBC there was 'a tight watch' being kept on video and pictures posted to MySpace, with civilian contractors monitoring the Internet on behalf of the Pentagon." The BBC has not been able to confirm that contractors are scouring the Internet for inappropriate material from the military, but reported that "US Central Command—which is responsible for troops in Iraq and Afghanistan—does have a team reading blogs and responding to what they consider inaccuracies about the so-called war on terror."[19]

Other reports stated that some liberal Web sites were blocked on military computers in Iraq as part of filtering nonconservative content. Examples include The Memory Hole[20] and Wonkette.[21] However, officials from the U.S. Defense Department denied that they block political Web sites from soldiers serving in Iraq.[22]

In January 2007, at the request of the British government, Google agreed to remove updated images that included British bases in Iraq from Google Earth after British divisional headquarters came under almost daily mortar barrages.[23]

Conclusion

Iraqi citizens have unfettered access to the Internet, but this fact is overshadowed by the ongoing security condition. The deadly conflict makes journalists and media professionals working in Iraq particularly vulnerable.

"Though Iraq's state of disorder has opened up a space of freedom, it has also produced serious fears. Living conditions continue to deteriorate. Owners of Internet centers close their stores at night out of fear—fear of both the occupying forces and those of the resistance."[24]

NOTES

1. *The Washington Post,* "Full Text of Iraqi Constitution," Article 26, October 2005, http://www.washingtonpost.com/wp-dyn/content/article/2005/10/12/AR2005101201450.html.
2. Human Rights Watch, "Iraq: New martial law powers threaten basic rights," February 23, 2007, http://hrw.org/english/docs/2007/02/23/iraq15393.htm.
3. Gilbert Burnham, Shannon Doocy, Elizabeth Dzeng, Riyadh Lafta, and Les Roberts, The Human Cost of the War in Iraq, 2006, web.mit.edu/CIS/pdf/Human_Cost_of_War.pdf.
4. Reporters Without Borders, "The war in Iraq," February 2007, http://www.rsf.org/special_iraq_en.php3.
5. Iraq Coalition Casualty Count, http://icasualties.org/oif (as of October 2, 2007).
6. The Arab Network for Human Rights Information, "Iraq: A look behind bars," http://www.hrinfo.net/en/reports/net2004/iraq.shtml.
7. International Telecommunication Union, http://www.ituarabic.org/arab_country_report.asp?arab_country_code=12.
8. *The New York Times,* "An Internet lifeline for troops in Iraq and loved ones at home," July 8, 2006, http://www.nytimes.com/2006/07/08/us/08FAMILY.html?ex=1310011200&en=4980662c7ca1ae71&ei=5088&partner=rssnyt&emc=rss.
9. http://www.asharqalawsat.com/details.asp?article=362940&issue=10028§ion=3.

10. http://www.uruklink.net/.
11. State Company for Internet Services, Baghdad Wireless Broadband Network, http://www.wbb-iraq.com/index.htm.
12. BBC News, "Iraqi women's blog adapted for stage," August 14, 2006, http://news.bbc.co.uk/2/hi/entertainment/4790577.stm.
13. See http://www.feministpress.org/Book/index.cfm?GCOI=55861100869560.
14. The Iraqi National Communications and Media Commission,http://www.ncmc-iraq.org/press%20Release%20.IQ.pdf.
15. Iraqi government Web site, http://www.iraqigovernment.org/Content/Biography/English/consitution.htm.
16. The Committee to Protect Journalists, *Attacks on the Press in 2006,* New York: Committee to Protect Journalists, February 28, 2007.
17. Freedom House, *Freedom of the Press, 2007 Edition.* New York: Freedom House, forthcoming.
18. Reporters Without Borders, "Internet under surveillance," http://www.rsf.org/article.php3?id_article=10735.
19. BBC News, "Pentagon keeps one eye on war videos," July 29, 2006, http://news.bbc.co.uk/2/hi/technology/5226254.stm.
20. Memory Blog, "The Memory Hole Banned in Iraq," May 28, 2004, http://www.thememoryhole.org/memoryblog/archives/000156.html.
21. Wonkette, "Freedom on the march," October 24, 2006, http://wonkette.com/politics/war/freedom-on-the-march-209861.php.
22. Josh Rogin, "Boutelle: Army not blocking political sites," FCW.com, October 26, 2006, http://www.fcw.com/article96599-10-26-06-Web. Lt. Gen. Steven Boutelle is the U.S. Army's chief information officer.
23. Thomas Harding, "Google blots out Iraq bases on internet," *The Telegraph,* January 21, 2007, http://www.telegraph.co.uk/news/main.jhtml?xml=/news/2007/01/20/wgoogle20.xml.
24. The Arab Network for Human Rights Information, "Iraq: A look behind bars," http://www.hrinfo.net/en/reports/net2004/iraq.shtml.

Israel

Israel is among the world's leading countries in broadband Internet penetration. Although the censorship of information considered vital to national security is a reality, Israel has yet to legally authorize or implement filtering of the Internet by law or voluntary pact.

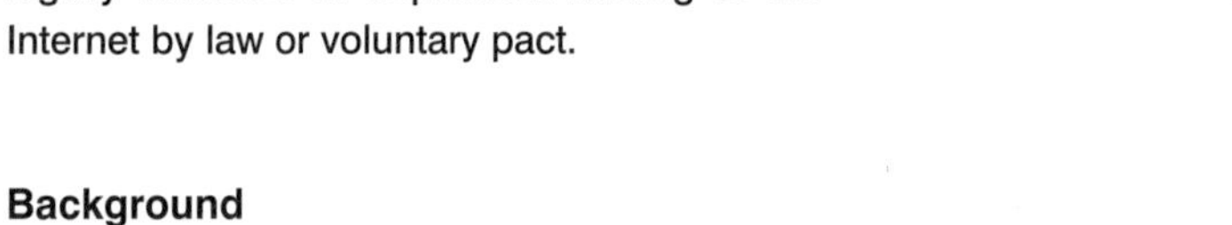

Background

Since its founding as a state in 1948, Israel has contended with the proper limits of security measures as a democracy under military threat. The Israeli Defense Forces' Military Censor decides what information should not be published, and both domestic journalists and foreign media organizations must comply as a condition of operating in Israel. This longstanding practice has been at the center of an ongoing debate about the curtailment of freedom of expression in order to protect national security and order.[1]

Internet in Israel

As a country self-described as always having to "depend on its intellectual resources for survival and development," Israel is home to one of the most vibrant technology centers in the world.[2] In 2003, the country drew USD1.1 billion in venture capital funding, placing it behind only Boston and Silicon Valley in attracting funding for start-ups.[3] Five major Internet service providers (ISPs) and approximately 70 smaller ISPs serve 3.6 million Internet users, about 60.6 percent of the total population in 2006.[4]

Israel ranks highest in the world in hours per user spent on the Internet, at 57.5 hours a month.[5] The vast majority of Israelis access the Internet from home, though many also do so at school, work, and other places.[6] Although blogs remain a relatively marginal activity in Israeli cyberspace, the Internet is now the main source of news for 26 percent of online users, second to

RESULTS AT A GLANCE

Filtering	No evidence of filtering	Suspected filtering	Selective filtering	Substantial filtering	Pervasive filtering
Political	●				
Social	●				
Conflict/security	●				
Internet tools	●				

Other factors	Low	Medium	High	Not applicable
Transparency				●
Consistency				●

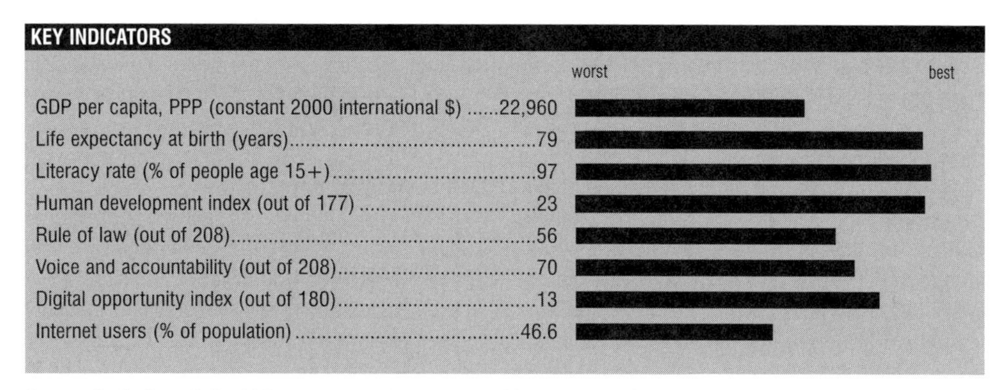

Source (by indicator): World Bank 2005, 2006a, 2006a; UNDP 2006; World Bank 2006c, 2006c; ITU 2006, 2004

television but surpassing print newspapers.[7] The Internet is also increasingly seen as a communication tool,[8] even a "new battleground," for vital Israeli interests and the national image.[9]

Initially, Internet penetration in Israel increased relatively slowly, because of the high cost of service—especially for broadband access.[10] Since 2001, however, the government has taken steps to allow more service providers to compete in Israel, reducing costs and dramatically increasing Internet use in general and broadband access in particular.[11] Bezeq, a formerly state-owned telecommunications giant that privatized in 2005, began offering ADSL service in 2001.[12] In large part because of the introduction of broadband cable modem access offered by cable companies in 2002, the percentage of households using broadband Internet increased from 4 percent in 2002 to 62 percent in 2006, with broadband service costing approximately USD9 a month.[13]

Legal and regulatory frameworks

The Ministry of Communications (MOC) regulates the Internet as part of the telecommunications sector.[14] Prior to the 1980s, the Israeli government controlled both telecommunications regulation and operations. In 1984 those functions were split, and all telecommunications facilities were transferred to Bezeq, a state-owned company.[15] Bezeq's monopoly on fixed-line transmissions within Israel led to a relatively high cost of Internet service in Israel.[16] After Bezeq's legal monopoly on fixed-line services expired in 1999, the MOC began liberally issuing licenses to competitors. This and other regulatory changes led to a burst of competition within the Internet sector, lowering prices and contributing to a large increase in Internet penetration after 2001.[17]

Israel's history as a state under constant military threat has strongly influenced its approach toward the control of information. Censorship of the media was "frozen" in law in 1945, when the military censor was authorized to ban the publication, printing, importing, and exporting of any material likely to cause damage to the security of Israel or public order.[18] Since then, censorship of sensitive, security- or military-related information has operated through voluntary agreements between military authorities and an editors' committee (the Israeli Committee of Daily Newspaper Editors). These agreements provide a platform for practical negotiation with a built-in arbitration body, and have been renewed periodically since

1949 with some significant amendments.[19] Despite the lack of full consent from all media, all such organizations operating in Israel, including foreign agencies, must agree to abide by the censor's rulings.[20]

The Directorate of Military Intelligence of the Israeli Defense Forces maintains the Military Censor unit that holds the authority to prevent reporting of information that may aid attacks on Israeli citizens. News outlets are prohibited, for instance, from revealing the exact location of enemy missile strikes, or stating that a high-ranking official is entering a threatened area.[21] After periods of more slack enforcement, the Censor has recently scaled up its efforts. During the 2006 war against Hezbollah in Lebanon, for instance, the government banned specific reports on troop movements, the location of Hezbollah rocket strikes, and other information that could be used to coordinate attacks or aim weapons.[22]

This regulatory structure has long been a source of controversy. In another example, the Military Censor blocked news about a National Security Council report on the vulnerability of an Israeli fuel depot. The censor was afraid the report might give terrorists ideas, but critics argue that such reports are necessary to spark public debate about security precautions.[23] A series of Supreme Court decisions limited the ban on publishing to content where there is a "tangible" and "near-certain" danger to the well-being of the public.[24] Over the decades, the Knesset has debated the role of the censors and the limits of free expression, especially in light of a changing media environment fueled by the growth of the Internet, but no legislation has been enacted to replace the current system.

Israel has yet to establish any explicit legal authority for filtering of the Internet. In 1998 the Knesset's Committee for Scientific and Technological Research and Development met to discuss the subject of Internet filtering.[25] Some groups in Israel, particularly the Orthodox community, were concerned over widespread pornography on the Internet, though the legislature seemed more worried about the availability of privileged information such as Israeli missile deployments.

ONI testing results

ONI testing in 2006 found no evidence of Internet filtering in Israel. In addition to the global list, ONI tested sites with content critical of the Israeli government or reflecting sensitive national security issues and state policies, from Palestinian groups such as Hamas, human rights organizations, and militant organizations (Hezbollah).

Conclusion

After years of somewhat stagnant growth, the Israeli Internet community is expanding rapidly. The country is likely to remain a center for the development of new Internet technologies, with widespread Internet access and deep broadband penetration. Israel does not filter the Internet, and in this respect maintains the freest Internet community in the Middle East. However, as proposed legislation to restrict access to pornography and violent content online continues to be debated, and as the space for online media increases, the Internet will likely challenge the bounds of the specific historical tradition and established practices of Israeli censorship.

NOTES

1. Matti Friedman, "Stop the press," *The Jerusalem Report,* April 4, 2005.
2. Israeli Ministry of Communications, Telecommunications in Israel 2006, November 6, 2006, p. 10, http://www.moc.gov.il/new/documents/broch_1.11.06.pdf.
3. Matthew Kalman, "Venture capital invests in Israeli techs: Recovering from recession, country ranks behind only Boston, Silicon Valley in attracting cash for startups," *San Francisco Chronicle,* April 2, 2004, http://www.sfgate.com/cgi-bin/article.cgi?file=/chronicle/archive/2004/04/02/BUG675V5L41.DTL.

4. Global Technology Forum, Doing Business in Israel, Economist Intelligence Unit, http://globaltechforum.eiu.com/index.asp?layout=newdebi&country_id=IL&channelid=6&country=Israel&title=Doing+e-business+in+Israel.
5. comScore Networks Press Release, "694 million people currently use the Internet worldwide according to comScore Networks," May 4, 2006, http://www.comscore.com/press/release.asp?press=849.
6. "The Information Technology Landscape in Israel," American University, http://www.american.edu/carmel/nk3791a/Internet.htm (accessed January 29, 2007).
7. Judy Siegel, "Web sites are a major source of news for a quarter of Internet users," *The Jerusalem Post,* November 14, 2006, reprinted by BBC Monitoring International Reports.
8. Michal Lando, "Ingathering of the intellectuals," *The Jerusalem Post,* January 9, 2007.
9. See Steve Linde, "Israel's newest PR weapon: The Internet Megaphone," *The Jerusalem Post,* November 29, 2006, describing a computer software tool called the "Internet Megaphone" that acts like a beeper alert system for citizens to bring a pro-Israel slant to public opinion polls, and so on.
10. See Israeli Ministry of Communications, Telecommunications in Israel 2006, November 6, 2006, http://www.moc.gov.il/new/documents/broch_1.11.06.pdf.
11. Ibid.
12. Global Technology Forum, Doing Business in Israel, Economist Intelligence Unit, http://globaltechforum.eiu.com/index.asp?layout=newdebi&country_id=IL&channelid=6&country=Israel&title=Doing+e-business+in+Israel.
13. Israeli High-Tech & Investment Report, "Overview of Israel's Internet & broadband," January 2007, http://www.ishitech.co.il/0107ar3.htm.
14. See the Israeli Ministry of Telecommunications Overview at http://www.moc.gov.il/new/english/index.html.
15. Israeli Ministry of Communications, Telecommunications in Israel 2006, November 6, 2006, http://www.moc.gov.il/new/documents/broch_1.11.06.pdf.
16. Ibid.
17. Ibid.
18. Articles 87(1), 88(1), Defense Regulations (State of Emergency), 1945, 1442 Palestine Gazette Index 855, (1945), cited in Hillel Nossek and Yehiel Limor, "Fifty years in a 'marriage of convenience': News media and military censorship in Israel," *Communication Law and Policy* 6(1): 1–35. Article 87(1) states that the "the censor is entitled, in general and in particular, to order the banning of the publication of material, the publication of which, will, or is likely, in his opinion, to harm the security of Israel or the well-being of the public or public order." Article 88(1) states that "the censor is entitled to order the banning of the importing and exporting, the printing and publishing, of every publication ... whose import or export, printing or publication, were or are likely to cause damage, in his opinion, to the security of Israel, to the well-being of the public or to public order."
19. See Hillel Nossek and Yehiel Limor, "Fifty years in a 'marriage of convenience': News media and military censorship in Israel," *Communication Law and Policy* 6(1): 1–35.
20. Ibid.
21. BBC Monitoring International Reports, "Analysis: Is Israel censoring media coverage of Lebanon war?" July 18, 2006.
22. Benjamin Harvey, "Israeli censor wields great power over coverage of rocket attacks," The Associated Press, July 19, 2006.
23. Lavie, Aviv, "Sensing the censor," *Haaretz,* May 27, 2002.
24. See Hillel Nossek and Yehiel Limor, "Fifty years in a 'marriage of convenience': News media and military censorship in Israel," *Communication Law and Policy* 6(1): 1–35.
25. Protocol no. 33 of the Knesset's Committee for Scientific and Technological Research and Development, The Internet, Freedom of Information and its Limits, February 24, 1998.

Jordan

Access to Internet content in the Hashemite Kingdom of Jordan remains largely unfettered, with filtering selectively applied to only a small number of sites. However, media laws and regulations encourage some measure of self-censorship in cyberspace,[1] and citizens have reportedly been questioned and arrested for Web content they have authored.[2]

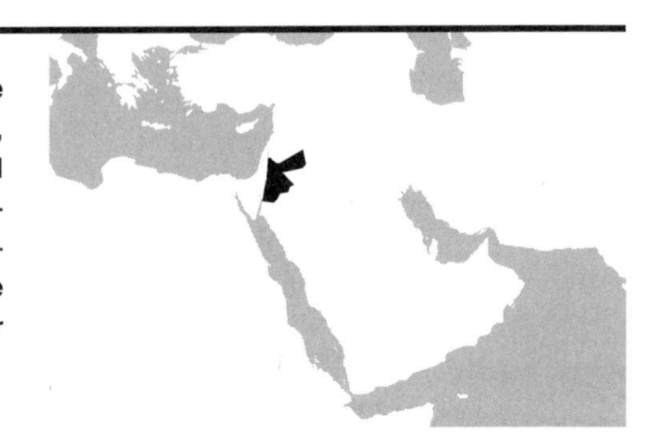

Background

Watchdog organizations continue to criticize the Jordanian government's record on human rights. In 2006, Human Rights Watch noted that "Jordanian authorities continued … to engage in practices that censor free speech," including charging journalists under controversial articles of the Penal Code.[3] Reports of prolonged detentions[4] of criminals and government harassment of opposition party members[5] have also surfaced. In June 2006 the government charged four parliamentarians from the Islamic Action Front (IAF) with fueling national discord and inciting sectarianism after the politicians visited the family of deceased al-Qaeda leader Abu Musab al-Zarqawi.[6] The constitutional monarchy has also demonstrated its willingness to silence and punish those critical of its allies.[7]

Internet in Jordan

Although the government provides schools with computers and encourages the growth of the Internet in Jordan,[8] connectivity prices remain prohibitively high for many Jordanians.[9] There are only five personal computers per hundred inhabitants, yet the country has achieved an Internet penetration rate of 12 percent—a relatively high figure for the region.[10] Most of this connectivity comes through the hundreds of Internet cafés[11] and community centers[12] in the country. A survey of 200 Jordanians, presented in

RESULTS AT A GLANCE

Filtering	No evidence of filtering	Suspected filtering	Selective filtering	Substantial filtering	Pervasive filtering
Political			●		
Social	●				
Conflict/security	●				
Internet tools	●				

Other factors	Low	Medium	High	Not applicable
Transparency	●			
Consistency			●	

KEY INDICATORS		worst — best
GDP per capita, PPP (constant 2000 international $)	4,585	
Life expectancy at birth (years)	72	
Literacy rate (% of people age 15+)	90	
Human development index (out of 177)	86	
Rule of law (out of 208)	79	
Voice and accountability (out of 208)	151	
Digital opportunity index (out of 180)	77	
Internet users (% of population)	11.2	

Source (by indicator): World Bank 2005, 2006a, 2006a; UNDP 2006; World Bank 2006c, 2006c; ITU 2006, 2004

2004, showed that the Internet serves as an important networking and communication tool, with all respondents using the Web to e-mail or chat.[13]

In an effort to further increase the Internet penetration rate, the government launched "Knowledge Stations" across the country in 2001. Jordanians in rural areas can access the Internet and attend courses in computers at these Stations.[14] An ongoing plan called the Jordan Broadband Learning and Education Network Project aims to create an extensive educational network by linking 8 public universities, 3,200 public schools, 23 community colleges, and 75 Knowledge Stations nationwide.[15]

Jordan has an advanced, though expensive, telecommunications infrastructure as compared with other countries in the region.[16] The telecom sector serves as a key industry for the Jordanian economy, accounting for 10 percent of the GDP.[17] Since the sector was liberalized in 2004,[18] private companies—in particular Wanadoo and Batelco, which together claim 83 percent of the Internet service provider (ISP) market—have overtaken the government's share of the market.[19] The government-owned National Information Technology Center (NITC) remains the exclusive registrar for the country code top-level domain (ccTLD) ".jo".[20]

Legal and regulatory frameworks

Established in 1995, the Telecommunications Regulatory Commission (TRC) regulates telecom and information technology services in Jordan.[21] Under the Telecommunications Law of 1995, the TRC is in charge of ISP licensing and telecommunications equipment.[22] Prospective ISPs must file a license application and document their financial resources, base prices, services and technologies, and geographical coverage areas.[23] ISPs must supply a mechanism for handling customer complaints and may increase user fees only after notification has run for at least one full month in two local newspapers.[24]

To obtain a license to open an Internet café, a party must submit an application with "an organizational site plan for the location to be used."[25] The Internet Café Regulations state that records of Internet use, including personal information, should be kept and that "a special technique should be provided to block and filter the sites"[26] that contain pornography or offensive religious material, that promote recreational drug use or gambling, or that show "the method of manufacturing of materials for military uses in an illegitimate manner."[27] Café patrons under thirteen years of age must be accompanied by a parent,[28] and managers cannot be younger than twenty-five.[29] Although the Regulations state that personal data should remain confidential,[30] a café must disclose such information when the government requests it.[31]

The Telecommunications Law instructs ISPs to withhold access from users who have "violated public morals"[32] or who use the Internet in a way that "endangers the national good,"[33] but leaves these stipulations undefined. The Law also stipulates that "Any person who originates or forwards, by *any* telecommunications means ... messages *contrary to public morals,* or forwards false information with *the intent to spread panic,* shall be punished by imprisonment of not less than one month or more than one year, or a fine ... or by both penalties."[34]

Article 5 of Jordan's Press and Publications Law (1998) prohibits journalists from publishing material that goes against "national obligation ... and Arab-Islamic values."[35] Article 7(e) is equally broad, forbidding the publication of anything that is "bound to stir violence or inflame discord of any form among the citizens."[36]

In January 2007, Jordan's Lower House National Guidance Committee began consultations with media experts and officials about a 2006 Press and Publications Draft Law.[37] Publishers and proponents of press freedom hope lawmakers will scrap provisions that set jail terms for journalists and amend articles they say restrict free expression.

ONI testing results

Testing conducted on three Jordanian ISPs—Batelco, Wanadoo, and Linkdotnet—showed no definitive blocking, though some filtering of political content is suspected given inconsistencies in the accessibility of certain sites.

Arab Times (www.arabtimes.com), a politically oriented news site that is sometimes critical of Arab leaders, was found to be inaccessible on Batelco and Wanadoo but was accessible via Linkdotnet. Though this finding does not constitute proof of filtering, it is worth noting, as the Web site was reportedly blocked in 2004.[38]

Conclusion

Jordanians appear to enjoy essentially unfiltered access to Internet content. However, the Press and Publications Law's broad provisions may lead some writers to engage in self-censorship. Although Jordan's government continues to develop initiatives to expand access to the Internet, laws restricting freedom of speech preserve an intimidating atmosphere that discourages free discourse on political and social issues.

NOTES

1. Reporters Without Borders, Internet Under Surveillance 2004: Jordan, http://www.rsf.org/article.php3?id_article=10737.
2. The Initiative for an Open Arab Internet, Implacable Adversaries: Arab Government and the Internet (2006): "Jordan," http://www.openarab.net/en/reports/net2006/jordan.shtml; Human Rights Watch, "Jordan: Rise in arrests restricting free speech," June 17, 2006, http://www.hrw.org/english/docs/2006/06/17/jordan13574.htm.
3. Ibid.
4. National Centre for Human Rights, The State of Human Rights in the Hashemite Kingdom of Jordan, May 31, 2005, http://www.nchr.org.jo/uploads/nchr-report.pdf.
5. U.S. Department of State, Country Reports on Human Rights Practices 2005: Jordan, http://www.state.gov/g/drl/rls/hrrpt/2005/61691.htm.
6. See "Jordan MPs face Zarqawi charges," BBC News, June 13, 2006, http://news.bbc.co.uk/2/hi/middle_east/5075422.stm.
7. In 2004, Ali Hattar was tried and sentenced to jail time for giving an "incendiary" anti-U.S. lecture at a conference. See Foreign Ministry of the Hashemite Kingdom of Jordan, "Hattar presents defense statement," January 19, 2005, http://www.mfa.gov.jo/events_details.php?id=8981.

8. Ministry of Information and Communications Technology, The National Broadband Network Briefing, http://www.moict.gov.jo/MoICT/MoICT_NBN.aspx.
9. The Arabic Network for Human Rights Information, "The Internet in the Arab world: A new space of repression?" http://www.hrinfo.net/en/reports/net2004/jordan.shtml.
10. See International Telecommunication Union, *World Telecommunication Indicators 2006*.
11. The Initiative for an Open Arab Internet, Implacable Adversaries: Arab Government and the Internet (2006): "Jordan," http://www.openarab.net/en/reports/net2006/jordan.shtml.
12. Deborah L. Wheeler, "The Internet in the Arab world: Digital divides and cultural connections," lecture presented June 16, 2004, at Jordan's Royal Institute for Inter-Faith Studies, http://www.riifs.org/guest/lecture_text/Internet_n_arabworld_all_txt.htm.
13. Ibid.
14. See Knowledge Stations Web site, http://www.ks.jo/. There are 132 Knowledge Stations across Jordan.
15. This project was slated for completion in 2006 but appears to be unfinished. See Ministry of Information and Communications Technology, "The national broadband network briefing," http://www.moict.gov.jo/MoICT/MoICT_NBN.aspx.
16. The Initiative for an Open Arab Internet, Implacable Adversaries: Arab Government and the Internet (2006): "Jordan," http://www.openarab.net/en/reports/net2006/jordan.shtml.
17. Ministry of Information and Communications Technology, E-readiness Assessment of the Hashemite Kingdom of Jordan, 2006, http://www.moict.gov.jo/MoICT/MoICT_Jordan_ereadiness.aspx.
18. In 1999, Jordan joined the WTO and fulfilled its telecom sector commitments in pursuit of full member status in 2000. See Telecommunications Regulatory Commission, Annual Report 2005, http://www.trc.gov.jo/Static_English/doc/annual%20report%2005.pdf.
19. Ministry of Information and Communications Technology, E-readiness Assessment of the Hashemite Kingdom of Jordan, 2006, http://www.moict.gov.jo/MoICT/MoICT_Jordan_ereadiness.aspx.
20. National Information Technology Center, http://www.nic.gov.jo/En/au.htm.
21. Telecommunications Regulatory Commission, http://www.trc.gov.jo/Static_English/main.shtm.
22. Telecommunications Law no. 13, 1995, http://www.trc.gov.jo/Static_English/doc/Telecom%20Law%20Translation%20(%20Final).pdf.
23. Ibid., Article 27.
24. See ibid., Articles 52, 53.
25. The actual location must fulfill certain conditions such as size. See Instructions for Regulating the Work of the Internet Centers and Cafés and the Bases for their Licensing for the Year 2001, http://www.reach.jo/Downloads/Legislative/Internet_Cafes_Regulations.pdf.
26. It is unclear what filtering technique should be used. See Instructions for Regulating the Work of the Internet Centers and Cafés and the Bases for their Licensing for the Year 2001, Article 6, http://www.reach.jo/Downloads/Legislative/Internet_Cafes_Regulations.pdf.
27. Ibid.
28. Ibid., Article 7.
29. Ibid., Article 17.
30. See ibid., Articles 6.3, 11.2.
31. Ibid.
32. Telecommunications Law no. 13, Article 58, 1995, http://www.trc.gov.jo/Static_English/doc/Telecom%20Law%20Translation%20(%20Final).pdf.
33. Ibid., Article 79.
34. Ibid., Article 75.
35. Press and Publications Law no. 8, Article 5, 1998. See Jordan Press Association, "Journalism laws," http://www.jpa.jo/all/english.htm.
36. Ibid., Article 7(e).
37. Ibtisam Awadat, "Press & Publications Draft Law 2006 under spotlight: Journalistic freedoms in the balance," The Star 18(133), February 1–7 2007, http://star.com.jo/viewnews/DetailNews.aspx?nid=3879.
38. Reporters Without Borders, Internet Under Surveillance 2004: Jordan, http://www.rsf.org/article.php3?id_article=10737.

Kazakhstan

Like many of the governments in the Commonwealth of Independent States (CIS), Kazakhstan has a conflicted position with regard to the Internet. The Kazakh government aims to make Kazakhstan the main IT portal in Central Asia. In this regard, the government has harnessed efforts to liberalize the IT sector, promote Internet use, and encourage e-government in order to spur social development. However, the government has also implemented a complex system that allows for state surveillance of Internet traffic that can be used to filter or suppress Internet content. Current rules require all Internet traffic to pass through state-owned channels, politically sensitive Internet content is selectively filtered, and opposition media and bloggers are said to practice self-censorship for fear of government reprisal.

Background

Kazakhstan is the largest country in Central Asia, covering a territory equivalent to the whole of Western Europe. An oil-rich country, Kazakhstan has recovered from the economic crises of the 1990s, and President Nursultan Nazarbayev is determined to turn Kazakhstan into an IT powerhouse in the region. An ambitious e-government project has been launched and the development of IT infrastructure is encouraged.

Politically Kazakhstan has become increasingly authoritarian. President Nazarbayev been the head of state since national independence in 1991, and he is widely alleged to have manipulated results of elections and suppressed opposition in order to remain in power.[1] Although press freedom is enshrined in the constitution,[2] the government controls most mass media outlets and exerts influence over most printing and distribution establishments. Anecdotal evidence

RESULTS AT A GLANCE

Filtering	No evidence of filtering	Suspected filtering	Selective filtering	Substantial filtering	Pervasive filtering
Political		●			
Social	●				
Conflict/security	●				
Internet tools	●				

Other factors	Low	Medium	High	Not applicable
Transparency				●
Consistency				●

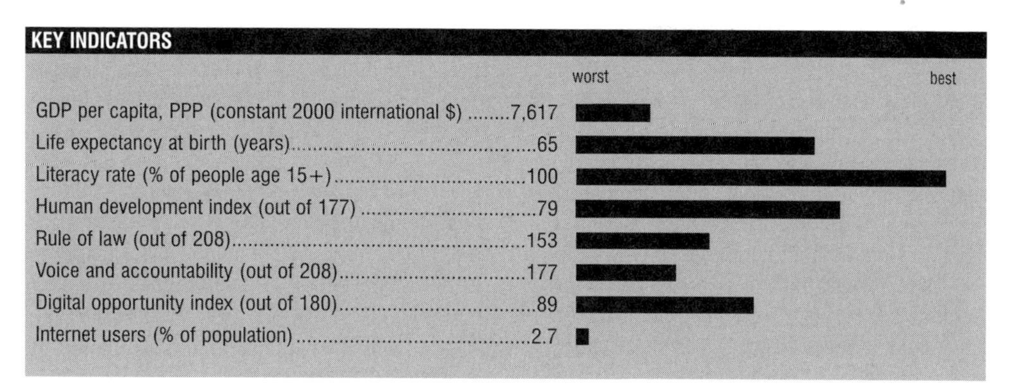

Source (by indicator): World Bank 2005, 2006a, 2006a; UNDP 2006; World Bank 2006c, 2006c; ITU 2006, 2004

points to online media and bloggers practicing self-censorship for fear of prosecution by the state under highly restrictive defamation laws.

Internet in Kazakhstan

The Kazakh Internet community is growing rapidly. Internet usage increased from 0.7 percent of the population in 2000 to 2.7 percent of the population in 2004.[3] Computer penetration is around fifteen to seventeen computers per 100 residents.

Because of its size and internal regional disparities (especially between rural and urban dwellers), Internet access remains beyond the reach of most Kazakhs, except for those living in major cities. Internet access is most popular among young urban dwellers, with a surprisingly high percentage of female users (44.1 percent).

Russian is the most popular language used on the Internet (94.1 percent), followed by Kazakh (4.5 percent) and English (1.4 percent), which may account for the high percentage of Kazakh Web sites hosted in Russia (including those on the ".kz" domain). Six percent of ".kz" Web sites are hosted in Kazakhstan, with the remainder hosted in Russia and elsewhere. Kazakhs use a wide range of search engines, including Russian, U.S., and Kazakh (www.rambler.ru, www.yandex.ru, www.yahoo.com, www.google.kz, www.site.kz).

Recent liberalization of the telecommunications market increased competition among the five licensed operators. These are Kazakhtelecom (the former state monopoly), Transtelecom, Kaztranscom, Arna [DUCAT], and Astel. Kazakstan also has five first-tier Internet service providers (ISPs) that possess independent channels to the Internet. These are Kaztelecom, Nursat, Astel, Telcom, and NIT. Some 100 second-tier providers lease access from the five first-tier ISPs.[4]

Kazakhtelecom is the operator of the national data transfer network, which connects the major cities of Kazakhstan with a total bandwidth of 665Mbit/s,[5] and carrying capacity of separate local segments of up to 10 GB/s.[6] Other leading first-tier ISPs (Nursat and Astel) also operate significant terrestrial and satellite based infrastructure.

Legal and regulatory frameworks

The Kazakh government exhibits an ambiguous and at times contradictory approach to the Internet. On one hand the "Development Strategy of Kazakhstan until 2030" demonstrates the government's strong commitment to create an independent and effective system of telecommunications services, which will be competitive with analogous infrastructures in more-developed countries. On the other hand the government follows a strong and multilevel information security policy, ensuring surveillance of telecommunications and Internet traffic in the country.

The Agency for Informatizaton and Communication (AIC), a central executive body in the IT field, is authorized to implement state policy in telecommunications and information technology development industries, carry out control in these sectors, and issue licenses to every type of telecommunications service.[7] The Security Council (SC), a body chaired by the president, is responsible for drafting decisions and providing assistance to the head of state on issues of defense and national security.[8] The SC also prepares a list of Web sites every six months that it wants to have blocked or forbidden from distribution. A 2005 SC decision legally forbade key national security bodies (namely the Ministries of Emergency Situations, of Internal Affairs, of Defense, and the National Security Committee) from connecting to the Internet. However, despite this prohibition, ONI researchers witnessed state officials accessing forbidden Web sites through an anonymizer.

The security system in Kazakhstan is complex and multilayered. The Inter-Departmental Commission is charged with coordinating and developing the national information infrastructure. The National Security Committee (NSC) monitors presidential, government, and military communications. The Office of the Prime Minister is an authorized state body responsible for the protection of state secrets and the maintenance of information security. "State secret" is broadly defined, encompassing various government policies as well as the president's private life, health, and financial affairs. The NSC has issued a general license to the private Agency on Information Security to establish and organize facilities for cryptographic protection of information, as well as to formulate proposals on information security for state organizations, corporate clients, banks, and other large commercial companies.

The information communications technology sector in Kazakhstan is highly regulated, as evidenced by some 300 legislative acts that expressly or implicitly control the information and telecommunications environment. All ISPs require a license from the AIC.[9] All telecommunications operators are legally obliged, as part of the licensing requirement, to connect their channels to a public network controlled by Kazakhtelecom. The so-called Billing Center of Telecommunication Traffic, established by the government in 1999, helps trace the activity of private companies and strengthen the monopolist position of Kazakhtelecom in the IT sphere. In practice, some telecommunications operators circumvent such regulations by using IP telephony to pass their interregional and international traffic.

The government has established systems to monitor and filter Internet traffic. Since the traffic of all first-tier ISPs goes through Kazakhtelecom's channels, filtering can be achieved using centralized resources. The ISPs may unknowingly receive filtered content because the main operator could install filters on any information that it deems inappropriate. ONI suspects that state officials informally ask Kazakhtelecom to filter certain content. Russian companies and Kazakhtelecom have openly signed an agreement to provide filtering, censorship, and surveillance on the basis of Security Council resolutions.

State regulations oblige Internet providers to register and maintain electronic records of customer Internet activity. ISPs are required to install special software and hardware equipment in order to create and store records for a specified amount of time, including log-in times, types of the connection, transmitted and received traffic between parties of the connection, identification number of the session, duration of time spent online, IP address of the user, and speed of data receipt and transmission. The ISPs are also required to prohibit their customers from disseminating (via Internet) pornographic, extremist, or terrorist materials or any other information not in accordance with the country's laws.[10]

The Kazakhstan Association of IT-Companies is the officially recognized administrator of the ".kz" domain. It is registered as a nongovernment organization but, in fact, it has 80 percent government ownership. The rules of registration and management of the ".kz" domain are issued by the AIC. The constitution guarantees freedom of speech and prohibits censorship, but the government often resorts to various mechanisms to suppress "inappropriate" information or to shut down oppositional domain names. These rules mean that an applicant may be denied registration if the resource server resides outside of Kazakhstan. Use of the Internet by political parties in Kazakhstan is limited, and few opposition or illegal parties have made the move to go online.

ONI testing results

ONI conducted testing on three main ISPs: Kazakhtelecom, Megaline, and Nursat. The evidence gathered from the testing is not sufficient to conclusively confirm the existence of a systematic filtering regime. However, a number of sites with sensitive political content, including locally sensitive topics and regional issues of concern to the Kazakh government, were inaccessible. Several of these inaccessible sites are hosted in Russia and Kyrgyzstan. ONI found some political sites were inaccessible for users of two ISPs (Kazakhtelecom and Megaline), while they remained accessible to Nursat users. Generally most of the inaccessible sites contained content related to political dissidents, allegations of government corruption, human rights issues, and strongly expressed criticism of the president.

Kazakh authorities also de-register Web sites that do not comply with its restrictive rules for registering domains within the ".kz" domain, and filters sites within this domain.[11] In 2005 Kazakh authorities de-registered a Web site created by the producers of *Borat,* (claiming that the site violated the rules by hosting the site outside of Kazakhstan and providing false contact information).[12]

ONI suspects that filtering practices in Kazakhstan have changed and are now performed at the network backbone. All traffic should pass through the Kazakhtelecom network and thus be subject to filters put in place by the state-controlled ISP. However, not all incoming and outgoing traffic passes through the network, which results in inconsistent patterns of blocking.

Most of the users are also on "edge" networks, such as cybercafés and corporate networks. Kazakhstan companies apply filtering mechanisms on a user level to prevent employees from accessing pornography, music, films, and dating Web sites. However, ONI testing found that Kazakhstan does not block any pornographic content or sites related to drug and alcohol use.

Conclusion

The Kazakh government has harnessed efforts to liberalize the IT sector, promote Internet use, and encourage e-government in order to spur social development. However, it has also put in place a complex security system that is capable of state surveillance of Internet traffic, and suppression of undesirable Internet content. Given government pressure on opposition media, self-censorship may also be an issue among online media publishers and bloggers. The technical sophistication of the Kazakhstan Internet environment and government's tendency toward stricter online controls warrant closer examination and monitoring.

NOTES

1. See Commission on Security and Cooperation in Europe, "Missed opportunity in Kazakhstan: Fraud and intimidation spoil election promised to be "free and fair," December 15, 2005, http://www.csce.gov/index.cfm?Fuseaction=ContentRecords.ViewDetail&ContentRecord_id=107&ContentType=G (accessed May 1, 2007).
2. Article 20, paragraphs 1 and 2 of the Constitution of Kazakhstan and the Law on Media and Telecommunications (with last amendments of Jan. 2006) article 2, pararagraph 1.
3. International Telecommunication Union, *World Telecommunication Indicators 2006*. The Kazakhstan Association of IT Companies estimated that 6.8 percent of the population used the Internet in 2005.
4. Source: The Agency for Informatizaton and Communication of Kazakhstan.
5. For comparison, by the end of 2002 the total Internet bandwidth capacity for Kazakhstan was 46Mb/s; by the end of 2003 it was 189 Mb/s.
6. Kazakhstan's investment in Internet capacity is part of the USD110 million loan from the European Bank for Reconstruction and Development. Lucent Worldwide Services and Winncom Technologies are providing support for the project. In 2006 Kazakhtelecom began construction of a next-generation network (NGN) and plans to deploy fixed wireless access (FWA) platforms such as Wi-Fi and WiMAX.
7. Resolution no. 724 of the Kazakh government, dated July 22, 2003.
8. Article 44, paragraph 20 of the Constitution of Kazakhstan.
9. Decree no. 998 of September 29, 2004, Concerning Question of Licensing in the Telecommunications Sphere.
10. Nursat Public Contract, http://www.nursat.kz/?72 (in Russian).
11. http://www.blokada.org/print.php?sid=1985.
12. http://www.rsf.org/article.php3?id_article=15919.

Kyrgyzstan

Recent liberalization of the telecommunications market in Kyrgyzstan has made Internet access affordable for the majority of the population. This access remains largely unfettered. However, an emerging regime shift toward more restrictive policy, dependence upon Russian and Chinese Internet connections, and political instability pose problems for clear and continual access to Internet in Kyrgyzstan.

Background

In 2005 Kurmanbek Bakiev won the presidential elections after the violent downfall of the fourteen-year authoritarian regime of former president Askar Akayev. The new head of state vowed to distribute more powers to the parliament, encourage free speech, fight corruption, and tackle poverty. However, this shift in power has yet to result in significant economic improvements in Kyrgyzstan, as two-thirds of the population remains below the poverty line. International observers predict that new civil conflicts may erupt if the country does not adopt urgent economic measures.[1] The Internet is one of the few free outlets for expressing public criticism in Kyrgyzstan, and has been used as an instrument to assemble people for protest against the government. Kyrgyzstan's U.N. global ranking for e-government (0.4417) has deteriorated; however, it remains in second place on the central Asian list, after Kazakhstan.[2]

Internet in Kyrgyzstan

Kyrgyzstan has one of the highest Internet penetration rates in Central Asia (5 percent in 2005).[3] Some local studies assert that the number of Internet users is two times higher than reported in the official data.[4] However, personal computers (PCs) remain unaffordable for the vast majority: only 2 percent of the population owns a PC.[5]

RESULTS AT A GLANCE

Filtering	No evidence of filtering	Suspected filtering	Selective filtering	Substantial filtering	Pervasive filtering
Political	●				
Social	●				
Conflict/security	●				
Internet tools	●				

Other factors	Low	Medium	High	Not applicable
Transparency				●
Consistency				●

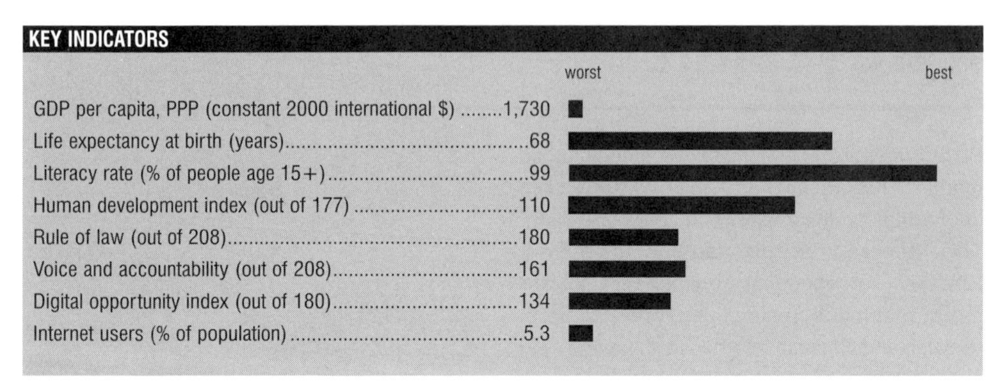

Source (by indicator): World Bank 2005, 2006a, 2006a; UNDP 2006; World Bank 2006c, 2006c; ITU 2006, 2005

Cybercafés are the main Internet access point in the country (for approximately 51 percent of all users). Other important venues for public access are workplaces (nearly 25 percent) and educational institutions (24 percent). There are approximately 150 public Internet access centers in the country, including cybercafés and free access centers sponsored by nongovernmental organizations (NGOs). Development of the Internet infrastructure targets only the two largest cities, Bishkek and Osh. There are slightly more female than male users, and 60 percent of all users are aged between fifteen and twenty-five, with an additional 20 percent aged twenty-six to thirty-five. Russian language sites remain the most visited among Kyrgyz Internet users (90 percent), compared with only 8 percent in Kyrgyz and 2 percent in English.

The privatization of both telecommunications and services, driven by the foreign investment and financial assistance, has resulted in an increasingly competitive Internet sector. This has caused access fees to decrease to USD0.30/hour, which in turn has made the Internet affordable for the average Kyrgyz. In 2005 the number of ISPs increased to thirty-eight, although only seven of these have an external Internet connection. Two of the seven ISPs—KygyzTelecom (KT) and SaimaTelecom—own the infrastructure they use. The others rent lines and cables from the state-controlled top-tier KT. The state has a major stake (50 percent) in Elcat, another top-tier ISP. The majority of ISPs connect by satellite to the Russian portion of the Internet. In addition to its major Russian connection, KygyzTelecom has built external connection ports to China and Kazakhstan.

The Internet Traffic Exchange Point (IXP),[6] shared by the ISPs with external Internet connection, runs the local traffic. The international Internet bandwidth in the country is 76 Mb/s,[7] and the most popular means for Internet access is through dialup connection. A private company, AsiaInfo, controls the country's top-level domain ".kg".[8] There are around 1,500 top-level domain names registered in the Kyrgyz Internet zone.

Legal and regulatory frameworks

Internet and ISP activities are not directly regulated by sector-specific laws in the communications sphere. Compared with its neighbor Kazakhstan, Kyrgyzstan does not compel local Internet providers to work with the state-owned provider. Therefore ISPs independently establish international connections. However, the state telecom continues to enjoy exclusive rights to national long-distance and international services, thus thwarting mobile operators and ISPs from entering the market. A licensing regime exists for providing Voice-over Internet Protocol (VoIP) services. To obtain a license, companies are required to contribute twenty million som (approximately USD517,000) to develop IT infrastructure. Once an applicant obtains the license, it may resell VoIP services to another company.

In 2002 the state declared ICT development to be a priority by way of the National Strategy on Information and Communication Technologies for Development of the Kyrgyz Republic.[9] Eager to harness Internet capabilities to stimulate economic growth, the government has encouraged e-government, e-education, and the e-economy.[10] For example, under a joint program between the government and international organizations, 95 percent of central government bodies, and 50 percent of local ones, have Internet access and provide online information about their services.[11] However, the cyber presence of political opposition is limited. ONI detected only three Kyrgyz Web sites belonging to political parties.[12]

The main institution responsible for the sector is the national ICT Council. The presidential administration has made efforts to introduce restrictive measures to control Internet content. In the spring of 2005 members of the government proposed amendments to the law on mass media that would have led to blocking all ".ru" domain sites containing offensive information on Kyrgyzstan. In turn, this would have limited Kyrgyz access to sources solely on the ".kg" domain, which is regulated by local authorities. Although this proposal was rejected, it revealed a shift in official attitudes toward Internet development in the country.

The National Communications Agency (NCA) is directly responsible to the President. It has taken over most of the responsibilities of the Ministry of Transportation and Communication in the telecommunication sector. The NCA regulates and supervises postal and electronic communication companies, issues licenses, and monitors the Internet.[13]

Kyrgyz security laws do not explicitly apply to Internet activities. However, the National Security Law of 2003 provides for the creation of specialized communication and information security bodies within the structures of the National Security Service. The Security Council will be *inter alia* responsible for examining internal and external policy questions in the field of information security. In 2005, a government resolution on the Program for Information Security was adopted, but it lacked precise definitions for what constitutes commercial secrets, state secrets, and private information. This absence of clear terminology may lead to variable interpretations, which could create space for potential abuse. Furthermore, the program does not exhaustively list what types of information can be limited, which again can allow for the broadening of the scope of restricted information.

There is no legislation allowing the national security services to organize surveillance over the Internet. In fact, KT itself launched a technical investigation to prevent "gray traffic." Possible surveillance exercised by state officials may take place at the ISP level. In July 2006, the State Agency for Intellectual Property proposed to create an Inter-Departmental Commission on State Regulation of the Kyrgyz Segment of Internet. This institution, which follows an existing Russian model, would coordinate the activity of the executive power bodies and organizations participating in the Kyrgyz segment of Internet. The implementation of restrictive measures by such an institution would deter further development of Internet in Kyrgyzstan.

ONI testing results

ONI conducted testing from various access points on all seven first tier ISPs: Aknet, AsiaInfo,

Elcat, KyrgyzTelecom, SaimaTelecom, Totel, and Transfer. The testing did not detect activity that is indicative of any deliberate or even selective pattern of filtering. Some U.S. military sites were inaccessible, but these are likely the result of "supply side" blocking by U.S. authorities or poor domain name propagation. Kyrgyzstan does not block the sites of religious or extremist groups.

Past work by ONI leads us to suspect that there may be just-in-time or event-based tampering applied during politically sensitive periods. This was the case during the 2005 parliamentary elections, when ONI documented the extensive use of DoS attacks against opposition and media Web sites and Kyrgyz ISPs.[14]

Blocking of voice traffic is carried out in order to limit access to non-Kyrgyz providers offering IP-telephony service, to thereby compel the use of local providers. Voice traffic is filtered in all the standard ports on all popular non-Kyrgyz providers of IP-telephony. Allegedly, Cisco (Pix) and Huawei (EuDemon) products are used for blocking voice content. Filtering also exists at the enterprise level (NGOs, corporate clients) in order to block access of content deemed irrelevant and to economize Internet traffic.

Conclusion

Kyrgyzstan does not officially engage in filtering of Internet content. Although the government generally encourages Internet development, a shift toward greater restriction may be emerging. The regime appears to be struggling to find a balance between maintaining control over the ICT sector and allowing the necessary freedom for spurring economic growth. Potential limits in Internet freedom are posed by generally poor access, the possibility of "in-stream filtering" resulting from dependence on Russian and Chinese connections, and the possibility of sporadic targeted filtering triggered by state instability. However, Kyrgyzstan is an aid-dependent country, and is therefore unlikely to pursue open filtering of Internet content.

NOTES

1. International Eurasian Institute for Economic and Political Research, http://www.iicas.org/libr_en/kg/libr_06_10_05kg.htm.
2. Department of Economic and Social Affairs; UN Global E-Government Readiness Report 2005, at 56, http://unpan1.un.org/intradoc/groups/public/documents/un/unpan021888.pdf.
3. International Telecommunication Union, *World Telecommunication Indicators 2006*.
4. According to the Expert Consulting Agency, the number of Internet users reached 550,000 in 2005, suggesting Internet penetration of more than 10 percent.
5. International Telecommunication Union, *World Telecommunication Indicators 2006*.
6. Telecoms Markets and Statistics, 2006.
7. Ibid.
8. For more information see http://www.domain.kg/.
9. http://www.ict.gov.kg/index.php?name=EZCMS&menu=2501&page_id=71.
10. See National ICT Action Plan, http://www.ict.gov.kg/index.php?name=EZCMS&menu=37&page_id=96.
11. The government gate portal is www.govservices.kg.
12. These are: the Moia Strana Party, the Democratic Party Turan, and the Ar-Namys Party, whose previous leader is today's Prime Minister F. Koulov.
13. Paul Budde Communication Pty Ltd., Telecoms Markets and Statistics, 2006.
14. See OpenNet Initiative, Special Report: Kyrgyzstan Election Monitoring in Kyrgyzstan (February 2005), http://www.opennetinitiative.net/special/kg/.

Libya

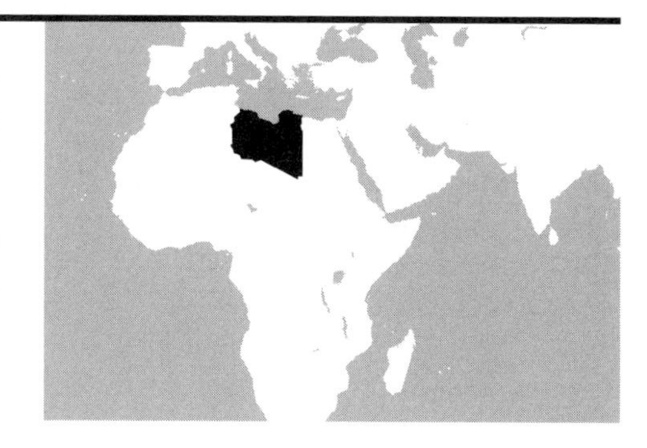

Libya continues to block Internet content related to political opposition, content critical of the government, and Web sites that advocate the rights of the minority group Amazigh (Berbers). This censorship of political content persists despite a trend toward greater openness and increasing freedom of the press.

Background

Libya has undertaken a radical shift in policies over the past few years. Formerly considered a state sponsor of terrorism and an international pariah, Libya moved to regain international acceptance by formally renouncing support of terrorism and dismantling their weapons of mass destruction development programs in 2003.[1] As a result, the United States, the European Union, and the United Nations lifted their respective embargoes on Libya soon after.[2] The United States recently established an embassy in Libya to further solidify relations between the two countries.[3]

Though much has changed, much has stayed the same. As Reporters Without Borders states, "despite Col. Muammar al-Gaddafi's recent pro-democracy pretensions, his regime still keeps a very tight rein on news."[4] Human rights watchdog groups still report serious violations, such as restriction of expression; prohibition of political parties and independent organizations; imprisonment of critics of the political system, the government, or its leader; torture; and unresolved disappearances from past years.[5] The press laws from 1972 and 1973 impose large fines and up to two year prison sentences for violations of a variety of press restrictions, including "doubting the aims of the revolution."[6] As a result self-censorship in the media is widespread. Reporters Without Borders reports that journalists rarely challenge the boundaries

RESULTS AT A GLANCE

Filtering	No evidence of filtering	Suspected filtering	Selective filtering	Substantial filtering	Pervasive filtering
Political				●	
Social	●				
Conflict/security	●				
Internet tools	●				

Other factors	Low	Medium	High	Not applicable
Transparency	●			
Consistency			●	

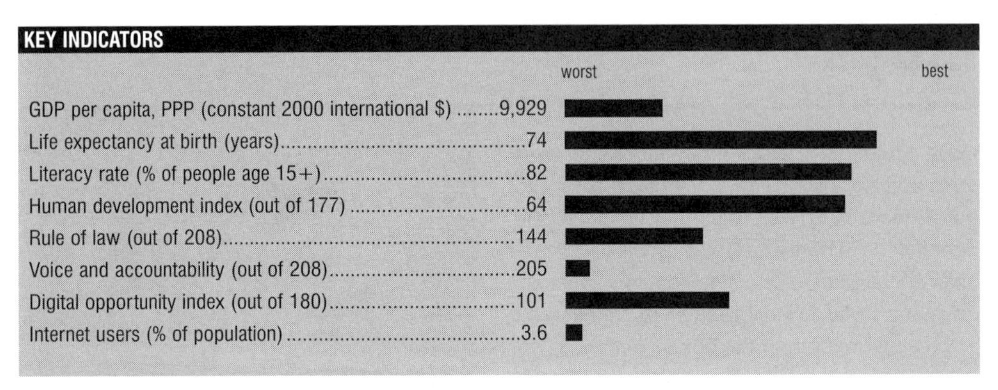

Source (by indicator): IMF 2006; World Bank 2006a, 2006b; UNDP 2006; World Bank 2006c, 2006c; ITU 2006, 2004

imposed by the government on content, especially those on topics relating to Gaddafi or his family or to the plight of the Berber minority.[7] The press laws also make the formation of private media illegal by restricting the right to publish to only two public organizations.[8]

At the same time as the Libyan opposition has increasingly used the Internet to spread its message,[9] the crackdown on journalistic freedom has moved into the realm of the Internet as well. In the country's most famous case, a fifty-one-year-old bookseller named Abdel Razak Al Mansouri was arrested in January 2005 and interrogated about a number his posts on the Akbar Libya Web site (www.akhbar-libya.com) that were critical of the government. Though never charged with a crime related to those posts, he was eventually charged, convicted, and sentenced to a year and a half in jail for possession of a gun without a license. He served a year before being granted amnesty.[10]

Internet in Libya

Internet access officially came to Libya at the end of 1998, but it was not widely available until early 2000.[11] Internet penetration remains low, at around 4 percent,[12] at least in part because of the long-term economic sanctions imposed on the country.[13] The primary means for people to connect is through Internet cafés.[14]

The state-owned General Post and Telecommunications Company (GPTC), run by Gaddafi's son, Mohamed al-Gaddafi,[15] regulates and operates Libya's telecommunications infrastructure, providing "international and local voice services, digital leased lines, telex, fax, mobile (through a partially owned subsidiary) and Internet services."[16] The GPTC also owns the country's primary ISP, Libya Telecom and Technology (LTT), which offers Internet services via dialup, DSL, broadband, and satellite,[17] though at least seven companies other companies are licensed. These competitors are effectively subordinated to LTT, however, as LTT maintains a monopoly over the country's international Internet gateway.[18]

In October 2006 the government of Libya reached an agreement with One Laptop per Child, a nonprofit United States group developing an inexpensive, educational laptop computer, with the goal of supplying machines to all 1.2 million Libyan schoolchildren by June 2008.[19] As the country only contained 130,000 computers in 2002,[20] this will be a major boost to the availability of information communications technology (ICT) technologies and the Internet.

Legal and regulatory frameworks

Libya continues to maintain strict limits on what can be said or written in the country. Libya's penal code, for example, punishes with life imprisonment or death anyone convicted of disseminating information that conflicts with the constitutional principles or the country's "fundamental social structures" or that tarnishes Libya's image abroad. Criticism of President Gaddafi is punishable by death.[21]

Further, by the press laws mentioned earlier, print and broadcast media are owned and strictly controlled by the government, and expression of opinions contrary to official policy is strictly forbidden. According to the Freedom House report, the pervasive use of secret police, informants, and arbitrary arrests intimidates citizens from speaking out and renders independent and critical journalism virtually impossible.[22]

With the rising threat from the Internet to government control over political information, the Libyan government appointed one of Gaddafi's closest friends to monitor and limit the growth of oppositional Web sites. Experts from Russia, Poland, and Pakistan were summoned to Libya to help handle the situation. One tactic that emerged was to force owners of Internet cafés to place stickers on computers that warn visitors from logging onto Web sites deemed oppositional.[23]

Beyond merely political content, the Libyan official ".ly" registry rules mandate that ".ly" domains "must not contain obscene, scandalous, indecent, or contrary to Libyan law or Islamic morality words, phrases or abbreviations."[24] ONI did not, however, find any social blocking in its tests.

In 2006 Reporters Without Borders removed Libya from their list of the Internet enemies after a fact-finding visit found no evidence of Internet censorship.[25] ONI's test results contradict that conclusion, however, as noted below.

ONI testing results

In 2007 ONI ran tests on Libya's three ISPs: Libya Telecom and Technology (LTT), Modern World of Communications (MWC), and Al-Falak. All three ISPs were found to block oppositional content such as the Web site of the Libyan Muslim Brotherhood (www.almukhtar.org) and the Libyan Constitutional Union (www.lcu-libya.co.uk and www.libyanconstitutionalunion.net).

The three ISPs also block Web sites containing information critical of the Libyan regime. For example, ONI found Libya for Ever (www.libya4ever.com), Libya al-Mostakbal (www.libya-almostakbal.com), and Libya Our Home (www.libya-watanona.com) to be blocked.

Access to Web sites containing information about the Amazigh (Berbers), including the preservation and teaching of the Tamazight language and culture, is restricted as well. Examples in this category found to be blocked are www.libyaimal.com, an Amazigh-related Web site, and www.tawalt.com, a site run by a Libyan Amazigh cultural foundation.

The filtering regime also targets content critical of the human rights situation in the country, notably the Web site of the Libyan Union for Human Rights Defenders (www.libyanhumanrights.com).

Evidence from ONI testing reveals that Libya employs IP blocking at the international gateway, carried out by Libya Telecom and Technology Company (LTT). Users who attempt to access banned content are not served with a blockpage, but rather encounter time-out messages.

Conclusion

Despite the general trend toward greater freedom and openness, Libya maintains an active Internet filtration regime focused on Web sites of political opposition groups, antigovernment news and views, and content related to the minority group Amazigh (Berbers). The filtering regime lacks transparency, as none of the three ISPs admits filtering or serves blockpages. If current trends hold, however, the government may decide to decrease their filtration efforts in the future.

NOTES

1. Elise Labott, "U.S. to normalize relations with Libya," CNN.com, May 15, 2006, http://www.cnn.com/2006/US/05/15/libya/index.html.
2. BBC News, "EU lifts weapons ban on Libya," October 11, 2004, http://news.bbc.co.uk/2/hi/europe/3732514.stm; Reporters Without Borders, "Libya: We can criticise Allah but not Gaddafi," October 2006, www.rsf.org/IMG/pdf/rapport_libye_gb.pdf.
3. U.S. Department of State, "About the Embassy," http://libya.usembassy.gov/history2.html.
4. Reporters Without Borders, Libya: Annual Report 2007, 2007, http://www.rsf.org/article.php3?id_article=20770&Valider=OK.
5. Human Rights Watch, Libya: Events of 2006, http://hrw.org/englishwr2k7/docs/2007/01/11/libya14712.htm.
6. BBC News, "EU lifts weapons ban on Libya," October 11, 2004, http://news.bbc.co.uk/2/hi/europe/3732514.stm; Reporters Without Borders, "Libya: We can criticise Allah but not Gaddafi," October 2006, www.rsf.org/IMG/pdf/rapport_libye_gb.pdf.
7. Reporters Without Borders, Libya: Annual Report 2007, 2007, http://www.rsf.org/article.php3?id_article=20770&Valider=OK.
8. United Nations Development Programme, Program on Governance in the Arab Region, Media and Government Regulations, http://www.pogar.org/countries/civil.asp?cid=10#sub5.
9. Islam Online, Breath for Youth, August 14, 2005, http://www.islamonline.net/Arabic/news/2005-08/14/article14.shtml (in Arabic).
10. Reporters Without Borders, "Imprisoned cyber-dissident in worrying condition after injury in fall from bunkbed," May 2005, http://www.rsf.org/article.php3?id_article=13890; Reporters Without Borders, Libya: Annual Report 2007, http://www.rsf.org/article.php3?id_article=20770&Valider=OK.
11. The Arabic Network for Human Rights Information, "Libya: The Internet in a conflict zone," http://www.hrinfo.net/en/reports/net2004/libya.shtml.
12. International Telecommunication Union, *World Telecommunication Indicators 2006*.
13. The Initiative for an Open Arab Net, Libya, http://www.openarab.net/en/reports/net2006/libya.shtml, (accessed March 19, 2007).
14. Ibid.
15. Economist Intelligence Unit, Libya: Privatisation Possiblities March 19, 2007, http://globaltechforum.eiu.com/index.asp?layout=rich_story&channelid=4&categoryid=31&title=Libya%3A+Privatisation+possibilities&doc_id=10336.
16. Internet Assigned Numbers Authority, IANA Report on the Redelegation of the .ly Top-Level Domain, October 2004, http://www.iana.org/reports/ly-report-05aug05.pdf.
17. LTT Co. Web site, http://ltt.ly/english/services.php.
18. Economist Intelligence Unit, Libya: Privatisation Possiblities, March 19, 2007, http://globaltechforum.eiu.com/index.asp?layout=rich_story&channelid=4&categoryid=31&title=Libya%3A+Privatisation+possibilities&doc_id=10336.
19. John Markoff, "U.S. group reaches deal to provide laptops to all Libyan children," The New York Times, October 11, 2006, http://www.nytimes.com/2006/10/11/world/africa/11laptop.html?ex=1318219200&en=84038e9ae540b091&ei=5088&partner=rssnyt&emc=rss.
20. International Telecommunication Union, *World Telecommunication Indicators 2006*.
21. International Press Institute, World Press Freedom Review, Libya 2005, http://www.freemedia.at/cms/ipi/freedom_detail.html?country=/KW0001/KW0004/KW0098/&year=2005.
22. Ibid.
23. The Arab Network for Human Rights Information, "Libya: The Internet in a conflict zone," http://www.hrinfo.net/en/reports/net2004/libya.shtml.
24. Official ".ly" registry Web site, http://www.nic.ly/regulations.php.
25. Reporters Without Borders, "The list of 13 Internet enemies," November 7, 2006, http://www.rsf.org/article.php3?id_article=19603.

Malaysia

Hoping that an Internet unencumbered by censorship will spur growth in domestic information technology industries, Malaysia has pledged not to censor the Internet. There is no evidence of technological Internet filtering in Malaysia. However, pervasive state controls on traditional media spill over to the Internet at times, leading to self-censorship and reports that the state investigates and harasses bloggers and cyber-dissidents.

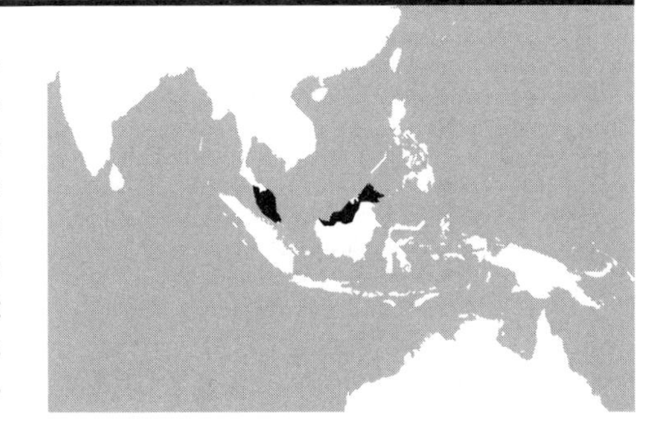

Background

Malaysia has a long history of state censorship and tight media controls. All four major newspapers are pro-state, and any oppositional and independent media outlets face the possibility of harassment by police, extended legal wrangling, detention, and imprisonment for publishing speech critical of the state.[1] As many as twenty different Malaysian laws restrict speech, and free speech activists contend that this leads to self-censorship by journalists.[2] The state also monitors the content of Web sites, and independent news Web site www.malaysiakini.com claims to have been the subject of several police investigations and an eviction notice as a result of publishing content deemed defamatory or offensive.[3]

Internet in Malaysia

Since 1996, Malaysia has embarked on an international public relations campaign to draw technology research and development to its Multimedia Super Corridor (MSC), a high-tech business center and communications infrastructure designed to help Malaysia become an international information technology leader.[4] Developing Internet infrastructure in Malaysia is a state priority, and consumers are encouraged to purchase PCs and Internet access. By 2005, Malaysia had approximately eleven million

RESULTS AT A GLANCE

Filtering	No evidence of filtering	Suspected filtering	Selective filtering	Substantial filtering	Pervasive filtering
Political	●				
Social	●				
Conflict/security	●				
Internet tools	●				

Other factors	Low	Medium	High	Not applicable
Transparency				●
Consistency				●

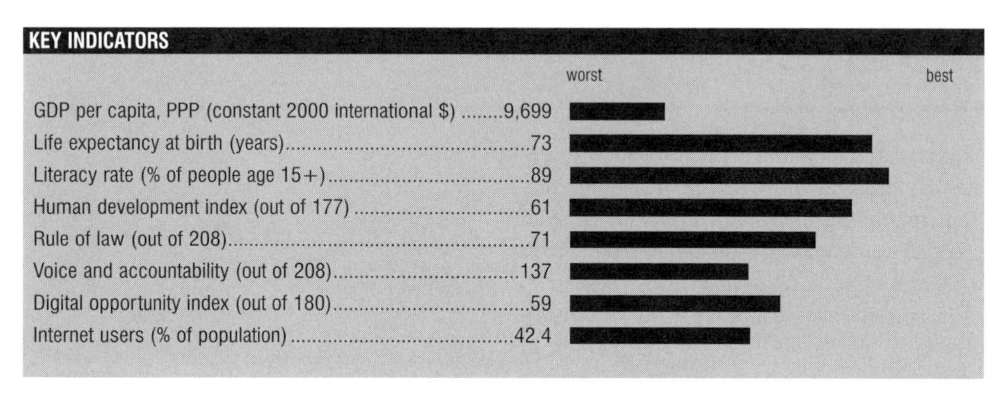

Source (by indicator): World Bank 2005, 2006a, 2006a; UNDP 2006; World Bank 2006c, 2006c; ITU 2006, 2005

Internet users, and with a national Internet penetration rate of 42 percent was third in Southeast Asia behind Hong Kong and Singapore.[5] The state, recognizing the opportunities for e-commerce and for individuals to exchange ideas and information,[6] has strongly encouraged adoption of broadband Internet throughout the country.[7] Nevertheless, uptake has been slow as dial-up remains the method by which most Malaysians access the Internet.[8] Broadband penetration reached a mere 2 percent in 2006, far behind other Southeast Asian regional leaders such as Singapore and Hong Kong, which had broadband penetrations of nearly 16 and 24 percent, respectively.[9]

Legal and regulatory frameworks

Malaysia's constitution guarantees every citizen the right of free speech and expression, but also sets significant limitations on that freedom, as Parliament may by law effect "such restrictions [on free speech] as it deems necessary or expedient in the interest of the security of the Federation ...."[10] Parliament has enacted numerous laws enabling broad state control over the media. Notable print and broadcast media regulations include the Printing Presses and Publications Act, which requires all print publishers to seek annual renewal of a publication license granted at the state's discretion, and the Sedition Act, which criminalizes the expression or publication of words that tend to incite hatred or contempt against any government.

The Communications and Multimedia Act of 1998 (CMA) and the Communications and Multimedia Commission Act of 1998 (CMCA) together directly govern Malaysia's telecommunications, broadcasting, and Internet sectors, including related facilities, services, and content.[11] The CMCA establishes the Malaysian Communications and Multimedia Commission, which is empowered to regulate the information technology and communications industries. The commission takes the position that Internet content must be regulated and controlled for "reasons of access, privacy and security and protection of individual rights."[12] The CMA empowers the commission with broad authority to regulate online speech, providing that "no content applications service provider, or other person using a content applications service, shall provide content which is indecent, obscene, false, menacing, or offensive in character with intent to annoy, abuse, threaten or harass any person."[13]

Publishers of media content in violation of this provision may face criminal penalties, including a fine of up to RM50,000 and/or a maximum of one year in prison.[14] The CMA also establishes the Content Forum, which formulates and implements the Content Code—voluntary guidelines for content providers concerning the handling of content deemed offensive and indecent.[15]

The CMA and other laws empower the state with extensive media controls. To foster the growth of the Internet market and the MSC, however, the state has generally refrained from directly censoring the Internet. In its "Bill of Guarantees" to approved MSC companies, the state pledges not to censor Internet content.[16] Nevertheless, Internet content publishers in Malaysia operate under constant risk that the CMA and numerous other laws regulating speech and content on traditional media will be interpreted or amended to extend to Internet publications.[17]

In January 2007, Malaysian Prime Minister Datuk Seri Abdullah Ahmad Badawi made a somewhat ineffectual distinction by stating that while the government policy is not to censor the Internet, bloggers are bound by laws on defamation, sedition, and other limits on speech.[18] Badawi's statement was an official restatement of the policy announced in August 2006 that bloggers who publish seditious, malicious, or defamatory content will be reported to the police.[19] In January 2007, the *New Straits Times* (NST) newspaper and several of its executives inaugurated the first known defamation suits against bloggers. Jeff Ooi (www.jeffooi.com) and Ahirudin Attan (www.rockybru.blogspot.com), both prominent bloggers and the latter the President of the National Press Club, were sued simultaneously for both blog posts and reader comments.[20] The allegedly libelous content included Jeff Ooi's blog coverage of NST and its editors' roles in misrepresenting facts, publishing a caricature of the Prophet Muhammad, and plagiarism in blog posts in 2006.[21] Ooi had previously been investigated by the Communications and Multimedia Commission and the police concerning comments a reader posted on his blog that were deemed offensive to the official version of Islam in Malaysia.[22]

ONI testing results

Testing was conducted during October and November 2006 on two of the largest Malaysian Internet service providers (ISPs), Jaring and TMNet, and also on Macrolynx, a smaller Malaysian ISP. The tests revealed no evidence of filtering for any of the categories tested.

Conclusion

Malaysia retains strict control over traditional broadcast and print media through a broad web of vaguely worded regulations.[23] To encourage growth of Internet and new media technologies and commerce in Malaysia, however, the state has promised Internet companies that it will not censor the Internet. ONI's testing revealed no evidence of technological Internet filtering. This does not necessarily mean, however, that the Internet environment in Malaysia is free of government influence and control. Bloggers and independent online news publishers report being investigated and harassed by police on several occasions for posting allegedly offensive or seditious content, and the state media frequently run articles and opinion pieces questioning whether the Internet should be subject to tighter state controls.

NOTES

1. See Reporters Without Borders, Malaysia: 2004 Annual Report, http://www.rsf.org/article.php3?id_article=10201.
2. Ibid.
3. See Steven Gan, "Yes, another police report," August 11, 2006, http://www.malaysiakini.com/editorials/55265; see also South East Asian Press Alliance, "World publishers and editors back Malaysiakini's non-disclosure policy," January 30, 2006, http://www.seapabkk.org/news/malaysia/20030130.html.
4. See http://www.msc.com.my/msc/msc.asp.

5. International Telecommunication Union, *World Telecommunication Indicators 2006*; Paul Budde Communication Pty Ltd., Asia: Internet, March 5, 2006, p. 3.
6. Communications and Multimedia Content Forum of Malaysia, "Broadband in Malaysia: More supply than demand?" 2006, http://www.cmcf.org.my/HTML/cmcf_industry_watch-12.asp; The Communications and Multimedia Content Forum of Malaysia, "Convergence," 2006, http://www.cmcf.org.my/HTML/cmcf_industry_watch_3.asp.
7. Paul Budde Communication Pty Ltd., Malaysia: Broadband Market, July 30, 2006, p. 1.
8. Ibid.
9. International Telecommunication Union, *World Telecommunication Indicators 2006*.
10. Constitution of Malaysia, Article 10.
11. See The Communications and Multimedia Content Forum of Malaysia, http://www.cmcf.org.my/.
12. Ibid.
13. Malaysian Communications Multimedia Act of 1998, §211(1).
14. Ibid., §233.
15. See The Communications and Multimedia Content Forum of Malaysia, http://www.cmcf.org.my/.
16. See MSC Malaysia National Rollout, http://www.msc.com.my/msc/rollout_status.asp.
17. See, for example, *Star Online,* "Government looking at gaps in printing Act," July 27, 2006, http://www.thestar.com.my/news/story.asp?file=/2006/7/27/nation/14961817&sec=nation ("The Government will study if the Printing Presses and Publications Act should be amended to include the electronic media and the Internet media").
18. *Malaysia General News,* "Gov't won't censor Internet bloggers but they must be responsible, says PM," January 23, 2007.
19. Reuters, "Malaysian leaders carry quarrel into cyberspace," August 11, 2006.
20. *South China Morning Post,* "Newspaper sues Internet bloggers for defamation," January 19, 2007, reprinted at http://www.asiamedia.ucla.edu/article.asp?parentid=61629.
21. See Jeff Ooi's blog *Screenshots* at http://jeffooi.com/.
22. See Ethan Zuckerman, "Global voices blogger Jeff Ooi questioned in Malaysia regarding Weblog," Global Voices, February 28, 2005, http://www.globalvoicesonline.org/2005/02/28/global-voices-blogger-jeff-ooi-questioned-in-malaysia-regarding-weblog-post/.
23. See, for example, http://www.kempen.gov.my/coci/mypress.htm.

Moldova

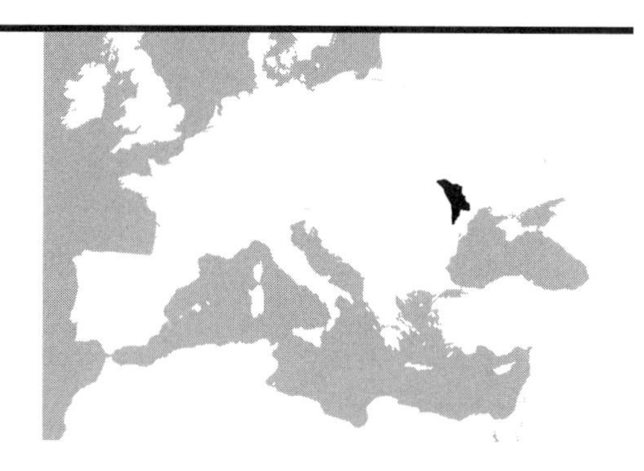

Internet users in Moldova enjoy largely unfettered access despite the government's restrictive and increasingly authoritarian tendencies overall. Development of the Internet has been rapid, propelled by a national ICT strategy that is harmonized with the European Union as well as the large diaspora population for whom telecommunications and the Internet are important channels of communication, and, possibly, for the transfer of remittances. Although filtering does not occur at the backbone level, both filtering and surveillance occur at the places where most Moldavians access the Internet: cybercafés and workplaces. Moldovan security forces have developed the capacity to monitor the Internet, and national legislation concerning "illegal activities" is strict.

Background

In the early 1990s, as a newly sovereign state, Moldova experienced both political and economical turmoil. Separatist movements erupted in two regions: Gagauzia, which later obtained autonomy, and the unrecognized breakaway state of Transdniester. The Transdniester region operates as an independent (albeit unrecognized) state with separate telecommunications and broadcasting networks.

Moldova has one of the lowest Internet development levels in Eastern Europe, and ranks 109th worldwide on the U.N. Global E-readiness Survey of 2005.[1] Yet the government has prioritized information communications technology (ICT) as means for national development and adopted a National ICT Strategy designed to align the sector with EU norms and standards via the EU-sponsored Electronic South Eastern Europe initiative (eSEE). Certain aspects of "e-government," such as the state registration

RESULTS AT A GLANCE

Filtering	No evidence of filtering	Suspected filtering	Selective filtering	Substantial filtering	Pervasive filtering
Political	●				
Social	●				
Conflict/security	●				
Internet tools	●				

Other factors	Low	Medium	High	Not applicable
Transparency				●
Consistency				●

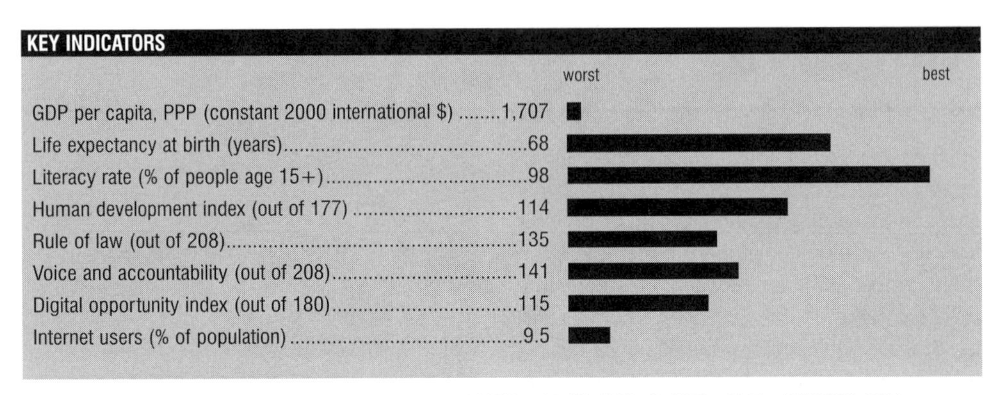

Source (by indicator): World Bank 2005, 2006a, 2006a; UNDP 2006; World Bank 2006c, 2006c; ITU 2006, 2004

database (registru.md) that acts as a central portal for all government department and services, are highly advanced and have been used as a model for other Commonwealth of Independent States (CIS) countries. Some human rights groups have voiced concerns that the database is too comprehensive and lacks oversight. Given that the legal basis for protecting citizens' rights to privacy is not yet defined, the information held in the database represents a risk for unwarranted (and unprecedented) surveillance. The president, a former internal ministry general, supports the register, which is not surprising given that it was originally developed within the Ministry of Internal Affairs. The telecommunications sector in Moldova is formally liberalized, but the government has faced problems privatizing the main operator.

Internet in Moldova

Internet use in Moldova has tripled since 2002 and penetration currently stands at 10 percent of the population.[2] However, development is constrained by a lack of quality infrastructure, low affordability, and the slow development of the telecommunications sector. A national survey indicates that 24.1 percent of the population claim the Internet is very expensive and difficult to afford.[3] Ownership of personal computers is low, with only 3 percent penetration (as of 2004). Nearly half of users access the Internet from their place of work, 33.6 percent use Internet at home, and 8.1 percent use public access points.[4]

Moldova has seven tier-one providers: Globnet, Moldtelecom, Telemedia, MDL.NET (MegaDat), Dynamic Network Technologies (DNT), Relsoft, and Riscom. A further eleven Internet service providers (ISPs) provide access to all regions of the country. International Internet traffic is routed by way of providers in Europe. The telecommunications market is dominated by Moldtelecom, which retains its near monopoly position in the market. Most of the other ISPs rent infrastructure from Moldtelecom. All ISPs exchange traffic via an Internet exchange point located at Moldtelecom.

Fixed-line and mobile teledensity remain underdeveloped, as do Internet and broadband penetration; however, all have recorded solid growth.[5] According to the national telecom regulator, dialup connections in the first nine months of 2006 rose by 88.9 percent (375,500), while broadband connections in this period tripled to 16,900. International Internet bandwidth in the country is 410 Mb/s for 2005.[6] In 2006, mobile phone ownership jumped to 32.2 percent. There are more than fourteen operators providing Voice-over Internet Protocol (VoIP) services on the international voice market, although Moldtelecom has retained the largest share.[7] Operators need to obtain a license in order to offer IP services.

Over 3,000 domain names are registered in the country code top-level domains (".md").[8] The most popular languages accessed by Internet users are Romanian, Russian, and English. The most-visited local Web site is the news agency site Azi (www.azi.md). The most-used search engines are www.ournet.md, www.super.md, and www.mail.ru.

Legal and regulatory frameworks

To meet requirements set by the World Trade Organization, the telecommunications market was liberalized on January 1, 2004. The main operator decreased its tariffs on average by 25 percent, allowing other providers into the market.[9] However, low computer penetration rates and inconsistent government policy remain major impediments to Internet growth. The state has officially committed to developing Moldova's information society, including promoting e-governance, although certain policies undermine with these objectives. The main telecommunications operator and top-tier ISP in the country, Moldtelecom, remains under state control despite large-scale criticism. ISPs rent access from Moldtelecom's well-developed infrastructure, which increases their costs and diminishes their competitiveness.

The ISPs are licensed by the National Agency for Telecommunications and Information Regulation (NATIR),[10] the main telecommunications regulator in Moldova. The law and corresponding regulation do not require special requirements for receiving a license. NATIR is responsible for issuing and suspending licenses, establishing license fees, and enforcing sanctions where necessary. In addition it regulates the management of the country's highest-level Internet domain (".md"). NATIR was established with an amendment to the Law on Telecommunications, which introduced a licensing regime for most Internet and telecommunications services. A new law on e-communications entered the parliament in 2006 as part of an effort to harmonize national legislation with European standards. The drafted law envisions broader rights for the final user and wider access to public networks, and provides for more-efficient market liberalization. The draft law also seeks to establish a new independent body to regulate telecommunications.

The Supreme Security Council (SSC), which oversees implementation of the president's decrees related to national security, monitors ministries' and state agencies' various activities to ensure national security. The SSC Ministry of Information Development carries out government policies related to information and communications and encourages collaboration between state and private organizations. The National Security and Information Service is empowered with broad authority to monitor and gather information on Internet usage.

ONI testing results

ONI carried out testing on three of the tier one ISPs: Globnet, Moldtelecom, and Telemedia. The tests revealed no evidence of filtering for any of the categories of content tested.

ONI research determined that some ISPs perform a differentiated multilevel filtering as a means of protection against various network attacks and spam or viruses. There was, however, no evidence of ISP-level filtering based upon sensitive political or social content. More commonly this type of filtering occurs at the level of business workplaces and cybercafés. ONI researchers performed a survey among more than 600 businesses to determine the level of filtering at work enterprises.[11] The results indicate that filtering or surveillance of Internet exists in all types of businesses. Some practice "sanitized" access to Internet, where employees can access only a limited number of sites directly related to the work they perform. Other enterprises allow employees to access the Internet, but filter out sexual, "harmful," and "entertainment" content.

In cybercafés access is limited more by surveillance than by direct filtering. Specific content is prohibited and, if accessed, the user is fined. Approximately 56 percent of cybercafés' administrators surveyed by ONI admitted to filtering and surveillance activities. Other administrators stated that they noted that some Web sites are inaccessible, but would not confirm that they used any specific filtering system in the cybercafé itself.

Conclusion

Despite increasingly authoritarian tendencies, the Internet in Moldova remains largely unaffected by filtering, at least at the backbone level. At "edge" locations, such as cybercafés and some enterprises, ONI research revealed filtering that restricted access to certain content and services. Given that over half (55 percent) of all Moldovans access the Internet through their workplace or cybercafés, this form of filtering has a significant impact on the way in which Moldovans "experience" the Internet. ONI research also suggests that Moldovan security forces have developed mechanisms to monitor Internet content. Given a relatively underdeveloped legislative base protecting citizen's rights and privacy, there are few checks and balances in place to prevent authorities from taking a more aggressive stance on policing Internet content.

NOTES

1. Department of Economic and Social Affairs, U.N. Global E-Government Readiness Report 2005, http://unpan1.un.org/intradoc/groups/public/documents/un/unpan021888.pdf.
2. International Telecommunication Union, *World Telecommunication Indicators 2006.*
3. A study conducted by the Centre of Sociological Politological and Psychological Investigation and Analysis CIVIS, 2005.
4. Centre of Sociological Politological and Psychological Investigation and Analysis CIVIS, 2005.
5. See Paul Budde Communication Pty Ltd., Moldova: Telecoms Market Overview & Statistics, April 2, 2006.
6. International Telecommunication Union, *World Telecommunication Indicators 2006.*
7. See Paul Budde Communication Pty Ltd., Moldova: Telecoms Market Overview & Statistics, April 2, 2006, p. 6.
8. Super.md, http://www.super.md.
9. See Paul Budde Communication Pty Ltd., Moldova: Telecoms Market Overview & Statistics, April 2, 2006.
10. Article 8 of the Law on Licensing Certain Types of Activities, http://www.anrti.md/en/acte/Leg_licente.htm#cap8 (last accessed April 30, 2007). The Law on Telecommunications and the Regulation of Licensing of Telecommunications and Informatics No. 5 of 2002 are also relevant to the licensing regime of IT services.
11. Less than half of the interviewed enterprises are connected to the Internet, and the majority of these have no more than two computers with access to the Net. The number of employees allowed to use computers with access is limited to nine in companies of medium size (from 50 to 249 employees). The use of Internet is three times more intensive in urban enterprises than rural ones. Dialup access is used by 52.6 percent of businesses, with 39.7 percent using broadband technologies.

Morocco

Internet access in Morocco is, for the most part, open and unrestricted. ONI testing revealed that Morocco filters only a small number of Web sites, mainly pro-Western Sahara independence sites. A small number of Weblog servers and anonymizers were also found to be blocked. The filtration regime is not comprehensive—similar content can be found on other Web sites that are not blocked.

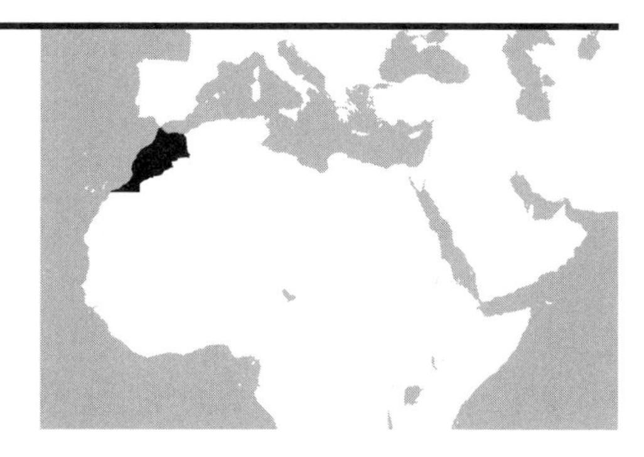

Background

Morocco faces two major issues that inform its actions regarding the press and human rights in general: the status of Western Sahara and terrorism. As to the first issue, Morocco has vied with the Polisario Front for control of Western Sahara ever since Spain pulled out of the region in 1976.[1] Morocco asserts a historical claim on the region,[2] while the Polisario Front asserts the right of self-determination.[3] After decades of fighting, both sides agreed to a UN-sponsored ceasefire in 1991 that required an eventual referendum on independence in the region.[4] As of yet, this referendum has not been held.[5] Despite the cease fire, reports of overzealous suppression of peaceful resistance to Moroccan rule persist.[6] Journalism on the subject has been restricted as well. In February 2006 a journalist from and the managing editor of *Le Journal Hebdomadaire* were fined 3.1 million dirhams (USD370,668) for questioning the objectivity of a study run by the European Strategic Intelligence and Security Center on the Polisario Front.[7] In October 2006 Morocco barred foreign journalists covering human rights issues from entering Moroccan-controlled Western Sahara,[8] and continued its crackdown on media coverage of the conflict by arresting and expelling a Swedish photographer found taking pictures of a pro-Polisario demonstration in Western Sahara in February 2007.[9]

RESULTS AT A GLANCE

Filtering	No evidence of filtering	Suspected filtering	Selective filtering	Substantial filtering	Pervasive filtering
Political	●				
Social	●				
Conflict/security			●		
Internet tools			●		

Other factors	Low	Medium	High	Not applicable
Transparency	●			
Consistency			●	

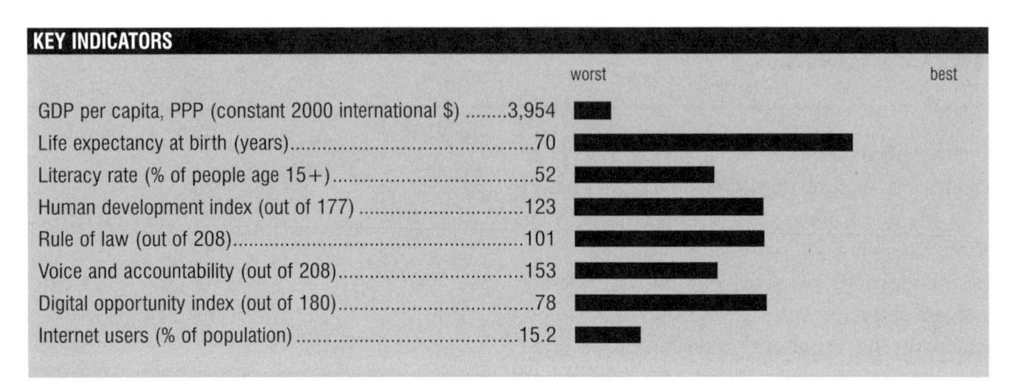

Source (by indicator): World Bank 2005, 2006a, 2006a; UNDP 2006; World Bank 2006c, 2006c; ITU 2006, 2005

As to the second issue, terrorism, Casablanca was the site of a major terrorist attack in May 2003 when suicide bombers detonated five bombs targeting a Jewish community center, a Spanish restaurant and social club, a hotel, and the Belgian consulate.[10] According to Human Rights Watch, "several hundred" suspects remain detained in connection with the incident and face "mistreatment, and sometimes torture, while under interrogation, and then convict[ion] in unfair trials."[11] An antiterrorism law passed soon after the attacks placed further restrictions on the press.[12]

Internet in Morocco

The Internet was first introduced in Morocco in 1995,[13] and the country now has one of the highest degrees of connectivity in Africa. Maroc Télécom is the largest Internet service provider (ISP) in Morocco, with an estimated market share in June 2006 in excess of 95 percent.[14] The majority of the remaining customers are with Maroc Connect, which provides an ADSL service through wholesale access agreements with Maroc Télécom.[15] Maroc Télécom offers wholesale services to other ISPs, following the Reference Access Offer approved by Morocco's telecom regulatory body, the National Telecommunications Regulatory Agency (ANRT), in October 2006.[16] In 2005 the total number of Internet users in Morocco was 4.6 million, up from 100,000 in 2000.[17] This constitutes approximately 15.1 percent of the country's population.[18] Penetration remains low in absolute terms as a result of cost, the country's low literacy rate, and a lack of infrastructure in rural areas. The mushrooming number of cybercafés, however, has expanded access from the capital city of Casablanca to Morocco's smaller cities and villages. As of mid-2006, ANRT had granted over 10,000 licenses to ISPs and cybercafés.[19] Internet service is also becoming faster and more reliable; ADSL service was launched in Morocco in 2003 and has attracted an increasing number of subscribers. The ADSL market grew by nearly 300 percent in 2005, and the number of subscribers topped 300,000 in mid-2006.[20] Morocco offers the cheapest DSL access in Africa; a monthly broadband connection starts at USD17 per month.[21]

The growing Moroccan blogosphere now includes over 10,000 blogs, mostly in French. As a result of the more widespread availability of the Internet, a Moroccan company recently launched the country's first Arabic-language blog platform.[22] Though largely free of filtration, posters generally avoid "red line" topics such as Western Sahara, defamation of royal authority, and defamation of Islam.[23]

In the last few years, the Moroccan government has reportedly begun to block access to Web sites run by fundamentalist Islamic groups and Web sites that advocate for the independence of Western Sahara.[24] This filtering regime appears to be less than comprehensive, however, as many related sites go unblocked. Furthermore, groups and individuals who are unable to express their views in the traditional media have increasingly turned to the Internet to voice their opinions.[25] However, there is growing religious pressure to block explicit material such as pornography.[26]

Legal and regulatory frameworks

The Moroccan government held a monopoly on the country's telecommunications sector until a privatization initiative that began in 1997.[27] The ANRT was founded in 1998 and has encouraged private companies to offer competitively priced services, including ISPs and cybercafés.[28] The ANRT grants licenses to companies that wish to run ISPs, but individuals who want to obtain an Internet account do not need to obtain approval.[29] The government does not yet appear to have taken action against ISPs for account holders' activities or to monitor sites accessed by account holders or customers of cybercafés.[30]

However, as mentioned earlier, Morocco is a consistent censor of the independent media, frequently fining newspapers and arresting journalists who report on human rights, politics, or Islam.[31] Current laws criminalize criticizing the monarchy or Morocco's claim to Western Sahara.[32] The antiterrorism bill that was passed following suicide bombings in Casablanca in 2003 grants the government sweeping legal power to arrest journalists or to filter Web sites that are deemed to "disrupt public order by intimidation, force, violence, fear or terror."[33] In recent years, the Moroccan government appears to be growing increasingly proactive about shutting down newspapers and imprisoning reporters; in January 2007, a newsweekly was ordered shut down for two months after publishing jokes about Islam and a reporter and editor were given three-year suspended prison sentences.[34]

Restrictions on freedom of expression reportedly have been extended to the Internet in recent years. According to Reporters Without Borders, in November 2005 the Moroccan government began blocking access to several major Web sites that support independence for the Western Sahara, a decision that was made either by the country's communications ministry or its interior ministry.[35] Shortly thereafter, Morocco reportedly cut off access to www.anonymizer.com, a Web site that allows Internet users to access banned sites from within the country. It has also been reported that the Web sites of Islamic fundamentalist organizations have been blocked, particularly those run by a fundamentalist group called the Justice and Charity Organization.[36]

ONI testing results

ONI carried out testing of Moroccan Internet service on the principal Internet provider, Maroc Telecom, and a smaller ISP, MTDS. The results of the testing found blocking of a small number of sites. Blocking was found primarily on sites promoting the independence of Western Sahara, such as www.saadasahara.com and www.sahara-occidental.com, as well as the Web sites for the Union of Sahrawi Journalists and Writers (www.upes.org), the Association of Families of Sahrawi Prisoners and Disappeared (www.afapredesa.org), and the General Trade Union of the Western Sahara Petition (www.umdraiga.com/eucoco2004/documentosytalleres/tallersindical.htm). Two blog hosting sites, www.haloscan.com and www.livejournal.com, were blocked as well. Finally, ONI also found that two anonymizer

Web sites, www.anonymizer.com and www.multi-proxy.org, were blocked.

However, a number of sites reported blocked in the past were found to be accessible. These include www.wsahara.net and www.arso.org, both pro-Western Sahara independence sites, as well as www.spsrasd.info, the Saharan press service's site, and www.aljamaa.info, Justice and Charity's site (though the title of the site is Justice and Spirituality).

Conclusion

Morocco's Internet filtration regime is relatively light and focuses on Western Saharan independence, a few blog sites, and highly visible anonymizers. The issues Morocco faces in Western Sahara's push for independence, the specter of Islamist terrorism, and the protection of the royal family and Islam from defamation have led Morocco to crack down on free speech and the press, but have not yet led it to significantly censor the Internet. As Internet users can find blocked material on other accessible sites, it is clear that Morocco's filtration regime is not comprehensive. Relative to the region, Moroccan Internet access is fairly free.

NOTES

1. BBC News, "Regions and territories: Western Sahara," February 2007, http://news.bbc.co.uk/2/hi/africa/country_profiles/3466917.stm.
2. United Nations Mission for the Referendum in Western Sahara, Western Sahara: MINURSO – Background, 2005, http://www.un.org/Depts/dpko/missions/minurso/background.html.
3. Western Sahara Online, "History," http://www.wsahara.net/history.html.
4. United Nations Mission for the Referendum in Western Sahara, Western Sahara: MINURSO – Background, 2005, http://www.un.org/Depts/dpko/missions/minurso/background.html.
5. Human Rights Watch, Morocco/Western Sahara: World Report 2007, January 2007, http://hrw.org/englishwr2k7/docs/2007/01/11/morocc14714.htm.
6. Ibid.
7. Reporters Without Borders, Morocco: Annual Report 2007, February 2007, http://www.rsf.org/article.php3?id_article=20772.
8. Ibid.
9. Reporters Without Borders, "Swedish photographer expelled from Western Sahara a day after his arrest," February 2007, http://www.rsf.org/article.php3?id_article=21059.
10. BBC News, "Terror blasts rock Casablanca," May 2003, http://news.bbc.co.uk/2/hi/africa/3035803.stm.
11. Human Rights Watch, Morocco/Western Sahara: World Report 2007, January 2007, http://hrw.org/englishwr2k7/docs/2007/01/11/morocc14714.htm.
12. Privacy International, "Internet censorship report 2003: Morocco," November 2004, http://africa.rights.apc.org/?apc=s21807e_1&x=28046.
13. United Nations Economic Committee for Africa, Morocco: Internet Connectivity, http://www.uneca.org/aisi/nici/country_profiles/Morocco/morocinter.htm.
14. Frontier Economics Ltd., London. Country Analysis 2007: A Report Prepared for the NATP II, January 2007, http://www.natp2.org/midtermnews/Country%20analysis%202007.pdf.
15. Ibid.
16. Ibid.
17. Internet World Stats, Internet Usage Statistics for Africa, http://www.internetworldstats.com/stats1.htm.
18. Ibid.
19. afrol News, "Moroccan ADSL market grows by 300%," November 7, 2006, http://www.afrol.com/articles/22425.
20. Ibid.
21. "Two major North African markets look set to become legal VoIP pioneers, says new report," *Al-Bawaba,* June 6, 2006.
22. Adam Mahdi, "Blogs becoming increasingly popular in Morocco," *Magharebia,* September 11, 2006, http://www.magharebia.com/cocoon/awi/xhtml1/en_GB/features/awi/features/2006/09/11/feature-01.
23. Human Rights Watch, The Internet in the Mideast and North Africa: Free Expression and Censorship – Morocco, June 1999, http://hrw.org/advocacy/internet/mena/morocco.htm.
24. Reporters Without Borders, Morocco: Annual Report 2007, February 2007, http://www.rsf.org/article.php3?id_article=20772; and Human Rights Watch, Morocco/Western Sahara: World Report 2007, January 2007, http://hrw.org/englishwr2k7/docs/2007/01/11/morocc14714.htm.
25. The Initiative for an Open Arab Internet, Implacable Adversaries: Arab Government and the Internet (2006): Morocco, http://www.openarab.net/en/reports/net2006/morocco.shtml.
26. Ibid.

27. United Nations Economic Commission for Africa, Morocco: NICI Policy, http://www.uneca.org/aisi/nici/country_profiles/Morocco/morocpol.htm.
28. ANRT, "L'ANRT en bref," 2006, http://www.anrt.net.ma/fr/.
29. Privacy International, "Internet censorship report 2003: Morocco," November 2004, http://africa.rights.apc.org/?apc=s21807e_1&x=28046.
30. Ibid.
31. Reporters Without Borders, Morocco: Annual Report 2007, February 2007, http://www.rsf.org/article.php3?id_article=20772.
32. Committee to Protect Journalists, Attacks on the Press in 2006: Morocco, http://www.cpj.org/attacks06/mideast06/mor06.html.
33. Human Rights Watch, Background: The State of Human Rights in Morocco, November 2005, http://hrw.org/reports/2005/morocco1105/4.htm; Mohammad Ibahrine, "Morocco: Internet making censorship obsolete," *Arab Reform Bulletin,* 3 (7): September 2005, http://www.carnegieendowment.org/files/ibrahine1.pdf.
34. Reporters Without Borders, "Three-year suspended sentences for editor and reporters called 'outrageous,'" January 15, 2007, http://www.rsf.org/article.php3?id_article=20414.
35. Reporters Without Borders, "Access to Sahrawi sites blocked within Morocco," December 2, 2005, at http://www.rsf.org/article.php3?id_article=15809.
36. Privacy International, "Internet censorship report 2003: Morocco," November 2004, http://africa.rights.apc.org/?apc=s21807e_1&x=28046.

Myanmar (Burma)

Myanmar's authoritarian military junta is slowly expanding access to the Internet while maintaining one of the world's most restrictive systems of control. Despite the fact that less than 1 percent of Myanmar's population access the Internet, the government has targeted online independent media and dissent with the same commitment it has demonstrated to stifling traditional media and voices for reform.

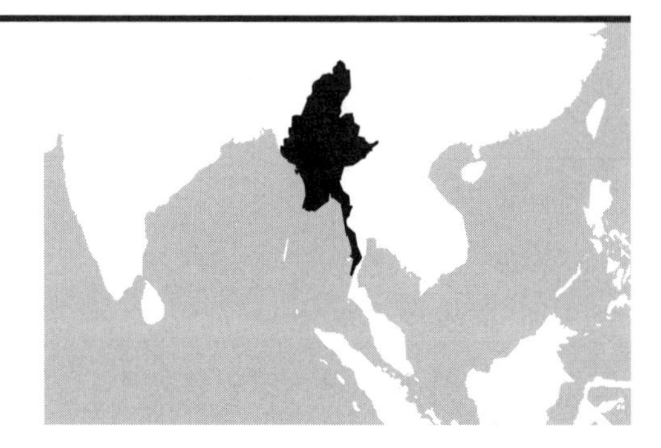

Background

Myanmar's abysmal human rights record worsened in 2006,[1] prompting increased pressure from the United States, the EU, and ASEAN for reform. In September the U.N. Security Council approved the U.S. government's proposal to put Myanmar formally on the Council's agenda.[2] Leaders from the State Peace and Development Council (SPDC) claim neocolonialists are infiltrating media technology on pretexts of protecting human rights and countering drug trafficking.[3] Other sensitive issues included political and constitutional reform, separatist movements, religious and ethnic minorities, forced and child labor, access by humanitarian organizations, and the country's first disclosed outbreak of bird flu. The government suppressed reports on a wide range of additional issues, from rising cement and fuel prices to restrictions on private banks,[4] and jailed two journalists who photographed the new, remote capital at Pyinmana.[5]

Internet in Myanmar

The reported number of Internet users in 2005 ranged from 78,000 to nearly 300,000, at the upper limit representing approximately 0.56 percent of Myanmar's population.[6] Myanmar remains one of thirty countries with less than 1 percent Internet penetration.[7] Most users access the Internet in cybercafés (starting at USD0.30

RESULTS AT A GLANCE

Filtering	No evidence of filtering	Suspected filtering	Selective filtering	Substantial filtering	Pervasive filtering
Political					●
Social				●	
Conflict/security				●	
Internet tools				●	

Other factors	Low	Medium	High	Not applicable
Transparency		●		
Consistency	●			

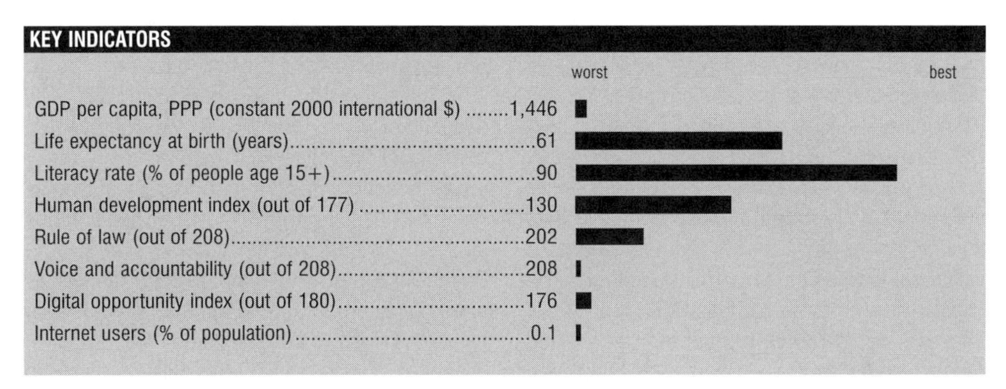

Source (by indicator): IMF 2006; World Bank 2006a, 2006a; UNDP 2006; World Bank 2006c, 2006c; ITU 2006, 2004

per hour, down from USD0.75 in 2004 and USD0.95–1.50 in 2003),[8] which are said to be present in five cities but are planned to reach 324 townships within three years.[9] Connection speeds are slow, however, as broadband is available primarily to government and businesses and used mostly for Internet telephony via Voice-over Internet Protocol (VoIP), though the government pledged to bring ADSL to every township by the end of 2006.[10] There are only two Internet service providers (ISPs) allowed in Myanmar: state-owned telecom Myanmar Posts and Telecom (MPT), which is the only source of new Internet services,[11] and Myanmar Teleport (MMT, formerly Bagan Cybertech), which is reportedly the infrastructure arm of Myanmar's Internet system and responsible for blocking content. In September 2005 the Ahaed Co. of Myanmar and the Canadian ICT company Teleglobe reportedly signed a memorandum of understanding to establish a private ISP.[12] Infrastructural reliability is also an issue: in May 2006 the entire country was disconnected for four days because of alleged damage to an undersea cable.[13]

Legal and regulatory frameworks

Myanmar heavily regulates online access and content via legal, regulatory, and economic constraints. As in other areas, however, the state's policies are difficult to assess because they are rarely published or explained.

Network-ready computers must be registered (for a fee) with the MPT; failure to do so can result in fines and prison sentences of seven to fifteen years.[14] Sharing registered Internet connections is also punishable by revocation of access and presumably similar "legal action."[15] Broad laws and regulations confer power upon the SPDC, which is also involved in all judicial appointments,[16] to punish citizens harshly for any activity deemed detrimental to national interests or security. Regulations issued in 2000 subjected online content to the same kind of strict filtering that the Press Scrutiny and Registration Division carries out (despite print media being almost exclusively state owned):[17] users must obtain MPT permission before creating Web pages, and they cannot post anything "detrimental" to the government or simply related to politics. The MPT can "amend and change regulations on the use of the Internet without prior notice."[18]

Costs indeed limit access significantly: even households that can afford a PC and long-distance connection fees outside the capital Yangon (Rangoon) and Mandalay cannot pay USD35/month[19] for a broadband account. Dialup access leaves them with state-monitored e-mail (free services are blocked)[20] and a small collection of pre-approved sites on the country's intranet, known as the Myanmar Wide Web.[21] As for cybercafés, promoted since 2002 by a "Public Access Centers" (PAC) program for e-mail and gaming purposes,[22] the government has been urging business owners to legally register as PACs. This requires them to log user identities and Web sites visited and send the information back to the state-owned Myanmar Info-tech.[23] There are reports, however, that many tech-savvy users risk connecting to proxy servers abroad and thereby access the entire Web undetected.[24]

ONI testing results

Testing was conducted on the two ISPs in Myanmar, Myanmar Teleport (MMT) and Myanmar Posts and Telecom (MPT). Both MMT and MPT filtered extensively and focused overwhelmingly on independent media, political reform, and human rights sites relating to Myanmar, as well as free Web-based e-mail services and circumvention tools.

Both ISPs blocked roughly the same number of circumvention tools, including Proxify, Guardster, and Anonymizer (although only MPT blocked www.anonymizer.com).

In June 2006 Gmail and Gtalk were made inaccessible and Skype was banned[25]—a reported attempt not only to censor communications but also to preserve the government's monopoly over telephone and e-mail services as MPT's revenues dipped.[26] ONI testing confirmed that although no search engines (MSN, Google, and so on) were blocked, Yahoo! Mail, Gmail, Hushmail, and mail2web were blocked by both ISPs, while MPT took the precaution of blocking thirteen additional e-mail sites, including Hotmail and Fastmail. Only MPT blocked Skype.

In addition to filtering Radio Free Asia (www.rfa.org) and OhmyNews (www.ohmynews.com), both MMT and MPT blocked many major independent news sites reporting on Myanmar. This included English language publications such as the *Irawaddy, Mizzima News,* and *BurmaNet News* (www.burmanet.org), as well as sites in the national language (www.burmatoday.net). Only MPT blocked the Voice of America Web sites (www.voanews.com) in English and Burmese, while MMT targeted regional news sites such as the Times of India and Asia Observer.

Sites containing content on human rights advocacy and democratic reform continued to be a priority for blocking. A number of nongovernmental organization (NGO) sites with different levels of involvement in Myanmar human rights issues were blocked (Open Society Institute at www.soros.org; www.burmacampaign.org.uk). Within this group were Web sites documenting the persecution of ethnic minorities and the personal Web site of Daw Aung San Suu Kyi. Other continuities in blocking included coalitions for democratic change in Myanmar, such as the Web site of the coalition government of the Union of Burma (www.ncgub.net), opposition movements (www.chinforum.org), and rights groups (www.womenofburma.org).

There were significant differences in filtering between the two ISPs. Of the sites found to be blocked in Myanmar, less than a third were blocked on both ISPs. The remaining blocked sites were blocked on one ISP or the other, but not both. MMT blocked almost exclusively sites with ties to Myanmar, where the term "Burma" in the URL was one of the common threads among the filtered sites, from human rights groups (www.burmawatch.org; www.hrw.org) critical of the government to peripheral personal sites (such as a site with photographs of Myanmar). MPT filtered many more sites from the global list, blocking a large majority of the pornography Web sites tested, while MMT filtered very few such sites.

Several curious results indicated that the Myanmar government does not take an entirely systematic approach to filtering. For example, Amnesty International (www.amnesty.org) was blocked entirely on MPT, but MMT filtered only several Amnesty reports on the country. Other significant variations among the ISPs, including the inconsistent blocking of pornography and gambling sites that suggest distinct filtering methods, are unusual given both ISPs are state-run.

Conclusion

Although Myanmar does not deploy its filtering regime with the same sophistication and breadth as other countries with similarly repressive online environments, the paranoid grip of the SPDC is felt in the restrictions on access, the high cost of services, and the frequently brutal clampdown on information and expression in all other spheres of Burmese life. This may be why there are not many known cases of cyber-dissidents in custody, given that people have been arrested for anything from publishing subversive poetry to listening to the BBC or Radio Free Asia in public.[27]

NOTES

1. U.S. Department of State, Country Reports on Human Rights Practices 2005: Burma, http://www.state.gov/g/drl/rls/hrrpt/2005/61603.htm.
2. Summary statement by the Secretary-General of the United Nations on matters of which the Security Council is seized and on the stage reached in their consideration, UN Doc. S/2006/10/Add.36, September 22, 2006.
3. Gen. Than Shwe, 85th Anniversary National Day Message, November 24, 2005, http://www.mofa.gov.mm/news/24nov05.html.
4. *Financial Times,* "Burma's privately owned presses are on a roll; Private sector journals are gaining popularity in spite of heavy pressure from state censors," December 8, 2005.
5. Reporters Without Borders, "Court upholds three-year sentences for journalists who photographed new capital," June 27, 2006, http://www.rsf.org/article.php3?id_article=16898.
6. Xinhua News Agency, "Internet users in Myanmar number nearly 300,000," November 8, 2006; International Telecommunication Union, *World Telecommunication Indicators 2006*.
7. International Telecommunication Union, ICT Statistics, http://www.itu.int/ITU-D/ict/statistics/ict/index.html.
8. *BBC Monitoring International Reports,* "Burma Internet users use proxy servers to visit blocked websites," October 17, 2006 (includes text from Ko Thet, "A hole in the Net," *The Irawaddy,* October 1, 2006); *The Guardian Online,* "The great firewall of Burma," July 22, 2003; and Reporters Without Borders, Internet: Burma, http://www.rsf.org/article.php3?id_article=10748&Valider=OK.
9. Xinhua News Agency, "Internet users in Myanmar number nearly 300,000," November 8, 2006.
10. Ibid.
11. Xinhua News Service, "Myanmar to grant foreign, local engagement in emerging cyber city," November 28, 2006.
12. http://www.burmanet.org/news/2005/09/12/xinhua-news-agency-myanmar-to-expand-internet-services/.
13. *BBC Monitoring International Reports,* "Burma's Internet link with outside world fails for fourth day" (text by official Chinese agency Xinhua), May 16, 2006.
14. Computer Science Development Law, sections 27, 28, September 20, 1996, http://www.myanmar.com/gov/laws/computerlaw.html.
15. Digital Freedom Network, "The new Net regulations in Burma," January 31, 2000, archived copy available at http://web.archive.org/web/20010220220441/http://dfn.org/voices/burma/webregulations.htm.
16. Ibid.
17. International Press Institute, 2005 World Press Freedom Review: Burma, http://www.freemedia.at/cms/ipi/freedom_detail.html?country=/KW0001/KW0005/KW0112/.
18. Digital Freedom Network, "The new Net regulations in Burma," January 31, 2000, archived copy available at http://web.archive.org/web/20010220220441/http://dfn.org/voices/burma/webregulations.htm.
19. BaganNet, "Access Services," October 30,2006, http://www.bagan.net.mm/products/access/broadband_ADSL.asp.
20. BaganNet, "About mail4u," October 2006, http://www.bagan.net.mm/products/services/aboutmail4u-e.asp.

21. Shawn W. Crispin, "A quantum leap in censorship," Asia Times Online, September 22, 2006, http://www.atimes.com/atimes/Southeast_Asia/HI22Ae01.html.
22. *BBC Monitoring International Reports,* "Burma Internet users use proxy servers to visit blocked websites," October 17, 2006 (includes text from Ko Thet, "A hole in the Net," *The Irawaddy,* October 1, 2006).
23. *The Myanmar Times,* "Burma enforces licensing of Internet cafes," March 20, 2006, (text of report in English by Khin Hninn Phyu, reprinted by the BBC) at http://www.burmanet.org/news/2006/03/31/myanmar-times-via-bbc-burma-enforces-licensing-of-internet-cafes/.
24. Democratic Voice of Burma, "Press freedom in Burma," May 7, 2006, http://english.dvb.no/news.php?id=7010; and Shawn L. Nance, "How to fool the cyber spooks," The Irrawaddy Online, March 27, 2005, http://www.irrawaddy.org/aviewer.asp?a=4504&z=104 (inset).
25. *Indo-Asian News Service,* "Google, Gmail banned in Myanmar: Surfers," June 30, 2006.
26. Reporters Without Borders, "Internet increasingly resembles an Intranet as foreign services blocked," July 4, 2006, http://www.rsf.org/article.php3?id_article=18202; *The Irrawaddy,* "Junta blocks Google and Gmail," June 30, 2006.
27. See, for example, *Mizzima News,* "Four dissidents sentenced up to 19 years in prison for anti-government poems," June 21, 2006, http://www.mizzima.com/MizzimaNews/AlertBurma/22-June-2006-02.html.

Nepal

Extremely unstable political conditions in Nepal have at times led to harassment of journalists and censorship of traditional media. In 2005 conditions deteriorated to the point where a week-long national media and Internet blackout was imposed. However, Nepal does not filter the Internet on an ongoing basis.

Background

Nepal is among the world's least-developed countries. It has endured extreme political instability in recent years because of its transition from absolute monarchy to democracy and because of its years of struggle between the state and militant Maoist insurgents, who control large portions of the countryside. Nepal was under the rule of an absolute monarch until 1990, when popular pressure forced the king to transition to a democratic system of parliamentary monarchy.[1] Since then, internal governmental collapse and parliamentary dissolution have been common occurrences.[2] During periods of extreme political volatility, the state has clamped down on the press and free expression. In 2005, citing deteriorating security conditions in Nepal from Maoist violence, the king imposed authoritarian rule and a week-long media blackout, during which the country was cut off from the Internet.[3] The state and Maoist rebels both have a history of harassing journalists and repressing media coverage.[4] Nevertheless, with the exception of King Gyanendra's authoritarian rule in 2005–06, Nepal has experienced tremendous growth of a "vibrant" and largely free independent media since parliament was established in 1990.[5]

Internet in Nepal

Although through 2005 less than 1 percent of Nepal's population of twenty-three million used the Internet, the Internet market in Nepal is

RESULTS AT A GLANCE

Filtering	No evidence of filtering	Suspected filtering	Selective filtering	Substantial filtering	Pervasive filtering
Political	●				
Social	●				
Conflict/security	●				
Internet tools	●				

Other factors	Low	Medium	High	Not applicable
Transparency				●
Consistency				●

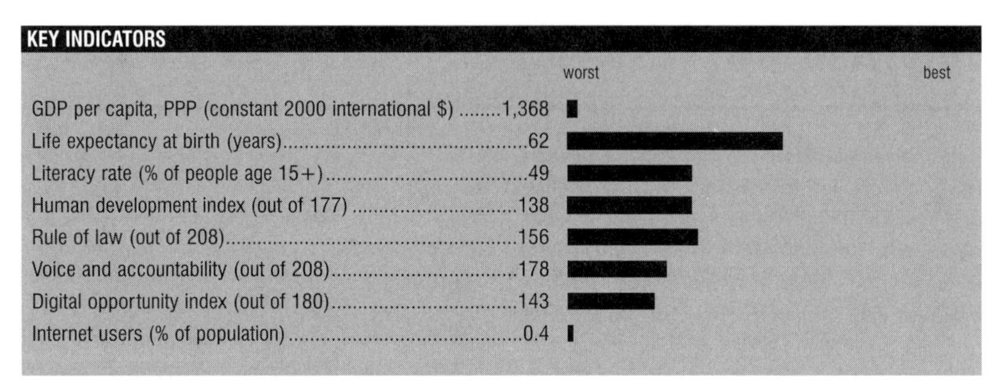

Source (by indicator): World Bank 2005, 2006a, 2006a; UNDP 2006; World Bank 2006c, 2006c; ITU 2006, 2005

growing rapidly—the result of a competitive Internet service provider (ISP) market and low Internet access prices.[6] Thirty-one private ISPs offer Internet access to businesses and consumers, though two, Worldlink and Mercantile, dominate the market with a combined share of more than 70 percent.[7] Cybercafés are important sources of Internet access for Nepalis; the country is believed to have the highest concentration of cybercafés in the world.[8] Much of Nepal's Internet access is concentrated in the more-developed Katmandu Valley region, as the mountainous terrain and low income in remote regions of the country make access more difficult. However, one effort to bring Internet access to rural populations—the Nepali Wireless Networking Project—has already wirelessly connected seven remote mountain villages to the Internet, with plans to network twenty-one villages in all.[9]

Although relatively few Nepalis presently get their news from the Internet, it has nevertheless become an important source of independent news in Nepal.[10] When King Gyanendra assumed authoritarian control in 2005, for example, traditional media were either shut down or heavily censored to ensure the publication of only favorable news about the monarch.[11] Nepali bloggers became an important political voice and source of information to the world about the situation unfolding inside the country.[12]

Legal and regulatory frameworks

Nepal's legal system is in flux because of its unstable political landscape and its new constitution. The most recent collapse occurred in February 2005, when the king assumed control of the government and armed forces.[13] Mass civilian protests followed, and he was forced to reinstate parliament and ultimately relinquish all official powers to the prime minister and parliament.[14] The king sought to stifle the independent media during his tenure, passing the repressive Media Law, which prohibited criticism of the king and royal family and the broadcast of news over independent FM radio stations (an important source of independent news in the country). The Media Law also increased the penalties for defamation tenfold. The law was repealed once parliament was reinstated.[15]

In December 2006, seven political parties and the Maoists agreed on a new interim constitution that paves the way for the Maoists to join the political mainstream and nationalizes royal properties,[16] leaving the fate of the monarchy up to a general election.[17] The interim constitution guarantees certain social freedoms including freedom of speech and expression, freedom to protest, and freedom to establish a political party, among others.[18] The constitution also guarantees the freedom to publish, including a specifically enumerated freedom to publish on the Internet.[19] It advises, however, that those who publish information that causes social disruption or disparages others may be subject to punishment under relevant laws.[20]

One such law is likely the Electronic Transaction and Digital Signature Act of 2004 (ETDSA), which regulates online commerce and financial transactions and criminalizes certain online behavior, including hacking and fraud. ETDSA also provides criminal penalties, including fines and up to five years in prison, for the publication of "illegal" content on the Internet (though it provides no definition of illegal content), or for the publication of hate speech or speech likely to trigger ethnic strife.[21] Similarly, the National Broadcasting Act of 1993 and the National Broadcasting Regulation of 1995 provide for fines and/or imprisonment for broadcasting content likely to cause ethnic strife or social unrest, undermine national security or moral decency, or conflict with Nepali foreign policy.[22]

However, the extent to which any previously existing laws will retain their force under the new government is unclear.

ONI testing results

Testing was conducted from October 2006 through January 2007 on six Nepali ISPs: Worldlink, Everest, Mercantile, Nepal Telecom, Speedcast, and Websurfer. The tests revealed no evidence of filtering for any of the categories tested.

Conclusion

Ongoing political instability remains a constant threat to independent media in Nepal, as there is a history of insurgents and the state harassing journalists and clamping down on media freedoms during times of political tension. In 2006 Nepal emerged from a particularly repressive period: the king's authoritarian rule was abolished, parliament was reinstated, and a new interim constitution was put into effect guaranteeing freedom of expression and of the press. These freedoms do not, however, extend to speech that is likely to incite social unrest or disparage others, which are sensitive issues for the state because of the ethnic and socioeconomic strife underlying the struggle with the Maoists. At present, however, Nepali journalists report virtually unconditional freedom of the press, including the Internet, and ONI's testing revealed no evidence that Nepal imposes technological filters on the Internet.

NOTES

1. See Maya Chadda, *Building Democracy in South Asia: India, Nepal, Pakistan,* Boulder, CO: Lynne Reinner Publishers (2000), pp. 113–20.
2. See generally John Whelpton, *A History of Nepal,* Cambridge University Press (2005), pp. 208–24.
3. BBC News, "Q&A: Nepal's future," November 8, 2006, http://news.bbc.co.uk/1/hi/world/south_asia/2707107.stm; see Mark Glaser, "Nepalese bloggers, journalists defy media clampdown by king," Online Journalism Review, February 23, 2005, http://www.ojr.org/ojr/stories/050223glaser/.
4. See, for example, Reporters Without Borders, "Maoists and government urged to respect press freedom undertakings," August 22, 2006, http://www.rsf.org/article.php3?id_article=18632; International Freedom of Expression Exchange, "Parliament abolishes repressive media law," http://www.ifex.org/fr/content/view/full/74580/; Committee to Protect Journalists, Press Release, "Over 200 journalists arrested, 31 in custody," April 20, 2006, http://peacejournalism.com/ReadArticle.asp?ArticleID=8550.
5. See, for example, World Association of Newspapers, "Founding father of independent media in Nepal remains hopeful, despite continued restrictions," June 2005, http://www.wan-press.org/article7574.html; BBC Online, "Nepal protests against media law," November 15, 2005, http://news.bbc.co.uk/2/hi/south_asia/4432882.stm.

6. Paul Budde Communication Pty Ltd., Nepal, Telecoms Market Overview and Statistics, July 30, 2006, p. 11.
7. Ibid., pp. 1, 14.
8. Ibid., p. 12.
9. Ibid., p. 13.
10. See Vincent Lim, "Blogging for democracy in Nepal," AsiaMedia, April 13, 2006, http://www.asiamedia.ucla.edu/article.asp?parentid=43000.
11. See Mark Glaser, "Nepalese bloggers, journalists defy media clampdown by king," Online Journalism Review, February 23, 2005, http://www.ojr.org/ojr/stories/050223glaser/.
12. Ibid.
13. BBC News, "Q&A: Nepal's future," November 8, 2006, http://news.bbc.co.uk/1/hi/world/south_asia/2707107.stm.
14. Charles Haviland, "Erasing the 'royal' in Nepal," BBC News, May 19, 2006, http://news.bbc.co.uk/1/hi/world/south_asia/4998666.stm.
15. International Freedom of Expression Exchange, "Parliament abolishes repressive media law," http://www.ifex.org/fr/content/view/full/74580/.
16. S. Chandrasekharan, "NEPAL: Interim constitution unveiled: Monarchy dumped," December 17, 2006, http://www.saag.org/%5Cnotes4%5Cnote354.html.
17. See Nepal Interim Constitution, http://www.nepalnews.com/archive/2007/jan/jan15/Constitution_2063.doc.
18. See Nepal Interim Constitution, Article 12.3, http://www.nepalnews.com/archive/2007/jan/jan15/Constitution_2063.doc.
19. Nepal Interim Constitution, Article 45.1.
20. Nepal Interim Constitution, Article 12.
21. Nepal Electronic Transaction and Digital Signature Act, Article 47, http://www.nta.gov.np/cyber_law.html.
22. National Broadcasting Act of Nepal, Articles 15–16, http://www.moic.gov.np/policy/pol_broad_act_2049.php; National Broadcasting Regulation, Article 9, http://www.nta.gov.np/national_broadcasting_regulation_2052.html.

North Korea

Government restrictions on online content and connectivity render the Democratic People's Republic of Korea (North Korea) a virtual "black hole" in cyberspace.[1] While shunning Internet accessibility and functionality, Pyongyang has opted for an isolated, domestic intranet consisting of approximately thirty Web sites approved by the government and available only to a privileged minority.

Background

The North Korean regime does maintain a nominal presence on the World Wide Web through sites promoting its ideology and agenda. As with print and broadcast media, these sites largely extol the nation's leader Kim Jong Il, his father Kim Il Sung, and the Juche Idea of national "self-reliance," while espousing the country's stance on reunification of the Korean Peninsula. Unlike other media, however, North Korean Web sites lie at some distance from Pyongyang. Because the state lacks an active top-level domain (TLD), it relies on servers in China, Japan, Germany, and even Texas to host its official pages, including www.korea-dpr.com (North Korea's Web page) and www.kcna.co.jp (the home page of the state-run Korean Central News Agency). North Korea has asked repeatedly for the country code TLD ".kp", but the U.S.-based Internet Corporation for Assigned Names and Numbers (ICANN) has yet to grant the request.

Internet and Intranet in North Korea

The community of Internet users in North Korea consists almost entirely of elites and foreigners. A select few, including members of Kim Jung Il's inner circle, enjoy unfiltered Internet access via satellite link to servers in Germany, thanks to a 2004 joint venture between Pyongyang's Korea Computer Center (KCC) and its Berlin-based counterpart KCC Europe. Most Internet users, however, are dependent upon Chinese service providers for connectivity—and thus are subject to China's filtering regime. For years, these providers could be reached only via international dialup from exclusive hotels in Pyongyang. In 2002, optical cable connections between the North Korean capital and Shanghai became operational at the Internet PC Room—the first Internet café in the country. Still, few North Koreans can afford the hourly fee of USD10, effectively limiting use of the PC Room to foreign diplomats, businesspeople, journalists, and tourists.

A growing segment of the North Korean population is gaining access to Chinese networks via Web-enabled mobile phones smuggled in from China and sold on the black market.[2] However, for most North Koreans, access to online content is exceedingly rare and limited to the few dozen Web sites that comprise Kwangmyong, the nation's domestic intranet.

ONI did not carry out empirical testing for Internet filtering in North Korea for this report.

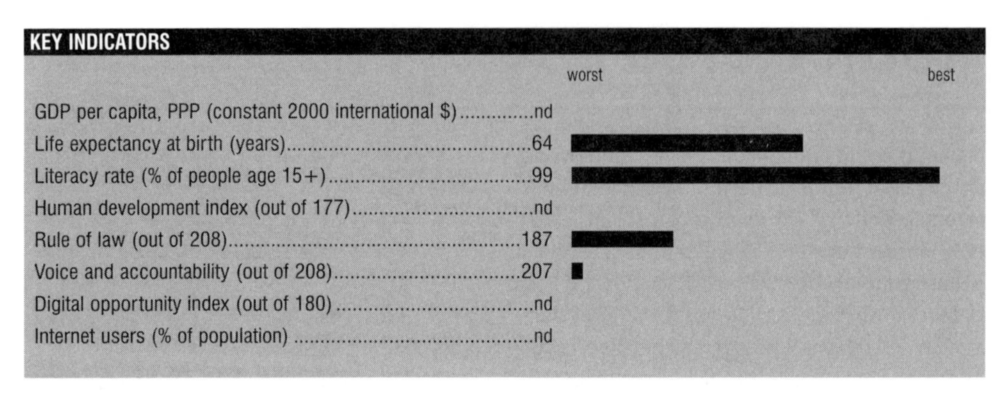

Source (by indicator): World Bank 2006a; US Department of State 2007a; World Bank 2006c, 2006c
nd = no data available

Content on Kwangmyong is chosen, and user conduct monitored, by the government. Information comes primarily from databases maintained by the Central Scientific and Technological Information Agency, the Grand People's Study House, and other repositories.[3] This content is intended for use at select research institutes, schools, and factories. Aside from these establishments, only government ministries and a handful of enterprises and individuals have the computers, telecommunications capacity, and the authorization needed to utilize the national intranet.

Small, government-sanctioned businesses offering public intranet access have been observed in urban areas, but user fees are likely prohibitive for the average North Korean. In 2005 human rights groups revealed photographs of one such venue—called the Information Technology Store—in the city of Chungjin. The facility houses several terminals with intranet connectivity and offers computer classes at the steep price of 20,000 won per month—seven to eight times the average monthly wage.[4] Such costs are believed to be mandated by the state so as to deter ordinary citizens from using the resources and services of these facilities.

Legal and regulatory frameworks

The near absence of connectivity, even to the isolated and heavily filtered Kwangmyong intranet, is consistent with the North Korean regime's efforts to regulate all information and communication in the country. There are no independent media in North Korea. Personal radios and televisions must be modified to receive only government stations and registered with the authorities. A nationwide ban on mobile phones has also been in place since May 2004.

It is the state's command of institutions and resources that allows it to achieve such pervasive control over online media. The government allocates available technologies to establishments and authorizes user access as it sees fit. Legal measures play only a subsidiary role in actualizing state control, and for citizens, they confer no actionable rights vis-à-vis the state. Thus, although Article 67 of the DPRK's Socialist Constitution guarantees freedom of speech and of the press, there is no means of instituting a legal challenge to the state's dominion over online access and expression.

NOTES

1. Tom Zeller, "The Internet black hole that is North Korea," *The New York Times,* October 23, 2006.
2. Rebecca MacKinnon, "Chinese cell phone breaches North Korean hermit kingdom," *YaleGlobal,* January 17, 2005, http://yaleglobal.yale.edu/display.article?id=5145.
3. Office of the National Counterintelligence Executive, "North Korea: Channeling foreign information technology to leverage IT development," December 2003, http://www.ncix.gov/archives/docs/NORTH_KOREA_AND_FOREIGN_IT.pdf.
4. A. Yang Jung, "Controlling the Internet café in North Korea," *The Daily NK,* July 13, 2005, http://www.dailynk.com/english/read.php?cataId=nk00300&num=206.

Oman

The Sultanate of Oman engages in extensive filtering of pornographic Web sites, gay and lesbian content, and anonymizer sites used to circumvent blocking. Although filtering of political content is highly selective, laws and regulations restrict free expression online and encourage self-censorship.

Background

Oman is a monarchy, with Sultan Qaboos bin Said exercising absolute power and the bicameral Majlis Oman (Council of Oman) acting in a mostly advisory position. Although the government is generally protective of human rights, it has been criticized by international groups for restricting free speech and assembly.[1] In early 2005, thirty-one Omanis were imprisoned for allegedly plotting to overthrow the government; all were granted royal pardons later that year.[2] In July 2005, two Omani human rights activists were arrested for criticizing the government: Taiba al-Mawali was jailed for six months and Abdullah Al-Riyami, who accused the police of torturing prisoners, was detained incommunicado for a week.[3]

Internet in Oman

Oman's communications infrastructure is well developed.[4] Oman Telecommunications Company (Omantel), the country's sole Internet service provider (ISP), is owned by the government. Omantel began providing full Internet service in early 1997. As of October 2006, Internet subscriptions numbered 92,126 (approximately 29 subscribers per 1,000 inhabitants).[5] Approximately 14 percent (12,900) of these subscriptions were to high-speed Internet (ADSL) services. The majority of subscribers continue to rely on dialup connections.[6]

The low number of Internet subscribers has been attributed to the paucity of personal

RESULTS AT A GLANCE

Filtering	No evidence of filtering	Suspected filtering	Selective filtering	Substantial filtering	Pervasive filtering
Political	●				
Social					●
Conflict/security	●				
Internet tools		●			

Other factors	Low	Medium	High	Not applicable
Transparency			●	
Consistency			●	

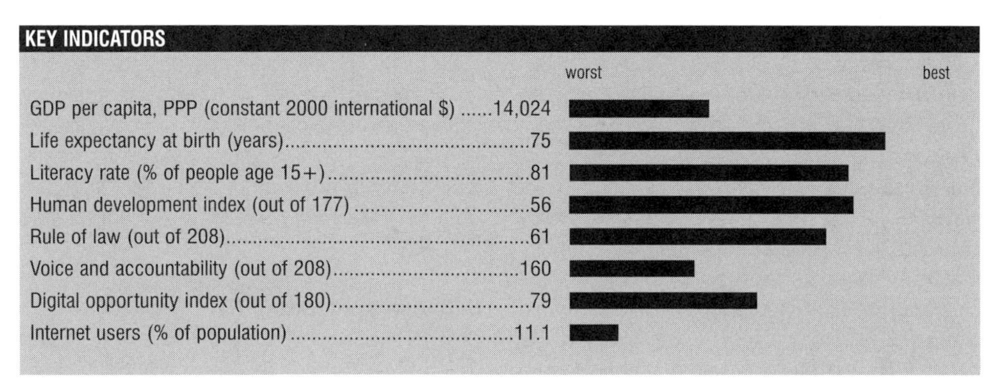

Source (by indicator): World Bank 2004, 2006a, 2006a; UNDP 2006; World Bank 2006c, 2006c; ITU 2006, 2005

computers, which in 2006 numbered 32 per 1,000 people.[7] Prepaid Internet service cards and Omantel's "Log 'N' Surf" service provide alternatives to regular subscriptions; neither requires prior registration.[8] Omantel is also working on a Wireless Local Loop (WLL) project to extend telecom services to rural areas where cable connectivity is either impossible or prohibitively expensive.[9]

Oman's government has plans to use the Internet to increase e-government and e-education. On January 9, 2007, Omantel launched the Easy Learning Service to provide hundreds of electronic training courses in accounting, sales, marketing, and customer services.[10] In February 2007 Omantel began offering sixty free hours of Internet access to new subscribers, describing the initiative as part of a larger plan to spread digital culture.[11]

The monarchy has also begun opening Oman's telecommunications sector to private investors and competitors. In 2005, the government sold a 30 percent stake in Omantel to local investors. The move came shortly after the country's second mobile-phone operator started operations, offering for the first time a choice to local consumers. The government has also announced plans to offer licenses for fixed-line telecom services in competition with Omantel.[12]

Legal and regulatory frameworks

On November 6, 1996, Sultan Qaboos bin Said issued the Basic Law of the State ("The White Book"), considered to be Oman's first constitution. Article 29 of the White Book guarantees "freedom of opinion and expression ... within the limits of the Law."[13] "Material that leads to public discord, violates the security of the State or abuses a person's dignity and his rights" may not be printed or published.[14] In July 2005 former parliamentarian Taiba al-Mawali was arrested and sentenced to eighteen months imprisonment for insulting a public official and sending allegedly libelous text messages that criticized government actions.[15] Her sentence was later reduced and she was released on January 31, 2006.[16]

Arrest and search warrants are not required by law, and the government can and does monitor both written and oral communications, including cell phone, e-mail, and Internet chat room exchanges. Publications that contravene cultural or political norms are subject to government censorship under the 1984 Press and Publication Law, and online forums admonish visitors that criticism of the sultan or government officials will be censored and could lead to police questioning.[17] Although some degree of criticism of the government has been tolerated in practice, especially on the Internet, writers and publishers generally exercise significant self-censorship. In November 2006, the Omani authorities briefly detained the administrator and a number of moderators of Oman's most popular online discussion forum (www.omania.net) after an article about corruption in the country was posted in the forum.[18] The administrator and moderators were banned from traveling outside the country, awaiting charges of defamation under the publication law, telecommunications law, and penal code.[19] In February 2007 a note in Arabic was posted on the Web site saying that the administrator was found innocent. No other details were mentioned.

Internet use in Oman is regulated by Omantel's Terms & Conditions, which mandates that users "not carry out any unlawful activities which contradict the social, cultural, political, religious or economical values of the Sultanate of Oman or could cause harm to any third party .... Any abuse and misuse of the Internet Services through e-mail or news or by any other means shall result in the termination of the subscription and may result in the proceedings of Criminal or Civil lawsuits against the Customer."[20]

To use the Internet, individuals, companies, and institutions are asked to sign an agreement not to publish anything that destabilizes the state; insults or criticizes the head of state or the royal family; questions trust in the justice of the government; creates hatred toward the government or any ethnicity or religion; promotes religious extremism, pornography, or violence; promotes any religious or political system that contradicts the state's system; or insults other states. Users must also agree not to promote illegal goods or prescription drugs over the Internet.[21]

Omantel imposes additional physical restrictions on Internet access in Internet cafés. Individuals or companies wishing to open an Internet café must submit a floor plan for the proposed site. The plan must be designed so that the computer screens are visible to the floor supervisor. No closed rooms or curtains are allowed that might obstruct view of the monitors.[22] Moreover, Internet café operators are asked to install proxy servers to monitor and log user activity.[23]

ONI testing results

Oman's exclusive ISP, Omantel, was tested using dialup, "Log 'N' Surf" service, and ADSL connections. As suggested by the text of Omantel's blockpage, results indicated extensive blocking of pornographic Web sites. Some Web sites featuring provocative attire were blocked as well.

There was also extensive blocking of gay and lesbian sites, though sites relating to gay civil rights and equality issues, such as www.glaad.com and www.hrc.org, were largely accessible.

Omantel also blocked some dating Web sites—probably because they contained either sexually explicit images (www.adultfriendfinder.com) or gay and lesbian content (www.gayromeo.com).

Anonymizing and proxy circumvention tools, such as Anonymizer and Proxify, were heavily blocked. Some Web sites dealing with hacking and cracking, such as keygencrack.com, were blocked. Only one of the many peer-to-peer Web sites tested was blocked (www.hypertorrent.com).

Although testing did not reveal the Web sites of Voice-over Internet Protocol (VoIP) services to be blocked, subscriber complaints suggest that the functionality of Skype, a popular VoIP application, has been crippled.[24] Omantel sources have reported to the media that past filtering of Skype was unintentional and would be remedied, but this statement has raised suspicions because SmartFilter categorizes www.skype.com as a "Web Phone" site, a category that Omantel would have had to specifically activate.[25] In March 2007 Oman's TRA openly banned the use of Internet telephony at Internet cafés and warned Internet café operators against providing basic voice service. The TRA also warned that violators face punishments that include imprisonment and fines.[26]

Although all blog sites tested were found to be accessible, some adult humor sites, such as www.collegehumor.com, were blocked. The Web site of the Arab-American newspaper *Arab Times* (www.arabtimes.com) was blocked, as were its Google cache copies. Unlike many states in the region, Oman does not appear to block Web sites that criticize Islam or that attempt to convert Muslims to other religions.

Omantel uses the American-made commercial filtering software SmartFilter. Omantel's blockpage states that the blocking of banned sites is not a unilateral decision taken by the ISP, but rather that "an overwhelming number of requests from the subscribers made [Omantel] rethink [its] strategy and conform to the popular demand to block pornographic and certain hacking sites that encourage hacking."[27] The blockpage also suggests that users submit an e-mail link to a site if they feel it has been blocked unfairly and that such a page should be re-categorized and unblocked.

Conclusion

Filtering of pornography, gay and lesbian content, and circumvention tools is pervasive in Oman. In addition to blocking Web sites, the authorities impose legal and physical controls to ensure that the Internet community does not access or publish objectionable or unlawful material. These laws and regulations give rise to self-censorship among writers and publishers, both off- and online.

NOTES

1. U.S. Department of State, Background Note: Oman, October 2006, http://www.state.gov/r/pa/ei/bgn/35834.htm.
2. Ibid.
3. Ibid.
4. The Political Risk Services Group, "Oman: Infrastructure," July 1, 2006 (accessed on LexisNexis).
5. Omantel News Room, "Net subscribers cross 92,000-mark," December 11, 2006 http://www.omantel.net.om/media_room/show_news.asp?news_id=322.
6. Ibid.
7. Telecommunications Regulatory Authority of Oman, Policy of Liberalization of the Telecommunication Sector, http://www.tra.gov.om/test1/lib.htm.
8. Omantel, "Internet Log'N'Surf," http://www.omantel.net.om/services/residential/internet/log_n_surf.asp.
9. *The Times of Oman,* "Omantel introduces WLL in 11 new rural villages," November 18, 2002, www.timesofoman.com/newsdetails.asp?newsid=36021&pn=local.
10. *The Times of Oman,* "Omantel to launch Easy Learning Service tomorrow," January 8, 2007, http://www.timesofoman.com/echoice.asp?detail=2005.
11. *The Times of Oman,* Times News Service, "Free surfing hours for new Omantel subscribers," February 4, 2007, http://www.timesofoman.com/inner_cat.asp?cat=1&detail=2887.
12. AMEInfo, "Oman's success in introducing cellular competition sets the stage for further market liberalization," November 28, 2005, http://www.ameinfo.com/72790.html.
13. The White Book: The Basic Law of the Sultanate of Oman, Article 29, http://www.omanet.om/english/government/basiclaw/overview.asp?cat=gov&subcat=blaw.
14. The White Book: The Basic Law of the Sultanate of Oman, Article 31, http://www.omanet.om/english/government/basiclaw/overview.asp?cat=gov&subcat=blaw.
15. U.S. Department of State, Country Report on Human Rights Practices: Oman, March 8, 2006, http://www.state.gov/p/nea/ci/79327.htm.
16. Amnesty International, "Oman: Taiba al Mawali," http://web.amnesty.org/library/Index/ENGMDE200012006?open&of=ENG-OMN.
17. U.S. Department of State, Country Report on Human Rights Practices: Oman, March 8, 2006, http://www.state.gov/p/nea/ci/79327.htm.
18. The Arabic Network for Human Rights Information, Internet and Freedom of Expression in the Sultanate of Oman, February 1, 2007, http://www.hrinfo.net/mena/achr/2007/pr0120.shtml (in Arabic).
19. Ibid.
20. Omantel, "Omantel Terms & conditions," http://www.omantel.net.om/policy/terms.asp.

21. Oman Telecommunications Company, Terms and Conditions, http://www.omantel.net.om/services/business/internet/Terms_and_condtions_internet_cafe.pdf.
22. Oman Telecommunications Company, Procedures for Internet Cyber Café Pre-Approval, http://www.omantel.net.om/services/business/internet/preapprovaleng.pdf.
23. Logs are to be kept for at least three months. See http://www.omantel.net.om/services/business/internet/policy_using_internet_cafe.pdf.
24. See, for example, Customer service enquiry submitted December 14, 2006, at Omantel Support Ticket Knowledge Base, http://www.omantel.net.om/help%5Fdesk/view_ticket.asp?ticket_id=1414.
25. Nart Villeneuve, "Target Skype?" Internet Censorship Explorer, http://ice.citizenlab.org/?p=123.
26. *The Times of Oman,* "TRA bans popular cyber cafe internet telephony," March 11, 2007, http://www.timesofoman.com/echoice.asp?detail=4227&rand=jx1GaXvoqloVJK99kPRjufZa4Z.
27. See Omantel's blockpage at http://www.omantel.net.om/new1.html.

Pakistan

Building on past attempts to filter blasphemous content, the Pakistan government expanded and intensified its Internet censorship campaign in February 2006, initiated in response to the Danish cartoons that depicted images of the Prophet Muhammad.[1] In addition to the Supreme Court ban on publishing or posting sites deemed to be presenting blasphemous material, the Pakistan Telecommunications Authority (PTA) has filtered content determined to be irredentist, secessionist, antistate, or antimilitary.

Background

Press freedom in Pakistan is restricted by the military-run government, headed by General Pervez Musharraf since 1999. In addition to applying military control over the judiciary and the ruling party in Parliament, print and electronic media have been censored where the content is deemed to be antigovernment or anti-Islamic. Government repression of media is particularly acute with regard to Balochi and Sindhi political autonomy, content considered blasphemous, and other antistate or antireligious content. A vibrant civil society movement working against Internet censorship continues to operate within Pakistan and monitors all developments in URL blocking.[2] International human rights groups have reported on the persecution of journalists at the hands of the Pakistani military intelligence agency.[3]

Internet in Pakistan

Internet usage in 2005 was reported to be 10.5 million users, with a 6.8 percent penetration rate.[4] According to September 2006 estimates, there are approximately twelve million Internet users in Pakistan, at a 7.2 percent penetration rate.[5] Pakistan has experienced considerable growth in its information communications technology (ICT)

RESULTS AT A GLANCE

Filtering	No evidence of filtering	Suspected filtering	Selective filtering	Substantial filtering	Pervasive filtering
Political			●		
Social				●	
Conflict/security					●
Internet tools			●		

Other factors	Low	Medium	High	Not applicable
Transparency			●	
Consistency		●		

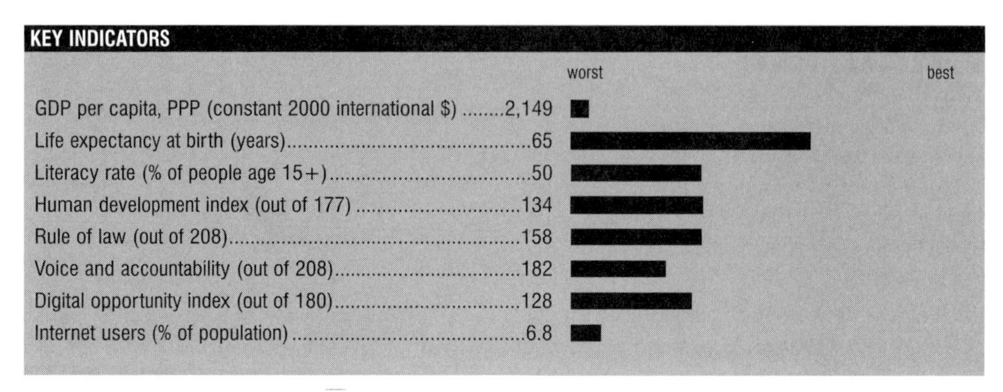

Source (by indicator): World Bank 2005, 2006a, 2006a; UNDP 2006; World Bank 2006c, 2006c; ITU 2006, 2005

sector; in 2003 the government deregulated its telecom market, opening itself up to corporate competition in telephone, mobile, and Internet services.[6] Internet access is widely available at cybercafés, which accommodate many lower-income and casual users. Rates for usage range from USD0.15/hour to upward of USD0.50/hour, depending on location and amenities. Although the Net Café Regulation Bill 2006 requires Internet cafés to monitor their patrons, there is currently no effective mechanism to verify compliance or enforce this law.[7] Athough Net café managers are expected to monitor the activities in their establishments, based on user experience these cafés appear to be unregulated by the regular police.

Since deregulation the market has become highly competitive, and there are currently over thirty Internet service providers (ISPs) in Pakistan of varying size and quality of service. The largest ISPs in the country include Cybernet, Comsats, Brainnet, Gonet, and Paknet (a subsidiary of the Pakistan Telecommunications Company Limited, or PTCL). Modem, DSL, and recently high-speed Internet service are all available in Pakistan, but the reliability of these connections remains low. The majority of home Internet users are connected by modem, while cybercafés tend to split one modem or DSL connection over many computers, reducing connection speed. High-speed Internet service is currently accessible only to wealthier patrons or businesses.

All Internet traffic in and out of Pakistan is routed by the PTCL through its subsidiary, the Pakistan Internet Exchange (PIE), with three international gateways at Islamabad/Rawalpindi, Lahore, and Karachi, and small/medium points of presence (POPs) in six other cities.[8] Currently, PIE handles 2,324 Mb/s of IP backbone traffic that comes to Pakistan using SMT4s on SMW4 and STM1s on SMW3 connected with BT, France Telecom, Telecom Italia, Verizon, and so on.[9] Bandwidth through FLAG Telecom in collaboration with PTCL is at 620 Mb/s. Domestic Internet traffic is peered at the PIE gateways within the country. The PIE's Karachi exchange reportedly processes at least 95 percent of Pakistan's Internet traffic.[10]

Bloggers across Pakistan objected to the intermittent block on www.blogspot.com and the temporary blocking of Wikipedia in 2006, and initiated a virtual civil society movement to repeal the orders.[11] This virtual civil society engages in awareness and advocacy work on Pakistan's

Internet censorship through up-to-date blogs, as well as by posting information on Wikipedia. Through these sites, users share a multitude of techniques to circumvent the URL block and continue to access their Web sites of choice. An example of this is the use of www.pkblogs.com to access and post on banned www.blogspot.com sites.

Legal and regulatory frameworks

Internet censorship in Pakistan is legally regulated by the PTA, under the directive of the government, the Supreme Court of Pakistan, and the Ministry of Information Technology and Telecommunications (MITT). The PTA implements its censorship regulations through directives handed down to the PTCL,[12] of which the Emirates Telecommunications Corporation (Etisalat) took majority control in 2006.[13]

In February 2002 the PTA challenged the legality of the use of Voice-over Internet Protocol (VoIP) as a replacement for long-distance calls. Because VoIP has achieved considerable popularity as a cost-effective alternative to long-distance calls, the PTCL banned VoIP and voice chat Web sites in early 2002; the service was undermining revenues for outgoing long-distance phone calls to the United States.[14]

In January 2003 the MITT directed the PTCL to block pornographic and blasphemous sites by placing content filters at all Internet exchanges,[15] an effort that was not entirely effective.[16] In March 2004 the Federal Investigation Agency also ordered all ISPs to block pornographic Web sites, a task beyond the technical capability of the ISPs at the time.[17]

On February 28, 2006, the PTCL issued a blocking directive banning a dozen URLs determined to have posted controversial Danish cartoons depicting images of the Prophet Muhammad.[18] Within two weeks in March, in a series of escalating instructions, the Supreme Court directed the government to block all Web sites displaying the cartoons; to explain why they had not been blocked earlier; to block all blasphemous content; and to determine how access to such content could be denied on the Internet worldwide.[19] The Supreme Court also ordered police to register cases of publishing or posting the blasphemous images under Article 295-C of the Pakistan Penal Code, where blasphemy or defamation of the Prophet Muhammad is punishable by death.[20] Desecration or derogation of the Quran is punishable by life imprisonment.[21]

On September 2, 2006, the MITT announced the creation of a committee to monitor content of offensive Web sites. According to the Ministry statement, "the committee, headed by the secretary of the MITT, will examine contents of websites reported or found to be offensive and containing anti-state material."[22] To address the grievances of Internet users with this censorship body, the government set up a Deregulation Facilitation Unit to deal with users' complaints.[23]

ONI testing results

ONI field testing was conducted on Brainnet, Cybernet, and Paknet ISPs. Testing results showed that blacklisted URLs were blocked at either the ISP or PIE level, or at both locations. The PTCL has implemented a limited, perhaps symbolic, block on pornography and religious conversion sites. However, more aggressive efforts have been made to target content regarding Balochi independence movements, Sindhi human rights and political autonomy movements, material considered blasphemous, antigovernment material, and anti-Islamic materials, though a clear pattern or criteria for what is filtered is lacking. Among these categories, Web sites depicting blasphemous content or addressing Balochi political independence were the most comprehensively blocked.

Because one of the twelve Web sites identified as depicting the Danish cartoons was hosted on Blogspot, the PTCL used a blocking mechanism that filtered the entire www.blogspot.com domain. As a result of this strategy, thousands of personal blogs hosted on www.blogspot.com were inadvertently filtered for most of 2006. Most material relating to the Danish cartoon incident was blocked by the ISPs; only one Web site containing the cartoons that was reportedly blocked (www.danishcartoons.ytmnd.com) was found to be fully accessible through the testing process.

By April 2006 the PTA extended their blocking to antistate Web sites as well as those promoting Balochi human rights and political autonomy.[24] ONI testing confirmed that internal security conflicts were a strong focus for filtering: all Web sites tested relating to independence (for example, www.balochunitedfront.org) and human rights (for example, www.balochestan.com) in the province of Balochistan were blocked, as well as selected sites promoting Sindhi political autonomy and human rights. Notably, though Balochi and Sindhi independence and human rights sites have been filtered, the few existing Web sites pertaining to Pashtun secessionism were fully accessible. This may be because the majority of Pashtuns are illiterate in their local language, and secessionist politics in the northwest frontier province are significantly less potent than in Balochistan and Sindh provinces. Therefore the more politically organized Balochi and Sindhi movements arguably pose a greater threat to the central government than these selected pro-Pashtunistan Web sites.

In addition to blasphemous, secessionist, and human rights Web sites, a variety of blogs and Web sites containing anti-Islamic and anti-Pakistani content were blacklisted, such as Indian militant extremist sites (www.hinduunity.com) and anti-Islamic blogs (www.jihadwatch.com). A number of less polemical Web sites, including personal blogs hosted on www.blogspot.com, and Web sites dedicated to promoting religious tolerance (www.faithfreedom.com) were also blocked.

ONI testing showed that the majority of newspapers and independent media, circumvention tools, international human rights groups, VoIP services, civil society groups, minority religious sites, Indian and Hindu human rights groups, Pakistani political parties, and sexual content (including pornography and gay and lesbian content) were accessible on all three ISPs. Pornographic content was largely accessible, with only symbolic blocking of selected sites. Civil society groups contend that all www.blogspot.com sites have been blocked; however, ONI testing found the site for the "Don't Block the Blog" campaign (www.help-pakistan.com) to be accessible on all three ISPs.

The lack of technical sophistication of the PTCL explains the comprehensive block on Blogspot. The PTCL lacks the capacity to target the specific URLs that contain offensive content and simply blocks the entire IP address on which the offending site was hosted. Although this filtering system has resulted in the collateral blocking of entire domains such as www.blogspot.com, the rudimentary nature of the blocking mechanism also makes it easier for users to circumvent the block using proxy servers or other bypassing methods.[25] Not only is the PTCL charged with blocking blacklisted URLs, but it also hands down blocking directives directly to the ISPs to implement. The ISPs then implement, or attempt to implement, the blocking orders; the results of the ONI testing show that this sometimes led to a redundancy in blocking at both the ISP level and the central Internet exchange point.

Conclusion

Currently Pakistanis have unimpeded access to most sexual, political, social, and religious content. However, the Pakistani government continues to use repressive measures against antimilitary, Balochi, and Sindhi political dissidents, and it blocks Web sites highlighting this repression. The government also filters high-risk antistate materials and blasphemous content.

The Pakistani government does not currently employ a sophisticated blocking system, nor does the government have a coherent policy on what sites should be blacklisted. The recently established ministerial committee will probably contribute to the development of a comprehensive framework for government censorship as methods for implementing blocking directives are refined. Civil society activists and cyber-dissidents continue to advocate for free expression and blogging rights, which are curtailed by crude blocking methods that have imposed blanket blocks on entire domains such as www.blogspot.com.

NOTES

1. Ammara Durrani, "Ban on 'blasphemous' websites: PTA blocks blogs not carrying profane material," *The International News,* March 6, 2006, http://www.jang.com.pk/thenews/mar2006-daily/06-03-2006/national/n8.htm.
2. Pakistan 451, http://pakistan451.wordpress.com/; and Don't Block the Blog campaign, http://help-pakistan.com/main/dont-block-the-blog/.
3. Reporters Without Borders, Pakistan: Annual Report 2007, February 19, 2007, http://www.rsf.org/article.php3?id_article=20794.
4. International Telecommunication Union, *World Telecommunication Indicators 2006*.
5. Internet World Stats, Asia Marketing Research, Internet Usage, Population Statistics and Information, http://www.internetworldstats.com/asia.htm.
6. Pakistan Telecommunication Company Limited, http://www.ptcl.com.pk/introduction.html.
7. Zulfiqar Ali, "Cabinet approves bill to regulate cyber-cafes: Bill makes cyber-cafe owners liable for prosecution if any 'unethical' or 'immoral' activities are detected within their premises," Asia Media, May 26, 2006, http://www.asiamedia.ucla.edu/article.asp?parentid=46879.
8. E-mail interview with Convener of Internet Service Provider Association of Pakistan (ISPAK).
9. Ibid.
10. Naveed Ahmad, "PTCL blocks vital Internet sites to comply with SC order," *The News,* http://www.thenews.com.pk/top_story_detail.asp?Id=6254.
11. Don't Block the Blog campaign, http://help-pakistan.com/main/dont-block-the-blog/.
12. Wikipedia, "Internet censorship in Pakistan," February 20, 2007, http://en.wikipedia.org/wiki/Internet_censorship_in_Pakistan.
13. DAWN, "Deal signed to give Etisalat PTCL control," March 13, 2006, http://www.dawn.com/2006/03/13/top5.htm.
14. *Daily Times,* "PTCL still directionless over VoIP issues," January 26, 2003, http://www.dailytimes.com.pk/default.asp?page=story_26-1-2003_pg7_30.
15. DAWN, "PTCL directed to block porno, blasphemous sites," January 29, 2003, http://www.dawn.com/2003/01/29/nat5.htm.
16. Imran Ayub, "PTCL to block all objectionable websites," The News International Pakistan, December 27, 2003, http://www.apnic.net/mailing-lists/s-asia-it/archive/2003/12/msg00028.html; World IT Report, "Pakistan faces difficulties to block porn sites, February 3, 2003, http://www.worlditreport.com/News/&mod=search&searchWords=Pakistan%20PTCL&st_id_search=93657&time=1&sub=1.
17. Reporters Without Borders, Pakistan: Annual Report 2004, http://www.rsf.org/article.php3?id_article=10794.
18. Ammara Durrani, "Ban on 'blasphemous' websites: PTA blocks blogs not carrying profane material," *The International News,* March 6, 2006, http://www.jang.com.pk/thenews/mar2006-daily/06-03-2006/national/n8.htm.
19. Ibid.
20. Article 295-C, Criminal Law (Amendment) Act, 1986, (Gazette of Pakistan, Extraordinary, part 1, October 12, 1986), http://www.thepersecution.org/50years/paklaw.html; *Daily Times,* "SC orders case against cartoon publishers," April 18, 2006, http://www.dailytimes.com.pk/default.asp?page=2006%5C04%5C18%5Cstory_18-4-2006_pg1_9; see also Akbar S. Ahmed, "Pakistan's blasphemy laws: Words fail me," *The Washington Post,* May 19, 2002, p. B01, http://www.washingtonpost.com/ac2/wp-dyn?pagename=article&node=&contentId=A36108-2002May17¬Found=true.
21. Article 295-B, Pakistan Penal Code (Act XLV of 1860), http://www.thepersecution.org/50years/paklaw.html.
22. Nasir Iqbal, "Body set up to block websites," *Dawn News,* September 3, 2006, http://www.dawn.com/2006/09/03/nat3.htm.
23. Ibid.
24. Pakistan Telecommunications Authority, Blocking of Websites Access, April 25, 2006, http://ice.citizenlab.org/blogimages/PTA_-_Blocking_of_website_25-4-06.pdf.
25. Khalid Omar, "Connecting the dots," Spider: Pakistan's Internet Magazine, April 2006, http://www.spider.tm/apr2006/main.html?pgsrc=szone&submenu=szone1&dirtarget=none.

Saudi Arabia

Saudi Arabia has filtered the Internet since its introduction into the kingdom less than a decade ago. The filtering regime most extensively covers religious and social content, though sites relating to opposition groups and regional political and human rights issues are also targeted.

Background

Saudi Arabia is a monarchy without elected political institutions.[1] The ruling Al Saud family has presided over the Islamic nation and accumulated a poor human rights record. At times there has been increased discussion of sensitive subjects, such as political reform and women's rights, but despite explicit promises to improve the human rights situation, the government continues to maintain that such rights are subordinate to Islamic law and tightly limits political and religious freedom.[2] Religious police, *Mutawwa'in,* are charged with enforcing public morality. A wide range of media, including books and films, are censored or banned. Arbitrary arrests, prolonged detentions of political prisoners, corporal punishment, and the denial of basic conditions for fair trials make for a bleak judicial landscape.[3] Journalism is strictly controlled and journalists must exercise self-censorship in order to avoid government scrutiny and dismissal.[4] Most Saudis get their information from foreign television and the Internet, and—though officially banned—dish receivers are becoming increasingly common.[5] Al-Jazeera, a Qatar-based Arab satellite television station, is banned in the country, and foreign journalists are rarely granted visas.[6]

Internet in Saudi Arabia

The belated arrival of the Internet in Saudi Arabia, several years after its introduction into other Arab

RESULTS AT A GLANCE

Filtering	No evidence of filtering	Suspected filtering	Selective filtering	Substantial filtering	Pervasive filtering
Political				●	
Social					●
Conflict/security			●		
Internet tools				●	

Other factors	Low	Medium	High	Not applicable
Transparency			●	
Consistency			●	

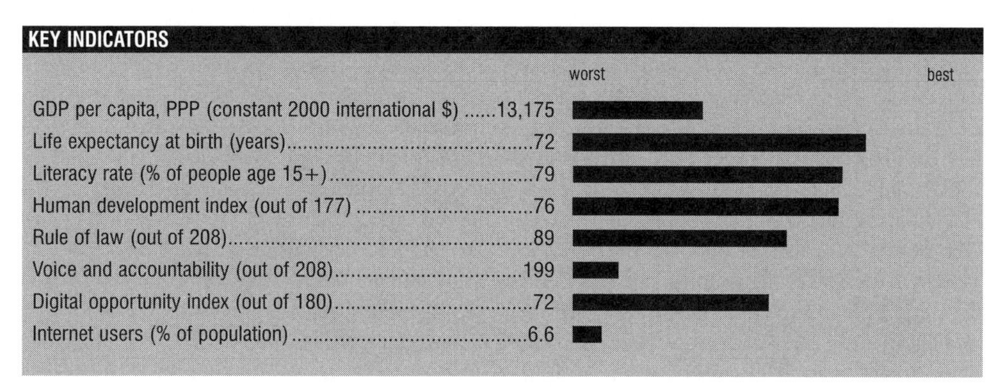

Source (by indicator): World Bank 2005, 2006a, 2006a; UNDP 2006; World Bank 2006c, 2006c; ITU 2006, 2004

countries, was largely the result of the government's concerns about regulating content. Since the year 2000, Internet usage has increased from less than 1 percent to more than 10 percent (over 2.5 million users) of the population.[7] Telecom companies have begun taking advantage of the still relatively low penetration rate by unveiling 3G networks in the country.[8] The government's Internet Services Unit (ISU), a department of the King Abdulaziz City for Science & Technology (KACST), has been responsible for overseeing Internet services in Saudi Arabia and for implementing government censorship.[9] As its Web site explains, twenty-one licensed Internet service providers (ISPs) and one more not yet in service connect users to the national network.[10] The ISU manages the link from the national network to international networks.[11] In accordance with a Council of Ministers decision, the Saudi Communications Commission was renamed the Communications and Information Technology Commission (CITC) and took charge of licensing and filtering processes previously managed by KACST.

Blogging has grown as a medium for expression in Saudi Arabia, with the number of bloggers tripling to an estimated 2,000.[12] Half of these bloggers are women.[13] In 2005 the government tried to ban the country's primary blogging tool, www.blogger.com.[14] However, after a few days the ban was lifted, with the censors choosing to block specific content on the blogging Web site instead.[15] Paralleling the increase in Internet use has been a proliferation of Internet cafés. As hourly rates can be too expensive for average Saudis, some Internet cafés offer monthly subscriptions that are more affordable.[16] From time to time, the authorities have shut down Internet cafés for reasons such as "immoral purposes."[17]

Legal and regulatory frameworks

The government of Saudi Arabia allowed public access to the Internet only after it was satisfied that an adequate regulatory framework could be put in place. The authorities use Secure Computing's SmartFilter software for technical implementation and to identify sites for blocking.[18] Furthermore, the expertise of local staff and input of ordinary citizens aid the filtering regime. The government makes no secret of its filtering, which is explained on a section of the ISU Web site.[19] According to this Web site, KACST is directly responsible for filtering pornographic content, while other sites are blocked upon request from "government security bodies." The Web site also has forms by which Internet users can request that certain sites be blocked or unblocked. It has been noted by a KACST official that the majority of blocked Web sites contain pornographic content, and over 90 percent of Internet users have tried to access a blocked Web site.[20]

In 2001 the Council of Ministers issued a resolution outlining content that Internet users are prohibited from accessing and publishing.[21] Among other things, it forbids content "breaching public decency," material "infringing the sanctity of Islam," and "anything contrary to the state or its system."[22] The resolution also includes approval requirements for publishing on the Internet and mechanical guidelines for service providers on recording and monitoring users' activities.[23]

A new law, approved by the Saudi Shoura (Advisory) Council in October 2006, criminalizes the use of the Internet to defame or harm individuals and the development of Web sites that violate Saudi laws or Islamic values, or that serve terrorist organizations.[24]

ONI testing results

Testing was conducted on two ISPs: National Engineering Services & Marketing (Nesma) and Arabian Internet and Communications Services (Awalnet). Both providers blocked the same Web sites, as expected given the centrally administered filtering system.

Testing indicates that the Web sites of Saudi political reformist and opposition groups, such as the Islah movement (www.islah.tv) and the Tajdeed movement (www.tajdeed.net), are targeted for blocking. In keeping with the Saudi government's emphasis on protecting the "sanctity of Islam"[25]—and the legitimacy of the regime—sites relating to minority faiths or espousing alternative views of Islam are blocked. These include the Web sites of a number of local Shiite groups.

The Web pages of a few global free speech advocates, such as Article19 (www.article19.org) and the Free Speech Coalition (www.freespeech coalition.com), are blocked. However, filtering of human rights content primarily targets Saudi or regional organizations. All Web pages of the Saudi Human Rights Center (www.saudihr.org) are blocked. Although the main pages of the Arab Human Rights Information Network and the Arabic rights organization Humum are accessible, the Saudi sections of the two sites, www.hrinfo.net/ifex/alerts/saudi and www .humum.net/country/saudi.shtml, are not.

Most global media sites tested, including Israel-based news outlets such as the daily Haaretz (www.haaretz.com), were accessible. However, the Arab-language news sites Al-Quds Al Arabi (www.alquds.co.uk) and Elaph (www .elaph.com) were blocked.

"Immoral" social content continues to be a priority target for Saudi censors. Over 90 percent of pornographic Web sites and most sites featuring provocative attire or gambling that were tested were blocked. Numerous sites relating to alcohol and drugs, gays and lesbians, and sex-education and family planning were also inaccessible. This pervasive filtering of social content is achieved through the use of SmartFilter software, which builds "blacklists" of sites from user-selected categories, such as Drugs, Gambling, Obscene, Nudity, Sex, and Dating. The substantial filtering of Internet tools, including anonymizers and translators, in Saudi Arabia is also achieved in this manner.

Conclusion

Saudi Arabia maintains a sophisticated filtering regime. Social content and Web-based applications are extensively filtered using commercial software. Additional political and religious sites are individually targeted for blocking. The result of this filtering system is consistent with the Saudi government's express commitment to censoring morally inappropriate and religiously sensitive material online. More generally, Internet filtering in Saudi Arabia mirrors broader attempts by the state to repress opposition and promote a single religious creed.

NOTES

1. U.S. Department of State, Country Reports on Human Rights Practices 2005: Saudi Arabia, March 8, 2006, http://www.state.gov/g/drl/rls/hrrpt/2005/61698.htm.
2. Amnesty International, Annual Report 2006: Saudi Arabia, http://www.amnestyusa.org/countries/saudi_arabia/document.do?id=ar&yr=2006.
3. U.S. Department of State, Country Reports on Human Rights Practices 2005: Saudi Arabia, March 8, 2006, http://www.state.gov/g/drl/rls/hrrpt/2005/61698.htm.
4. Reporters Without Borders, Saudi Arabia: Annual Report 2007, http://www.rsf.org/article.php3?id_article=20775&Valider=OK.
5. Reporters Without Borders, Saudi Arabia: Annual Report 2006, http://www.rsf.org/article.php3?id_article=17185.
6. Ibid.
7. Internet World Stats, Internet Usage in the Middle East, http://www.internetworldstats.com/stats5.htm#me.
8. *Wireless World Forum,* "Telecom industry sees 3G boom in Saudi Arabia," http://www.w2forum.com/i/Telecom_industry_sees_3G_boom_in_Saudi_Arabia.
9. Internet Services Unit, King Abdul Aziz City for Science & Technology, http://www.isu.net.sa/.
10. Ibid.
11. Ibid.
12. Faiza Saleh Ambah, "New clicks in the Arab world," *The Washington Post,* November 12, 2006, http://www.washingtonpost.com/wp-dyn/content/article/2006/11/11/AR2006111100886.html.
13. Ibid.
14. Reporters Without Borders, Saudi Arabia: Annual Report 2007, http://www.rsf.org/article.php3?id_article=20775&Valider=OK.
15. Ibid.
16. The Baheyeldin Dynasty, "Saudi Arabia's ISPs: Internet/Cyber Cafes," http://baheyeldin.com/saisp/0006-cafes.phtml.
17. Human Rights Watch, Saudi Arabia: Human Rights Developments 2001, http://www.hrw.org/wr2k1/mideast/saudi.html.
18. The OpenNet Initiative, Internet Filtering in Saudi Arabia in 2004, http://www.opennetinitiative.net/studies/saudi/#toc1c.
19. Internet Services Unit, "Introduction to Content Filtering," http://www.isu.net.sa/saudi-internet/contenet-filtring/filtring.htm.
20. Raid Qusti, "Most of Kingdom's Internet users aim for the forbidden," *Arab News,* October 2,2005, http://www.arabnews.com/?page=1§ion=0&article=71012&d=2&m=10&y=2005.
21. Council of Ministers Resolution, Saudi Internet Rules, February 12, 2001, http://www.al-bab.com/media/docs/saudi.htm.
22. Ibid.
23. Ibid.
24. Raid Qusti, "Shoura approves law to combat e-crimes," *Arab News,* October 10, 2006, http://www.arabnews.com/?page=1§ion=0&article=87941&d=10&m=10&y=2006.
25. Council of Ministers Resolution, Saudi Internet Rules, February 12, 2001, http://www.al-bab.com/media/docs/saudi.htm.

Singapore

The government of the Republic of Singapore engages in minimal Internet filtering, blocking only a small set of pornographic Web sites as a symbol of disapproval of their contents. However, the state employs a combination of licensing controls and legal pressures to regulate Internet access and to limit the presence of objectionable content and conduct online.

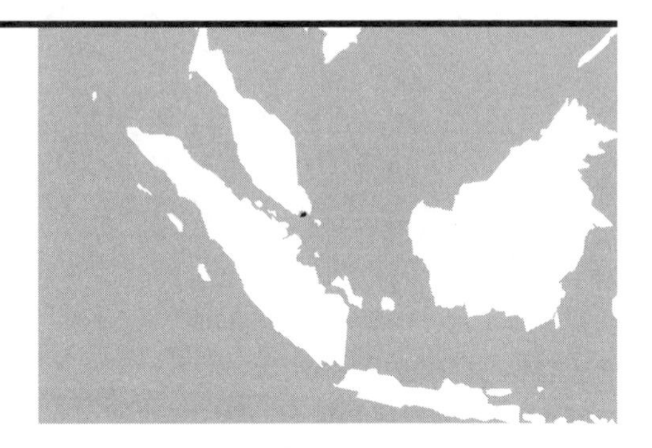

Background

Singapore's government uses restrictive laws, political ties to the judiciary, and ownership and intimidation of the media to suppress dissenting opinion and opposition to the ruling People's Action Party (PAP). Provisions of the Internal Security Act (ISA), the Criminal Law (Temporary Provisions) Act (CLA), the Undesirable Publications Act (UPA), and other statutes prohibit the production and possession of "subversive" materials and permit the detention of suspected offenders without judicial review.[1] Citizens, including Singapore Democratic Party (SDP) leader Chee Soon Juan, have been arrested for speaking publicly without a permit,[2] and foreign activists from civil society organizations have been detained, interrogated, and deported.[3] Government plaintiffs have been able to levy civil liability and heavy damages through defamation suits against independent and critical voices, including those of opposition politicians and of regional publications with domestic circulation.[4] Moreover, virtually all domestic newspapers and television and radio stations are owned by corporations with economic ties to the government; hence they adhere closely to the PAP line when reporting on sensitive issues.[5] Taken together, these economic and legal controls contribute to a climate of pervasive self-censorship of political commentary. These mechanisms of control and influence allow the Singapore government to

RESULTS AT A GLANCE

Filtering	No evidence of filtering	Suspected filtering	Selective filtering	Substantial filtering	Pervasive filtering
Political	●				
Social			●		
Conflict/security	●				
Internet tools	●				

Other factors	Low	Medium	High	Not applicable
Transparency			●	
Consistency			●	

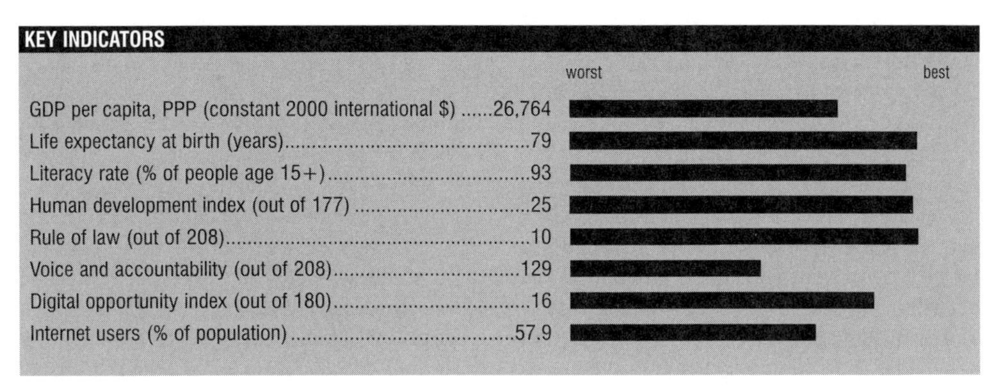

Source (by indicator): World Bank 2005, 2006a, 2006a; UNDP 2006; World Bank 2006c, 2006c; ITU 2006, 2004

cripple basic freedoms of expression and assembly under the guise of protecting public security and preserving order.

Internet in Singapore

In 2005, the number of Internet users in Singapore reached 2.42 million, or 67.2 percent of the population,[6] giving the country one of the highest Internet penetration rates in the world. Home access is commonplace, with residential dialup and broadband subscriptions totaling more than 2.1 million.[7] Over 70 percent of businesses use the Internet,[8] and public access is widespread and expanding. In December 2006, a three-year national wireless service was launched, providing laptop users with free Wi-Fi Internet access in high-traffic areas across the island.[9] Terminals in cybercafés and libraries supply the public with additional connectivity.

Three main Internet Access Service Providers (IASPs)—SingNet, StarHub, and Pacific Internet—serve as the "gateways" to the Web, providing access to Internet service resellers (ISRs) for sale to the public.[10] Though all three IASPs are public corporations, Temasek Holdings (the government's holding company) remains the majority shareholder in SingNet and StarHub.[11]

Legal and regulatory frameworks

Singapore's Media Development Authority (MDA) claims to have instituted a "light-touch" regulatory framework for the Internet, promoting responsible use while giving industry players "maximum flexibility."[12] In addition to promoting self-regulation and public education, the MDA maintains license and registration requirements that subject Internet content and service providers to penalties for noncompliance with restrictions on prohibited material. The MDA is charged with ensuring that "nothing is included in the content of any media service which is against public interest or order, or national harmony, or which offends good taste or decency."[13] The core of this framework is a class license scheme stipulated by national statute (the Broadcasting Act)[14] and by industry policies and regulations issued by the MDA.

Under the class license scheme, all Internet service providers (ISPs) and those Internet content providers (ICPs) determined to be political parties or persons "engaged in the propagation, promotion or discussion of political or religious issues relating to Singapore" must register with the MDA.[15] As licensees, ISPs and ICPs are also bound by the MDA's Internet Code of Practice. The Code defines "prohibited material" broadly, specifying only a few standards for sexual, violent, and intolerant content.[16] Where filtering is not mandated at the ISP level, the Code requires that ICPs deny access to material if so directed by the MDA. Licensees that fail to comply with the Code may face sanctions, including fines or license suspensions or terminations, as authorized under the Broadcasting Act. In 2005, one Web site titled "Meet Gay Singapore Friends" was reportedly fined USD5,000 by the MDA for being in violation of the Code.[17]

Threats of civil and criminal liability under other laws further deter Internet users from posting comments or content relating to sensitive issues. In May 2005 the state-funded agency A*STAR accused Jiahao Chen, a Singaporean doctoral student in the United States, of posting "untrue and serious accusations against A*STAR, its officers and other parties," and threatened Chen with "legal consequences unless the objectionable statements were removed and an acceptable apology published."[18] Chen complied with A*STAR's demands and replaced the posts with an apology, thereby avoiding a potential defamation suit.[19] The high-profile case prompted caution[20] in the Singapore blogosphere and discussion[21] on how to avoid suit under the nation's defamation laws.[22]

In October 2005 two men were jailed under the Sedition Act[23] for the first time in nearly forty years. One received a one-month sentence and the other a nominal one-day sentence and a USD5,000 fine for posting racist remarks denigrating Muslims and Malays.[24] In January 2006, a twenty-one-year-old was also charged with violating the Sedition Act after he posted four cartoons of Jesus on his blog. The charges were eventually dropped, but not before Singaporean authorities had confiscated the individual's computer and removed the cartoons from his blog.[25]

In November 2006 SDP activist Yap Keng Ho was sentenced to ten days in jail after he refused to pay a fine for speaking at an illegal SDP rally, held in April 2006. Yap had posted a video of the speech on his blog and was ordered to remove it by a judge.[26]

The above incidents appeared to presage further repressive legislation and policies against Singaporean Internet users. In 2007 the Ministry of Home Affairs (MHA) is expected to table before parliament a slate of amendments to the Penal Code. The proposed amendments expand the scope of nineteen offenses to cover acts perpetrated via electronic media, including "uttering words with deliberate intent to wound the religious feelings of any person" (§298); defamation (§499); and making "statements conducing to public mischief" (§505).[27] Section 298 is being modified further to cover "the wounding of racial feelings," so that offenders may be prosecuted under the Sedition Act or the Penal Code.[28] The MHA amendments also introduce nineteen new offenses, including abetting "an offense which is committed in Singapore, even if any or all of the acts of abetment were done outside Singapore," as via Internet or mobile phone (§108B).[29]

ONI testing results

ONI conducted testing on Singapore's two major IASPs, SingNet and StarHub, and on a third ISP, SysTech. A common perception of the Singaporean Internet community points to the existence of a list of 100 banned Web sites purportedly maintained by the Media Development Authority (MDA). ONI found that only seven Web sites tested, all relating to pornography, were blocked, including www.sex.com, www.playboy.com, and www.penthouse.com. The blocking of only these high-profile sites suggests that filtering is indeed mandated for symbolic, rather than preventative, purposes. Moreover, the seven sites blocked on SingNet and StarHub were all accessible on SysTech.

Conclusion

The Singapore government implements a limited filtering regime, relying mainly on nontechnological measures to curb online commentary and content relating to political, religious, and ethnic issues. The purported purpose of these measures is "to promote and facilitate the growth of the Internet while at the same time safeguarding social values and racial and religious harmony."[30] The threats of lawsuits, fines, and criminal prosecution inhibit more open discourse in an otherwise vibrant Internet community.

NOTES

1. U.S. Department of State, Country Reports on Human Rights Practices 2006: Singapore, at 1.e., http://www.state.gov/g/drl/rls/hrrpt/2006/78790.htm; see also Singapore Statutes Online, http://statutes.agc.gov.sg/.
2. Human Rights Watch, "Singapore: Release opposition leader," December 8, 2006, http://hrw.org/english/docs/2006/12/08/singap14792.htm.
3. Asian Forum for Human Rights and Development, "Detained civil society activists suffer cruel, inhuman and degrading treatment before deportation," October 15, 2006, http://www.forum-asia.org/index.php?option=com_content&task=view&id=83&Itemid=32.
4. U.S. Department of State, Country Reports on Human Rights Practices 2006: Singapore, at 1.e., 2.a., 2.d., 3, http://www.state.gov/g/drl/rls/hrrpt/2006/78790.htm.
5. U.S. Department of State, Country Reports on Human Rights Practices 2006: Singapore, at 2.a., http://www.state.gov/g/drl/rls/hrrpt/2006/78790.htm.
6. Asia Internet Usage and Population Statistics, http://www.internetworldstats.com/stats3.htm (citing International Telecommunication Union data).
7. InfoComm Development Authority (IDA), Statistics on Telecom Services for 2006 (Jul. – Dec.), http://www.ida.gov.sg/Publications/20061205181639.aspx (listing 1,489,500 residential dialup subscriptions and 657,900 residential broadband subscriptions as of October 2006).
8. InfoComm Development Authority (IDA), Measuring Infocomm Usage by Companies, 2005, http://www.ida.gov.sg/Publications/20061205110753.aspx.
9. InfoComm Development Authority (IDA), "Media Release: Free access to Wireless@SG Network extended from two to three years," http://www.ida.gov.sg/News%20and%20Events/20061204103552.aspx?getPagetype=20.
10. See Media Development Authority (MDA), Internet Industry Guidelines, http://www.mda.gov.sg/wms.file/mobj/mobj.496.internet_industry_guide.pdf.
11. See SingTel, Board & Management, http://home.singtel.com/about_singtel/board_n_management/default.asp; StarHub, Investor Relations – Stock Information, http://www.starhub.com/corporate/investorrelations/index.html.
12. See Media Development Authority (MDA), Internet, http://www.mda.gov.sg/wms.www/devnpolicies.aspx?sid=161.
13. Media Development Authority Act of Singapore, Part III, 11.1(h), http://statutes.agc.gov.sg/.
14. Broadcasting Act (Cap. 28), http://statutes.agc.gov.sg/.
15. Broadcasting (Class License) Notification 2001, July 15, 1996, http://www.mda.gov.sg/wms.file/mobj/mobj.487.ClassLicense.pdf.
16. "Prohibited material is material that is objectionable on the grounds of public interest, public morality, public order, public security, national harmony, or is otherwise prohibited by applicable Singapore laws." Media Development Authority (MDA), Internet Code of Practice, http://www.mda.gov.sg/wms.file/mobj/mobj.497.internet_code.pdf.
17. See "MDA bans gay website and fines another one," *The Straits Times,* October 28, 2005.
18. A*STAR, Press Statement, May 6, 2005, http://www.a-star.edu.sg/astar/about/action/pressrelease_details.do?id=0fc8783e9bYl.
19. See *Singapore News,* "Student shuts down blog after A*Star threatens to sue," May 6, 2005.
20. See, for example, "A sad day for the Singapore blogosphere," April 25, 2005, http://singaporeangle.blogspot.com/2005/04/sad-day-for-singapore-blogosphere.html (expressing views of over thirty bloggers).
21. See *South China Morning Post,* "Bloggers cautiously test the limits of free speech," July 18, 2005.
22. See Defamation Act (Cap. 75), http://statutes.agc.gov.sg/.
23. See Sedition Act (Cap. 290), http://statutes.agc.gov.sg/.
24. See "Racist bloggers jailed," *The Straits Times,* October 8, 2005.
25. Reporters Without Borders, "Government drops charges against blogger who posted Jesus cartoons," July 20, 2006, http://www.rsf.org/article.php3?id_article=18106.

26. Reporters Without Borders, "Blogger belonging to democratic party jailed for ten days," November 24, 2006, http://www.rsf.org/article.php3?id_article=19702.
27. See Ministry of Home Affairs, Consultation Paper on the Proposed Penal Code Amendments, p. 6, http://www.agc.gov.sg/publications/docs/Penal_Code_Amendment_Bill_Consultation_Paper.pdf.
28. Ibid, p. 9.
29. Ibid, p. 17.
30. See Media Development Authority (MDA), Internet, http://www.mda.gov.sg/wms.www/devnpolicies.aspx?sid=161.

South Korea

Although South Korea has one of the highest Internet penetration rates in the world, the state imposes substantial legal and technological controls over online expression. South Korea filters a large amount of content that supports or praises North Korea, South Korea's historical political adversary, as well as a small number of sites devoted to gambling and pirated software.

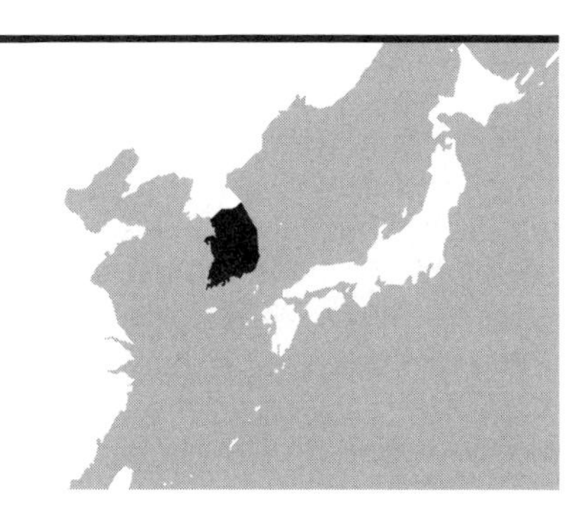

Background

The Republic of Korea (also known as South Korea) was established in 1948 and spent four decades under authoritarian rule until a democratic system emerged in 1987.[1] South Korean foreign relations remain dominated by the state's relationship with its traditional adversary, the Democratic People's Republic of Korea (or North Korea), with which South Korea has technically been at war since the two sides fought to a stalemate in 1953.[2] Since that time, South Korea has been largely intolerant of dissident views and those espousing communism or supporting North Korea; publicly praising North Korea has been, and remains, illegal. Human rights groups charge that, since its enactment in 1948, thousands of South Koreans have been arrested under the state's anti-communist National Security Law (NSL).[3] Those arrested over the years include students, publishers, trade unionists, political activists, professors, and Internet surfers.[4] Many have been arrested and jailed for peacefully expressing their political views.[5] Some prisoners arrested under the NSL were allegedly held for three to four decades, ranking them among the world's longest-held political prisoners.[6]

Despite South Korea's current "sunshine policy" of diplomatic engagement with North Korea, investigations and arrests continue for

RESULTS AT A GLANCE

Filtering	No evidence of filtering	Suspected filtering	Selective filtering	Substantial filtering	Pervasive filtering
Political	●				
Social			●		
Conflict/security					●
Internet tools	●				

Other factors	Low	Medium	High	Not applicable
Transparency			●	
Consistency			●	

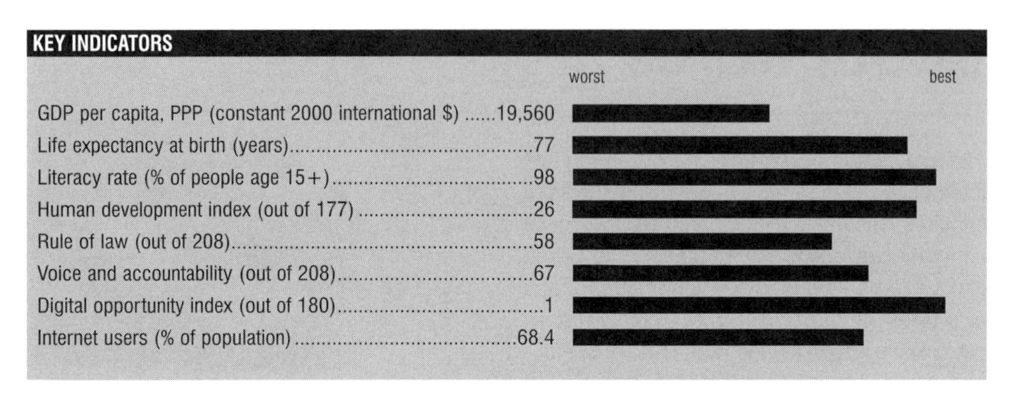

Source (by indicator): World Bank 2005, 2006a; U.S. Department of State 2007b; UNDP 2006; World Bank 2006c, 2006c; ITU 2006, 2005

those publicly supporting North Korea and its policies.[7] In a recent celebrated case, a sociology professor at Dongguk University was investigated by authorities and suspended by the university for posting an article on the Internet in which he argued that North Korea's invasion of the South in 1950 should be interpreted as an attempt to reunify the two Koreas.[8] Overall, however, Korea's human rights record has steadily and markedly improved since the 1990s.[9]

Internet in South Korea

South Korea is one of the most connected countries in the world. By 2005 more than 89 percent of South Korean households had Internet access; 75 percent of these households used broadband.[10] South Koreans are connected to the most advanced national network infrastructure in the world. Following the Asian financial crisis in the late 1990s, South Korea invested heavily in its broadband infrastructure, providing its citizens with a national network that carries data at speeds up to 50 megabits per second.[11] A majority of South Korean Internet users use the Internet more than once per day.[12] The vast majority of users access the Internet from home.[13] Even so, playing video games and chatting online remains a popular pastime in the approximately 30,000 broadband "PC bangs" (Internet cafés) throughout South Korea.[14] Online gaming, fueled by South Korea's ultra high speed broadband infrastructure, is a national obsession, with as much as 35 percent of the population playing online games regularly.[15]

By 2004, seventy-six different Internet service providers (ISPs) were providing connection services to South Korean Internet users.[16] But three South Korean ISPs control nearly 85 percent of the market for Internet access, the largest of which—KorNet—provides about half the ADSL lines in the country, making it the largest ADSL supplier in the world.[17]

In accordance with state ethics guidelines, most South Korean search engines require users to verify they are at least nineteen years old (using a national identification number) before allowing access to porn sites.[18] Peer-to-peer file sharing is a popular online activity in South Korea,[19] though authorities have begun to crack down on peer-to-peer services and monitor them for pornography and other content deemed harmful. Anecdotally, however, many users appear able to circumvent the various technological restrictions on Internet use and have unrestricted access to pornography and other sites that the state deems harmful or offensive.

Online citizens' media has played an important role in Korean politics and Internet culture in recent years, led by www.ohmynews.com, a popular Seoul-based online newspaper that mostly publishes articles written and submitted by ordinary citizens.[20] OhmyNews has been widely acknowledged as strongly influencing the 2002 election of Korean President Roh Moo-hyun.[21]

Legal and regulatory frameworks

The primary regulation governing Internet speech in South Korea is the NSL. First promulgated in 1948, the NSL was designed to prevent communist ideology and pro–North Korea sentiment from penetrating South Korean society.[22] The NSL punishes pro–North Korea activists by criminalizing "antistate" activities.[23] The statute provides for up to seven years' imprisonment for "those who praise, encourage, disseminate or cooperate with anti-state groups ... being aware that such acts will endanger the national security and the democratic freedom."[24] The NSL provisions are vague, permitting state actors broad discretion in their application. The statute governs both print and online media, and has been invoked against individuals attempting to engage with North Korea or promote North Korea's political views. It has therefore been cited as having a chilling effect on free expression in the media.[25] Citing the NSL, the Ministry of Information and Communication in 2004 instructed ISPs in South Korea to block access to thirty-one Web sites considered to be North Korean propaganda.[26]

The NSL is immensely controversial in South Korean society and is a focal point of intense debate between conservative leaders, who argue the law is necessary to protect the nation from threats posed by North Korea, and liberal politicians, who argue the law is repressive, dictatorial, and outdated, and should therefore be repealed.[27] In 2004, the Korean Constitutional Court upheld Article 7 of the NSL, which criminalizes the act of publicly praising and supporting North Korea, as a constitutionally permissible restriction on speech.[28]

Several other laws and decrees extend legal liability to content posted on the Internet, including the Telecommunications Business Act, which makes it illegal to transmit over telecommunications lines any content that compromises public safety, order, or morals;[29] and the Election Law, amended in 2004 to illegalize Internet dissemination of information that defames politicians during their election campaigns and to empower authorities to review ISP records containing information about suspected violators.[30]

The Korean Internet Safety Commission (KISCOM), formerly the Information and Communications Ethics Committee (ICEC), is an independent body established in 1995 under the Telecommunications Business Act to formulate a code of communications ethics and inform state policy aimed at "eradicating subversive communications and promoting active and healthy information."[31] KISCOM is empowered to define harmful content and recommend which Web sites should be blocked.[32] KISCOM also employs a system to monitor the circulation of "illegal and harmful contents on the Internet."[33] In addition, KISCOM formulates and administers a voluntary "Internet Content Rating Service" permitting Web sites to self-evaluate their level of appropriateness for minors, and provides to parents and schools filtering software and related technologies compatible with the rating service.[34]

ISPs have become increasingly responsible for policing content on their networks. In 2001, the state promulgated the Internet Content Filtering Ordinance,[35] which requires ISPs to block as many as 120,000 Web sites on a state-compiled list, and requires Internet access facilities that are accessible to minors, such as public libraries and schools, to install filtering software.[36] The Youth Protection Act of 1997[37] makes ISPs officially responsible, as "protectors of juveniles," for making inappropriate content inaccessible on their networks.[38]

The 2001 ordinance also classified homosexual Internet content as "harmful and obscene" under the Youth Protection Act.[39] The Ministry of Information and Communications formally adopted this classification and immediately ordered a large South Korean Web site devoted to issues of homosexuality to classify itself in ICEC's content rating system as harmful and block minors from accessing the site or face fines and imprisonment.[40] Homosexual rights advocates challenged the order in court as an illegal restriction on free speech. Although the court ruled in favor of the ICEC, it seriously questioned the constitutionality of ICEC's ordinance classifying homosexual content as harmful to minors.[41] In 2003, the Korean National Youth Protection Committee removed homosexuality from the categories of "harmful and obscene." The reversal came in response to a Korean National Human Rights Protection Committee resolution finding that classifying homosexual content as harmful and obscene is an unconstitutional restriction on individuals' rights of expression and pursuit of happiness.[42]

ONI testing results

In 2001, South Korea reportedly required its ISPs to block as many as 120,000 sites on an official list.[43] When ONI conducted its testing at the end of 2006, however, the evidence indicated that Internet filtering in South Korea, although present, is not as extensive as reports have suggested. Testing was conducted through residential Internet access inside South Korea on the two of the largest South Korean ISPs—KorNet and HanaNet—between October 2006 and January 2007. The testing revealed that South Korea filters political and social content, specifically targeting sites containing North Korean propaganda or promoting the reunification of North and South Korea, as well as a handful of sites devoted to gambling and two sites devoted to pirated software (www.mscracks.com and www.kickme.to/fosi).

ONI determined that a large majority of pro-North Korea or pro-unification Web sites on ONI's testing list were blocked,[44] along with a selected number of gambling-related sites. The blocking was extremely consistent across the two ISPs tested, as in virtually every instance a Web site that registered as blocked on HanaNet registered as blocked on KorNet as well. On each ISP, ONI detected two methods of blocking: IP (Internet Protocol) blocking and domain name server (DNS) tampering. IP blocking occurs at the router level, between the South Korean ISP and the Internet. The routers are programmed to stop information coming from certain IP addresses. DNS tampering prevents Internet domain names from resolving to their proper IP addresses. Sites blocked by KorNet through DNS tampering resolve to a blockpage hosted by the police at http://211.253.9.250/, which states that the page has been lawfully blocked and lists the user's own IP address.

ONI's tests suggested there is little blocking of sensitive social content in South Korea, despite KISCOM's focus on cleansing the Web of "harmful" social content. Besides two sites devoted to pirated software, ONI's testing registered no blocks in other social categories, including pornography and gay and lesbian content. South Korea does, however, attempt to restrict minors' access to pornography by requiring age identification for entry to Korean porn sites.

Conclusion

Although South Korea is the world leader in Internet and broadband penetration, its citizens do not have access to a free and unfiltered Internet. The state imposes a substantial level of filtering for a free and democratic society. It requires ISPs to block sites on government lists and fosters a culture of self-censorship through broadly worded laws that make individuals criminally liable for posting "antistate" content. The state also encourages Korean Web site operators to engage in a self-rating system, and requires ISPs and other Internet access facilities, such as cybercafés and schools, to self-police for content deemed harmful to youths. Despite reports that the South Korean government has considered discontinuing its filtering of pro–North Korean Web sites,[45] ONI's testing indicated that the government still filters a large amount of content related to North Korea, as well as a handful of Web sites devoted to gambling and pirated software.

NOTES

1. BBC Online, Country Profile: South Korea, November 10, 2006, http://news.bbc.co.uk/1/hi/world/asia-pacific/country_profiles/1123668.stm#media.
2. See Time, "North, South Korea to restart talks," February 15, 2007, http://www.time.com/time/world/article/0,8599,1590185,00.html.
3. See, for example, Asian Human Rights Commission, "South Korea: Seven activists detained and charged with violating the National Security Law," August 31, 2001, http://www.ahrchk.net/ua/mainfile.php/2001/160/.
4. Amnesty International, Republic of Korea (South Korea): Time to Reform the National Security Law, February 1, 1999, http://web.amnesty.org/library/Index/ENGASA250031999?open&of=ENG-KOR.
5. Ibid.
6. Ibid.
7. See Human Rights Watch, South Korea: Events of 2006, in World Report 2007, http://hrw.org/englishwr2k7/docs/2007/01/11/skorea14758.htm.
8. Cho Chung-un, "Kang case rekindles debate on National Security Law," *Korea Herald,* October 17, 2005, http://www.asiamedia.ucla.edu/article.asp?parentid=31651.
9. See Amnesty International Australia, "South Korea: Moving forward on human rights," February 12, 2007, http://www.amnesty.org.au/Act_now/campaigns/asia_pacific/features/south_korea_moving_forward_on_human_rights.
10. International Telecommunication Union, *World Telecommunication Indicators 2006*.
11. See Kristin Kalning, "Forget reality TV. In Korea, online gaming is it," MSNBC.com, February 21, 2007, http://www.msnbc.msn.com/id/17175353/.
12. See South Korea Internet Statistics Information System, http://isis.nida.or.kr/.
13. Ibid.
14. Paul Budde Communications Pty Ltd., Asia: Broadband: Market Overview, 2006, p. 12.
15. See Kristin Kalning, "Forget reality TV. In Korea, online gaming is it," MSNBC.com, February 21, 2007, http://www.msnbc.msn.com/id/17175353/.
16. Paul Budde Communications Pty Ltd., South Korea: Key Statistics, Telecom Market Overview & Analysis – 2005, 2006, p. 1.
17. Paul Budde Communications Pty Ltd., South Korea: Broadband Market: Overview & Statistics, 2006, pp. 1–2.
18. See, for example, Kim Tae-Gyu, "Is Google becoming victim of its own success? Strong search capability allegedly used for identity theft, porn site harvesting," *Korea Times,* July 27, 2006, http://times.hankooki.com/lpage/200607/kt2006072718493610230.htm.
19. See, for example, Cho Jin-So, "Korea: Online music sharing flourishes," *Korea Times,* August 29, 2006, http://www.asiamedia.ucla.edu/article.asp?parentid=51923.
20. See http://ohmynews.com; see also http://en.wikipedia.org/wiki/OhmyNews; Christopher M. Schroeder, "Is this the future of journalism?" Newsweek, June 18, 2004, http://www.msnbc.msn.com/id/5240584/site/newsweek/.
21. See Christopher M. Schroeder, "Is this the future of Journalism?" Newsweek, June 18, 2004, http://www.msnbc.msn.com/id/5240584/site/newsweek/.
22. *The Washington Post,* "South Korea weighs allowing once-taboo support for the North," November 22, 2004, http://www.washingtonpost.com/wp-dyn/articles/A2477-2004Nov21.html.
23. John Kie-chiang Oh, *Korean Politics: The Quest for Democratization and Economic Development,* pp. 36–7, Ithaca and London: Cornell University Press (1999).
24. National Security Law of South Korea, Article 7, (unofficial English translation at http://www.kimsoft.com/Korea/nsl-en.htm).
25. See, for example, Sandra Coliver, Paul Hoffman, Joan Fitzpatrick and Stephen Bowen, eds., Secrecy and Liberty: National Security, Freedom of Expression and Access to Information, pp. 420–23, The Hague and Boston : M. Nijhoff Publishers (1999).
26. See Nart Villeneuve, The Filtering Matrix: Integrated Mechanisms of Information Control and the Demarcation of Borders in Cyberspace, January 2006, available at http://www.firstmonday.org/issues/issue11_1/villeneuve/.
27. *Korea Herald,* "A nation-splitting law," September 8, 2004, http://www.asiamedia.ucla.edu/article.asp?parentid=14429.
28. Ibid.

29. Telecommunications Business Act of the Republic of Korea, Article 53.
30. See Public Official Election Act of the Republic of Korea, Articles 8-5, 8-6, 272-3, English translation available at http://www.nec.go.kr/english/res/Public_Official_Election.pdf.
31. See Korea Internet Safety Commission Web site, http://www.icec.or.kr/.
32. Ibid.
33. Ibid.
34. Ibid; see also 2600 News, "Internet censorship in South Korea," June 5, 2002, http://www.2600.com/news/view/article/1184.
35. Electronic Frontiers Australia, Internet Censorship: Law & Policy Around the World, 2002, http://www.efa.org.au/Issues/Censor/cens3.html#sk.
36. See, for example, http://web.skku.edu/~sktimes/251/society.html.
37. See Youth Protection Committee Web site (in Korean), http://youth.go.kr/.
38. See *Korea Times,* "Teenagers to be blocked from cyber pornography," August 19, 2004, http://search.hankooki.com/times/times_view.php?term=cyber+pornography++&path=hankooki3/times/lpage/tech/200408/kt2004081918161611810.htm&media=kt.
39. See Han Chae-yun and Yi Huso, Sungkyun Times, "On-again and off-again: Korean on/off-line LGBTQ/Iban community blocked," September 2002, http://web.skku.edu/~sktimes/251/society.html; http://www.gaylawnet.com/news/2002/ce02.htm.
40. Ibid.
41. Chosun.com, "Homosexual Web site ruled constitutional," December 22, 2003, http://english.chosun.com/w21data/html/news/200312/200312220010.html.
42. See Homosexuality Removed from Classification of "Harmful and Obscene" in Youth Protection Law, Sodomy Laws, April 22, 2003, http://www.sodomylaws.org/world/south_korea/sknews001.htm.
43. See, for example, Reporters Without Borders, "Internet under surveillance," at http://www.rsf.org/article.php3?id_article=7248.
44. The blocked pro–North Korea sites on ONI's testing list include: http://osaka.korea-htr.com/koreakokoku.html, http://www.baekdoonet.has.it/, http://www.bommin.net/, http://www.chongryon.com/index.html, http://www.cnet-ta.ne.jp/juche/defaulte.htm, http://www.dprk-book.com/, http://www.dprk-stamp.com/, http://www.jpth.net/, http://www.kancc.org/, http://www.kcckp.net/, http://www.kcna.co.jp/, http://www.korea-dpr.com/, http://www.korea-np.co.jp/main/main.aspx, http://www.krbook.net/index-k.htm, http://www.krsrt.com/, http://www.minjok.com/, http://www.ournation-school.com/, http://www.silibank.com/silibank/korea/, http://www.uriminzokkiri.com/, and http://www.worldcorea.net/.
45. See, for example, Chosun.com, "S. Korea to lift ban on pro–North Korean Web sites," January 5, 2005, http://english.chosun.com/w21data/html/news/200501/200501050015.html.

Sudan

Sudan openly acknowledges filtering content that transgresses public morality and ethics or threatens order.[1] The state's regulatory authority has established a special unit to monitor and implement filtration; this primarily targets pornography and, to a lesser extent, gay and lesbian content, dating sites, and provocative attire.

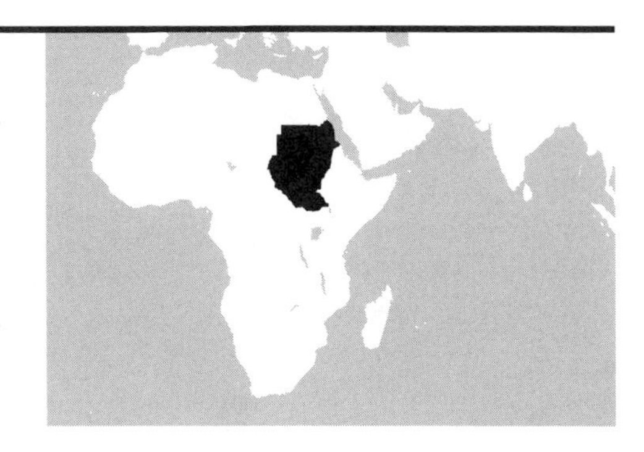

Background

Since gaining independence from the UK in 1953, Sudan has been plagued by constant strife and civil war, which have stunted the development of both the economy and the government.[2] Previously an authoritarian state with all effective power vested in the president, Sudan is currently in a period of transition following the historic signing of the Comprehensive Peace Agreement (CPA) in 2005.[3] The CPA requires the sharing of power and wealth between the rebel Sudan People's Liberation Movement/Army (SPLM/A) and the Government of Sudan.[4] The CPA has prompted the drafting of an interim national constitution that affords basic rights, including freedom of religion and of the press, and that prohibits human rights abuses, including torture and cruel punishment. In practice, however, violations of these provisions by the government and its security forces are numerous.[5] Non-Muslims, non-Arab Muslims, and Muslims from sects unaffiliated with the ruling party face discriminatory policies and practices, as evidenced in the allocation of government jobs.[6] Killings of civilians in conflict, abductions, life-threatening prison conditions, arbitrary arrests and detentions (of political opponents as well as journalists), and human trafficking (often for sexual exploitation, forced labor, or military conscription) constitute additional human rights violations.[7]

RESULTS AT A GLANCE

Filtering	No evidence of filtering	Suspected filtering	Selective filtering	Substantial filtering	Pervasive filtering
Political	●				
Social					●
Conflict/security	●				
Internet tools				●	

Other factors	Low	Medium	High	Not applicable
Transparency			●	
Consistency			●	

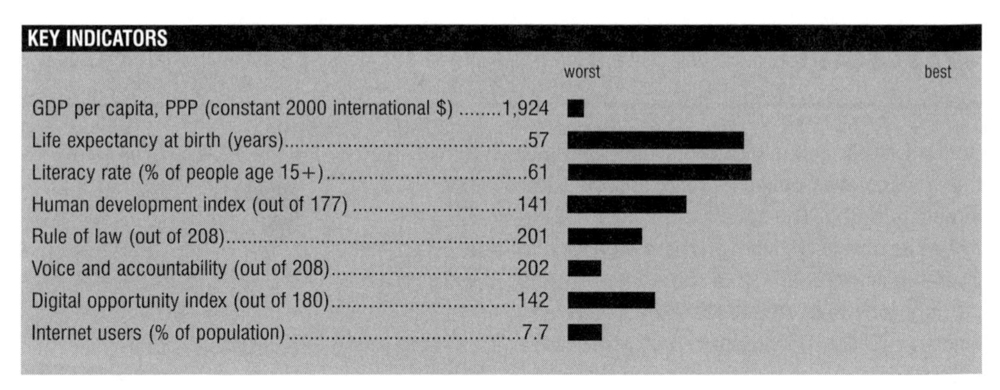

Source (by indicator): World Bank 2005, 2006a, 2006a; UNDP 2006; World Bank 2006c, 2006c; ITU 2006, 2005

Violence and human rights abuses continue in the Darfur region of western Sudan, in a conflict that has spread across the Chad border. In February 2007, the government of Sudan denied the U.N. Human Rights Council visas to enter Darfur to conduct an impartial review.[8] U.N. officials say that conflict in the region has resulted in over 400,000 deaths and displaced approximately two million people.[9]

Internet in Sudan

Internet usage in Sudan is limited. Where infrastructure does exist, access can be prohibitively expensive. There are few locally produced Web pages.[10]

The infrastructure in Sudan is not optimized for high-speed data communications services, and both the capability and reliability of domestic data networks need improvement. Fifteen Internet service providers (ISPs) operate in Sudan (2006), but only two have direct connectivity to the global Internet; the rest are considered by the Sudanese government to be operating illegally.[11]

The number of home Internet subscriptions increased by a factor of ten between 2001 and 2005, rising from 50,000 to 500,000. During the same period, the number of Internet cafés more than doubled. However, Internet usage remains concentrated in Khartoum, accounting for 95 percent of Internet users. The majority of Internet users in Sudan rely on dialup connections (59 percent), and very few have high-speed Internet (19 percent). While 81 percent of universities in Sudan are Internet-equipped, most (65 percent) still use dialup connections.

The information and telecommunications sector in Sudan is regulated by the National Telecommunication Corporation (NTC). In 1993, the state-owned Public Telecommunication Corporation was transformed into the Sudan Telecommunication Company (Sudatel), allowing private investors to purchase a share in the enterprise. However, two-thirds of the shares of the company remained in government hands while it assumed exclusive operational control of the sector.[12]

In 2001, the Sudanese government adopted the National Strategy for Building the Information Industry, with the goal of enabling "all sectors of society to access information media in a way leading to the widest dissemination and utilization of information, all of which shall contribute to achieve an appreciated economic growth, wealth development, job opportunities, enhancement of all-sector production rates and eradication of poverty."[13] As a result of the Strategy, Sudatel's monopoly over mobile telephony ended in 2002 and competitive operators—including several ISPs—in telecommunications were licensed.[14]

Legal and regulatory frameworks

Article 39 of the 2005 interim national constitution of the Republic of Sudan states that "[e]very citizen shall have an unrestricted right to the freedoms of expression, reception and dissemination of information, publication, and access to the press without prejudice to order, safety or public morals as determined by law."[15] The same article also states that the "state shall guarantee the freedom of the press and other media as shall be regulated by law in a democratic society."[16] However, in practice, these rights have been severely restricted.[17] Since emergency laws (which had provided for official censorship) were lifted on July 9th, 2005, the government has continued to censor print media.[18] In 2006, the New York–based Committee to Protect Journalists (CPJ) voiced alarm over "increasing censorship of opposition and independent newspapers in Sudan."[19] Additionally, fear of reprisals has led to self-censorship among journalists.[20]

The 2001 National Strategy for Building the Information Industry called for filtering Internet content that is "morally offensive and in violation of public ethics and order, [and] that may promote corruption and deface traditional identity."[21] The NTC declares that, although it targets several categories, "[t]he most important is the pornographic material, which accounts for over 95 percent of the total volume of the censored materials. Other categories include pages related to narcotics, explosives, alcohols, sacrilege, blasphemy, and gambling."[22] Interestingly, the NTC uses Western peer-reviewed research to support its decision to block these materials in defense of the public good. The NTC states that "[t]here is no political site among the list of blocked sites," and admits that "some translation sites were blocked as they were used to circumvent filtering."[23]

The NTC has set up a special filtering unit to screen Internet media before it reaches users in Sudan. The NTC asserts that sites are filtered based on their contents rather than their names, and that filtering is needed "to preserve noble values and . . . safeguard the society against evil."[24] According to the NTC, the Internet Service Control Unit receives daily requests to add Web sites to, or remove them from, the blacklist. The NTC makes available on its Web site an e-mail address for such requests.[25]

ONI testing results

Testing was conducted on two ISPs in Sudan, Sudanet and Zina Net. Their blocking behavior was identical.

Pornography was extensively filtered. However, some online discussion groups that facilitate the exchange of Arabic sex materials were found to be accessible. There was also some blocking of gay and lesbian, dating, and provocative-attire Web sites. Those dating Web sites that were blocked were those likely to host sexually explicit (for example, www.adultfriend finder.com) or gay and lesbian (www.gay romeo.com) content. Other blocked gay and lesbian Web sites included a site addressing domestic violence (www.lesbians-against-violence.com) and a search portal (www.bglad.com), which were filtered due to being miscategorized as pornography by the commercial software Smart Filter.

Also blocked were health-related sites pertaining to the alteration of body parts, such as www.circumcision.org and www.breast enlargementmagazine.com. Similarly, most of the miscellaneous sites blocked—such as www.collegehumor.com, www.metacafe.com, and www.bootyologist.com—probably contain sexually explicit content.

Access to the feminist Web site www.feminista.com was blocked.

Many of the tested sites that facilitate anonymous Web surfing or circumvention of Internet filters were blocked. Additionally, some Web sites with hacking, cracking, or WAREZ content were blocked.

A small number of translation Web sites—which the NTC argues are used to circumvent filtering[26]—were blocked.

Only one tested blog, Boingboing, was blocked. This may have been an unintentional artifact of Smart Filter, which categorizes Boingboing as a pornographic Web site.[27] Still, blogging is subject to scrutiny and can incur serious consequences. In October 2006, Sudan expelled Jan Pronk, a top U.N. official, from the country after he posted in his blog (www.janpronk.nl) sensitive statements relating to the conflict in Darfur.[28] ONI has monitored and verified the blog's accessibility from Sudan.

Some Web sites discussing Christianity or criticizing Islam, such as www.islamreview.com, were blocked.

The Arab Network for Human Rights Information (www.hrinfo.org) reported that the NTC blocked access to the Web site www.sudaneseonline.com in 2004.[29] This site was not found to be blocked during ONI testing.

Conclusion

Online pornography is extensively blocked in Sudan, as the government openly acknowledges. Many anonymizer and proxy Web sites are blocked, as are some sites related to provocative attire, dating, and gay and lesbian interests. Sudan is relatively transparent in its filtering of the Internet compared with other Arab states, and even provides an appellate process for challenging the blocking of a site.

NOTES

1. National Telecommunication Corporation, Information Society in Sudan: National Strategy for Building the Information Industry, http://www.ntc.org.sd/english/ntc/management%20telecom%20sector/information%20technology/info_society.htm.
2. U.S. Department of State, Bureau of African Affairs, Background Note: Sudan, November 2006, http://www.state.gov/r/pa/ei/bgn/5424.htm.
3. Ibid.
4. Ibid.
5. U.S. Department of State, Country Reports on Human Rights Practices 2005: Sudan, http://www.state.gov/g/drl/rls/hrrpt/2005/61594.htm.
6. U.S. Department of State, International Religious Freedom Report 2005: Sudan, http://www.state.gov/g/drl/rls/irf/2005/51497.htm.
7. U.S. Department of State, Country Reports on Human Rights Practices 2005: Sudan, http://www.state.gov/g/drl/rls/hrrpt/2005/61594.htm.
8. Sean McCormack, spokesman, U.S. Department of State, Bureau of Public Affairs, Press statement, Human Rights Council: Sudan Assessment Mission Denied Visas, February 16, 2007, http://www.state.gov/r/pa/prs/ps/2007/february/80619.htm.
9. U.N. News Centre, "Annan welcomes extension of African Union mission in Darfur," September 21, 2006, http://www.un.org/apps/news/story.asp?NewsID=19948.
10. National Information Centre, Republic of Sudan Ministry of the Cabinet, Sudan E-Readiness Assessment Report 2006.
11. Ibid.
12. National Telecommunication Corporation, Establishments, Objectives & Powers, http://www.ntc.org.sd/english/ntc/estb-%20obj-func.htm.
13. National Telecommunication Corporation, Information Society in Sudan: National Strategy for Building the Information Industry, http://www.ntc.org.sd/english/ntc/management%20telecom%20sector/information%20technology/info_society.htm.
14. Ibid.
15. The Republic of the Sudan, The Draft Constitutional Text, March 16, 2005, http://www.mfa.gov.sd/arabic/contry_leader/20060902130908.doc (in Arabic), translated at http://www.reliefweb.int/library/documents/2005/govsud-sud-16mar.pdf.
16. Ibid.
17. U.S. Department of State, Country Reports on Human Rights Practices 2005: Sudan, http://www.state.gov/g/drl/rls/hrrpt/2005/61594.htm.
18. Ibid.

19. Committee to Protect Journalists, "Sudan: Authorities intensify newspaper censorship and seizures," September 14, 2006, http://www.cpj.org/news/2006/mideast/sudan14sept06na.html.
20. U.S. Department of State, Country Reports on Human Rights Practices 2005: Sudan, http://www.state.gov/g/drl/rls/hrrpt/2005/61594.htm.
21. National Telecommunication Corporation, Information Society in Sudan: National Strategy for Building the Information Industry, http://www.ntc.org.sd/english/ntc/management%20telecom%20sector/information%20technology/info_society.htm.
22. National Telecommunication Corporation, Filtering of Information and Blocking of Offensive Sites on the Internet, http://www.ntc.org.sd/english/filtering/filtering.htm.
23. Ibid.
24. Ibid.
25. Ibid.
26. Ibid.
27. Xeni Jardin, "Exporting censorship," *The New York Times,* March 9, 2006.
28. See Jonathan Steele, "Sudan expels UN official for blog revealing Dafur military defeats," *The Guardian,* October 23, 2006, http://www.guardian.co.uk/sudan/story/0,,1929019,00.html.
29. Arab Network for Human Rights Information, "Sudan: Banning SudaneseOnline, A violation of freedom of expression," July 10, 2004, http://www.hrinfo.net/en/reports/2004/pr040710.shtml.

Syria

In addition to filtering a range of Web content, the Syrian government monitors Internet use closely and has detained citizens "for expressing their opinions or reporting information online."[1] Vague and broadly worded laws invite government abuse and have prompted Internet users to engage in self-censoring and self-monitoring to avoid the state's ambiguous grounds for arrest.[2]

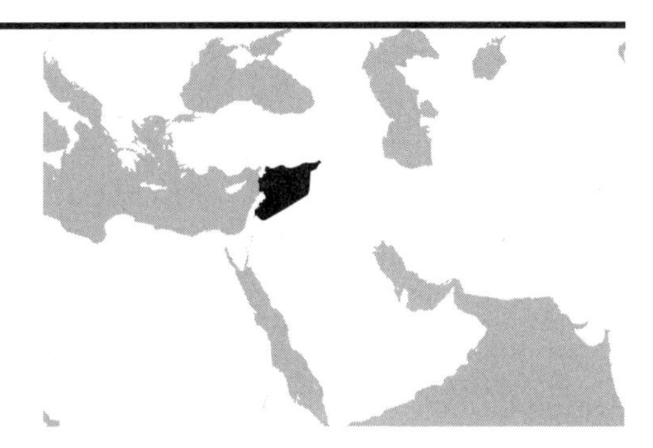

Background

Syria is among the most repressive countries in the world with regard to freedom of expression and information. Criticisms of the president and reports on the problems of religious and ethnic minorities in Syria remain particularly sensitive topics.[3] Human rights organizations have reported exhaustively on political arrests and detentions.[4]

In 2006 Reporters Without Borders ranked Syria among the thirteen "enemies of the Internet."[5] Although the government does recognize the importance of the Internet as a source of economic growth, it also admits to automatically blocking pornographic Web sites[6] and to censoring "pro-Israel and hyper-Islamist" Web sites, such as "those run by the illegal Muslim Brotherhood, and those calling for autonomy for Syrian Kurds."[7] In defense of these practices, Minister of Technology and Communications Amr Salem said, "Syria is currently under attack … and if somebody writes, or publishes or whatever, something that supports the attack, they will be tried."[8]

Internet in Syria

With a literacy rate of 80 percent,[9] Syria's main barriers to Internet access are economic. Only 4.2 percent of the population own personal computers, with just 1 percent of Syrians

RESULTS AT A GLANCE

Filtering	No evidence of filtering	Suspected filtering	Selective filtering	Substantial filtering	Pervasive filtering
Political					●
Social			●		
Conflict/security			●		
Internet tools				●	

Other factors	Low	Medium	High	Not applicable
Transparency		●		
Consistency			●	

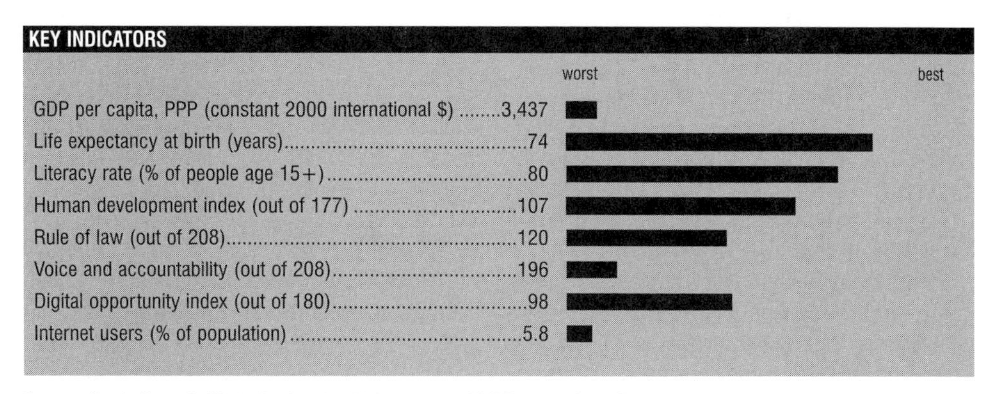

Source (by indicator): World Bank 2005, 2006a, 2006a; UNDP 2006; World Bank 2006c, 2006c; ITU 2006, 2005

subscribing to Internet services.[10] The proliferation of Internet cafés[11] has helped raise the Internet penetration rate to approximately 6 percent,[12] but many Syrians still find the cost of these cafés prohibitive.[13]

In recent years, the government has endeavored to expand Internet access by installing hardware and telecommunications capabilities in schools, by subsidizing the cost of personal computers, and, most recently, by fostering competition among Internet service providers (ISPs).[14]

There are four ISPs that are neither owned nor funded by the government. Still, the two government-affiliated ISPs,[15] Syria Telecommunication Establishment (STE) and SCS-net (now Aloola), continue to occupy the majority of the market.[16] Aya, one of the privately owned ISPs, has close ties to the government.[17]

Legal and regulatory frameworks

In addition to maintaining regulatory control over ISPs, the Syrian government imposes financial and technical constraints on Internet users. Syrian Internet subscribers wishing to use ports other than port 80—the port most often used for Web browsing—must apply for a special service and pay a small monthly fee.[18] Aya and other ISPs offer plans that allow users to access the Internet with a fixed IP address, which is necessary to host sites, to use Virtual Private Networks, and to bypass the ISP's proxy server. They may also pay for a special plan that allows them to open otherwise blocked ports, such as those used for Voice-over Internet Protocol (VoIP) and video chat.[19]

Points of Internet access are also strictly regulated and sometimes monitored. To open an Internet café an owner must obtain a license from the Telecommunications Department's office in the local governorate. To acquire a license, the owner must follow the regulations in the Conditions Manual, which include specifications on the spacing between computers.[20] Though users at Internet cafés are not required to show ID or give their names, some Syrians have reported that plainclothes officials watch Internet cafés and take note of the users.[21]

The Constitution of the Arab Republic of Syria affords every citizen "the right to freely and openly express his views in words, in writing, and through all other means of expression," while also guaranteeing "the freedom of the press, of printing, and publication in accordance with the law."[22] In actuality, these freedoms are limited by other legislative provisions. Article 4(b) of the 1963 Emergency Law authorizes the government to monitor all publications and communications.[23] That law also allows the government to arrest those who commit "crimes which constitute an overall hazard" or other vaguely defined offenses.[24]

The Press Law of 2001 subjects all print media—from newspapers, magazines, and other periodicals to books, pamphlets, and posters—to government control and censorship.[25] Printing "falsehoods" or "fabricated reports" is a criminal offense under the Press Law, and writing on topics relevant to "national security [or] national unity" is forbidden.[26] Violators may be penalized with hefty monetary fines, lengthy prison terms, or license suspensions or revocations.[27] Furthermore, "periodicals that are not licensed as political publications [are prohibited] from publishing 'political' articles"—a provision that "amounts to blanket government censorship."[28] Thus, although the Internet has facilitated access to unofficial information, that information is limited by the controls and threats codified in Syrian law.

The government has demonstrated its willingness to punish Syrians for writing and transmitting information online.[29] Authorities have detained individuals for e-mailing an image or article produced by another party, for voicing complaints about the government, and for posting original photographs of police crackdowns on the Web.[30] These incidents have engendered caution and self-censorship across the Syrian Internet as a whole and within the Syrian blogosphere, which nonetheless continues to grow and to become more vibrant.[31]

ONI testing results

Testing was conducted on one of the main ISPs in Syria, Aloola (formally SCS-Net). Although the tests indicate that Syria now blocks fewer Web sites than it has in the past, many sites remain blocked.

The Web site of the Syrian branch of the Muslim Brotherhood, www.jimsyr.com, was blocked, though the Web site of the Egyptian branch, the region's largest, was available. Two Kurdish Web sites, www.tirej.net and www.amude.net, were blocked, as was the Web site of the United States Committee for a Free Lebanon (www.freelebanon.org), which campaigns for an end to Syrian influence in Lebanese politics. The Arabic- and English-language sites of the unrecognized Reform Party of Syria were filtered, along with the Web sites of the *Hizb al-Tahrir* (Liberation Party)—an Islamist group that seeks to restore the Caliphate and that remains banned in many countries.

ONI's tests found that 115 Syrian blogs hosted on Google's popular blogging engine, www.blogspot.com, were blocked, strongly suggesting that the ISP had blocked access to all blogs hosted on this service, including many apolitical blogs. www.freesyria.wordpress.com, a blog created to campaign for the release of Michel Kilo, a prominent Syrian journalist imprisoned for his writings, was also blocked.

In the past, Syria has reportedly filtered access to popular e-mail sites. ONI testing found www.hotmail.com to be blocked, along with two, relatively small Web-based e-mail sites, www.address.com and www.netaddress.com. None of the Arabic-language e-mail sites ONI tested were blocked, though the Arabic-language hosting site www.khayma.com was.

Nearly one-third of the anonymizer sites tested were blocked, indicating some measure of effort to preempt circumvention.

Though most foreign news sites were accessible, Web sites of some important Arabic newspapers and news portals were found to be blocked. Examples include the pan Arab, London-based, Arabic-language newspapers, *Al-Quds al-Arabi* (www.al-quds.co.uk) and *Al-Sharq al-Awsat,* (www.asharqalawsat.com), the news portal www.elaph.com, the Kuwaiti newspaper *Al Seyassah* (www.alseyassah.com), the U.S.-based Web site of the *Arab Times* (www.arabtimes.com), and the Islamically oriented news and information portal www.islamonline.net. These publications frequently run articles critical of the Syrian government.

Web sites of human rights organizations were generally available. Sites associated with the London-based Syrian Human Rights Committee (SHRC) marked an important exception; all URLs on the www.shrc.org.uk domain were found blocked in this round of testing. As indicated above, some blogs that criticize the human rights record of Syria were also blocked.

Only three Web sites of the Web sites tested with pornographic content were blocked: www.playboy.com, www.sex.com, and www.netarabic.com/vb (this last is a message board with pornographic content).

Web sites that focus on lesbian, gay, bisexual, and transgendered issues were generally available. One site, www.gaywired.com, was an exception.

Unfortunately, an insufficient number of Israeli Web sites were tested to confirm whether or not Syria blocks the entire ".il" domain, as past reports have suggested.[32] However, the fact that the Institute for Counter Terrorism's Israeli Web site (www.ict.org.il) was blocked—while the Institute's alternate URL (www.institutefor counterterrorism.org), lacking the ".il" suffix, was not—lends credence to such reports. Furthermore, the Web site for the World Zionist organization (www.wzo.org.il) was blocked.

Conclusion

The Web sites blocked in Syria span a range of categories, with the most substantial filtering occurring among sites that criticize government policies and actions or espouse oppositional political views. Repressive legislation and the imprisonment of journalists and online writers for their activities online have led many Syrians to engage in self-censorship. Meanwhile, the government continues to promote the growth of the Internet throughout the country.

NOTES

1. Human Rights Watch, False Freedom: Online Censorship in the Middle East and North Africa: Syria, November 2005, http://hrw.org/reports/2005/mena1105/.
2. Guy Taylor, "After the Damascus spring: Syrians search for freedom online," *Reason Online: Free Minds and Free Markets,* February 2007, http://www.reason.com/news/show/118380.html.
3. U.S. Department of State, Country Reports on Human Rights Practices 2005: Syria, March 8, 2006, http://www.state.gov/g/drl/rls/hrrpt/2005/61699.htm.
4. See, for example, Syrian Human Rights Committee, Annual Report on Human Rights Situation in Syria 2006 (Covering the period from June 2005 to May 2006), June 2006, http://www.shrc.org.uk/data/pdf/ANNUALREPORT2006.pdf.
5. Reporters Without Borders, "List of the 13 Internet enemies 2006," November 7, 2006, http://www.rsf.org/article.php3?id_article=19603.
6. Guy Taylor, "Syrians search for online freedom," podcast audio of interview with Amr Salem, *World Politics Watch,* January 13, 2007. Excerpts available at http://www.worldpoliticswatch.com/blog/blog.aspx?id=470.
7. Guy Taylor, "After the Damascus spring: Syrians search for freedom online," *Reason Online: Free Minds and Free Markets,* February 2007, http://www.reason.com/news/show/118380.html.
8. Ibid.
9. World Bank, "Syrian Arab Republic Data Profile 2004," *World Development Indicators database* (April 2006), http://devdata.worldbank.org/external/cpprofile.asp?ccode=syr&ptype=cp.
10. OneWorld.net, OneWorld Country Guide: Syria, http://us.oneworld.net/guides/syria/development#top.
11. The Arabic Network for Human Rights Information, "Syria Internet under siege," *The Internet in the Arab World: A New Space of Repression?* June 2004, http://www.hrinfo.net/en/reports/net2004/syria.shtml.
12. International Telecommunication Union, *World Telecommunication Indicators 2006*.
13. The Initiative for an Open Arab Internet, Implacable Adversaries: Arab Governments and the Internet: Syria, December 2006, http://www.openarab.net/en/reports/net2006/syria.shtml.

14. Human Rights Watch, False Freedom: Online Censorship in the Middle East and North Africa: Syria, November 2005, http://hrw.org/reports/2005/mena1105/.
15. Jihad Yazigi, "Syrian telecom to be gradually liberalized," *The Syria Report,* January 10, 2005, http://www.syria-report.com/article2.asp?id=1657 (paid subscription required).
16. The Initiative for an Open Arab Internet, Implacable Adversaries: Arab Governments and the Internet: Syria, December 2006, http://www.openarab.net/en/reports/net2006/syria.shtml.
17. Human Rights Watch, False Freedom: Online Censorship in the Middle East and North Africa: Syria, November 2005, http://hrw.org/reports/2005/mena1105/.
18. OpenNet Initiative interview with a Syrian computer consultant who requested anonymity, November 13, 2006.
19. Aya's list of "VIP plans," for example, is available at http://aya.sy/?id=vip.
20. The Initiative for an Open Arab Internet, Implacable Adversaries: Arab Governments and the Internet: Syria, December 2006, http://www.openarab.net/en/reports/net2006/syria.shtml.
21. Human Rights Watch, False Freedom: Online Censorship in the Middle East and North Africa: Syria, November 2005, http://hrw.org/reports/2005/mena1105/.
22. The Constitution of the Arab Republic of Syria, Article 38, http://www.oefre.unibe.ch/law/icl/sy00000_.html.
23. Syrian Human Rights Committee, Special Report: Repressive Laws in Syria, February 19, 2001, http://www.shrc.org.uk/data/aspx/d4/254.aspx#D3.
24. Ibid.
25. Human Rights Watch, Memorandum to the Syrian Government, Decree No. 51/2001: Human Rights Concerns, January 31, 2002, http://hrw.org/backgrounder/mena/syria/.
26. Ibid.
27. Ibid.
28. Ibid.
29. Human Rights Watch, False Freedom: Online Censorship in the Middle East and North Africa: Syria, November 2005, http://hrw.org/reports/2005/mena1105/.
30. Ibid.
31. Megan K. Stack, "Arabs take byte at regimes," *Democracy Council,* September 12, 2005, http://www.democracycouncil.org/ArabsTakeByte.cfm.
32. The Initiative for an Open Arab Internet, Implacable Adversaries: Arab Governments and the Internet: Syria, December 2006, http://www.openarab.net/en/reports/net2006/syria.shtml.

Tajikistan

Internet access in Tajikistan remains largely unfettered, although the run-up to the December 2006 presidential elections produced a documented case of event-driven filtering of a political Web site. Overall Internet penetration remains weak, and the telecommunications sector is relatively unencumbered by regulation—a consequence of the decentralized nature of the government (itself a result of the compromise that ended the civil war in 1997).

Background

The Internet in Tajikistan emerged as the country was ending a bloody civil war that followed the demise of Soviet rule in the early 1990s. The resulting fragmentation of power also meant that Internet services were developed largely without state interference and the Ministry of Communications played a weak role in the development of the sector. Internet as well as telecommunications services remained fragmented up until the end of the 1990s, with several companies failing to interconnect because of fierce (and at times violent) competition. During the period of instability, Internet service providers (ISPs) were aligned with feuding political and economic interests that spilt over to the competition among the ISPs themselves.

Internet use among Tajiks has been increasing, but remains relatively low (1.19 percent) despite government efforts to make information communications technology (ICT) a pillar of national development. Opposition media are not actively exploiting the Internet's potential largely because of the low levels of penetration and the lack of a mature critical mass of Internet users. None of the registered opposition parties have domain names registered in the ".tj" Internet zone, and only one has its Web site available in Tajik.[1] The incumbent president, who recently

RESULTS AT A GLANCE

Filtering	No evidence of filtering	Suspected filtering	Selective filtering	Substantial filtering	Pervasive filtering
Political			●		
Social	●				
Conflict/security	●				
Internet tools	●				

Other factors	Low	Medium	High	Not applicable
Transparency		●		
Consistency	●			

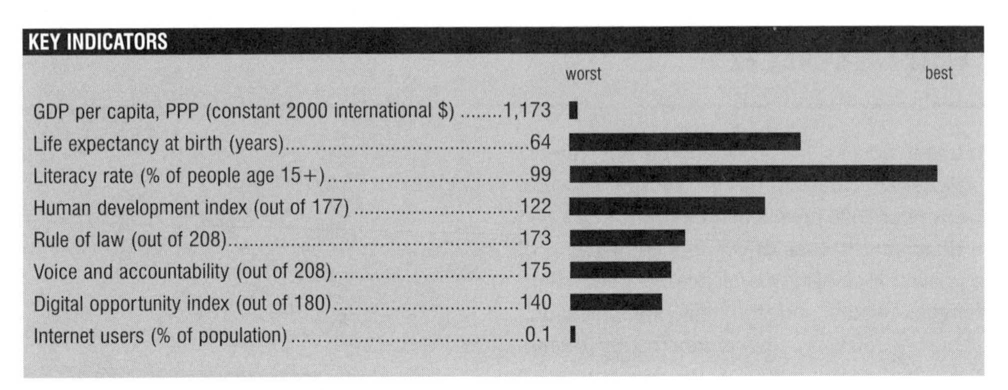

Source (by indicator): World Bank 2005, 2006a, 2006a; UNDP 2006; World Bank 2006c, 2006c; ITU 2006, 2004

started his third seven-year mandate,[2] exerts a degree of control over the independent media while suppressing the opposition with prosecutions based on broad and inconsistent interpretations of the relevant laws.[3]

Internet in Tajikistan

The rate of Internet penetration in Tajikistan is estimated between 0.075 percent[4] and 1.19 percent.[5] Empirical data show the number of active Internet users to be growing rapidly, with estimated total numbers that are higher than the Ministry of Communications' estimate of 26,000.[6] The discrepancy in figures probably arises from the hesitancy of most commercial ISPs to disclose accurate user statistics in order to avoid a per user charge. No official data exist on the number of personal computers in Tajikistan. Khoma,[7] a local nongovernmental organization (NGO), estimates that 1 percent of households own personal computers and over 1 percent of those access Internet from home, mostly using dialup technology. Access via DSL and Wi-Fi technologies is affordable only to a handful of companies. Satellite connection is widely used and few ISPs use Mach 6 technology to connect mountain towns and district regions. The state-owned telecommunications company Tajiktelecom built a connection to the Trans Asia Europe (TAE) fiber-optic highway passing through Uzbekistan; however, ISPs prefer to use their own infrastructure. The Internet exchange point in Tajikistan, managed by the Association of Tajikistani ISPs, connects only four of the eleven ISPs.

Internet access remains largely unaffordable, as the average monthly salary in the country amounts to USD30–40, while the minimum salary drops to USD7. The price for one hour of Internet access in cybercafés is USD0.41; unlimited monthly traffic by dialup access costs USD29.41 and limited ADSL access costs USD25.[8]

Most Internet users are young and access the Internet through cybercafés close to schools and universities. In January 2006, the Ministry of Communications estimated that some 400 cybercafés existed, most concentrated in large cities. The cybercafés, operating as second-tier ISPs, need to obtain licenses before starting their activity. Although over 70 percent of the population resides in rural areas, the Internet is mainly accessible in urban areas because of poor infrastructure and low affordability. A 2005 study by CIPI shows that over three-quarters of Internet users are male.[9]

Tajik is the official national language. However, Russian is the most popular language for Internet use. The most-visited Web site in Tajikistan is www.mail.ru, and the most popular search engines are www.rambler.ru, www.google.com, www.yahoo.com, and www.yandex.ru.

Legal and regulatory frameworks

The Tajik top-level domain name was registered with Internet Assigned Numbers Authority (IANA) in 1997 but the domain name was later suspended as it was used mainly for registering pornography sites. In 2003 the domain name registration was delegated to the Information and Technical Centre of the President of Tajikistan Administration, a state entity that now supervises registrations within the ".tj" domain.[10]

The Ministry of Communications requires all ISPs to obtain licenses in order to operate. Currently eleven first-tier ISPs are actively providing Internet service in the country.[11] The ISPs do not reveal information about their bandwidth because these data are a legally protected commercial secret. This protection extends to the countries from which the connection originates. ONI data reveal that most ISPs have two points of access, one located in Russia and the other in Western Europe. The majority of ISPs are eligible to provide Voice-over Internet Protocol (VoIP) services under an IP-telephony license.[12] Recent amendments require VoIP service providers to obtain a special license from the Ministry of Communications.

The main state entities regulating Internet in Tajikistan are the Security Council (SC), the ICT Council, and the Ministry of Communications. The president of the republic, however, remains the key authority, ratifying the main legal documents in the IT sector and directing ICT policy in the country. The SC controls the implementation of the State Strategy on Information and Communication Technologies for Development of the Republic of Tajikistan (E-Strategy). The SC monitors telecommunications, including Internet, for national security reasons. The ICT Council[13] is responsible for implementing and coordinating work under the E-Strategy and advising the president. The Ministry of Communications is the main regulator in the telecommunications industry and is empowered to issue licenses for any related activities. The government adopted the Conception on the Information Security,[14] which serves as a platform for proclaiming official views, principles, and policy directions to preserve state information security.

The government restricts the distribution of information that contains state secrets and other privileged data that intend to "discredit dignity and honor of the state and the President," or that contain "violence and cruelty, racial, national and religious hostility..., pornography... and any other information prohibited by law."[15] The provisions of this regulation are broad, allowing state agencies wide discretion in their application. The control over information security is assigned to the Main Department of State Secrets and the Ministry of Security.

Tajikistan does not have an official policy on Internet filtering. However, state authorities have been known to restrict access to some Web sites at politically sensitive times by communicating their "recommendations" to all top-level ISPs. Prior to the 2006 presidential elections, the Communications Regulation Agency issued a "Recommendation on Filtering" advising all ISPs that "for the purpose of information security" they should "engage in filtering and close access to those Internet sites that are directed to undermining the state policy on information sphere."[16] As a result, several oppositional news Web sites hosted in Russia or Tajikistan were inaccessible to Tajik users for several days.[17] Although the officials offered unclear reasons for shutting down the Web sites, independent media foresee that the list of affected sites might grow in the future.[18]

ONI testing results

ONI tested in Tajikistan on three key ISPs: Babilon-T, Tajiktelecom, and Telecomm-Technology. The tests revealed no direct evidence of filtering for any of the selected categories. Nevertheless, ONI did document the sporadic filtering of political content during the 2006 presidential election.

Considered the most conservative Central Asian country, with a predominantly Muslim population, Tajikistan does not technically filter access to pornography sites. However, accessing such sites in public centers is illegal. Any such access may be penalized with a fine ranging from USD15 to 100 as provided in the Administrative Code and may be prosecuted under the Criminal Code. Based on ONI's investigation, we concluded that currently most cybercafés do not employ any filtering applications to limit access to information. However, cybercafés routinely monitor users to ensure they do not visit forbidden sites.

Conclusion

Although the government has adopted a strategy aimed at developing information society and employing ICT potential for spurring economic growth, it does not seek to encourage independent online publishers, journalists, and bloggers. Media freedom is widely challenged and subject to *de facto* censorship, although the constitution provides that "state censorship and prosecution for criticism are forbidden."[19] Filtering is unlikely to be declared as an official policy since Tajikistan depends on international aid. The Tajik government, however, has in place policies and instruments to maintain firm control over the distribution of information, particularly before elections. The government is engaged in developing programs aimed at restricting citizens' Internet access, following on from President Rahmonov's message that "Western values aren't always applicable" to Eastern countries.[20]

NOTES

1. See SNGNews.ru, http://sngnews.ru/articles/5/67577.html (in Russian) (accessed May 3, 2007).
2. See Joanna Lillis, "Tajikistan: No surprises in presidential elections," Eurasia Insight, November 6, 2006, http://www.eurasianet.org/departments/insight/articles/eav110606a.shtml (accessed May 2, 2007); and Nigora Buhari-zade, "The opposition raises protests," Deutsche Welle, August 29, 2006, http://www.dw-world.de/dw/article/0,2144,2150509,00.html (in Russian).
3. In 2005 the State Licensing Commission formally denied BBC a license, basing its argumentations on a complex interpretation of the Law on Licensing Certain Types of Activities. See Eurasia Insight, "Tajik government "tightening the screws" on independent media," August 26, 2006, http://www.eurasianet.org/departments/insight/articles/eav082506a.shtml. In addition, in 2005 the leader of the main opposition party, Iskandarov, was convicted on terrorism and corruption charges and sentenced to a twenty-three-year prison term.
4. See InternetWorldStat, http://www.internetworldstat.com/global_internet_stats.htm (accessed May 2, 2007).
5. Estimate by the Civil Initiative on Internet Policy (CIPI), Civil Initiative on IT Tajikistan.
6. Data of the Ministry of Communications of RT, AsiaPlus 30 (313) from January 24, 2006, http://www.asiaplus.tj.
7. Internews Network, http://www.khoma.tj.
8. See State Statistics Committee, http://www.stat.tj; Internet access tariffs of ISP Intercom, http://www.intercom.tj; and ISP Babilon-T, http://www.tojikiston.com.
9. 2005 study conducted by the Civil Initiative on Internet Policy (CIPI), http://www.cipi.tj.
10. See the Tajikistani TLD hosting organization, Information and Technical Centre of the President of Tajikistan Administration, http://www.nic.tj.
11. A joint Tajik-American company, TACOM, stopped providing Internet service in the summer of 2006, but it is still a licensee.
12. See the Ministry of Communications, http://www.mincom.tj, Law on Telecommunications, law no 56 of 2002, May 10, 2002, http://www.tajik-gateway.org/index.phtml?lang=en&id=1414
13. The ICT Council was established by presidential decree no. 1707 of February 27, 2006.
14. The Conception was ratified by presidential decree no. 1175 of November 7, 2003.
15. Points 2 and 3 of regulation no. 389, "On Creating a Republican Network of Data Transfer and Measures to Order Access to Global Information Networks," August 8, 2001, (unofficial translation from Russian).

16. Recommendation on Filtering sent to ISPs by the Communications Regulation Agency (unofficial translation), obtained by ONI researchers, December 2006.
17. See Deutsche Welle, "Access to opposition media websites is forbidden in Tajikistan," October 8, 2006, http://www.dw-world.de/dw/article/0,2144,2198763,00.html; Fergana News, "The websites officially blocked in Tajikistan were announced. Among them – Ferghana.Ru," October 9, 2006, http://www.ferghana.ru/news.php?id=3633&mode=snews; and SNGnews.ru, "In Tajikistan is closed the access to some websites," October 7, 2006, http://sngnews.ru/articles/5/68007.html (accessed May 2, 2007).
18. See at SNGnews, "Internet Service Providers in Tajikistan are prepared for filtering of 'unsafe' Web sites," http://sngnews.ru/articles/5/68051.html (accessed May 3, 2007).
19. Article 30 of the Constitution of the Republic of Tajikistan, 1994.
20. Joanna Lillis, "Tajikistan: No surprises in presidential elections," Eurasia Insight, November 6, 2006, http://www.eurasianet.org/departments/insight/articles/eav110606a.shtml.

Thailand

Amidst lingering political uncertainty, Thailand's censorship of the Internet continues to be a contested and controversial policy because the legal basis for filtering and actual filtering practices are not transparent.

Background

In the aftermath of a military coup that followed years of heightened fear and self-censorship, the Internet community in Thailand continues to face uncertainties created by censorship policies, antiquated laws, regulatory reform, and the privatization of state-owned telecoms. Considered by many to have inaugurated Internet filtering in Thailand, former Prime Minister Thaksin Shinawatra pursued aggressive censorship policies and, through his family-owned Shin Corporation, orchestrated a series of defamation suits against his critics.[1] After Thaksin was deposed in a military coup on September 19, 2006, the interim government abrogated the 1997 Constitution, abolished the Constitutional Court, and imposed a series of restrictions on news reporting and political activity that threatened national solidarity.[2]

Internet in Thailand

Internet usage in Thailand began with a small base and has increased sixfold over the past five years.[3] Initially, rather than encouraging growth of the Internet for all people, the government used and developed it only for state academic institutions and government agencies.[4]

The total number of Internet users in 2005 was estimated at 12,500,000, representing an Internet penetration rate of approximately 19 percent.[5] However, homes and businesses in Bangkok and other major cities make up most of the penetration rate, and there is little Internet connectivity in surrounding areas.[6] In 2004, about 15 percent of schools had access to the

RESULTS AT A GLANCE

Filtering	No evidence of filtering	Suspected filtering	Selective filtering	Substantial filtering	Pervasive filtering
Political			●		
Social				●	
Conflict/security	●				
Internet tools			●		

Other factors	Low	Medium	High	Not applicable
Transparency		●		
Consistency		●		

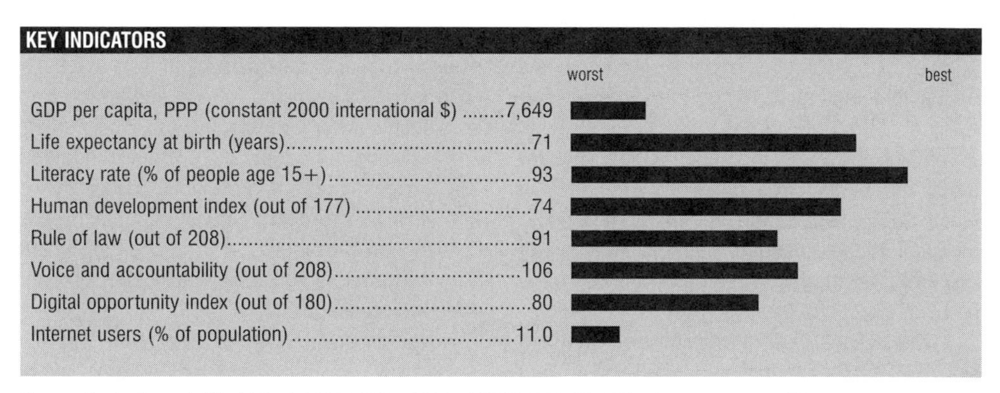

Source (by indicator): World Bank 2005, 2006a, 2006a; UNDP 2006; World Bank 2006c, 2006c; ITU 2006, 2005

Internet.[7] It is believed that more people may use the Internet as content becomes available in local languages rather than English.[8] Although no significant gender divide has emerged, over half of Thai Internet users are between fifteen and twenty-four years old.[9] Of this group nearly 27 percent use the Internet at cybercafés while 18 percent access from home.[10] Broadband Internet access is available, but it is still undeveloped at less than 2 percent household penetration.[11]

Internet connectivity in Thailand is built around education/research networks, commercial networks (Internet service providers, or ISPs), and government networks.[12] CAT Telecom (CAT, formerly the Communications Authority of Thailand) and the Telephone Organization of Thailand (TOT), the two big state-owned telecoms, each operate an international Internet Gateway (IIG) as well as one each of three domestic exchanges for twenty-one licensed ISPs and four noncommercial Internet hubs.[13]

Legal and regulatory frameworks

The Ministry of Information and Communications Technology (MICT) and its subordinate bodies, including the National Information Technology Committee (NITC), CAT, TOT, and the National Electronics and Computer Technology Center (NECTEC), all regulate the Internet.[14]

Prior to the coup the constitution provided a nominal legal basis for censorship, although the precise authority for filtering Internet content remains unclear. Under the abrogated 1997 Constitution, Thai citizens were guaranteed the rights to express opinions; to communicate by "lawful" means; and to access information with certain limitations for state security, maintaining public order or morals, and safeguarding others' right to privacy and reputation.[15] It remains unconstitutional to criticize or level accusations against the king.[16]

Broad claims associating criticism of government with injury to the king, or *lèse majesté,* have also been used to enforce censorship. Thailand is one of the few remaining countries in the world to prosecute crimes of *lèse majesté,* where individuals who insult, defame, or threaten the Thai royal family can be sentenced to three to fifteen years of imprisonment. Such allegations, in spite of King Bhumibol's own sanction of public criticism of the Thai crown, are leveled infrequently but have targeted independent media voices[17] and been used as a "political tool to discredit opponents."[18] *Lèse majesté,* which in

Thailand involves a scope of expression far broader than the actions of the king himself, has begun to form the basis for the blocking and removal of Web sites.[19]

In July 2003 Thailand became the first country to impose a curfew on online gaming.[20] In March 2006 a regulation enforced by the Culture Ministry forbade persons under eighteen years of age from entering Internet cafés between the hours of 10pm and 2pm.[21]

The National Telecommunications Commission (NTC) was brought into operation in late 2004 as an independent telecom regulator and given the exclusive authority to grant licenses for telecom or IT services.[22] Previously, an ISP could obtain a concession contract only by giving a free equity stake of about 35 percent to CAT Telecom (formerly the Communications Authority of Thailand) in exchange for a share of the profits from the networks these companies built and paid for.[23] In March 2005 the NTC announced that it would grant free licenses once permanent guidelines were in place.[24]

In August 2003, Thaksin's government ordered ISPs to begin blocking a list of Web sites that were compiled by CAT and hosted on its server.[25] The MICT's Cyber Inspector team was also charged with rooting out gambling and sex sites.[26] In late 2005 the government announced its plans to block over 800,000 pornographic and violent Web sites; ISPs would be ordered to take down the sites, and those that did not follow the order would have their licenses revoked.[27] The prime minister also formed a nine-member Internet inspection committee, which met online each morning to compile a list of sites for ISPs to block.[28] Although citizens were encouraged to submit sites for blocking through various forums,[29] there has been a marked lack of transparency in the government's decision-making process and execution of filtering. As a new constitution is slated for 2007, the legal authority for Internet filtering continues to be contested.

In the first days of martial law after the coup, military leaders issued orders intended to restore "normalcy," demanding all political parties to stop their activities, banning new political parties, and requiring the cooperation of news media to discourage the reporting of public opinion.[30] The MICT followed suit, enforcing a temporary ban on political text-messaging and phone-ins, where ISPs and authors would be held responsible for offensive messages.[31]

Not yet enacted at the time of the coup, a revised law laying out the terms and penalties of computer crimes was approved in principle by the newly installed National Legislative Assembly on November 15, 2006. Sponsored by the MICT and the interim military government, this bill in its current form would punish the forwarding of a pornographic e-mail with up to three years imprisonment and the posting of online activity posing a threat to "national security" as an offense under the national security law.[32]

ONI testing results

The stated goal of blocking 800,000 pornographic and violent Web sites as a result of Thaksin's policy is only one of many reported figures of the number of blocked sites in Thailand. For example, in 2004 there were reportedly 1,247 blocked URLs, most of which were pornographic sites, along with a few sites devoted to online gaming and one site belonging to a separatist movement.[33] This proportion remained relatively intact in other accounts. Before it took down its public reports, the Police Bureau on High Tech Crime claimed to have blocked all of the over 34,000 "illicit" Web sites reported since April 2002, with Thai and foreign pornography sites at about 56 percent of the total, sites that sell sex equipment 12 percent, and sites with content posing a "threat to national security" at 11 percent.[34] From 2002 to 2005 the MICT also blocked over 2,000 sites, reportedly mostly pornography sites.[35] In addition, multiple alleged block lists containing a majority of pornography sites were "leaked." It was common for prominent sites to be made inaccessible, only to be unblocked after a period of time.

ONI conducted testing after the coup on three major ISPs: KSC, LoxInfo, and True. Of the sites tested, only a small percentage were actually blocked. The Thai government does implement filtering and primarily blocks access to pornography, online gambling sites, and circumvention tools. Outside these categories, only a few sites were blocked by all three ISPs. Two of these sites were inaccessible and suspected to be blocked. One of these sites, the anti-coup Web site www.19sep.com received significant media coverage for being blocked six times over a period of three months.[36] The other, the Web site of the Patani United Liberation Organisation (www.puloinfo.net) considered by the government to be a Malay Muslim separatist group, appears to be a recent incarnation of the site www.pulo.org that was also blocked and has since been taken down.

Although it has long been declared a top priority of filtering in Thailand, a minority of the Thai-related pornography sites ONI tested were actually blocked by all three ISPs. Only one pornography site (www.sex.com) on the global list was blocked by all three ISPs.

Filtering is demonstrated by redirection to an MICT blockpage. Although it has been reported that ISPs are required to block a list of banned Web sites distributed by the NITC, ONI testing found that filtering varies across ISPs. LoxInfo and True showed significant overlap in sites filtered, blocking a substantial number of circumvention tools and anonymous proxies (www.guardster.com; www.stayinvisible.com), as well as pornography and gaming sites. A few sites promoting human rights, such as the Patani Malay Human Rights Organisation (www.pmhro.org), were also blocked by both ISPs.

Only KSC appeared to address the issue of *lèse majesté,* blocking a number of pages on Amazon.com and other commerce sites featuring biographies of the king. These present an example of URL filtering in Thailand, as various Amazon.com URLs were blocked but the domain (www.amazon.com) remained available on all ISPs tested.

Conclusion

The current official approach toward filtering is in flux, especially in the face of questions about the legal authority and procedures for censorship after the abolishment of the 1997 Constitution. However, evidence from ONI testing suggests that targets for blocking have remained consistent, with a strong focus on pornography and lesser priorities made of gaming and circumvention tools. Only a small number of sites with sensitive political content, particularly about the Thai monarchy and insurgents in the south, continue to be inaccessible. It remains to be seen whether the harsh legacy of censorship of all media created by the former prime minister's government will be carried forward in post-coup Thailand.

NOTES

1. For example, the Shin Corporation sought four hundred million baht (USD10 million) in defamation suits from the Thai Post newspaper and the first defendant, Supinya Klangnarong, a media freedom activist with the NGO Campaign for Popular Media Reform. See also Reporters Without Borders, Thailand:Annual Report 2006, http://www.rsf.org/article.php3?id_article=17364.
2. Council for Democratic Reform, Announcement No. 3, September 19, 2006.
3. Paul Budde Communication Pty Ltd., Thailand-Internet, 2006, p. 1.
4. Steven Huter, Sirin Palasri, and Zita Wenzel, *The History of the Internet in Thailand.* Network Startup Resource Center: University of Oregon Books, http://www.nsrc.org/case-studies/thailand/english/conclusion.html.
5. Paul Budde Communication Pty Ltd., Thailand-Internet, 2006, p. 1.
6. National Statistical Office: Thailand, 2005 Information and Communication Technology Survey, http://web.nso.go.th/eng/en/stat/ict/ict05_rep.pdf.

7. National Electronics and Computer Technology Center (NECTEC), Thailand MICT Indicators 2005, February 2005, pp. 28, 41, http://iir.ngi.nectec.or.th/download/indicator2005.pdf.
8. Paul Budde Communication Pty Ltd., Thailand-Internet, 2006, p. 1.
9. National Electronics and Computer Technology Center (NECTEC), Thailand MICT Indicators 2005, February 2005, http://iir.ngi.nectec.or.th/download/indicator2005.pdf.
10. Ibid.
11. Paul Budde Communication Pty Ltd., Telecommunication Sector Snapshot: Thailand, 2006.
12. National Electronics and Computer Technology Center (NECTEC), Internet Network Infrastructure:Thailand's Perspective, January 10, 2002, http://unpan1.un.org/intradoc/groups/public/documents/APCITY/UNPAN012808.pdf.
13. Internet Connectivity in Thailand, December 2006, http://iir.ngi.nectec.or.th/internet/map/current.html.
14. Reporters Without Borders, Internet Under Surveillance 2004: Thailand, http://www.rsf.org/article.php3?id_article=10777.
15. Articles 37, 39, 58, 59. See Article 19, Freedom of Expression and the Media in Thailand, December 2005, p. 38.
16. Article 8, Constitution of the Kingdom of Thailand, adopted October 11, 1997; Article 1, Constitution of the Kingdom of Thailand (Interim Edition), adopted October 1, 2006.
17. Asia Human Rights Commission, Update on Urgent Appeal, April 10, 2006, http://www.ahrchk.net/ua/mainfile.php/2006/1651/.
18. Article 19, Freedom of Expression and the Media in Thailand, December 2005, p. 75.
19. For example, the Midnight University Web site and forum (www.midnightuniv.org) was blocked in July 2006 on claims of *lèse majesté,* although it was accessible at time of testing. See *Bangkok Post*, "Web board banned, claim of lese majesty," July 28, 2006, reprinted at http://www.asiamedia.ucla.edu/06thailandcoup/article.asp?parentid=49971. The site for discussing political and social issues was also blocked by the MICT on September 29, 2006, the day after scholars at Chiang Mai University affiliated with the Web site tore up mock copies of the interim military government's constitution. *Bangkok Post*, "Thai university website closed after protest over interim charter," October 1, 2006, reprinted at http://www.asiamedia.ucla.edu/article.asp?parentid=54251.
20. Phermsak Lilakul, "Ragnarok curfew starts tonight," *The Nation*, July 17, 2003.
21. Anchalee Kongrut, "16-hour Internet cafe curfew for under-18s," *Bangkok Post*, April 25, 2006.
22. Telecommunications, IT and E-Commerce, http://bia.co.th/027.html.
23. Paul Budde Communication Pty Ltd., Thailand-Internet, 2006, p. 5.
24. Ibid.
25. *Bangkok Post*, "Govt forces ISPs to block `inappropriate' web sites," September 7, 2003.
26. *Bangkok Post*, "PM wants tighter curbs on internet," December 18, 2003.
27. *Agence France-Presse,* "Thailand to block over 800,000 sites," November 28, 2005.
28. Ibid.
29. See, for example, Thaisnews, "New program will be launched on Radio Thailand," January 31, 2006, http://www.thaisnews.com/news_detail.php?newsid=160479.
30. Announcement by the Council for Democratic Reform No. 10: Request for Cooperation in News Reporting, September 20, 2006; Announcement by the Council for Democratic Reform No. 15: Ban on Meetings and other Political Activities by Political Parties, September 21, 2006, http://www.MICT.go.th/cdrc/read_all.asp?cid=1; *Bangkok Post*, "Coup leaders authorise press censorship," September 20, 2006.
31. *Financial Times,* Thai Press Reports, "MICT imposes temporary ban on political text-messaging and phone-in," September 26, 2006.
32. Thai Press Reports, "National Legislative Assembly approves computer crime bill in principle," November 20, 2006.
33. Miles Ignotus, "Censoring the Web," *Bangkok Post*, February 15, 2004.
34. *Bangkok Post,* "Censors busy on the Internet," November 24, 2006.
35. Kavi Chongkittavorn, "Stop messing with Internet access and free debate," *The Nation*, November 20, 2006.
36. *The Nation,* "Freedom of speech: Anti-coup website blocked again without notification," December 30, 2006, http://nationmultimedia.com/2006/12/30/politics/politics_30022916.php.

Tunisia

Although Tunisia has actively sought to develop its information and communications technology (ICT) infrastructure, the government blocks a range of Web content and has used nontechnical means to impede journalists and human rights activists from doing their work. This pervasive filtering of political content and restrictions on online activity has prompted frequent criticism from foreign governments and human rights organizations.[1]

Background

The Tunisian government curtails dissent, free expression, and the flow of information into and out of the country. The government relies on legal and economic means to maintain effective control over the press and the broadcast media.[2] State interference in assemblies is commonplace. In 2005 the government banned the first congress of the Union of Tunisian Journalists and shut down the offices of the Association of Tunisian Judges.[3] The government has dispatched the police to surround and disrupt meetings of the National Council for Liberties in Tunisia,[4] and leveraged the courts to enjoin the Tunisian Human Rights League from preparing for its national congress.[5] The government has also reportedly threatened judges with assignments to remote locations; tortured prisoners; and arrested, harassed, and intimidated human rights activists.[6] In March 2005, for instance, lawyer and human rights activist Radhia Nasraoui was beaten by police on the way to a demonstration.[7] Despite the release of eighty political prisoners in March 2006, more than two hundred are believed to remain in custody.[8]

Internet in Tunisia

The Tunisian Ministry of Communications established the Tunisian Internet Agency (ATI) to

RESULTS AT A GLANCE

Filtering	No evidence of filtering	Suspected filtering	Selective filtering	Substantial filtering	Pervasive filtering
Political					●
Social					●
Conflict/security			●		
Internet tools				●	

Other factors	Low	Medium	High	Not applicable
Transparency	●			
Consistency			●	

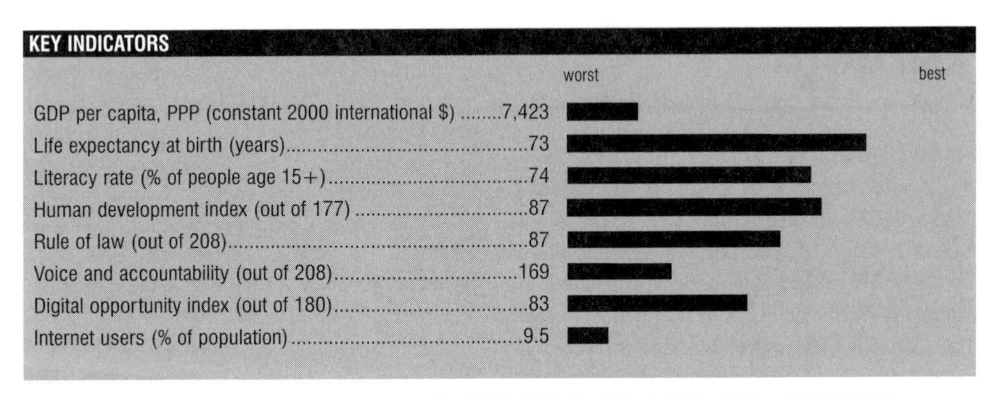

Source (by indicator): World Bank 2005, 2006a, 2006a; UNDP 2006; World Bank 2006c, 2006c; ITU 2006, 2005

regulate the country's Internet and domain name system (DNS) services, which had formerly fallen under the Regional Institute for Computer Sciences and Telecommunications (IRSIT)'s purview.[9] The ATI is the gateway from which all of Tunisia's twelve Internet service providers (ISPs) lease their bandwidth.[10] Seven of these ISPs are publicly operated; the other five—Planet Tunisie, 3S Global Net, Hexabyte, Tunet, and Topnet—are private.[11] These ISPs offer a range of options, including hourly dialup access,[12] broadband access (with prices starting at less than USD25 per month),[13] and satellite-based Internet.[14]

The government has energetically sought to spread access to the Internet. The ATI reports connectivity of 100 percent for universities, research laboratories, and secondary schools, and 70 percent for primary schools.[15] Government-brokered "Free Internet" programs that provide Web access for the price of a local telephone call and increased competition among ISPs have significantly reduced the economic barriers to Internet access. Those Tunisians for whom personal computers remain prohibitively expensive may also access the Internet from more than 300 cybercafés set up by the authorities.[16]

Tunisia's rapid growth in Internet capacity is reflected in an increase in Internet use. In just five years, Tunisia's Internet penetration rate rose from 1 percent (2001) to 9.3 percent (2006),[17] and today there are roughly one million Internet users in the country.[18]

Legal and regulatory frameworks

In addition to filtering Web content, the government of Tunisia utilizes laws, regulations, and surveillance to achieve strict control over the Internet.

The Tunisian External Communication Agency (ATCE), the government body responsible for media regulation, contends that fewer than 10 percent of newspapers are under state ownership and editorial control.[19] However, the ATCE uses its regulatory powers to help government supporters and hamper detractors seeking advertising space in the print media.[20] The state maintains direct ownership of all three of the country's television stations and all but one radio station (which does not air news). Although the Internet has unquestionably made it easier for Tunisians to read news and opinions not found in the country's monolithic press and broadcast media, legal threats exert pressures on content providers operating within the country.

In 2001 President Zine el-Abidine Ben Ali removed prison sentences from the Press Law, which criminalizes criticism of the government.[21] However, rights groups have pointed out that imprisonment and other harsh penalties are preserved in the Penal Code.[22]

ISPs are required to send the Ministry of Telecommunications a current list of their subscribers each month. [23] ISPs, Internet service subscribers, Web page owners, and Web server owners are responsible for ensuring that the content of the pages and Web servers that they host conform to the Press Code's prohibitions against publications "likely to upset public order."[24] In addition, these parties are "obliged to constantly monitor the content of web servers operated by the service provider so as to not allow any information contrary to public order and good morals to remain on the system."[25]

These regulations also apply to Publinets, government-sponsored Internet cafés. Café owners are responsible for the activities of their patrons.[26] Computer monitors in Publinet cafés visited by an ONI researcher were angled so that the café owner could see the screens, and in one case, the café owner commented when the researcher attempted to access blocked sites.[27]

Tunisia achieves its filtering through the use of a commercial software program, SmartFilter, sold by the U.S. company Secure Computing. Because all fixed-line Internet traffic passes through facilities controlled by ATI, the government is able to load the software onto its servers and filter content consistently across Tunisia's twelve ISPs. Tunisia purposefully hides its filtering from Internet users. SmartFilter is designed to display a 403 "Forbidden" error message when a user attempts to access a blocked site; the Tunisian government has replaced this message with a standard 404 "File Not Found" error message, which gives no hint that the requested site is actively blocked.[28]

ONI testing results

ONI testing in Tunisia revealed pervasive filtering of Web sites of political opposition groups such as the Al-Nadha Movement (www.nahdha.info) and Tunisian Workers' Communist Party (www.albadil.org). Web sites that contain oppositional news and politics were also blocked. Examples include www.perspectivestunisiennes.net, www.nawaat.org, www.tunisnews.com, and www.tunezine.com.

Web sites that publish oppositional articles by Tunisian journalists were also blocked. For example, ONI verified the blocking of the French daily *Libération* Web site in February 2007 because articles by Tunisian journalist Taoufik Ben Brik critical of President Zine el-Abidine Ben Ali appeared on the site.[29]

Also blocked are Web sites that criticize Tunisia's human rights records. For example, the Web sites of the League for the Defense of Human Rights (www.ltdh.org) and the Congrés Pour la République (www.cprtunisie.net) were blocked, along with the Web sites of Reporters Without Borders (www.rsf.org), the International Freedom of Expression eXchange (www.ifex.org), the Islamic Human Rights Commission (www.ihrc.org), and the Arabic Network for Human Rights Information (www.hrinfo.org). Although the home page of Human Rights Watch (HRW) was accessible, the Arabic- and French-language versions of an HRW report on Internet repression in Tunisia were blocked.

Pornographic sites and anonymizers and circumvention tools, such as Anonymizer (www.anonymizer.com) and Guardster (www.guardster.com), were filtered extensively. Indeed, almost all of the tested sites belonging to these categories were blocked.

A few sites that criticize the Quran (www.thequran.com) and Islam (www.islameyat.com) or encourage Muslims and others to convert to Christianity (www.biblicalchristianity.freeserve.co.uk) were blocked, though their small number points to limited filtering of religious content in Tunisia.

Other blocked sites included several gay and lesbian information or dating pages, sites containing provocative attire, hacking Web sites, and several online translation services.

Conclusion

Tunisia's government continues to suppress critical speech and oppositional activity, both in real space and in cyberspace. Unlike other states that employ filtering software, Tunisia endeavors to conceal instances of filtering by supplying a fake error page when a blocked site is requested. This makes filtering more opaque and clouds users' understanding of the boundaries of permissible content. Tunisia maintains a focused, effective system of Internet control that blends content filtering with harsh laws to censor objectionable and politically threatening information.

NOTES

1. Tunisia has regularly been labeled an "enemy of the Internet"; see Reporters Without Borders, "List of the 13 enemies of the Internet in 2006 published," November 7, 2006, http://www.rsf.org/article.php3?id_article=19603.
2. Amnesty International, "Tunisia: Human rights abuses in the run up to the WSIS," http://web.amnesty.org/library/index/engmde300192005.
3. Ibid.
4. Human Rights Watch, "Tunisia: Police use force to block rights meeting," http://hrw.org/english/docs/2004/12/14/tunisi9841.htm.
5. Amnesty International, "Tunisia: Fear for safety/Intimidation," http://web.amnesty.org/library/Index/ENGMDE300222005?open&of=ENG-TUN.
6. Amnesty International, "Tunisia: Human rights abuses in the run up to the WSIS," http://web.amnesty.org/library/index/engmde300192005.
7. Amnesty International, Tunisia – Report: 2006, http://web.amnesty.org/report2006/tun-summary-eng#1.
8. afrol News, "Tunisia still holds some 200 political prisoners," March 1, 2006, http://www.afrol.com/articles/18285.
9. Tunisia Online, "Internet in Tunisia: History," June 25, 2002, http://www.tunisiaonline.com/internet/history.html.
10. Network Startup Resource Center, Tunisia and the state of the Internet (e-mail from Lamia Chaffai of ATI to Dolores Lizarzaburu of NSRC), November 14, 2002, http://www.nsrc.org/db/lookup/report.php?id=1037285984211:488846420&fromISO=TN.
11. Tunisian Internet Agency, http://www.ati.tn/Defaulten.htm.
12. See, for example, Hexabyte's Free Internet FAQ, http://www.zerodinar.com/faq.php (French language only).
13. See, for example, Topnet, http://www.topnet.tn/ (French language only).
14. Tunet.tn, "L'accés Internet haut dèbit par satellite (TUNET VSAT)," http://www.tunet.tn/?item=solutions&sp=Satellite (French language only).
15. Tunisian Internet Agency, http://www.ati.tn/Defaulten.htm.
16. Reporters Without Borders, "Tunisia," http://www.rsf.org/article.php3?id_article=7271.
17. Internet World Stats, "Tunisia: Internet usage and population growth," http://www.internetworldstats.com/af/tn.htm, citing data from the International Telecommunication Union.
18. Internet World Stats cites ITU data, which place the number of Internet users in Tunisia at 953,000 (http://www.internetworldstats.com/af/tn.htm). The Tunisian government's estimate of 1.14 million is slightly higher (http://www.ati.tn/Defaulten.htm).
19. International Freedom of Expression eXchange (IFEX), The IFEX Tunisia Monitoring Group: Media Censorship, http://campaigns.ifex.org/tmg/censorship.html.
20. Ibid.
21. Tunisia Online, Government, http://www.tunisiaonline.com/government/index.html.
22. Human Rights Watch, False Freedom: Online Censorship in the Middle East and North Africa: Tunisia, http://hrw.org/reports/2005/mena1105/7.htm, citing Ligue Tunisienne pour la Défense des Droits de l'Homme, Report on the Freedom of Information in Tunisia, http://www.iris.sgdg.org/actions/smsi/hr-wsis/ltdh03-press-en.pdf.
23. Decree of the Ministry of Telecommunications of March 22, 1997, Article 8, translated by Harvard Law School Langdell Library.
24. Decree of the Ministry of Telecommunications of March 22, 1997, Article 9, and Code de la Presse, Article 49, translated by Harvard Law School Langdell Library.
25. Decree of the Ministry of Telecommunications of March 22, 1997, Article 9, translated by Harvard Law School Langdell Library.

26. International Freedom of Expression eXchange (IFEX), The IFEX Tunisia Monitoring Group, http://campaigns.ifex.org/tmg/about.html; IFEX, Tunisia: Freedom of Expression Under Siege, February 2005, http://www.ifex.org/download/en/FreedomofExpressionunderSiege.doc.
27. Human Rights Watch, The Internet in the Mideast and North Africa: Free Expression and Censorship, http://hrw.org/advocacy/internet/mena/tunisia.htm.
28. Internet Censorship Explorer, "Tunisia: Internet filtering," June 7, 2005, http://ice.citizenlab.org/?p=115.
29. Reporters Without Borders, "French media censored in Tunisia because of articles by Tunisian journalist Taoufik Ben Brik," http://www.rsf.org/article.php3?id_article=21119.

Ukraine

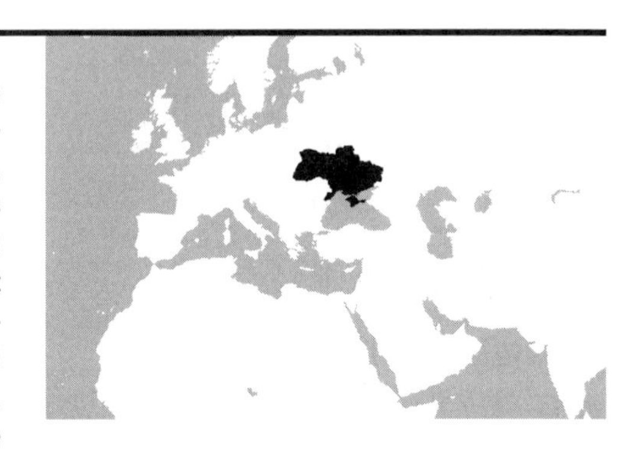

Access to Internet content in Ukraine remains largely unfettered. However, despite a generally liberal media and telecommunications policy, the authorities have enlisted special bodies and regulations to survey Internet content in order to "protect national security" and limit other forms of "undesirable" information content. These regulations embody the potential for expanded formal and informal controls, although such constraints are unlikely in the near future.

Background

Among the Commonwealth of Independent States (CIS) countries, Ukraine is second only to Russia in the size and strength of its IT establishment. Ukraine was the birthplace of Soviet computing and Kyiv remains a major center for IT development. The county was an early adopter of policies to support information communications technology (ICT) for development as a pillar of national development, and the government has invested in building out the country's ICT infrastructure.

The Ukrainian government recognizes the significance of the Internet for economic development and for the development of information society. The state has demonstrated the political will to undertake vital reforms in the telecommunications sector, although much remains to be done to promote a favorable environment for developing the Internet, fostering e-commerce, and introducing e-governance. The World Economic Forum ranks Ukraine 76th out of 115 countries for 2005–2006 in the Internet readiness index.[1]

The January 2005 "orange revolution"—when opposition groups successfully challenged the outcome of the November 2004 presidential

RESULTS AT A GLANCE

Filtering	No evidence of filtering	Suspected filtering	Selective filtering	Substantial filtering	Pervasive filtering
Political	●				
Social	●				
Conflict/security	●				
Internet tools	●				

Other factors	Low	Medium	High	Not applicable
Transparency				●
Consistency				●

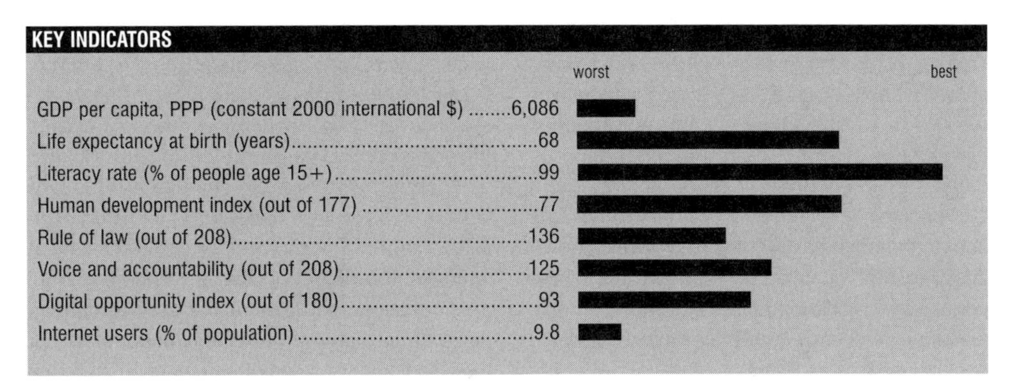

Source (by indicator): World Bank 2005, 2006a, 2006a; UNDP 2006; World Bank 2006c, 2006c; ITU 2006, 2005

elections that were thought to be unfair—highlighted the latent political power resulting from the "convergence" of information infrastructures (cell phones, Internet, and independent media) and political mobilizaton. The opposition made full use of these technologies to mobilize and direct supporters in acts of civil disobedience, sit-ins, and general strikes. Although the Internet did not play a determining role in the success of the "orange revolution," its use by the opposition helped to foster the perception that these technologies served an important strategic role in organizing political opposition (which observers have termed "hyper-democracy"). This perception, in turn, prompted neighboring authoritarian governments such as that of Belarus to crack down on Internet openness.

Internet in Ukraine

The partly liberalized Ukrainian telecommunications market is relatively undeveloped. Fixed-line penetration remains low and the telephone system requires modernization. The demand for mobile services has expanded rapidly, to reach a penetration of nearly 50 percent. The largest telecom and top-tier Internet service provider (ISP), Ukrtelecom, has 92.9 percent state ownership. The parliament has legalized its privatization,[2] but this has been delayed in anticipation of the company increasing in market value.[3] The state monopolies Ukrtelecom and Utel, which is controlled by Ukrtelecom, together own 95 percent of the long-distance and international calls market.[4]

State-owned Ukrtelecom is the largest ISP in the country, but does not decisively control the countries other major ISPs. As of June 2006 some sixty ISPs connected to six Internet traffic exchange points.[5] Recently the number of ISPs offering broadband access services has rapidly increased.[6] The government, recognizing the need for attracting foreign investment and stimulating favorable Internet environment, has also announced plans to introduce Wi-Fi and WiMAX technologies.[7]

The Ukrainian national country code top-level domain (".ua") is administered by the Hostmaster Company, a specialized nonprofit organization.

Internet penetration in the country was estimated at 9.8 percent in 2005,[8] well below the European average of 36.1 percent. Several obstacles compromise expansion, including high access costs, poor infrastructure in the regions, high call rates, and low levels of personal computer (PC) ownership. The International Telecommunication Union (ITU) 2005 estimates show that only 4 percent of the population owns a PC.[9] Although ISPs have considerably reduced their access costs (for example, by leasing outdated or redundant infrastructure from Ukrtelecom) and a few providers offer free access during the night, most Ukrainians cannot afford to use the Internet: 46.8 percent of the population identified themselves as poor.[10] Men are more frequent users than women (at 59.3 percent), and most users access Internet at the office, cybercafés, or home. The most popular search engines in Ukraine are Ukrainian www.BigMir.net and www.Ukr.net, Russian Yandex and Rambler, and Google.

Legal and regulatory frameworks

The 2003 Law on Communications established the National Communication Regulation Commission, which regulates the IT and telecommunications market. Under this law telecommunications operators require a license before starting activity.[11] With the present government, the Internet enjoys a high degree of freedom. Internet activity is not subject to licensing or other forms of regulation. Liberalization of the market has led to a rapid increase in the number of ISPs, which numbered 260 in 2006.

At present there are no controls on Internet access or content. However, this may be changing as government figures have made public calls for stricter regulation of the internet, citing national security concerns.[12] Suggested measures include licensing ISPs, registering Internet resources, and monitoring content related to obscene or harmful material. The threat of Internet censorship was raised in 2005 when the Ministry of Transport and Telecommunications introduced, and subsequently withdrew, a decree regulating registration of Web sites hosted in Ukraine for the purposes of national security.[13] An earlier Act to introduce mechanisms for Internet monitoring (the 2002 Order of the State Committee on Communications) required ISPs to install a state monitoring system in order to provide Internet access to state organizations. The purpose of this monitoring was to control unsanctioned transmission of data containing state secrets. However, a "state secret," as provided in current regulations, lacks concrete definition, allowing authorities broad discretion in interpretation. The difficulties in separating state from nonstate users expose the latter to monitoring. Human rights groups have suggested that the Security Service has been intercepting messages and carrying out surveillance on over approximately 50 percent of Ukrainian traffic.[14]

The Council of National Security and Defense is the main governmental body responsible for national security and defense; this body is chaired by the president. The Council monitors information security policy and coordinates the work of the other executive bodies in this field. The Security Service of Ukraine is empowered to initiate criminal investigations and use wiretapping devices on communications. Legislation has not made clear either the circumstances that justify interception of information from communication channels, or the time limits of any such interception.[15] The recently established State Service for Special Communications and Information Protection Service implements governmental policy on protecting state information and confidential communication, and exercises control over cryptographic and technical information security.[16]

The Law on Protection of Public Morals of November 20, 2003, enacted during the term of the previous government, is still effective. It prohibits production and circulation of pornography; dissemination of products that propagandize war or spread national and religious intolerance; humiliation or insult to an individual or nation on the grounds of nationality, religion, or ignorance; and the propagation of "drug addition, toxicology, alcoholism, smoking and other bad habits." The National Expert Council for the Protection of

Public Morals has authority to inspect media, including the Internet, in order to start a procedure for revocation of the license in case of violation. The National Expert Council, however, has not issued any decision yet because it lacks legal mechanisms for enforcement.

ONI testing results

ONI conducted testing on seven ISPs: Adamant, Cornel, Elvisti, Ukrtelecom, Volia, Goldentelecom, and Ukr.net. The testing did not detect any filtering, although a few Web sites with content related to alcohol and drugs, public health, human rights, and minority faiths were temporarily inaccessible.

Conclusion

Citizens of Ukraine enjoy an unfettered access to the Internet. The country has an Internet infrastructure oriented toward European providers, and thus the ISPs are not influenced by the policies of Russian providers. However, the country has built up an intricate system of bodies and regulations that could be geared to surveillance of information carried on telecommunications networks, including the Internet.

NOTES

1. World Economic Forum, Global Information Technology Report 2005–2006, http://www.weforum.org/en/initiatives/gcp/Global%20Information%20Technology%20Report/index.htm (last accessed May 2, 2007).
2. For more about Urktelecom's privatization and the alternatives to public sale, see "Ukrtelecom's policy" (in Russian), http://proit.com.ua/telecom/2006/05/15/114501.html (last accessed May 2, 2007).
3. Serhey Malyhin, Events Digest (Review of Events in ICT Policy in Ukraine over July: First part of August 2005), Global Internet PolicyInitiative, 2005, http://gipi.internews.ua/eng/events_digest/digest_events_july_eng.pdf (last accessed May 2, 2007).
4. Andriy Vorobyov, Ukraine Telecommunications Market Report, BISNIS, 2005, http://www.bisnis.doc.gov/bisnis/bisdoc/0602UkraineTelecomReport.htm (last accessed May 2, 2007).
5. For further information about the UA-IX Internet Traffic Exchange, see Ukrainian traffic exchange network at http://www.ua-ix.net.ua/eng.
6. News Wire Feed, "Ukrtelecom uses Cisco routers: Light reading: IP & convergence," March 23, 2005, http://www.lightreading.com/document.asp?doc_id=70731; Thomson Press Release, "Thomson Pioneers Next Generation Telecoms in Ukraine and Estonia," May 19, 2006, http://www.thomson.net/EN/Home/Press/Press+Details.htm?PressReleaseID=ca229b68-c6ba-4e69-9b95-bbc7dd8540f0 (accessed May 2, 2007); News@Cisco News Release, "DataGroup to deliver DWDM network in Ukraine with Cisco Optical Technology," February 15, 2006, http://newsroom.cisco.com/dlls/2006/prod_021506.html (accessed May 2, 2007).
7. A statement by Victor Bondar, Minister of Transport and Telecommunication, quoted in "Minister promises to cover Ukraine with Internet" Ukrainskaya Pravda, January 18, 2006 (in Russian), http://www.pravda.com.ua/ru/news/2006/1/18/36834.htm (last accessed May 2, 2007).
8. See International Telecommunication Union, *World Telecommunication Indicators 2006*. For comparison, Internetworldstat estimates the data at 11.5 percent for 2006; see Internet World Stats, Usage and Population Statistics, 2007, http://www.internetworldstats.com/stats4.htm (last accessed May 2, 2007).
9. International Telecommunication Union, *World Telecommunication Indicators 2006*.
10. United Nations Development Programme in Ukraine, Prosperity Through Poverty Alleviation, http://www.undp.org.ua/?page=areas&area=2 (accessed May 2, 2007).
11. Law on Telecommunications of November 18, 2003. The law abolished the provisions of the 1995 Communication Law, including the charges for incoming calls for all kinds of telephone communications. See the text of the law at http://ilaw.org.ua/ (in Ukrainian).
12. The Director of Ukraine's Security Service Konstantyn Boyko pointed out the imminent danger that Internet may cause to the country, stating: "Foreign political forces, intelligence departments and extremist organizations, which are able to direct resources and endowments of the Internet to harm our nation," See "Security service to take totalitarian control over Internet," Ukrainskaya Pravda, May 27, 2006 (in Russian), http://www.pravda.com.ua/ru/news/2006/5/27/41096.htm (last accessed May 2, 2007).

13. The decree asked for compulsory registration of Web sites, and specified criteria that sites had to respect before being launched. International Press Institute, 2005 World Press Freedom Review:,Ukraine, http://www.freemedia.at/cms/ipi/freedom_detail.html?country=/KW0001/KW0003/KW0087/&year=2005 (last accessed May 2, 2007).
14. Kharkiv Human Rights Protection Group, Human Rights in Ukraine: IV. Right to Privacy, February 7, 2006, http://www.khpg.org/en/index.php?id=1151854687 (last accessed May 2, 2007).
15. This justification in the draft decree was in part drawn from an existing law, the "Law on Operative Investigative Activity" of February 18, 1992, No. 2135-XII.
16. The agency is established by the Law of February 23, 2006, No. 3475-IV. For excerpts of the law translated into English, see Yaroslav the Wise Institute of Legal Information, The Law of Ukraine: On the State Service for Special Communications and Information Protection of Ukraine, http://www.welcometo.kiev.ua/pls/ili/docs/a_law_eng/E3475-IV.html (last accessed May 2, 2007).

United Arab Emirates

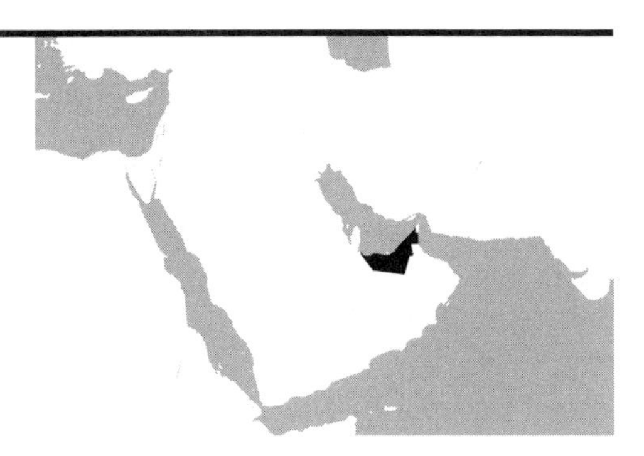

The government of the United Arab Emirates (UAE) pervasively filters Web sites that contain pornography or relate to alcohol and drug use, gay and lesbian issues, or online dating or gambling. Web-based applications and religious and political sites are also filtered, though less extensively. Additionally, legal controls limit free expression and behavior, restricting political discourse and dissent online.

Background

The United Arab Emirates is ruled chiefly by a Federal Supreme Council, consisting of the hereditary leaders of the seven individual emirates. Although the UAE Constitution provides for judicial independence and guarantees freedom of speech, of the press, and of assembly, political interference and legal constraints undermine these provisions. Rulings by Islamic and civil courts are scrutinized by the government, and the foreign nationals occupying most judicial seats can be deported.[1] Print and electronic media are subject to the Press and Publications Law, which permits censorship of content by the state Media Council and prosecution under the Penal Code—for example, for publishing material that causes someone moral harm (Article 372), or for defaming someone without concrete evidence (Article 373). Journalists practice self-censorship, and newspapers often rely on the state's Emirates News Agency for material.[2] Although citizens can voice their concerns through channels such as the open councils (*majlis*), they do not have the power to transform the government or national law and are prohibited from criticizing their leaders.[3] Human rights activists have been detained and academics and critics barred from making their views public.[4] The government also controls Sunni and Shi'a

RESULTS AT A GLANCE

Filtering	No evidence of filtering	Suspected filtering	Selective filtering	Substantial filtering	Pervasive filtering
Political			●		
Social					●
Conflict/security			●		
Internet tools				●	

Other factors	Low	Medium	High	Not applicable
Transparency		●		
Consistency	●			

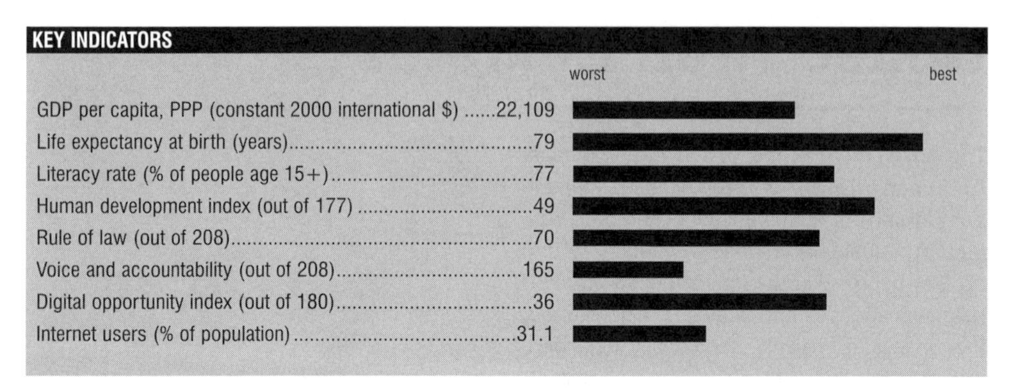

Source (by indicator): World Bank 2004, 2006a, 2006b; UNDP 2006; World Bank 2006c, 2006c; ITU 2006, 2005

mosques and monitors sermons for political commentary.[5]

Internet in the United Arab Emirates

The UAE is among the most highly Internet-connected countries in the Middle East. The UAE *Yearbook* for 2007 states that there are more than 578,000 Internet subscribers in the country.[6] September 2006 figures on Internet penetration place the number of Internet users at 1.40 million, or 35 percent of the population,[7] though Etisalat (Emirates Telecommunications Corporation)—the nation's primary service provider—estimates that more than 51 percent of the country is online.[8]

Since 1976, nearly all telephone and Internet service in the UAE has been furnished by the government-owned Etisalat—either through direct sale of subscriptions to customers or through commercial resale of Etisalat services via providers such as Dubai Internet City (DIC) telecom. With the release of the General Policy for the Telecommunication Sector (GTP) in 2006,[9] the UAE government moved to liberalize the telecommunications market, though it remains to be seen how competition will affect Etisalat's monopoly on telecom operations. The UAE Telecommunications Regulatory Authority (TRA) granted a twenty-year license to the Emirates Integrated Telecommunications Company (EITC)—more widely known under its traded name "du"—to offer fixed-line, mobile, and Internet services. Prior to liberalization, du served the Dubai free zone and a few affiliated residential complexes, providing unfettered Internet access in those areas, but acquisitions and partnerships with other telecom companies have expanded du's capabilities and customer base. du now aims to capture 30 percent of the UAE telecom market within three years.[10]

Legal and regulatory frameworks

Controls on Internet content in the UAE, actualized through filtering and other forms of enforcement, are geared toward safeguarding political, moral, and religious values. According to Etisalat, there is some evidence that these controls enjoy popular support. A 2002 survey found that 60 percent of Etisalat subscribers surveyed favored retaining the ISP's automatic filtering system, with 51 percent saying that it protected family members from objectionable content. In 2004, the UAE cited the survey as indicating that the role of filtering "in protecting users from offensive material is considered to be an acceptable form of censorship."[11]

The mandate for technical filtering in the UAE derives from the TRA and is executed at the ISP level. Etisalat prohibits the use of its services for any "criminal or unlawful purpose such as but not limited to vice, gambling or obscenity, or for carrying out any activity which is contrary to the social, cultural, political, economical or religious values of the UAE."[12] Emerging competitor du is also moving toward compliance with the TRA's filtering policies. In January 2007, the company defended its decision to block the use of Voice-over Internet Protocol (VoIP) in the free zone, saying that TRA rules and guidelines mandated the ban.[13] This decision heralded a comprehensive plan to implement technical filtering throughout the Dubai free zone in 2007.[14]

The UAE government has also issued a federal law on combating cybercrimes. Cyber-Crime Law No. 2 of 2006 considers any intentional act that abolishes, destroys, or reveals secrets, or that results in the republishing of personal or official information, to be a crime.[15] Individuals may be imprisoned for using the Internet to abuse Islamic holy shrines and rituals, insult any recognized religion, incite or promote sins, or oppose the Islamic religion.[16] Anyone convicted of "transcending family principles and values"[17] or setting up a Web site for groups "calling for, facilitating and promoting ideas in breach of the general order and public decency"[18] may be jailed.

ONI testing results

ONI conducted tests on the UAE's two ISPs: Etisalat, which services most of the country; and du, which (at the time of testing) serviced only Dubai Media City, Dubai Internet City, and some residential areas associated with the free zone. To conduct the tests, ONI used dialup, broadband, and wireless connections. Access in the Dubai free zone was unfettered, while considerable filtering behavior was exhibited on the Etisalat ISP.

Testing in the UAE points to selective filtering of Web sites that express alternative political or religious views. www.UAEprison.com, a site hosting testimonials of former prisoners and critiques of the government's human rights practices, was blocked, as was the site of the U.S.-based *Arab Times* (www.arabtimes.com). Several sites presenting unorthodox perspectives on Islam (www.thekoran.com, www.islamreview.com, www.secularislam.org) were blocked, along with a handful of sites promoting minority faiths (www.albrhan.org, www.ansarweb.net). Among the few extremist sites filtered in the UAE were www.hinduunity.org, a site advocating Hindu solidarity and resistance to Islam, and www.kahanetzadak.com, a site devoted to the founder of the militant Jewish Defense League. Meanwhile, the state continued to deny access to all sites on the Israeli country code top-level domain ".il."

Testing revealed pervasive filtering of pornographic and gay and lesbian sites, which were extensively blocked. Web pages relating to sexual health (www.circumcision.org) and education (www.sexualhealth.com) or containing provocative attire (www.lingerie.com) were filtered to lesser degrees. Sites promoting alcohol and drug use or facilitating online gambling or dating were also blocked in large numbers.

ONI found substantial filtering of Internet tools in the UAE, including translation (www.systranbox.com), hacking (www.thesecretlist.com) and anonymizer (www.surfsecret.com) sites. Numerous VoIP sites (www.skype.com, www.pc2call.com) were blocked in accordance with the national ban on such applications. In October 2006, the UAE unblocked access to social networking and multimedia sharing sites, including www.youtube.com, www.flickr.com, www.metacafe.com, and www.myspace.com. However, sections of these sites containing objectionable material remain unavailable.

Conclusion

The UAE prevents its citizens from accessing a significant amount of Internet content spanning a variety of topics, though the majority of sites filtered appear to be those deemed obscene. Outside the free zones, the state employs SmartFilter software to block content such as nudity, sex, dating, gambling, cults/occult, religious conversion, and drugs. Sites containing anonymizer, hacking, translation, and VoIP applications are also filtered in this manner. The manual blocking of the entire Israeli domain is indicative of the government's political opposition to the Israeli state, rather than to the particular contents of the Web sites hosted there. Though most political sites and news sources are accessible throughout the country, a handful are blocked. It remains to be seen how severely the enforcement of TRA policies in the free zone and affiliated residential clusters will hamper access to Internet content and transform the traditionally unrestricted information environment in those areas.

NOTES

1. U.S. Department of State, Country Reports on Human Rights Practices 2006: United Arab Emirates, at 1.e., http://www.state.gov/g/drl/rls/hrrpt/2006/78865.htm.
2. Ibid., at 2.a.
3. Ibid., at 2.a.3.
4. Human Rights Watch, World Report 2007: United Arab Emirates, http://hrw.org/englishwr2k7/docs/2007/01/11/uae14724.htm.
5. U.S. Department of State, Country Reports on Human Rights Practices 2006: United Arab Emirates, at 2.c., http://www.state.gov/g/drl/rls/hrrpt/2006/78865.htm.
6. United Arab Emirates Yearbook 2007: Infrastructure at 190, http://www.uaeinteract.com/uaeint_misc/pdf_2007/English_2007/eyb6.pdf.
7. Internet World Stats "Middle East Internet usage and population statistics," http://www.internetworldstats.com/stats5.htm (citing International Telecommunication Union data).
8. See Etisalat, About Us: Corporate Information, http://www.etisalat.co.ae/.
9. See General Policy for the Telecommunication Sector (GTP), http://www.tra.ae/NationalTelecomPolicyofUAE.pdf.
10. See United Arab Emirates Yearbook 2007: Infrastructure at 193, http://www.uaeinteract.com/uaeint_misc/pdf_2007/English_2007/eyb6.pdf.
11. United Arab Emirates Yearbook 2004: Information and Culture at 254, http://www.uae.org.ae/uaeint_misc/pdf/English/Culture_&_Information.pdf.
12. Etisalat (ecompany) Policies: Terms & Conditions, "Condition of Use" at 1(v), http://ecompany.ae/eco/isp/english/cs/policies/terms.html.
13. Gulf News, "du defends blocking of VoIP calls in free zones," January 29, 2007, http://archive.gulfnews.com/articles/07/01/29/10100242.html.
14. *Gulf News,* "Free zones to be put under official web filters," February 10, 2007.
15. Article 2, Cyber-Crime Law No. 2 of 2006, printed in *Gulf News,* February 13, 2006, http://archive.gulfnews.com/uae/uaessentials/more_stories/10018507.html.
16. Article 15, Cyber-Crime Law No. 2 of 2006.
17. Article 16, Cyber-Crime Law No. 2 of 2006.
18. Article 20, Cyber-Crime Law No. 2 of 2006.

Uzbekistan

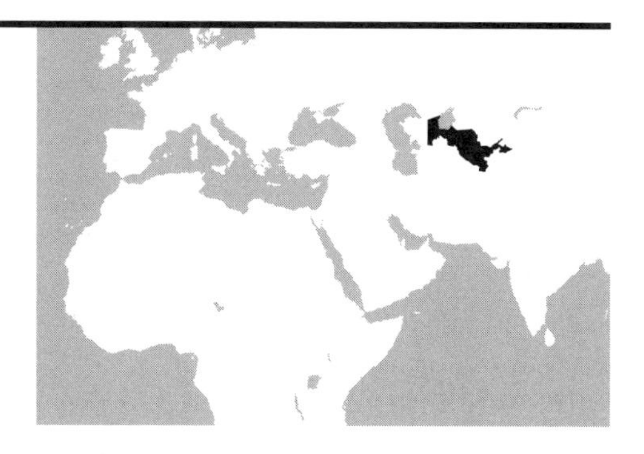

Among the Commonwealth of Independent States (CIS) countries, Uzbekistan is the undisputed leader in applying Internet controls. Filtering is comprehensive and, until 2006, largely undeclared with the government denying the existence of these practices. At present, the government employs sophisticated multilayered mechanisms to exercise control over the Internet, including adopting restrictive policies, applying technological measures, and compelling self-censorship on the media.

Background

At present, and in spite of the formal separation of powers enshrined in the Constitution of the Republic of Uzbekistan, virtually all power is invested in President Islam Karimov. During his extended authoritarian rule the president has demonstrated an active commitment to controlling the information environment in the country and constraining the expression of dissident viewpoints. The active opposition has been forced to leave the country. For them, the Internet often remains the only way to communicate with Uzbek society. The complex series of laws and regulations have resulted in self-censorship of online publishers, independent journalists, and bloggers. This, complemented with a restrictive Internet filtering regime, significantly stifles public discourse on political and human rights topics.

Uzbekistan's control of the Internet embodies the most pervasive regime of filtering and censorship in the CIS. It stands in stark contrast to the government's official enthusiasm for information and communications technology (ICT) development and the Internet. Until 2001 Uzbekistan was a regional leader in the adoption of the Internet and the prioritization of ICTs as a mechanism for national development. Uzbekistan was among the first of the post-Soviet

RESULTS AT A GLANCE

Filtering	No evidence of filtering	Suspected filtering	Selective filtering	Substantial filtering	Pervasive filtering
Political				●	
Social			●		
Conflict/security	●				
Internet tools			●		

Other factors	Low	Medium	High	Not applicable
Transparency	●			
Consistency			●	

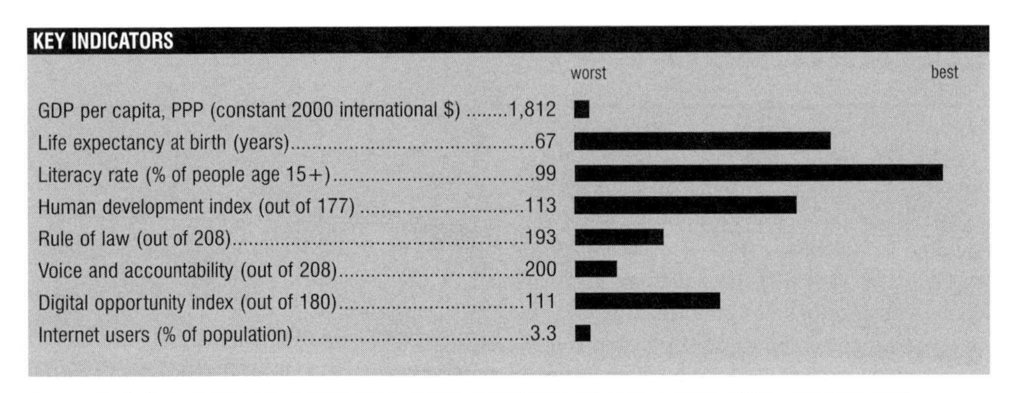

Source (by indicator): World Bank 2005, 2006a, 2006b; UNDP 2006; World Bank 2006c, 2006c; ITU 2006, 2004

republics to establish a national agency responsible for ICT development (UzInfoCom), to contribute state resources to building a sizable academic and research network (UzSCINET), and to launch an ambitious project to provide Internet to the main government institutions (Cabinet of Ministers and Presidency). After 2001 Uzbekistan continued to receive sizable foreign support aimed at developing its ICT infrastructure, including a large network of Internet access points in the regions. Uzbek government officials at all levels were sent abroad to study e-government systems, and ICT was prioritized as a means for national development. Until 2001–02 the Internet remained open and free from filtering with the exception of some limited filters for pornography that were implemented on the academic and research network (UzSCINET).

The turning point in the state's relationship to Internet freedom began following a series of attacks in Tashkent in 2004 blamed on the Hizb-ut-Tahrir (Hit) and the Islamic Movement of Uzbekistan. These attacks have been generally associated with a deepening crackdown in Uzbek society and encompass all forms and channels of dissent, including the Internet.

Internet in Uzbekistan

In 2004 the International Telecommunication Union estimated some 880,000 regular Internet users in Uzbekistan, or a 3 percent Internet penetration rate.[1] According to local surveys the total Internet audience is approximately 1,820,000 as of June 2006. In contrast to neighboring countries, Uzbek women use Internet at an almost equal rate to their male counterparts, with a difference of only 3 percent.[2] About 41.3 percent of Internet users are sixteen to twenty years old.[3] Access is most common from homes (42.7 percent) and work (44.6 percent). Approximately 30 percent of Internet users also visit cybercafés.[4] As of January 2005 there were 463 Internet access centers in Uzbekistan; in January 2006 the number dropped to 344.[5]

Residential Internet services are unaffordable for the majority of the population. The cost of dialup services is USD0.37 per hour and unlimited access is USD67.14 per month. The cost of ADSL access is significantly lower: on average, it does not exceed USD15 per month and offers speed of 128 Kbit/second. The quality of Internet access and communications services in Uzbekistan is rapidly improving.[6] The bandwidth capacity of external channels has shown steady growth: as of June 2006 it totaled 160.2 Mb/s, up from 44 Mb/s in July 2004.

The number of Internet service providers (ISPs) in Uzbekistan has grown considerably, from 25 in 1999 to 539 in 2005. Because of increased licensing requirements the number of ISPs dropped to 430 in 2006. There are seven top-tier ISPs with connections to China, Russia, Italy, Germany, and the Netherlands. The country also has a network of microwave radio relay lines that provide for high-speed data transmission. The sole Internet exchange point, used by the seventeen aggregator ISPs, is located in Uzbek Central Telegraph's premises.[7]

The domain registration of the national ".uz" zone was decentralized in December 2005 when five operators were granted the status of registrars. Created with foreign organizations' support, the Computerization and Information Technology Developing Center (UzInfoCom) is a quasi-nongovernmental organization[8] that develops computer and information technologies and administers the country code top-level domain name (".uz").[9] According to the data of UzInfoCom, as of October 2005 there were 2,704 second-level domains.

Russian is the most popular language among Internet users (up to 70 percent), followed by Uzbek and English. The most visited Web sites in Uzbekistan are media sites and search engines located in the Russian Internet zone (".ru").

Legal and regulatory frameworks

Although the constitution of Uzbekistan guarantees freedom of expression and prohibits censorship,[10] the Central Inspection on Protecting State Secrets in the Press officially censored media until 2002. Since then the government has increasingly compelled self-censorship on online media publishers, bloggers, and opposition leaders through a variety of means.[11] A recent example is the newly adopted Mass Media Law.[12] Discussions of its drafts were closed to the public to minimize media criticism of restrictive provisions. The Law holds media owners, editors, and staff members responsible for the "objectivity" of the published materials.[13] Independent and foreign media, including online publishers, need to register with the Cabinet of Ministers in Uzbekistan. In addition, the Law forbids entities with 30 percent or more foreign participation to establish their own media outlets in the country. Online versions of newspapers also fall within the Law's scope, and as such are subject to registration if their content differs from the printed publication. In order to gain more control over Internet content, the government has stated that subsequent regulations will specify the type of Web sites that need to be registered.[14]

The formal regulation of the Internet and electronic mass media commenced with the adoption of regulation no. 52 by the Cabinet of Ministers,[15] which established a National Network of Information Transmission (UzPAK) and ensured its monopoly on international Internet connectivity for the purposes of preserving the national information security. The government's strict enforcement of this regulation resulted in several Web sites becoming temporarily inaccessible.[16] Regulation no. 352 abolished UzPAK's monopoly on the international connections and fostered a decentralization process in the field of Internet providers. [17] However, more than 80 percent of ISPs still run their connection through UzPAK despite the high tariffs. A few ISPs have their own international satellite connections, which provide better service than UzPAK, for lower fees. A growing trend among ISPs is to use UzPAK's lines to send messages and satellite networks to view or download information. This solution allows the providers to circumvent UzPAK's monitoring network and the channels' low capacities.

UzPAK was set up within the Communications and Information Agency (UzACI),[18] which is the principal state agency regulating services in the area of communications, including the

Internet.[19] Under Resolution of the Cabinet of Ministers No. 232 of 2002, UzACI provides information security and coordinates providers' activities in this field. All Internet service providers and operators must obtain a license from UzACI.[20] Under order no. 216 Internet providers and operators cannot disseminate information that *inter alia* calls for the violent overthrowing of the constitutional order of Uzbekistan, instigates war and violence, contains pornography, or degrades and defames human dignity.[21] UzbekTelecom, the national telecommunications operator, has discretionary power to oversee the ISPs' observance of this order.[22] In 2005 the ISPs in Uzbekistan faced another regulatory hindrance in the form of resolution no. 155 (Cabinet of Ministers), which stipulated that only legal entities should be entitled to provide licensed telecommunication services. Individuals have to register as legal entities and obtain new licenses before continuing to provide Internet services.

In 2004 the Cabinet of Ministers adopted regulation no. 555, establishing a Center for Mass Media Monitoring within UzACI. The Center's key objectives are to analyze the content of information disseminated through the Internet and ensure its compliance with existing laws and regulations.[23] Another regulatory body, the Uzbek Agency for Press and Information (UzPIA), monitors the observance of media law, issues registrations and licenses for media outlets.[24] This agency has the power to suspend media licenses for "systematic" breaches of Uzbekistan's restrictive media and information laws.

The 2002 Law on Principles and Guarantees on Access to Information reserves the government's right to restrict access to information when necessary to protect the individual "from negative informational psychological influence."[25] The government further controls information streams by authorizing the use of political, economic, or other measures when necessary to counteract "threats in the sphere of information security" or "ideas of terrorism and religious extremism."[26]

Uzbekistan's principal intelligence agency, the National Security Service (SNB), monitors the Uzbek sector of the Internet and thereby compels ISPs, including cybercafés, to self-censor. Soviet-style censorship structures were replaced by "monitoring sections" that basically work under the SNB's guidance. There is no mandatory government pre-publication review, but the ISPs risk having their licenses revoked if they post "inappropriate" information. On some occasions, the SNB has ordered ISPs to block access to opposition or religious Web sites.[27] The SNB's censorship is selective and often targets articles on government corruption, violations of human rights, and organized crime. Usually this censorship affects specific pages instead of top-level domain names. The SNB regularly exchanges data with Russian intelligence sources and allegedly collaborates with the Russian Foreign Intelligence Academy.

Paradoxically, Internet filtering in Uzbekistan did not begin with the security forces but rather with the academic and research network, whose existence was funded with foreign development assistance.[28] UzSCINET was the first Uzbek ISP to implement a filtering policy, using an open source filtering product (Squid Guard) and publicly available list of pornographic sites. UzSCINET justified its position of filtering pornography on the basis of being a provider to schools and universities, as well as the need to conserve bandwidth. However, UzSCINET lacked formal legal status in Uzbekistan and as a result was dependent on UzInfoCom, a quasi-government agency for maintaining its license as a service provider. As it happened, the formal "head" of UzSCINET was also the director of UzInfoCom and a deputy director of UzASCI, the government communications agency and regulator. Simultaneously, he was also acting as an adviser to the presidential Security Council. As a result pressure was exerted on UzSCINET to cooperate with authorities, and over time the network became a "testing ground" that security forces used to develop a system for selecting and blocking unwanted Web sites. As late as 2005 the system was far from comprehensive, with previous ONI research showing a great deal of divergence among the access available on various ISPs, where some comprehensively blocked content while others allowed unfettered access. The suspicion is that some commercial ISPs had close connections with Karimov's inner circle and hence were able to withstand pressure to implement filtering, which gave them a commercial advantage (as users who wished to access such content would pay to access the Internet through these ISPs).

ONI testing results

Testing was conducted on five of the largest ISPs in Uzbekistan: ROL, Sarkor, SHARQ, TPS, and UzPAK. ONI detected a consistent and substantial filtering system that re-directs users to another Web site (www.live.com). Blocked sites included numerous political sites and a wide range of sites with human rights contents from both the local and regional list. In general, online publications tackling political issues deemed subversive or sensitive to the government were heavily filtered. These Web sites are hosted outside of Uzbekistan (www.ferghana.ru) because the ones based in the country have been already forced to shut down (www.uznews.net). Selective filtering of Web sites displaying social topics was also detected, including sites with religious, extremist, porn, gay, and lesbian content. U.S. military Web sites were largely inaccessible on some of the ISPs, although this appears to be the rest of "supply-side" blocking by U.S. authorities. Several anonymizers, a few host URLs, and one e-mail site were also within the list of blocked Web sites.

Most of the cybercafés surveyed by ONI researchers have announcements cautioning users against visiting Web sites containing extremist, obscene, sexually explicit, or pornographic content, and some cybercafé administrators do carry out surveillance on a regular basis. However, observations demonstrate that this is unevenly applied. In some cases, users enjoy relatively unfettered Internet access. In others, notably during two visits by ONI researchers, accessing an "unauthorized site" led to a swift arrest by security forces who were summoned by the Internet café owner. Regular visits by SNB officers are reported at cybercafés in the Fergana valley where they are said to manually check to see if certain sites are accessible. Most cybercafés use commercially available software that allows them to manage and bill clients remotely for time spent online. This software is easily adapted to warn administrators when unauthorized content is being accessed, and also to block access.

Conclusion

Uzbekistan maintains the most extensive and pervasive filtering system among tested CIS countries. Although expressly banned in Uzbek law, filtering is widespread and apparently growing. A large number of sites with political and human rights content sensitive to the government remain inaccessible to Internet users. The security forces in Uzbekistan manually check Internet access at "edge locations" (such as cybercafés) and monitor users' activities. The regulatory framework is so intricately woven that, in most cases, ISPs and Internet publishers are unaware of the governing law. To avoid inflicting the wrath of authorities, Internet actors frequently undertake self-censorship.

NOTES

1. See International Telecommunication Union, *World Telecommunication Indicators 2006*.

2. U.N. Development Programme and Agency for Communication and Information of Uzbekistan (2005), Review of Information and Communication Technologies Development in Uzbekistan, http://ru.ictp.uz/downloads/annual_review_2005eng.pdf (accessed March 15, 2007).
3. Survey conducted by U.N. Development Programme (UNDP) Digital Development Initiative, a joint project between the UNDP and the government of Uzbekistan (UzASCI) (2004).
4. Opinion poll conducted by the joint UNDP, UzASCI, Information and Communication Policy project (ICTP) (see information about this project at http://en.ictp.uz). The total percentage exceeds 100 percent because respondents provided more than one answer.
5. See UNDP and Agency for Communication and Information of Uzbekistan (2005), Review of Information and Communication Technologies Development in Uzbekistan, http://ru.ictp.uz/downloads/annual_review_2005eng.pdf (accessed March 15, 2007).
6. According to UzACI's data, the total modems' capacity in 2006 reached 17,000, which is twice that of the analogous indicators of 2004.
7. For more information on the amount of traffic run through the IXP, see Infocom, Results of Tashkent Internet Exchange in 2006, January 13, 2007, http://ru.infocom.uz/more.php?id=A2109_0_1_0_M (accessed February 25, 2007).
8. In reality, UzInfcom is effectively part of the Uzbek Communications and Information Agency (UzASCI); its director is a deputy director of UzASCI and it possesses no formal autonomy or independent decision-making capacity. UzASCI, while having the status of a government agency, is chaired by a deputy prime minister who acts as the de facto minister of communication.
9. Internet Assigned Numbers Authority, Agreement Between Communications and Information Agency of Uzbekistan and UzInfoCom, October 12, 2002, http://www.iana.org/cctld/uz/govt-UzInfoCom-agmt-18oct02.htm (accessed January 16, 2007).
10. Article 29 and 67 of the Constitution of the Republic of Uzbekistan.
11. Alisher Taksanov, Between Scylla and Charybdis: Uzbek Press in Recent Years in OSCE, Sixth Central Asian Media Conference, September 23–24, 2004, *21st Century Challenges for the Media in Central Asia: Dealing with Libel and Freedom of Information* 47 (2005), http://www.osce.org/publications/rfm/2005/10/18583_576_en.pdf.
12. The new Mass Media Law entered into force on January 15, 2007.
13. Institute for War and Peace Reporting, "Internet hit by media law change," January 30, 2007, http://www.iwpr.net/?p=buz&s=b&o=328926&apc_state=henh (accessed January 30, 2007).
14. See the statement of Utkir Zakirov, Head of the Coordination of Media Activities and Publishing Houses Department within the Uzbek Agency for Press and Information, http://www.cctld.uz/news/?detail=92 (in Russian).
15. Paragraph 1, regulation no. 52, On the Establishment of the National Network of Information Transmission and Streamlining the Access to the World Information Networks, adopted on February 5, 1999.
16. David Stubbs, "American aid could worsen Internet restrictions in Uzbekistan," March 30, 2002, http://www.eurasianet.org/departments/rights/articles/eav033002.shtml (accessed January 16, 2007).
17. Paragraph 1, regulation no. 352, On the Decentralization of Access to the World Computer Networks, adopted by the Cabinet of Ministers of Uzbekistan on October 10, 2002.
18. The UzACI was established under regulation no. 215, On the Measures of Improving the Activity of the Uzbek Agency for Communications and Information of 2004.
19. RESEA Republic of Uzbekistan Portal of the State Authority, Communication and Information Agency of Uzbekistan, http://www.gov.uz/en/section.scm?sectionId=2762 (accessed January 18, 2007).
20. Order no. 285, approved by the Head of the Uzbek Agency for Communications and Information on August 25, 2004.
21. Provision 12, paragraph 2, order no. 216, approved by the Head of the Uzbek Agency for Communications and Information on July 23, 2004.
22. Regulation no. 221, adopted by the Cabinet of Ministers of Uzbekistan on October 6, 2005.
23. Paragraph 1, regulation no. 555, On the Measures of Improving the Organizational Structures in the Sphere of Mass Telecommunications, adopted by the Cabinet of Ministers of Uzbekistan on November 24, 2004.
24. Uzbekistan Agency for Press and Information, Republic of Uzbekistan Portal of the State Authority, http://www.gov.uz/en/section.scm?sectionId=2759.
25. Inera Safargalieva, Uzbek Media and the Authorities: A Strange Relationship in OSCE, Fifth Central Asian Media Conference, September 17–18, 2003, Central Asia: In Defense of the Future: Media in Multi-Cultural and Multi-Lingual Societies, p. 263 (2003), http://www.osce.org/publications/rfm/2004/02/12243_101_en.pdf.
26. See Article 15 of the 2002 Law on Principles and Guarantees on Access to Information, Information Security of the State.

27. Omar Sharifov, "Review of 30 Uzbekistan-related Websites," Committee for Freedom of Speech and Expression, June 3, 2005, http://www.freeuz.org/eng/analysis/?id1=298 (accessed January 16, 2007).
28. Significant donors included the U.S. government (IREX), Soros foundation, NATO, and UNDP.

Venezuela

Internet use in Venezuela is currently not subject to extensive content restrictions, and ONI testing found no evidence of Internet censorship. However, the announced nationalization of the country's largest telecommunications company, CANTV, and the restrictive general media policies are fueling concerns that the Chávez administration could institute Internet filtering in the near future.

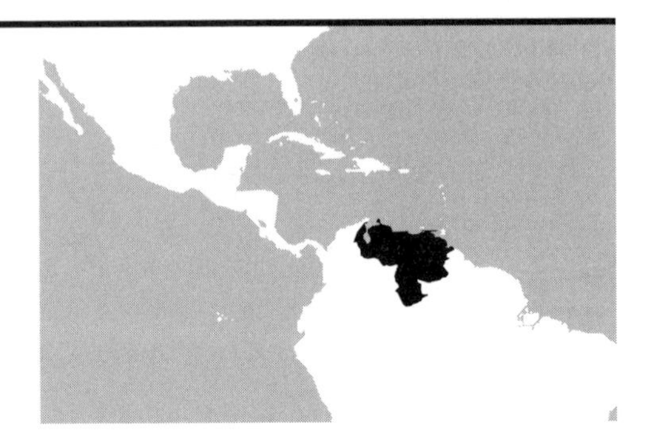

Background

The government of Hugo Chávez is in the process of consolidating power after a number of electoral victories and a failed coup in 2002. This process has taken two forms: undermining judicial independence and wresting greater control over the media. Judges on the First and Second Administrative Courts—the courts with jurisdiction over complaints relating to the government's administrative actions—are kept as provisional appointees. In 2005, six judges and their replacements were fired from the two courts for reportedly not passing performance tests.[1] As a result, these judges continue to be unable to pass judgment without fear of government retribution. As to the media, the Chávez government has recently passed two laws meant to restrict freedom of the press and of expression: the Law of Social Responsibility in Radio and Television of 2004 and the Criminal Code Reform Law of 2005. The first law delineates the standards for what is acceptable to be aired on radio and television within the country. Stations are threatened with large fines and broadcasting license suspensions for broadcasts that "condone or incite" public disturbances or carry messages "contrary to the security of the nation."[2]

In January 2006, a Venezuelan court accused ten media outlets of "obstruction of justice" and banned them from reporting on the

RESULTS AT A GLANCE

Filtering	No evidence of filtering	Suspected filtering	Selective filtering	Substantial filtering	Pervasive filtering
Political	●				
Social	●				
Conflict/security	●				
Internet tools	●				

Other factors	Low	Medium	High	Not applicable
Transparency				●
Consistency				●

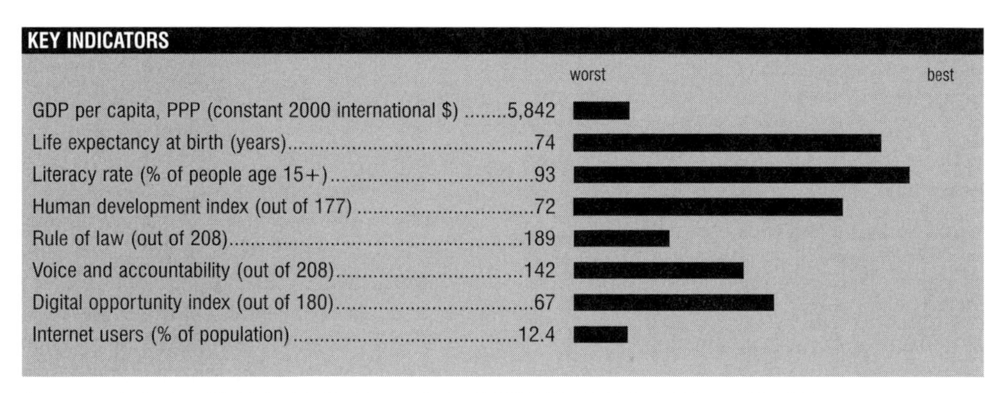

Source (by indicator): World Bank 2005, 2006a, 2006a; UNDP 2006; World Bank 2006c, 2006c; ITU 2006, 2005

investigation into the murder of Danilo Anderson, the lead prosecutor in the investigation of the failed 2002 coup. The state's lead witness, a psychiatrist, was labeled an imposter by members of the media.[3] The second law, the Criminal Code Reform Law, expanded sections of the criminal code relevant to "disrespect" of the government and increased their penalties. A television journalist critical of Chávez, José Ovidio Rodríguez Cuesta, was prosecuted in February 2006 under these newly expanded statutes. One judge rejected the case, but a Caracas court has since reopened it.[4]

Internet in Venezuela

Between 1998 and 2002, the number of Internet users in Venezuela grew from 207,000 to 1,585,000, but then decreased to 1,365,000 in 2003 for a current Internet penetration rate of 12.4 percent.[5] The vast majority of personal computers are not connected to the Internet.[6] The Venezuelan government estimates that 50.4 percent of the population have never used the Internet and would not be interested in doing so, while 28.9 percent are possible future Internet users, primarily young, educated, middle-class individuals.[7] Though there are sixty licensed ISPs, CANTV Servicios and Telcel control over 90 percent of the Internet market.[8]

Internet use is strongly concentrated among young, educated city residents, with 76 percent of users younger than thirty-five,[9] 67 percent having schooling beyond high school,[10] and more than 60 percent of users coming from Caracas.[11] Approximately 26 percent of Internet users log on daily. These users tend to be upper-class individuals using home connections for educational or work research and downloading. Over half of Internet users connect between once and five times per week, using cybercafés for e-mailing and chatting. This group is generally male and represents all socioeconomic levels with the exception of the lowest income segment. A smaller portion of users, 16.9 percent, connect between once every other week and once per month. These light users come from all economic strata except the lowest class, and they almost exclusively use cybercafés for job search purposes.[12] Hotmail, Google, and Yahoo are by far the most popular sites, followed by news sites and other search engines.[13]

Despite programs promoting Internet use by poor and rural Venezuelans, access for this segment of the population, about 60 percent of the total, is essentially nonexistent, and basic public education does not incorporate Internet technologies.[14]

In 2000, Venezuela had approximately 240 dot-com businesses, mostly business-to-business rather than business-to-consumer.[15] The government has been attempting to automate its processes and put its agencies and services online, assisted by a newly created agency for information technology,[16] but these attempts have not been consistent or thorough.[17]

Legal and regulatory frameworks

Venezuelan President Chávez has decreed the promotion of Internet use as essential to development.[18] Correspondingly, the government promotes use of information and communication technologies (ICT) through a regulatory framework designed to promote competition among ICT businesses, but no special programs encourage such businesses directly.[19] Personal Internet use appears to be essentially unrestricted by current law and regulation. Despite an erroneous press release listing Venezuela among countries with Internet censorship,[20] the U.S. State Department Report on Human Rights in Venezuela states that "there were no government restrictions on the Internet or academic freedom."[21] Individual reports of suspected filtering are not backed by substantial evidence. [22]

Fear of Internet regulation stems from broader Venezuelan law restricting freedom of press and speech. The Social Responsibility Law opens citizens to punishment for disrespecting authority and endangering children with improper content; these laws have led to censorship in the general news media.[23] President Chávez's announcement on January 8, 2007, of re-nationalization plans for CANTV has heightened fears of expanded regulation and content restrictions as the government assumes more control of Internet media.[24] A recent article notes that CANTV has held 83 percent of the Internet market since the market's privatization,[25] so any changes in filtering through a nationalized CANTV will have a strong impact on Internet users.

ONI testing results

Tests of Internet censorship were carried out in late 2006 on the two major ISPs in Venezuela. The testing covered a wide range of potentially sensitive content, including sites dedicated to political opposition, freedom of expression, and anti-Chávez media, as well as sites centered on controversial social issues such as minority religions, indigenous peoples, gambling, and pornography. This assessment turned up no evidence of filtering.

Conclusion

Despite fears to the contrary, ONI results give no indication of Internet censorship. The nationalization of CANTV and past censorship of different media are causes for concern about future filtering. However, current evidence indicates that Venezuelan Internet access is restricted only by initial socioeconomic, cultural, and geographic barriers to entry and not by any subsequent restraints on content once users are online.

NOTES

1. Human Rights Watch, World Report 2007: Venezuela, http://hrw.org/englishwr2k7/docs/2007/01/11/venezu14888.htm.
2. Ibid.
3. Ibid; Reporters Without Borders, Venezuela: Annual Report 2007, http://www.rsf.org/article.php3?id_article=20544&Valider=OK.
4. Human Rights Watch, World Report 2007: Venezuela, http://hrw.org/englishwr2k7/docs/2007/01/11/venezu14888.htm.
5. Cámara Venezolana de Comercio Electronico – Tendencias Digitales, http://www.cnti.gob.ve/cnti_docmgr/sharedfiles/indicadores.penetracion.Internet.vzla.pdf.
6. TILAN at the University of Texas's Latin American Network Information Center, http://lanic.utexas.edu/project/tilan/countries/ven/.

7. Cámara Venezolana de Comercio Electronico – Tendencias Digitales, http://www.cnti.gob.ve/cnti_docmgr/sharedfiles/indicadores.penetracion.Internet.vzla.pdf.
8. Global Competitiveness Report 2001–2002, Harvard Center for International Development, http://www.cid.harvard.edu/cr/profiles/Venezuela.pdf.
9. Cámara Venezolana de Comercio Electronico – Tendencias Digitales, http://www.cnti.gob.ve/cnti_docmgr/sharedfiles/indicadores.penetracion.Internet.vzla.pdf.
10. Ibid.
11. Global Competitiveness Report 2001–2002, Harvard Center for International Development, http://www.cid.harvard.edu/cr/profiles/Venezuela.pdf.
12. Cámara Venezolana de Comercio Electronico – Tendencias Digitales, http://www.cnti.gob.ve/cnti_docmgr/sharedfiles/indicadores.penetracion.Internet.vzla.pdf.
13. Tendencias Digitales, Indicadores de Uso de Internet en Venezuela, http://www.tendenciasdigitales.com/td/indicadores_uso.htm.
14. Harvard Center for International Development, Global Competitiveness Report 2001–2002, http://www.cid.harvard.edu/cr/profiles/Venezuela.pdf.
15. Ibid.
16. Lentro Nacionál de Technologías de Información, http://www.cnti.gob.ve/.
17. Harvard Center for International Development, Global Competitiveness Report 2001–2002, http://www.cid.harvard.edu/cr/profiles/Venezuela.pdf
18. Decreto no. 825, May 10, 2000, http://www.cnti.gob.ve/cnti_docmgr/sharedfiles/decreto825.pdf.
19. Harvard Center for International Development, Global Competitiveness Report 2001–2002, http://www.cid.harvard.edu/cr/profiles/Venezuela.pdf.
20. Caroline Walker, "Online filtering and censorship at issue on the Internet," http://usinfo.state.gov/dhr/Archive/2006/May/23-489666.html.
21. U.S. Department of State, Country Reports on Human Rights Practices 2005: Venezuela, http://www.state.gov/g/drl/rls/hrrpt/2005/61745.htm.
22. See Alexander Boyd, "Venezuela's Regime: Initial Stages of Internet Control," http://www.vcrisis.com/?content=letters/200504261545 for the only example found in preliminary research.
23. See http://hrw.org/english/docs/2005/03/24/venezu10368.htm, http://www.cidh.org/annualrep/2005eng/chap.4d.htm, and http://sipiapa.com/ (search country listings for Venezuela).
24. http://www.cnn.com/2007/WORLD/americas/01/08/chavez.media.ap/index.html.
25. http://www.lared.com.ve/archivo/telco12-01-07.html.

Vietnam

Vietnam currently regulates access to the Internet extensively, both in the management of Internet infrastructure as well as by restricting access to country- and language-specific content.

Background

Now a member of the World Trade Organization (WTO), the Socialist Republic of Vietnam is attempting simultaneously to promote the development of information communications technology (ICT) and e-commerce while struggling to limit access to content that might destabilize the communist state and undermine its control. Although citizens are legally allowed to question corruption, economic policy, and government deficiencies, the line is drawn at political criticism involving government leaders, political parties and multiparty democracy, and sensitive social and diplomatic issues. After a period of relative openness and tolerance of independent voices and criticism in 2006, where liberal publications were established, the government clamped down on what it considers unlawful usage of the Internet. Authorities continue to detain a number of individuals for Internet activities, such as discussing political reform over Voice-over Internet Protocol (VoIP).[1]

Internet in Vietnam

Vietnam's Internet system is growing and changing rapidly, and it is difficult to describe the situation "on the ground" with complete accuracy. From 2005 to 2006, the number of Internet users reportedly jumped from 9.2 million to 14.5 million, yielding an Internet penetration rate of 17 percent.[2] Because more than half of the population is under thirty and a significant portion of individual users use cybercafés for online gaming and access to the Internet, control over these venues

RESULTS AT A GLANCE

Filtering	No evidence of filtering	Suspected filtering	Selective filtering	Substantial filtering	Pervasive filtering
Political					●
Social			●		
Conflict/security	●				
Internet tools				●	

Other factors	Low	Medium	High	Not applicable
Transparency	●			
Consistency	●			

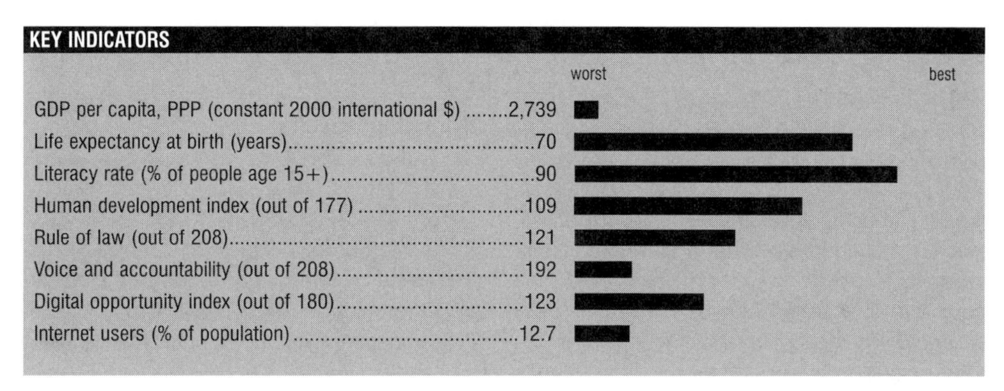

Source (by indicator): World Bank 2005, 2006a, 2006a; UNDP 2006; World Bank 2006c, 2006c; ITU 2006, 2005

is an important priority for the state.[3] Postal offices are also important providers of Internet access. VoIP is an increasingly popular means of communication.[4] Although the state affirmatively seeks to enhance the competitive edge of domestic enterprises, Vietnamese online services are still nascent, and there are few search engines currently available to Vietnamese users.

State regulation determines how Internet connectivity in Vietnam is organized and managed, and facilitates Internet content filtering by limiting external access points that must be controlled.[5]

Only Internet exchange points (IXPs) can connect to the international Internet, while online service providers (OSPs) and Internet content providers (ICPs) may connect to ISPs and IXPs.[6] At the edge of the network, Internet agents, such as cybercafés, connect to their contracted ISP.[7] ISPs may connect with each other and with IXPs, but private ISPs may not connect with each other in peering arrangements.[8] Currently, IXPs can theoretically maintain independent connections to the international Internet, but it is not clear how many do so in practice. Vietnam controls the allocation of domain names under the country code top-level domain, ".vn," through the Vietnam Internet Center,[9] and is also planning to implement a state-controlled Vietnamese-language second-level domain.[10]

Legal and regulatory frameworks

Vietnam's legal regulation of Internet access and content is multilayered and complex, and can occur at the level of National Assembly legislation, ministerial decisions, or through VNPT rules created for the management of the Internet infrastructure. Although Vietnam nominally guarantees freedom of speech, of the press, and of assembly through constitutional provisions,[11] state security laws and other regulations trump or eliminate these formal protections. Media in Vietnam are state-owned, and they are under increasingly tight control by the state. Effective July 1, 2006, the Decree on Cultural and Information Activities subjects those who disseminate "reactionary ideology" including revealing secrets (party, state, military, and economic), who deny revolutionary achievements, and who do not submit articles for review before publication to fines of up to thirty million dong (USD2,000).[12] These regulations appear to target journalists, as criminal liability already exists for some of the proscribed activities, including the dissemination of state secrets.

All information stored on, sent over, or retrieved from the Internet must comply with Vietnam's Press Law, Publication Law, and other laws, including state secrets and intellectual property protections.[13] All domestic and foreign individuals and organizations involved in Internet activity in Vietnam are legally responsible for content created, disseminated, and stored. New monitoring software issued by the Ministry of Posts and Telematics in July 2006 requires ISPs to record the identity and Internet behavior of users at Internet kiosks, and to store the information on their servers for one year.[14] Relevant legislation and administrative decrees may not be consistently enforced—such as the requirements to track IDs and record personal information as a condition for access in cybercafés that appears to be largely ignored in cities such as Ho Chi Minh City and Hanoi.[15] However, they do provide the state with considerable authority and discretion to control how citizens get online.

Just as ISPs and cybercafés are required to install monitoring software and store information on users, all users are also formally deputized to report content that opposes the state or threatens state security to the relevant authorities.[16] It is unlawful to use Internet resources or host material that opposes the state; destabilizes Vietnam's security, economy, or social order; incites opposition to the state; discloses state secrets; infringes organizations' or individuals' rights; or interferes with the state's Domain Name System (DNS) servers.[17] Those who violate Internet use rules are subject to a range of penalties, from fines to criminal liability for offenses such as causing chaos or "security disorder."[18] The National Assembly enacted the Law on Information Technology on June 22, 2006.[19]

Regulatory responsibility for Internet material is divided along subject-matter lines in Vietnam. While the Ministry of Culture and Information focuses on sexually explicit, superstitious, or violent content, the Ministry of Public Security monitors customers who access politically sensitive sites.[20]

ONI testing results

Testing was conducted from various access points (including hotel, cybercafé, and wireless connections) on two ISPs: FPT and VNPT. VNPT returns a "blockpage" indicating that the requested site was prohibited; FPT indicates that the filtered site does not exist, suggesting a form of DNS tampering where the listings for filtered sites had been removed from its DNS server. Our testing of Vietnam's Internet filtering found that the state concentrates its blocking on content about overseas political opposition, overseas and independent media, human rights, and religious topics. Proxies and circumvention tools, the use of which is illegal,[21] were the major exception and a substantial number were inaccessible on both ISPs.

A large majority of blocked and inaccessible content was specific to Vietnam—either in the Vietnamese language or related to Vietnamese issues, with a significant number of filtered sites operating out of California. Sites only in English or French, or from the global list, were rarely blocked. For example, the domain for Radio Free Asia (www.rfa.org) was blocked only on FPT, although RFA's Vietnamese-language home page (www.rfa.org/vietnamese) was blocked by both ISPs. At the same time, however, sites only tangentially or indirectly critical of the government, such as content focusing on local communities (www.nguoidan.net; www.vietnamdaily.com) or world news aggregation (www.thongluan.org; www.danchimviet.com/php/index.php) were also blocked, along with sites voicing strong anticommunist sentiments (www.conong.com; www.vietnamvietnam.com). Although a large number of overseas sites focusing on political opposition and reform (such as the Free Vietnam Alliance at www.lmvntd.org) were filtered, the only human rights Web site on the global list to be blocked by either (and in this case both) ISP belonged to the NGO Human Rights Watch (www.hrw.org).

Certain religious content, such as pages on religious freedom, Buddhism, and Caodai (www.caodai.net) are blocked to a limited degree. Some topics, such as the Montagnard people who assisted the United States during the war with Vietnam and who are commonly Christians, overlap multiple categories (such as the Montagnard Human Rights Organization Web site www.mhro.org) and are filtered accordingly.

Surprisingly, Vietnam does not block any pornographic content (though it does filter one site ONI tested with links to adult material), despite the state's putative focus on preventing access to sexually explicit material. The state's filtering practices are thus in obvious tension with the purported justification for these actions.

ONI has concluded that commercial filtering lists are not being used in Vietnam for several reasons: the pattern of blocking does not conform to any software product that ONI has studied, the observed pattern of deleting DNS records for prohibited sites is inconsistent with using Web filtering software, and the greater filtering of Vietnamese-language sites on a given topic compared with English-language sites. However, VNPT may be using a commercial product for filtering. Through multiple rounds of testing, inconsistencies in filtering persisted and evolved, also indicating that the Vietnamese state or Vietnamese ISPs are compiling their own block lists. For example, the news site www.saigonbao.com, blocked earlier in 2006 by both FPT and VNPT, was inaccessible only on VNPT when tested at the end of the year. VNPT also filtered a range of sites that were accessible on FPT, primarily independent media, human rights (from the Vietnam Human Rights Network to the International Criminal Tribunal for Rwanda Web sites), and overseas community (the Vietnam American National Gala awards) and political content.

Conclusion

Vietnam's filtering regime is multilayered, relying not only on computing technology but also on threats of legal liability, state-based and private monitoring of users' online activities, and informal pressures such as supervision by employees or other users in cybercafés. Over time, the state's online filtering has expanded, both in the content blocked for a given topic and the number of content categories that are targeted. Although purporting to protect national security and block obscene content, Vietnam actually focuses on blocking access to sites within an expansive definition of political "opposition" that includes the activities of Vietnamese communities overseas. Although the Vietnamese state's blocking of access to certain content on the Internet can be circumvented by users with technical knowledge, ordinary users will likely continue to find that filtering distorts their information environment.

NOTES

1. Deutsche Presse-Agentur, "Vietnam youths arrested over internet chats released after 9 months," August 16, 2006.
2. Thai Press Reports, "E-commerce services extended in 2006," January 4, 2007.
3. Paul Budde Communication Pty Ltd., 2006, Vietnam: Internet, 10, July 30, 2006. See also John Boudreau, "Bay Area entrepreneur leads way in online gaming in Vietnam," The Mercury News, January 17, 2007, http://www.mercurynews.com/mld/mercurynews/business/16475618.htm.
4. Paul Budde Communication Pty Ltd., 2006, Vietnam: Internet, 15, July 30, 2006.
5. See Articles 27–38, decree no. 55/2001/ND-CP of the Government on the management, provision and use of the Internet services, issued on August 23, 2001.
6. Article 27, decree no. 55/2001/ND-CP.
7. Ibid.
8. Ibid.
9. See Article 3, decision no. 27/2005/QD-BBCVT; IANA, .vn – Vietnam, http://www.iana.org/root-whois/vn.htm.
10. See Article 7, decision no. 27/2005/QD-BBCVT.

11. Article 69, Constitution of the Socialist Republic of Vietnam, adopted April 15, 1992, amended December 25, 2001, http://www.oefre.unibe.ch/law/icl/vm00000_.html.
12. Southeast Asian Press Alliance (SEAPA), "Vietnam readies stricter press laws to rein back aggressive journalists," June 16, 2006, http://www.seapabkk.org/newdesign/newsdetail.php?No=485.
13. Article 6(1), decree no. 55/2001/ND-CP of the Government on the management, provision and use of the Internet services, issued on August 23, 2001.
14. U.S. Department of State, Country Reports on Human Rights Practices 2006: Vietnam, March 6, 2007, http://www.state.gov/g/drl/rls/hrrpt/2006/78796.htm.
15. OpenNet Initiative, Internet Filtering in Vietnam in 2005–2006: A Country Study, 19, August 2006, http://www.opennet.net/studies/vietnam/.
16. Article 6(2), decree no. 55/2001/ND-CP of the Government on the management, provision and use of the Internet services, issued on August 23, 2001; Article III, Joint Circular no. 02/2005/TTLT-BCVT-VHTT-CA-KHDT of July 14, 2005, on management of Internet agents, issued by the the Ministry of Post and Telematics, the Ministry of Culture and Information, the Ministry of Public Security, and the Ministry of Planning and Investment on July 14, 2005.
17. Article 2(2), Regulation on Management and Use of Internet Resources, decision no. 27/2005/ QD-BBCVT, issued by the Ministry of Post and Telematics on August 11, 2005.
18. Article 11(3), decree no. 55/2001/ND-CP.
19. Thai Press Reports, "Vietnam post and telecommunication sector records outstanding achievements in 2006," January 1, 2007.
20. BBC Monitoring International Reports, "Vietnamese Security Ministry establishes special unit to tackle Internet crime," August 4, 2004.
21. Article I s.3, Joint Circular no. 02/2005/TTLT-BCVT-VHTT-CA-KHDT.

Yemen

Internet filtering in the Republic of Yemen is relatively broad in scope, with pornography a principal target for blocking. Despite the wide range of content censored, however, the depth of filtering in Yemen is inconsistent; many users of Yemen's primary Internet service provider (ISP) are not filtered when the user licensing quota in the filtering software agreement is exceeded.

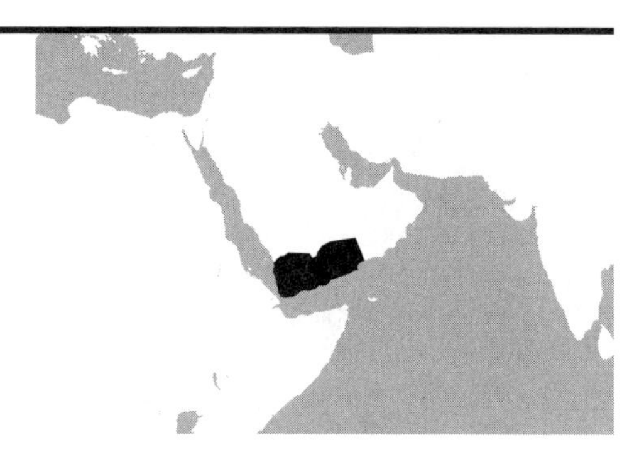

Background

The press in Yemen operates under the careful eye of a government hostile to independent reporting on political and social issues. Newspapers have been closed and journalists have been arrested, interrogated, imprisoned, fined, and banned from publication for their coverage of sensitive topics; reports of threats and physical attacks are also numerous.[1]

In 2005, the government and unidentified parties thought to be associated with government security forces intensified harassment of journalists and political critics. Human rights problems include limitations of citizens' ability to change the government, acknowledged torture, significant restrictions on freedom of press and assembly, and some restrictions on speech.[2]

Internet in Yemen

Yemen lacks a robust telecommunications and information communications technology (ICT) sector. The International Telecommunication Union (ITU) estimates that less than 1 percent of Yemen's population uses the Internet (0.87 users per 100 inhabitants) and that only 300,000 PCs exist in the country (1.5 per 100 inhabitants).[3] Many cannot afford and are simply unfamiliar with the equipment and services needed to access the Internet.[4] Only 9 out of every 100 inhabitants is a telephone subscriber.[5]

RESULTS AT A GLANCE

Filtering	No evidence of filtering	Suspected filtering	Selective filtering	Substantial filtering	Pervasive filtering
Political			●		
Social					●
Conflict/security			●		
Internet tools				●	

Other factors	Low	Medium	High	Not applicable
Transparency		●		
Consistency			●	

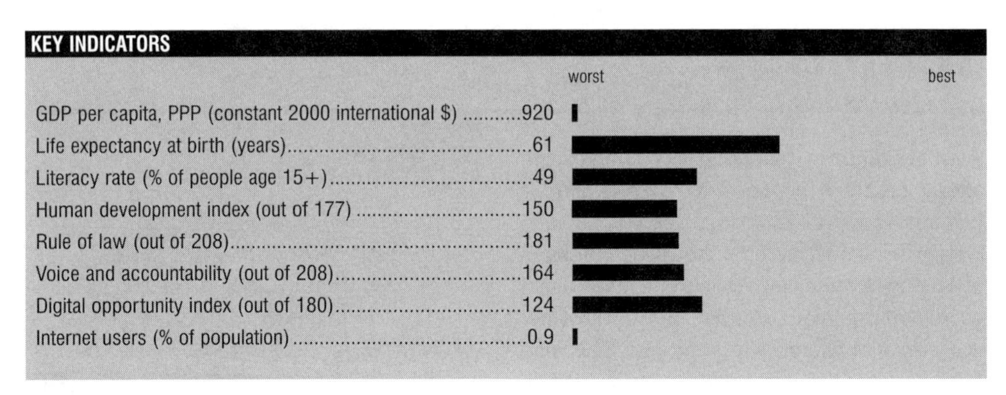

Source (by indicator): World Bank 2005, 2006a, 2006b; UNDP 2006; World Bank 2006c, 2006c; ITU 2006, 2004

Yemen is serviced by two ISPs: YemenNet, which is a service of the government's Public Telecommunication Corporation (PTC),[6] and TeleYemen's Y.Net, which is part of the government's PTC but is managed by FranceTelecom.[7]

Businesses own 60 percent of Internet subscriber accounts, while government and educational institutions own only 3 percent of subscriber accounts.[8] Far fewer women than men access the Internet, which may be because the primary Internet access locations are Internet cafés (61 percent) and work (24 percent), with home Internet availability considerably less frequent (13 percent). Only 2 percent access the Internet from schools.[9] By mid-2005, the number of Internet cafés in Yemen reached 736.[10]

Legal and regulatory frameworks

The Ministry of Telecommunication and Information Technology (MTIT) grants ISP licenses;[11] PTC, a branch under the MTIT, is responsible for the management and growth of telecommunications in Yemen.[12]

ISPs impose restrictions on the use of Internet services, preventing subscribers from accessing or transmitting certain content. The terms and conditions set by TeleYemen (aka Y.Net) state: "Access to applications which transmit or receive live video or audio, or make similar demands on the capacity of the network, constitutes an unreasonable usage which may affect the performance of the network, and is not permitted."[13] Also covered are customer responsibilities, including prohibitions on "sending any message which is offensive on moral, religious, communal, or political grounds" (6.1.1).[14] Additionally, TeleYemen reserves the right to control access "and data stored in the Y.Net system in any manner deemed appropriate by TeleYemen" (7.1).[15] Finally, section 6.3.3 admonishes subscribers that TeleYemen will "report to the competent authorities, any use or attempted use of the Y.Net service which contravenes any applicable Law of the Republic of Yemen."[16]

Yemen's Press and Publications Law, passed in 1990, subjects publications and broadcast media to broad prohibitions and harsh penalties.[17] This law theoretically establishes a press that "shall be independent and shall have full freedom to practice its vocation," but that must operate "within the context of Islamic creed, within the basic principles of the Constitution, goals of the Yemeni Revolution, and the aim of solidifying national unity."[18]

The Press and Publications Law further states that local journalists must be Yemeni citizens and must obtain Press Cards from the Ministry of Information. Foreign journalists must be accredited to receive Press Cards. Press Cards can be revoked by the Ministry of Information without any reason given, and this revocation requires the former holder to leave Yemen unless they have an independent reason for residency.[19]

A recent example of the implementation of this law is the conviction and fine handed down in December 2006 to the editor of the Yemen Observer for reprinting the Danish cartoons of the Prophet Muhammad. The Yemen Observer's license was revoked and the newspaper was closed down in February 2006 for three months after republishing fragments of the Danish cartoons.[20] Interestingly, the Web site of the newspaper was not targeted or blocked by the authorities.

A new draft of the law, proposed in 2005, was rejected by the Yemen Journalists Syndicate (YJS) as even more repressive than the existing 1990 law.[21] Despite a promise by the Yemeni president to reform the media laws and abolish imprisonment penalty in publishing offenses, Yemeni journalists are subject to violation by the government, the ruling party, opposition parties, and religious groups alike.[22]

The draft law "ignored the question of the electronic media freedom, putting an end to the state ownership and monopoly over broadcast media. Rather, it went on controlling the websites just like print media."[23]

ONI testing results

ONI ran in-country tests in 2006 on Yemen's two ISPs, YemenNet and TeleYemen/Y.Net. We found significant differences between the two. Interestingly, YemenNet, the primary ISP, was found to block very few Web sites. Because these results were contrary to previous information and ONI studies, we repeated the test runs from different locations using different connections but got the same results, which showed that YemenNet no longer filters as extensively as it did in the past. We investigated further and found that the ISP uses a Blue Coat integrated cache/filter appliance to run Websense but possesses a limited number of concurrent user licenses—not nearly enough to cover the 150,000-plus Internet users in the country. Thus, when the number of subscribers accessing the Internet at a given time exceeds the limited number of user licenses, the requests of users circumvent the filtering software.

The second ISP, TeleYemen/Y.Net, also obtains its filtering software from U.S.-based Websense. However, Y.Net was found to block almost all of the Web sites containing pornography, provocative attire, sex education materials, and anonymizing and privacy tools. Search strings containing the word "sex" are blocked, as are some sites hosting gay and lesbian content, hacking information, and non-erotic nudity. The ISP also filters some religious conversion sites and a limited number of Voice over Internet Protocol (VoIP) and circumvention sites.

The only political Web site found to be blocked by Y.Net is www.soutalgnoub.com, which is run by a Yemeni opposition group. Other than this Web site, neither provider blocked any of the other politically-related sites on the testing lists. However, ONI monitored Web access in Yemen during Yemen's September 2006's presidential election and found that the government-owned YemenNet did block access to several independent news and political opposition sites, including Nass Press (www.nasspress.com), Al-Mostakela Forum (www.mostakela.com), and the Yemeni Council (www.al-yemen.org).[24]

Conclusion

Extensive testing and analysis revealed no evidence that the Yemeni state is currently preventing citizens from accessing news or political content online. The availability of such content should not, however, suggest tolerance for criticism or dissent, as attested by the state's treatment of journalists and its timely blocking of oppositional media sites during the 2006 presidential elections. The failures of the filtering system installed on Yemen's principal ISP likewise hint at the state's limited capacity to control content, rather than any willingness to allow information to flow freely. In essence, the breadth of content filtered should temper any optimism about the evident ineffectiveness of filtering in Yemen witnessed in this round of testing.

NOTES

1. See, for example, Reporters Without Borders, "Yemen press release," December 14, 2005, http://www.rsf.org/article.php3?id_article=15713; Yemen: Annual Report 2006, http://www.rsf.org/article.php3?id_article=17212.
2. U.S. Department of State, Country Reports on Human Rights Practices 2005: Yemen, http://www.state.gov/g/drl/rls/hrrpt/2005/61703.htm.
3. International Telecommunication Union, *World Telecommunication Indicators 2006*.
4. Adnan Hizam, "IT market in Yemen experiences rapid growth," Yemen Observer, July 29, 2006, http://www.yobserver.com/cgi-bin/2007/exec/view.cgi/22/10622/printer.
5. International Telecommunication Union, *World Telecommunication Indicators 2006*.
6. YemenNet Web site, Background information, http://www.yemen.net.ye/index.php?q=background (in Arabic).
7. Library of Congress: Federal Research Division, Country Profile: Yemen, December 2006, http://lcweb2.loc.gov/frd/cs/profiles/Yemen.pdf.
8. Helmi Noman, An Overview of the Demographics and Usage Patterns of Internet Users in Developing Countries: Yemeni Internet Population as a Case Study, United Nations Development Programme, http://www.undp.org.ye/ict.php.
9. Ibid.
10. Yemen News Agency (Saba) http://www.sabanews.net/view.php?scope=f69b5&dr=&ir=&id=103280 (in Arabic).
11. Ministry of Telecommunication and Information Technology, http://www.mtit.gov.ye/ (in Arabic).
12. *AME Info,* "Yemen's Internet market registers high growth rates," March 14, 2006, http://www.ameinfo.com/80390.html.
13. Terms and conditions for Y.Net Service, http://www.y.net.ye/support/rules.htm.
14. Ibid.
15. Ibid.
16. Ibid.
17. Yemen News Agency (Saba) Press and Publications Law, Law http://www.sabanews.net/view.php?scope=319c3e9&dr=&ir=&id=44000.
18. Ibid.
19. Ibid.
20. Reporters Without Borders, Yemen: Annual Report 2007, http://www.rsf.org/country-43.php3?id_mot=157&Valider=OK.
21. See *Yemen Times,* "Journalists reject draft press law," April 28 to May 1, 2005, http://yementimes.com/article.shtml?i=837&p=front&a=2; IRIN, "Yemen: Journalists still targeted despite draft law,", April 30, 2006, http://www.irinnews.org/report.aspx?reportid=26327; International Press Institute, 2005 World Press Freedom Review, http://www.freemedia.at/cms/ipi/freedom_detail.html?country=/KW0001/KW0004/KW0108/.
22. Arab Press Freedom Watch, More Press Freedom Violation Recorded in Yemen, http://www.apfw.org/indexenglish.asp?fname=news%5Cenglish%5C2007%5C01%5C13203.htm.See full report at http://www.apfw.org/indexenglish.asp?fname=report%5Cenglish%5C2007%5C01%5C1014.htm.
23. Mohammed Al-Qadhi, "Journalists reject draft press law," *Yemen Times,* http://yementimes.com/article.shtml?i=837&p=front&a=2.
24. See Arabic Network for Human Rights Information, "To exit integral competitive presidential elections: The Yemeni government restricts freedom of expression by blocking independent websites," September 11, 2006, http://www.hrinfo.net/en/reports/2006/pr0911.shtml; *NewsYemen,* "Press and freedom of expression: The victims of Yemeni elections," September 30, 2006, http://www.newsyemen.net/en/view_news.asp?sub_no=4_2006_09_30_6360.

Zimbabwe

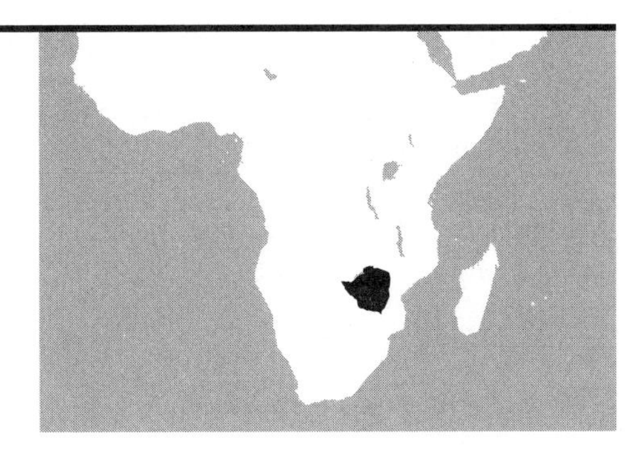

Despite the country's highly repressive regime, ONI found no evidence of Web site filtration in Zimbabwe. Because of limited Internet access and usage, the country's efforts have centered on regulating email.

Background

Zimbabwe's government is tightly controlled by President Robert Mugabe and the ruling Zimbabwe African National Union-Patriotic Front (ZANU-PF). They have dominated the political landscape since the country's independence from Great Britain in 1980 and have manipulated political structures to ensure that they stay in control.[1] The ZANU-PF–controlled government is known for its brutal repression and continuing violations of human rights. The best example of this is 2005's "Operation Murambatsvina," or "Operation Tsunami," as it is called locally.[2] Officially described as an effort to eliminate illegal housing and commerce, the "mass evictions and demolitions" were,[3] as reported by the U.N., "carried out in an indiscriminant and unjustified manner, with indifference to human suffering, and, in repeated cases, with disregard to several provisions of national and international legal frameworks."[4] Though the actual motivations are unknown, one theory is that the operation was meant to be retribution toward regions in which voters for opposition parties lived.[5] Free assembly is dramatically curtailed as the government often violently breaks up peaceful protests under the Public Order and Security Act.[6] There have been allegations of police abuse and the torture of detainees.[7] A severe press law passed in 2002 allows the Media and Information Commission to crack down on dissent within the media by controlling the licensing of journalists.[8] And, finally, the government jams a number of radio stations critical of the government, such as Voice of

RESULTS AT A GLANCE

Filtering	No evidence of filtering	Suspected filtering	Selective filtering	Substantial filtering	Pervasive filtering
Political	●				
Social	●				
Conflict/security	●				
Internet tools	●				

Other factors	Low	Medium	High	Not applicable
Transparency				●
Consistency				●

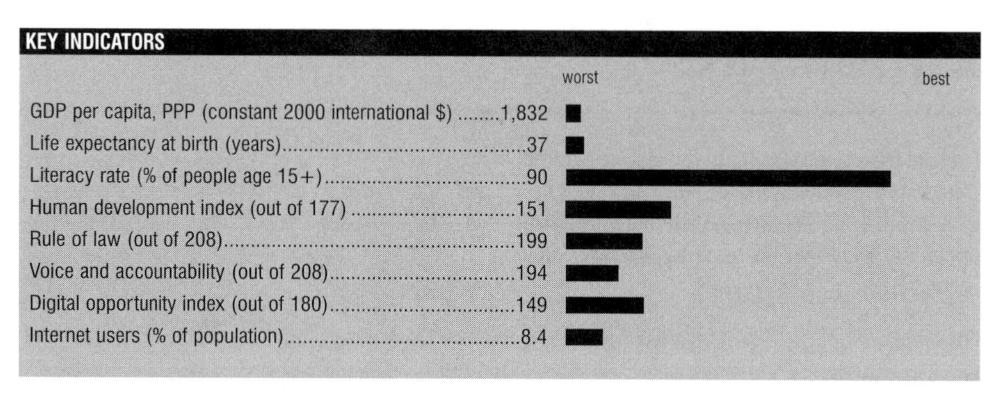

Source (by indicator): World Bank 2005, 2006a, 2006b; UNDP 2006; World Bank 2006c, 2006c; ITU 2006, 2005

America, Voice of the People, and SW Radio Africa.[9]

Internet in Zimbabwe

The number of Internet users in 2005 was reportedly 1,000,000, or approximately 8 percent of Zimbabwe's population.[10] The number of Internet service providers (ISPs) has risen from six in 2003 to twenty-seven in 2004, due to growing demand.[11] The *Business Mirror* in 2005 performed a survey showing that Harare has over thirty Internet cafés, up from about twenty, two years prior.[12] The Internet is a less expensive means of communication than the telephone service in Zimbabwe, fueling its growth. In 2004, electronic messaging cost between ZIM $200 (USD0.04 by 2004 exchange rate) and ZIM $250 (USD0.05 by 2004 exchange rate) per minute, while international telephone calls cost between ZIM $3,800 (USD0.72 by 2004 exchange rate) and ZIM $5,800 (USD1.10 by 2004 exchange rate) per minute.[13] However, because of limited awareness of its capabilities, Internet use is mostly limited to e-mail.[14] The low level of Internet penetration overall is likely the result of the increasingly rapid decline of the economy and quality of life in the country over the past seven years. In January 2007, inflation rates reached a staggering 1,593.6 percent.[15] The government is bankrupt, eight in ten Zimbabweans are destitute, and workers in Harare see their bus fares to and from work take up their entire salaries.[16] In such an environment, demand for luxury goods such as computers and Internet use is low. In September 2006, a large majority of the Internet went offline when the international satellite communications provider, Intelsat, cut service to the country, following the failure of government-owned telecommunications company, TelOne, to pay its debts to the company. Service was restored after the reserve bank paid the outstanding debt.[17]

Legal and regulatory frameworks

Zimbabwe's government mainly focuses its regulation of Internet use on e-mail.[18] The Post and Telecommunications Act of 2000 allows the government to monitor e-mail usage and requires ISPs to supply information to government officials when requested.[19] The Supreme Court, however, ruled in 2004 that the sections of the law that permit monitoring Internet users violated the constitution.[20]

The government struck back with an initiative in 2004 that requires ISPs to renew contracts with TelOne, the government-owned telecommunications company, with the stipulation that they report any e-mail with "offensive or dangerous" content.[21] In essence, this requires ISPs to do what the Supreme Court has ruled is unconstitutional. As of yet, no ISPs have signed new agreements.[22]

The government responded again with the Interception of Communications Bill of 2006. Under its provisions, the government would establish a telecommunications agency called the Monitoring and Interception of Communications Center to monitor, among other things, all telecommunications.[23] The government withdrew the bill in November 2006 over constitutionality objections from the Parliamentary Legal Committee and plans to revise it.[24] Even without explicit powers, the authorities appear to be pursuing a crackdown on e-mail dissent unabated. In 2005, for example, authorities arrested forty people in a raid on a local Internet café because an e-mail insulting President Robert Mugabe allegedly was sent from the location.[25]

ONI testing results

ONI testing of two Zimbabwean ISPs, Econet and YoAfrica, revealed no evidence of a filtration regime in the country. Despite the ZANU-PF regime's record of repression, this is not an unexpected finding. Internet use in Zimbabwe is extremely low and, as mentioned earlier, is generally limited to e-mail rather than Web browsing. As a result, Zimbabwe's main efforts toward control of the Internet are e-mail focused. A large-scale Internet filtration system in all likelihood does not hold much value to the Zimbabwean government relative to the price of its implementation.

Conclusion

Zimbabwe is a highly repressive country with a failing economy and a poverty-stricken population. Internet penetration is extremely low and the Internet is mainly used for e-mail. As a result, the government restricts its efforts toward Internet control to e-mail monitoring and censorship. Though its legal authority to pursue such measures is contested, the government appears to be following through on its wishes to crack down on dissent via e-mail. If Internet usage were to rapidly expand and increasingly spill over to Web browsing, it is likely, given its history, that Zimbabwe would move to Web site filtration. Given the state of the country, however, this does not appear imminent.

NOTES

1. Freedom House, Zimbabwe: Country Report, 2005, http://www.freedomhouse.org/template.cfm?page=22&year=2005&country=6866.
2. U.N. Special Envoy on Human Settlements Issues in Zimbabwe, Report of the Fact Finding Mission to Zimbabwe to Assess the Scope and Impact of Operation Murambatsvina, 2005, http://www.zimbabwesituation.com/zimbabwe_rpt.pdf.
3. Human Rights Watch, World Report 2007: Zimbabwe, 2007, http://hrw.org/englishwr2k7/docs/2007/01/11/zimbab14720.htm.
4. U.N. Special Envoy on Human Settlements Issues in Zimbabwe, Report of the Fact Finding Mission to Zimbabwe to Assess the Scope and Impact of Operation Murambatsvina, 2005, http://www.zimbabwesituation.com/zimbabwe_rpt.pdf.
5. Ibid.
6. Human Rights Watch, World Report 2007: Zimbabwe, 2007, http://hrw.org/englishwr2k7/docs/2007/01/11/zimbab14720.htm.
7. Ibid.
8. Reporters Without Borders, Zimbabwe: Annual Report 2007, 2007, http://www.rsf.org/article.php3?id_article=20744&Valider=OK.
9. Ibid.
10. International Telecommunication Union, *World Telecommunication Indicators 2006*.
11. Zimbabwe Internet Usage and Marketing Report, 2005, http://www.internetworldstats.com/af/zw.htm.
12. Ibid.
13. Ibid.
14. Ibid.

15. Reuters, "Zimbabwe inflation surges to new record," February 12, 2007, http://www.zimbabwesituation.com/feb13_2007.html#Z1.
16. Michael Wines, "Economic free fall in Zimbabwe," February 6, 2007, http://www.iht.com/articles/2007/02/06/news/zim.php?page=1.
17. BBC News, "Zimbabwe internet link restored," September 26, 2006, http://news.bbc.co.uk/2/hi/africa/5382518.stm.
18. Freedom House, Zimbabwe: Country Report, 2006, http://www.freedomhouse.org/template.cfm?page=22&year=2006&country=7092.
19. Reporters Without Borders, "Zimbabwe," 2004, http://www.rsf.org/article.php3?id_article=10710.
20. Ibid.
21. Andrew Meldrum, "Mugabe introduces new curbs on Internet," June 2004, http://www.guardian.co.uk/international/story/0,3604,1230096,00.html.
22. *ZimObserver News,* "ISP(s) block e-mails with political content," June 2006, http://www.zimobserver.com/index.php?mod=article&cat=Technology&article=1.
23. Clemence Manyukwe, "Zimbabwe: Government drops snooping bill," November 3, 2006, http://www.allafrica.com/stories/200611030016.html.
24. Ibid.
25. APC Blogs, Expression under Repression: WSIS and the Net, 2005, http://blog.apc.org/en/index.shtml?x=2503009.

Contributors

Ross Anderson is professor of security engineering at the Computer Laboratory of the University of Cambridge.

Malcolm Birdling is a researcher at the Oxford Internet Institute at the University of Oxford.

Ronald Deibert is associate professor of political science at the University of Toronto and director of the Citizen Lab at the Munk Centre for International Studies, University of Toronto.

Robert Faris is research director for the Internet filtering project at the Berkman Center for Internet & Society.

Steven J. Murdoch is a researcher in the Security Group of the University of Cambridge.

Helmi Noman is a research affiliate of the Berkman Center for Internet & Society.

John Palfrey is the executive director of the Berkman Center for Internet & Society and a Clinical Professor of Law at Harvard Law School.

Rafal Rohozinski is the former director of the Advanced Network Research Group at Cambridge University (Cambridge Security Programme). He is a principal with the SecDev Group.

Mary Rundle is a fellow at the Berkman Center for Internet & Society and a nonresident fellow at the Center for Internet & Society at Stanford Law School.

Nart Villeneuve is the director of technical research at the Citizen Lab at the Munk Centre for International Studies, University of Toronto.

Stephanie Wang is a research fellow at the Berkman Center for Internet & Society.

Jonathan Zittrain is professor of Internet governance and regulation at Oxford University and the Jack N. and Lillian R. Berkman Visiting Professor for Entrepreneurial Legal Studies at Harvard Law School.

Index